An Introduction to Signals and Systems

An Introduction to Signals and Systems

John Alan Stuller
The University of Missouri-Rolla

Australia   Canada   Mexico   Singapore   Spain   United Kingdom   United States

An Introduction to Signals and Systems, First Edition
by John Alan Stuller

Associate Vice President and Editorial Director:
Evelyn Veitch

Publisher: Chris Carson

Developmental Editor:
Hilda Gowans

Permissions Coordinator:
Vicki Gould

Production Services:
RPK Editorial Services, Inc.

Copy Editor:
Fred Dahl

Proofreader:
Erin Wagner

Indexer:
Shelly Gerger-Knechtl

Production Manager:
Renate McCloy

Creative Director:
Angela Cluer

Interior Design:
Carmela Pereira

Cover Design:
Andrew Adams

Compositor:
International Typesetting and Composition

Printer:
Thomson/West

Cover Image Credit:
MAZAAKI KAZAMA/Getty Images

North America
Nelson
1120 Birchmount Road
Toronto, Ontario M1K 5G4
Canada

Asia
Thomson Learning
5 Shenton Way #01-01
UIC Building
Singapore 068808

Australia/New Zealand
Thomson Learning
102 Dodds Street
Southbank, Victoria
Australia 3006

Europe/Middle East/Africa
Thomson Learning
High Holborn House
50/51 Bedford Row
London WC1R 4LR
United Kingdom

Latin America
Thomson Learning
Seneca, 53
Colonia Polanco
11560 Mexico D.F.
Mexico

Spain
Paraninfo
Calle/Magallanes, 25
28015 Madrid, Spain

To Sandy
and John, Pete, Michael, and David

Contents

7   DT Processing of CT Signals   426

8   Introduction to Random Signals   491

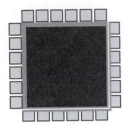

Preface

This book provides a concise and clear introduction to signals and systems theory with emphasis on fundamental analytical and computational techniques. It can be used for one-semester introductions to either continuous-time (CT) or discrete-time (DT) signals and systems, and for one- or two-semester introductions to both CT and DT signals and systems.

Signals and systems theory has strong roots in electrical engineering: Signals, regardless of their origins, are typically converted to their electrical counterparts and processed by electronic systems. Electrical engineering students traditionally study signals and systems after one or two semesters of circuits courses that include both transient and AC circuit analysis. This background is very valuable, but not necessary, for understanding this book. The book is accessible to a broad range of undergraduate engineering and science students as well as to electrical engineering students. The book is also valuable to practicing engineers and scientists who seek an insightful review. Prerequisites for the book are first year calculus and familiarity with complex arithmetic.

The book has the following features:

- The development of the core material is academically sound, concise, and easy to understand. New derivations not found in other texts and a modernized placement of topics helps students to learn faster and appreciate the unity of the overall theory.
- DT and CT topics are described in parallel chapters. The parallel development brings to light the similarities and the differences between DT and CT. Either CT or DT can be covered first, or CT and DT can be covered in parallel, chapter by chapter or section by section.
- The book uses plain language and emphasizes core issues. Discussions of noncentral issues and excessive detail that can overwhelm students are excluded.
- To accommodate students with different backgrounds, the technical depth starts at an elementary level in the first few chapters and increases gradually.
- Theory is presented in a way that builds upon and strengthens students' intuition.
- The book includes an assessable, heuristic introduction to random signals (Chapter 8), a topic that is central to signal processing but is usually omitted in an introduction because of its advanced nature. The material in this chapter conveys an *intuitive* understanding of probability, expectation, correlation functions, and power spectrums. This chapter provides students with a welcome preparation for more rigorous courses.
- A practical audio signal processing problem is introduced in Chapter 1. This problem, which is familiar to all students, is referred to in several subsequent

chapters in increasing depth as new analytical tools are developed. The problem serves as a unifying point of reference for the developing theory, and it helps students appreciate their progress.

Chapter-by-Chapter Prerequisites

The flow graphs in Figure P.1 show the chapter-to-chapter prerequisites of the book. In the figure, the left-hand and right-hand columns pertain to chapters that deal with exclusively CT and DT topics, respectively. The middle column pertains to chapters that deal with both DT and CT. Prerequisites are indicated by the arrows. The main thread of the development follows.

Figure P.1

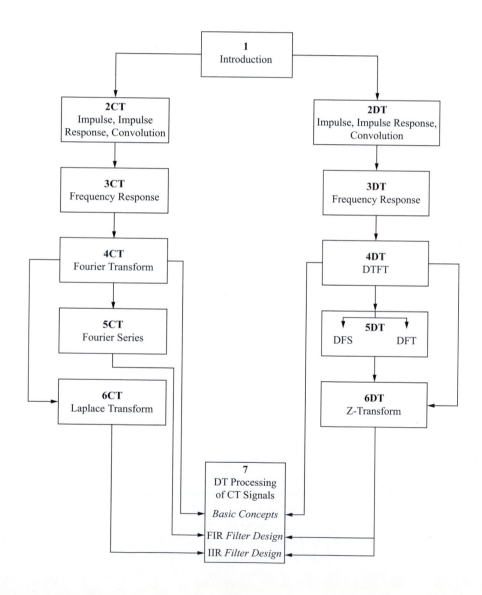

Chapter 1

Chapter 1 provides a brief introduction to CT and DT signals and systems. It defines the concepts of stability, causality, linearity, and time invariance, and it introduces the system models that are basic to signals and systems theory. It also introduces a practical audio signal processing problem that is referred to in subsequent chapters.

Chapters 2CT–6CT

Chapter 2CT introduces the unit impulse and shows that every waveform can be represented as an integral of impulses. The impulse response $h(t)$ of a linear time invariant (LTI) system is introduced and illustrated with several examples. The response of an LTI system for any input $x(t)$, given by the convolution $x(t) * h(t)$, is shown to follow directly from the integral-of-impulses representation of $x(t)$, time invariance, and linearity. A necessary and sufficient condition for an LTI system to be bounded input-bounded output (BIBO) stable is obtained from the convolution integral.

Chapter 3CT evaluates the convolution $x(t) * h(t)$ for a complex sinusoidal input waveform $x(t)$. A one-step evaluation establishes that (1) the output is a complex sinusoid having the same frequency as the input, (2) the input and output sinusoids are related by the system's frequency response characteristic $H(F)$, and (3) $H(F)$ is the Fourier transform of the impulse response $h(t)$. The derivation shows that $H(F)$ is a natural and fundamental descriptor of any LTI system. This descriptor is shown to apply not only to purely sinusoidal inputs, but also to inputs composed of multiple sinusoidal components. Many examples and properties are included.

Chapter 4CT again evaluates the convolution $x(t) * h(t)$, this time with $h(t)$ expressed as the inverse Fourier transform of $H(F)$. By a simple algebraic step, the Fourier transform $X(F)$ of $x(t)$ is derived along with the property that convolution in the time domain transforms into multiplication in the frequency domain. The Fourier transform is also shown to be a limiting case of multiple sinusoids introduced in Chapter 3CT. Thus, $X(F)$ is established as a natural and fundamental descriptor of any signal that is applied to an LTI system. Chapter 4CT continues with the derivation of important Fourier transform pairs and properties and the sampling theorem. Convergence properties of the Fourier transform are described in detail, but this topic can be omitted without loss in continuity.

Chapter 5CT presents the Fourier series and its applications as a special case of multiple sinusoidal component waveforms introduced in Chapter 3CT. It shows that the formulas for the Fourier series coefficients can be derived either from first principles, as in the historical derivation, or from the Fourier transform of a periodic waveform. The book departs from the historical placement of Fourier series before the Fourier transform because the Fourier transform is the central analytical tool in modern signals and systems theory. Moreover, the placement provides substantial insight into the relationship between the Fourier transform and Fourier series. Several examples of Fourier series are presented, including an analysis of the harmonic distortion introduced by a saturating amplifier. The convergence properties of the Fourier series are described in detail. An appendix to this chapter strengthens the connection between the Fourier transform and the Fourier series by deriving the Fourier transform from the Fourier series.

Chapter 6CT introduces the Laplace transform as an analytical extension of the Fourier transform. The unilateral Laplace transform is emphasized because of its widespread use in engineering practice for determining LTI system response to suddenly applied inputs. Systems characterized by ordinary nth-order differential equations are emphasized.

Standard topics, such as poles, zeros, and partial fraction expansions, are developed gradually and with many examples. The chapter shows the relationship between transient and sinusoidal steady-state response and illustrates this relationship with first- and second-order system examples.

Chapters 2DT–6DT

The developments of Chapters 2DT–5DT run parallel to those of 2CT–5CT: Chapter 2DT introduces the DT unit impulse and shows that every sequence can be represented as a sum of impulses. The impulse response $h[n]$ of an LTI system is defined and illustrated with several examples. The response of an LTI system for any input $x[n]$, given by the convolution $x[n] * h[n]$, is shown to follow directly from the sum-of-impulses representation of $x[n]$, shift invariance, and linearity. A necessary and sufficient condition for an LTI system to be BIBO stable is obtained from the convolution sum.

Chapter 3DT introduces sinusoidal sequences as sampled versions of sinusoidal waveforms. After describing the aliasing introduced when a rotating phasor is sampled below the Nyquist rate, the chapter then evaluates the convolution $x[n]*h[n]$ for a complex sinusoidal input sequence $x[n]$. A one-step evaluation establishes that (1) the output is a complex sinusoid having the same frequency as the input, (2) the input and output sinusoids are related by a multiplication with the system's frequency response characteristic $H(e^{j2\pi f})$, and (3) $H(e^{j2\pi f})$ is the discrete-time Fourier transform (DTFT) of the impulse response $h[n]$. The derivation shows that $H(e^{j2\pi f})$ is a natural and fundamental descriptor of any LTI system. This descriptor is shown to apply not only to purely sinusoidal inputs, but also to inputs composed of multiple sinusoidal components. Many examples and properties are included.

Chapter 4DT again evaluates the convolution $x[n]*h[n]$, this time with $h[n]$ expressed as the inverse DTFT of $H(e^{j2\pi f})$. By a simple change in the orders of summation, the DTFT $X(e^{j2\pi f})$ of $x[n]$ is derived along with the property that convolution in the time domain transforms into multiplication in the frequency domain. Thus, $X(e^{j2\pi f})$ is established as a natural and fundamental descriptor of any signal that is applied to an LTI system. Chapter 4DT continues with the derivation of important DTFT pairs and properties, including upsampling and downsampling and the sequence sampling theorem. Convergence properties of the DTFT are described in detail, but this topic can be omitted without loss in continuity.

Chapter 5DT introduces the *discrete Fourier series* (DFS) and the closely related *discrete Fourier transform* (DFT). As indicated by the flow graph, the developments of the DFS and DFT are independent: They can be covered in either order. Chapter 5DT presents the DFS and its applications as a special case of multiple sinusoidal component sequences introduced in Chapter 3DT. The DFS is also derived from the DTFT of a periodic sequence. The bulk of Chapter 5DT is devoted to the properties and applications of the DFS and DFT. The relation between circular and linear convolution is described in detail using multiple graphical aids. An introduction to spectrum estimation describes the picket fence effect, spectrum leakage, frequency resolution, and windowing. This chapter makes frequent reference to computational tools and includes many MATLAB problems.

Chapter 6DT introduces the z-transform as an analytical extension of the discrete-time Fourier transform. The unilateral z-transform is emphasized because of its widespread use in engineering practice for determining LTI system response to suddenly applied inputs. Systems characterized by ordinary nth-order difference equations are emphasized. Standard topics, including poles, zeros, and partial fraction expansion, are developed gradually along with corresponding examples and references to pertinent software. The chapter shows

Figure P.2

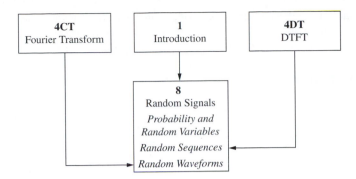

the relationship between transient and sinusoidal steady-state response and illustrates this relationship with first- and second-order system examples.

Chapter 7

Chapter 7 describes the bridge between continuous time and discrete time in the context of a hybrid system designed for DT processing of CT waveforms. It further develops the relationship between the Fourier and discrete-time Fourier transforms and derives the conditions for which DT and DT signal processing are equivalent. The chapter includes design techniques for IIR and FIR filters and presents the basic filter structures for realizing the filters.

Chapter 8

Virtually all modern applications of signals and systems theory involve random processes. Therefore, we believe that an introduction to random processes should be included in *An Introduction to Signals and Systems*. Chapter 8 presents the topics of probability, random variables, random sequences, expectation, autocorrelation, and power spectrums in a highly readable way. The emphasis is on developing students' intuitive understanding through examples. The chapter does not attempt to present the material with the completeness and mathematical depth found in higher-level courses. Prerequisites for Chapter 8 are shown in Figure P.2.

Problems

Each CT and DT chapter concludes with analytical problems keyed to the sections in the text. Learning engineering theory and how to apply it is similar to learning a foreign language or a musical instrument. A student cannot learn these skills without practice. The purpose of the problems is to provides a means for this practice. The problems range from straightforward to very challenging. The most challenging problems are indicated with a **tiger icon** .

MATLAB®

Hand led MATLAB problems are included with the analytical problems in appropriate DT chapters, Chapter 6CT, and Chapters 7 and 8. The substantial majority of these problems appear in Chapters 2DT, 5DT, 7 and 8 and are labeled with the MATLAB® icon. The intent is to help students get started using MATLAB and its help files. The problems frequently include suggested MATLAB code. For a MATLAB tutorial, one cannot do better than the one included in MATLAB itself. Students should be advised that facility at programming a computer is never an adequate substitute for understanding theory.

Computer Problems

The book contains additional computer problems labeled 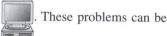. These problems can be performed using any software, including MATLAB. The COMPUTER problems appear primarily in the CT chapters.

Acknowledgments

The topic of signals and systems is founded on the pioneering work of many people over a long period of time. It has been my primary technical interest since 1963. This book is an outgrowth of this interest and of several years of experience in academia, governmental research facilities, and private industry. It is my pleasure and privilege to acknowledge the thoughtful and constructive suggestions of outstanding reviewers. These reviewers include Parham Aarabi, University of Toronto; Rocio Alba-Flores, University of Minnesota, Duluth; Nirmal K. Bose, Pennsylvania State University; Michael T. Johnson, Marquette University; Kurt Kosbar, University of Missouri, Rolla; Fabrice Labeau, McGill University; Thuy T. Le, San Jose State University; Aryan S. Mehr, University of Saskatchewan; and Andreas Spanias, Arizona State University. It is my further privilege to acknowledge the invaluable help of the staffs and contractors of Thomson Engineering and RPK Editorial Services, Inc., who worked closely with me from the start to produce the best possible book. My special thanks go to to Evelyn Veitch, Christopher T. Carson, Hilda Gowans, Rose Kernan, Andrew Adams, Angela Cluer, and Renate McCloy. Finally, some of the insights, personalities, and smiles of people I have be honored to know in my lifetime are also in this book. A list would consume many pages. I am grateful to all.

JOHN ALAN STULLER
Gaiberg–Heidelberg, Germany

1 Introduction

1.1 Signal and Systems Theory

Signals and systems theory can be defined as the theory addressing the analysis and development of systems that process signals. The theory necessarily includes an understanding of the signals and how they interact with the systems. The signals may originate from audio, video, or other sources. They may be naturally processed by the environment. Because they are typically processed by electronic systems, the signals often appear as voltages and currents in electronic systems. The electronic systems include transducers, such as microphones and scanners; analog systems, such as amplifiers and filters; digital systems, such as microprocessors; and converters that provide the interface between the analog and digital systems.

Applications of signals and systems theory are numerous and diverse. They include audio and video signal processing, remote robot control, cellular communication, weather forecasting, and processing the electric impulses from human nerves, to name a few. A representative application is illustrated in Figure 1.1. The figure depicts a singer whose voice (an acoustic waveform, or *signal*) $x(t)$ is picked up by a microphone, amplified, and broadcast through a speaker to an audience. Acoustic noise $\xi(t)$ from the audience also reaches the microphone. The amplifier has a graphic equalizer and a set of reverb "ambiance" options. The sound waves bounce off the walls, creating distinct echoes and house reverberation (multiple overlapping echoes). The naturally processed reflected sound $r(t)$ is fed back into the microphone. A line-output jack on the amplifier feeds the amplifier output $e(t)$ into an analog tape recorder.

After the performance, audio engineers process the tape's audio offline to improve the sound quality (Figure 1.2). The processing task involves reducing the noise, compensating for the effects of feedback, removing the echoes, and retuning the reverberation for the best effect. The offline processor consists of a prefilter that eliminates noise outside the audio frequency range, a sampler that converts the filtered audio waveform into a sequence of voltage samples, an analog-to-digital (A/D) converter that converts the audio samples to binary voltages (equivalent to ones and zeros), a digital processor (a computer) that improves the audio, and digital memory to store the processed audio. For playback, a digital-to-analog (D/A) converter transforms the binary voltages from the memory back to voltage samples, and an interpolator converts the output sequence of samples back to a waveform, $o(t)$, that is applied to the speaker.

Figure 1.3 shows a *block diagram* engineers might use to model the components of Figure 1.1. Signals and systems theory deals with the relationship between the input and

Figure 1.1
Singer performing on
stage.

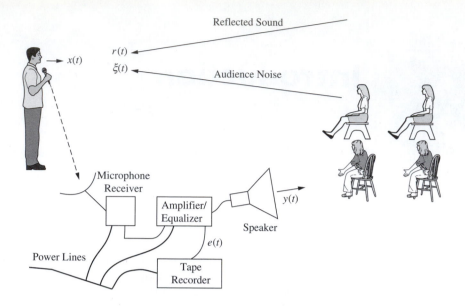

output of each block in Figure 1.3 as well as the system as a whole. Block diagrams are
quite intuitive: Arrows indicate the one-way flow of the signals into and out of the system
blocks, and the meaning of the plus signs (+) is obvious.

Figure 1.4 shows a *flow graph* that provides another way to represent the various system
components and their interconnections. The flow graph contains the same information as
the block diagram but shows it in a somewhat more compact manner.

Figure 1.5 shows a block diagram engineers might use to model their offline signal
processing task from the input $e(t)$ of Figure 1.2 to the output, $o(t)$ of the interpolator. Here,
the hardware that transforms $e(t)$ to $o(t)$ is modeled as a sampler, a discrete-time processor,
and an interpolator. The A/D and D/A converters and the memory have been omitted. The
omission of the A/D and D/A converters is valid if the distortion introduced by representing

Figure 1.2
Offline processing of
audio.

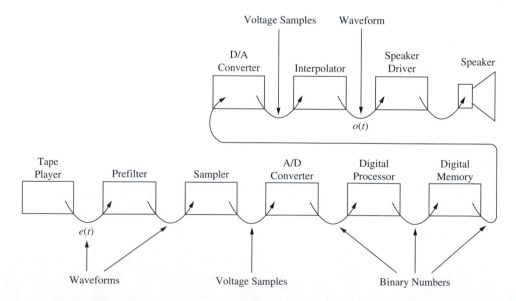

Figure 1.3
Block diagram for
Figure 1.1.

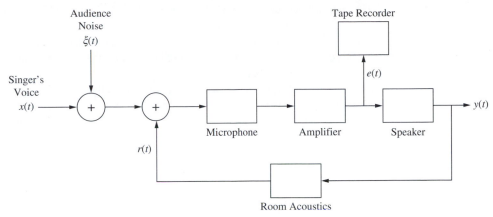

Figure 1.4
Flow graph for
Figure 1.1.

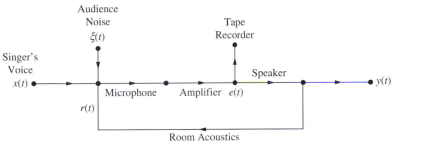

Figure 1.5
Block diagram for
Figure 1.2.

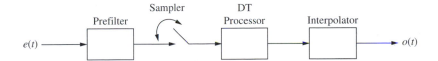

analog voltages by finite-length binary numbers can be neglected. We have identified the components in the physical system by their names in Figures 1.3–1.5. Later we will show how to describe and analyze these components mathematically.

You may have noticed that we called the singer's voice a signal and that we also called the waveforms in Figures 1.3–1.5 signals. Engineers routinely call any waveform or data sequence they work with or observe a signal, even when noise is present. For example, they might call the waveform $e(t)$ a signal. They would call the singer's voice waveform $x(t)$ the "signal" and $\xi(t)$ the "noise" when their attention is directed toward removing or reducing the effects of $\xi(t)$.

1.2   Waveforms and Data

As illustrated in Figures 1.1 and 1.2, we work with two kinds of signals: *waveforms* and *data sequences*. Please take careful note of the notation used to distinguish between the waveforms and the data. Waveforms are functions of real arguments, which are contained in parentheses. Thus, waveforms are denoted as $x(t)$, $y(t)$, and so forth, where t is a real variable. Data sequences are indexed using integer arguments, contained in brackets. Thus, data is denoted as $x[n]$, $y[n]$, and the like, where n is an integer variable. Data is not defined for noninteger arguments. For example, $x[3.5]$ is not defined. In this book we assume that the argument of a waveform t denotes time.[1] We generally refer to waveforms as *continuous-time (CT) signals* and data as *discrete-time (DT) signals*. CT signals are processed by waveform or analog processors (CT systems), and data signals are processed by data processors (DT systems).

Waveforms and data can be mathematically defined functions and sequences used for testing or otherwise understanding a signal or a system. The real-life waveforms and data the engineers must usually work with, however, are not the sine and cosine functions found in AC circuit theory. For example, in Figure 1.1, no rational audience would pay to listen to a sine wave (a single pure tone) for two hours. This book introduces you to the fundamental tools that enable you to work with real-life waveforms and sequences. But, as you progress through the chapters, you will also gain a deep appreciation for sinusoidal waveforms. This is because the sinusoidal waveforms of circuit theory provide the best known foundation for understanding and evaluating real-life signals and systems.

1.3   SLTI Systems

This book deals almost exclusively with *stable, linear, time-invariant (SLTI)* systems. The technical definitions of these terms are given shortly, but let's first consider what the terms mean in everyday language.

Stable means that the output variable does not "blow up" (go to infinity) if the input variable does not "blow up." For example, consider reverberation in the room in Figure 1.1. Reverberation is created by multiple reflections of sound off walls, ceiling, and other surfaces. The returned reverberated sound wave cannot have more energy than the incident waves. The room acoustics that produce the reverberation therefore comprise a stable system. For another example, consider the amplifier. A finite volume (amplitude) of music into the microphone produces a finite (not infinite) volume of music out of the speakers. Therefore the amplifier comprises a stable system. An *unstable* system does the opposite: For an unstable system, there exists some finite input for which the output is infinite. You have heard the loud screech due to too much feedback in a system setup like that in Figure 1.1. Too much feedback can make the overall response *unstable* even when each subsystem is stable. A tiny input can get amplified in the feedback loop and come out as a loud roar. (It is not infinite in a real setup only because the amplifier's output stage finally saturates and limits the output amplitude.[2]) When the signal processing engineers attempt to improve the sound using a DT processor, they use a stable processing algorithm. In Chapter 2, we find necessary and sufficient conditions for DT and CT systems to be stable.

Linear means that superposition applies. Consider again the reverberation of the room in Figure 1.1. An echo from two people clapping is the sum (or *superposition*) of the echoes of each person clapping. Because reverberation is the sum of many overlapping echoes, the reverberation from two people clapping is the superposition of the reverberation of each person clapping. Echoes and reverberation are both linear processes. Consider the amplifier/equalizer in Figure 1.1. The amplifier output due to the sum of the singer's voice and the noise from the audience is the sum of the amplified and equalized outputs from the two sources acting individually. Amplification and equalization are linear

---

[1] A waveform could be a function of some other variable such as space. For example, the waveform on the tape in Figure 1.1 is a function of the spatial distance along the tape.

[2] Real-life unstable systems must ultimately reach a saturation limit because there are no infinite physical variables. Once saturation is reached, the system is no longer linear.

processes. Because the physical systems affecting the quality of the sound in Figure 1.1 are all linear, the signal processing engineers find it natural to compensate for them using linear processing.[3]

Time invariance means that a system's response to a specific input does not depend on when the input is applied. The time-invariance property applies to echoes, reverberation, and the amplifier response. Because these physical systems affecting the quality of the sound in Figure 1.1 are all time invariant, signal processing engineers find it natural to compensate and improve the audio using time-invariant processing.

Sinusoidal inputs occupy a unique place in describing how SLTI systems work. Consider again the system of Figure 1.1: A pure sinusoidal acoustic tone, sung into the microphone, reproduces itself, except for a change in amplitude and phase, throughout the room *in spite of* the echoes. The sinusoidal tone retains its sinusoidal form and frequency everywhere. Any other input, however, does not retain its functional form in the presence of echoes and reverb. This fundamental property, sinusoid in/sinusoid out, is established in Chapters 2CT and 2DT and forms the foundation for many subsequent developments in this book.

Let's now proceed with the technical definitions.

1.3.1 Stability

The technical definition of a stable system uses the idea of a bounded signal. A signal is *bounded* if its absolute value is never infinite. The magnitude of a bounded signal is, by definition, never greater than some finite number B. In technical terms, *a system is stable if every bounded input produces a bounded output*. This definition of stability is called BIBO (bounded input/bounded output) stability. It applies to both CT and DT systems.

1.3.2 Linearity

Suppose we have a CT system and apply an input $x_1(t)$. We denote the output as $y_1(t)$. Using flow graph notation, we write this input-output (I-O) relation as

$$x_1(t) \rightarrow y_1(t) \tag{1}$$

Similarly, if we apply an input $x_2(t)$, the output is $y_2(t)$:

$$x_2(t) \rightarrow y_2(t) \tag{2}$$

The system is *linear* if, for every $x_1(t)$ and $x_2(t)$ and for all constants, a and b, the input $ax_1(t) + bx_2(t)$ produces the output $ay_1(t) + by_2(t)$:

$$ax_1(t) + bx_2(t) \rightarrow ay_1(t) + by_2(t) \tag{3}$$

Similarly, suppose that we have some DT system and apply an input $x_1[n]$. The output is $y_1[n]$. Similarly, if we apply an input $x_2[n]$, the output is $y_2[n]$:

$$x_1[n] \rightarrow y_1[n] \tag{4}$$

$$x_2[n] \rightarrow y_2[n] \tag{5}$$

The DT system is linear if, for every a and b and for every $x_1[n]$ and $x_2[n]$, the input $ax_1[n] + bx_2[n]$ produces the output $ay_1[n] + by_2[n]$:

$$ax_1[n] + bx_2[n] \rightarrow ay_1[n] + by_2[n] \tag{6}$$

Superposition applies for linear systems.

---

[3] The problem of optimum signal processing, including both linear and nonlinear processors, is the subject of higher-level texts dealing with detection and estimation theories. A linear system is often, but not always, optimum.

Figure 1.6
Example of time shift:
(a) original waveform;
(b) time-shifted
waveform.

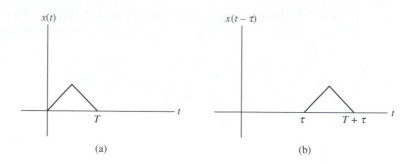

(a)   (b)

1.3.3
Time
Invariance

To define the concept of time invariance mathematically, we need to know how to mathematically describe a time-shifted waveform. We time-shift a waveform when we replace its argument t by $t - \tau$ where τ is the amount of shift. Assume for illustration that τ is positive. Then the value of $t - \tau$ is smaller than t by exactly τ. Therefore, $x(t - \tau)$ has the value that $x(t)$ had τ seconds earlier. Time shift is illustrated in Figure 1.6.

Consider a CT system having some input $x(t)$ and output $y(t)$.

$$x(t) \rightarrow y(t) \tag{7}$$

The system is time invariant if, for every $x(t)$ and every τ, a time shift of the input by τ produces a time shift of the output by τ

$$x(t - \tau) \rightarrow y(t - \tau) \tag{8}$$

Notice that in a time-invariant system, I-O relation (8) applies for every τ, including negative as well as positive values.

Similarly, to understand the concept of time invariance for a DT system, we need to know how to shift a data sequence. Shift for a DT signal is illustrated in Figure 1.7. We shift data when we replace the index n by $n - m$, where m is the amount of shift. The reason that the substitution $n - m$ for n shifts the sequence by m is simple. Assume for illustration that m is positive. Then the value of $n - m$ is smaller than n by exactly m. Therefore, $x[n - m]$ has the value that $x[n]$ had m index units earlier.

Consider a DT system having some input $x[n]$ and output $y[n]$.

$$x[n] \rightarrow y[n] \tag{9}$$

Figure 1.7
Shift for sequence:
(a) original data;
(b) shifted data.

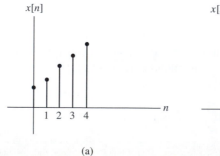
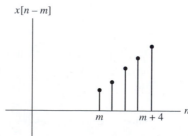

(a)   (b)

The system is time invariant if a shifted input produces the shifted version of the same output

$$x[n - m] \rightarrow y[n - m] \tag{10}$$

for every $x[n]$ and every m. Notice that in a time-invariant system, I-O relation (10) applies for every m, including negative as well as positive values.

1.3.4 Causality

The real-life systems that engineers work with have the property of cause and effect. For a CT system, this means that for every input, the output $y(t_o)$ at any time t_o does not depend on input values $x(t_i)$ that occur at times after t_o, i.e., for $t_i > t_o$. For a DT system cause and effect means that for every input, the output, $y[n]$ at any n is not affected by input data values $x[m]$ for $m > n$. All the systems in Figure 1.1 are causal. The microphone amplifier, and tape recorder do not produce an output before there is an input. You cannot get an echo or reverberation from the walls of the room before the originating sound occurs.

1.3.5 Noncausal Signal Processing

Some thought shows that signal processing engineers *can* process signals in a way that mimics a noncausal system. Consider the tape in Figure 1.2. Clock time in Figure 1.1 is represented by distance on the tape. Pick some point near the middle of the tape as a point of reference. If we look one way on the tape, we have audio that is in the "past" compared to the point of reference, and if we look the other way, we have audio that is in the "future" compared to the point of reference. There is no reason why we cannot use the future as well as the past part of the tape to improve or otherwise modify the waveform's value at the point we call the reference. We can do this to all points on the tape, except at the very ends where there is no future or past left. When the sound engineers try to improve the quality of the sound, they may well decide to mimic noncausal processing. Of course, they cannot do this without access to the future data, and this access is made possible only by the overall time delay of the recorded audio. Some advantages and disadvantages of noncausal signal processing, including specific examples on how it is done, are discussed in Chapter 7.

1.4 Engineering Models

The method of signals and systems theory, as in other branches of engineering, is based on idealized *models* that (approximately) describe the observed world. We define the models and proceed to analyze them. If the results of our analysis do not agree with our physical observations, either our models are wrong, or we made a mistake in our analysis, or the equipment we are using to make the observations is at fault.

The models include both system models and signal models.

1.4.1 System Models

The CT system models include the ideal amplifier, integrator, differentiator, and delay element shown in Figures 1.8–1.11. They are all linear, time-invariant (LTI) models. The *ideal amplifier*, depicted by either the system of Figure 1.8(a) or Figure 1.8(b), is defined by the I-O equation

$$y(t) = Ax(t) \tag{11}$$

where A is the amplification.

Figure 1.8
Ideal amplifier.

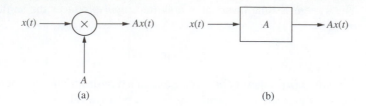

(a)   (b)

Figure 1.9
Integrator.

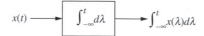

Figure 1.10
Differentiator.

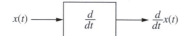

Figure 1.11
Delay element.

The *ideal integrator*, illustrated in Figure 1.9, is defined by the I-O equation

$$y(t) = \int_{-\infty}^{t} x(\lambda)\, d\lambda \tag{12}$$

The *ideal differentiator*, illustrated in Figure 1.10, is defined by the I-O equation

$$y(t) = \frac{dx(t)}{dt} \tag{13}$$

The *ideal delay element*, illustrated in Figure 1.11, is a CT system defined by the I-O equation

$$y(t) = x(t - \tau) \tag{14}$$

The DT system models include the ideal amplifier, accumulator, first differencer, and unit delay element shown in Figures 1.12–1.15. They are all LTI models. The *ideal amplifier*, depicted by either the system of Figure 1.12(a) or Figure 1.12(b) is defined by the I-O equation

$$y[n] = Ax[n] \tag{15}$$

where A is the amplification.

The *ideal accumulator*, illustrated in Figure 1.13, is defined by the I-O equation

$$y[n] = \sum_{k=-\infty}^{n} x[k] \tag{16}$$

The *ideal unit delay* is defined by the I-O equation

$$y[n] = x[n - 1] \tag{17}$$

The ideal unit delay system element, is denoted by a D, as illustrated in Figure 1.14.

Figure 1.12
Ideal amplifier.

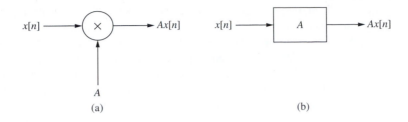

(a)   (b)

Figure 1.13
Accumulator.

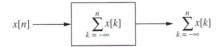

Figure 1.14
Unit delay.

Figure 1.15
First differencer.

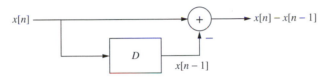

The *ideal first differencer*, illustrated in Figure 1.15, is a DT system defined by the I-O equation

$$y[n] = x[n] - x[n-1] \tag{18}$$

As we shall see in subsequent chapters, a model for a more complex CT system is obtained by interconnecting the individual CT elements. These interconnections may include (ideal) adders in which the outputs of two or more ideal elements are additively combined. Similar considerations apply to DT elements. In referring to either the CT or DT models, engineers routinely drop the word "ideal" to shorten the terminology.

Additional ideal system models will be introduced in subsequent chapters. For example, we describe the amplifier of Figure 1.1 using an LTI system called an ideal low-pass filter in Chapter 3CT. We define an ideal sampler and an ideal interpolator in Chapter 4CT and use them in Chapter 7 to model part of the A/D and D/A conversion depicted in Figure 1.1.

Remember that a DT system is itself commonly used as a model for the operations performed by a digital computer. Digital computers cannot exactly represent real numbers like π or $\cos(2\pi 0.1)$ or real waveforms like $e(t)$ of Figure 1.1. The DT system models used in this book are idealizations because they omit the errors (called quantization noise) introduced when a real number is expressed as a finite-length binary word. Normally, the system is designed so that quantization noise can be neglected.

1.4.2
Signal Models

The signal models we use in signal and systems theory are mathematical idealizations of the signals we observe. Unlike the system models, however, the signal models are too numerous to delineate. We will see many of them in subsequent chapters.

Signal models are organized into broad classes. For example, every signal is either *real* or *complex*. Waveform and sequence models that are *real* have values specified by the set of real numbers. These models correspond directly to signals we observe and measure physically. Waveforms and sequences that are *complex* have values taken from the set of complex numbers. We use complex signal models because they help efficiently describe, in mathematical terms, how the real signal models interact with LTI system models. We first use complex signal models in Chapter 3. If you are unsure of your complex arithmetic, you should study Appendix A before you get to Chapter 3. It is a good idea to test your knowledge of complex arithmetic by trying the problems included in that appendix.

We introduce other broad classes of signals in subsequent chapters as we need them.

1.5   Problems

The following problems are labeled CT and DT. If you intend to study CT signals and systems, you should do the CT problems. If you intend to study DT systems, you should do the DT problems.

1.2   Waveforms and Data

CT1.2.1 Consider the waveform

$$h(t) = \begin{cases} 0 & t < 0 \\ t & t \geq 0 \end{cases} \tag{19}$$

Plot:
a) $h(t)$ versus t.
b) $h(t - \lambda)$ versus t, where $\lambda = 4$.
c) $h(t - \lambda)$ versus t, where $\lambda = -4$.
d) $h(-t)$ versus t.

CT1.2.2 Consider the waveform $h(t)$ of Problem CT1.2.1. Plot:
a) $h(-\lambda)$ versus λ.
b) $h(t - \lambda)$ versus λ where $t = 5$.
c) $h(t - \lambda)$ versus λ where $t = -5$.

CT1.2.3 The *time-scaled* version of a waveform $x(t)$, is defined as $x(at)$, where a is a nonzero constant. Consider the waveform

$$x(t) = \begin{cases} 0; & t < 0 \\ t; & 0 \leq t < 1 \\ 0; & 1 \leq t \end{cases} \tag{20}$$

Plot:
a) $x(t)$ versus t.
b) $x(at)$ versus t, where $a = 0.5$.
c) $x(at)$ versus t, where $a = 2$.
d) $x(at)$ versus t, where a is shown as an arbitrary parameter on the plot. Assume that $a > 0$.
e) State in your own words how $x(at)$ depends on a. Then answer: (i) Does $x(at)$ become more compressed or more expanded in time as a is increased? (ii) Does the "ramp" part of $x(at)$ become more or less steep as a is increased? (iii) If $x(t)$ were recorded on an analog tape, would a speedup in the tape player correspond to an increase or a decrease in the value of a?

CT1.2.4 Consider again the waveform of Problem CT1.2.3.
Plot:
a) $x(t)$ versus t.
b) $x(at)$ versus t, where $a = -0.5$.
c) $x(at)$ versus t, where $a = -2$.
d) $x(at)$ versus t, where a is shown as an arbitrary parameter on the plot. Assume that $a < 0$.
e) State in your own words how $x(at)$ depends on a when a is negative. Then answer: (i) Does $x(at)$ become compressed or expanded in time as $|a|$ is increased? (ii) Does the "ramp" part of $x(at)$ get more or less steep as $|a|$ is increased? (iii) If $x(t)$ were recorded on an analog tape, would a backward speedup in the tape player correspond to an increase or a decrease in the value of $|a|$?

DT1.2.1 Consider the sequence $h[n]$

$$h[n] = \begin{cases} 0; & n < 0 \\ n; & 0 \leq n < 11 \\ 0; & 11 \leq n \end{cases} \qquad (21)$$

Plot:
a) $h[n]$ versus n.
b) $h[n - m]$ versus n, where $m = 4$.
c) $h[n - m]$ versus n, where $m = -4$.
d) $h[-n]$ versus n.

DT1.2.2 Consider the sequence $h[n]$ of Problem DT1.2.1. Plot
a) $h[-m]$ versus m.
b) $h[n - m]$ versus m, where $n = 5$.
c) $h[n - m]$ versus m, where $n = -5$.

DT1.2.3 The *downsampled* or *decimated* version of a sequence $x[n]$ is defined as

$$x_d[n; M] = \mathcal{DOWN}\{x[n]; M\} = x[nM]$$

where M is a positive integer. Consider the sequence

$$x[n] = \begin{cases} 0; & n \leq 0 \\ n; & 1 \leq n \leq 16 \\ 0; & 17 \leq n \end{cases} \qquad (22)$$

Plot:
a) $x[n]$ versus n.
b) $x_d[n; 2]$ versus n.
c) $x_d[n; 3]$ versus n.
d) State in your own words how $\mathcal{DOWN}\{x[n]; M\}$ depends on M. Does $\mathcal{DOWN}\{x[n]; M\}$ appear more compressed or more expanded as M is increased?

DT1.2.4 The *upsampled* version of a sequence $x[n]$ is defined as

$$x_u[n; M] = \mathcal{UP}\{x[n]; M\} = \begin{cases} x[\frac{n}{M}]; & \text{for } n = lM, \ l = \pm 1, \pm 2, \ldots \\ 0; & \text{otherwise} \end{cases} \qquad (23)$$

where M is a positive integer. By definition, upsampling consists of inserting $M-1$ zero samples between the values of a sequence $x[n]$. Consider the sequence $x[n]$ of Problem DT1.2.3.
Plot:
a) $x[n]$ versus n.
b) $x_u[n; 2]$ versus n.
c) $x_u[n; 3]$ versus n.
d) State in your own words how $\mathcal{UP}\{x[n]; M\}$ depends on M. Does $\mathcal{UP}\{x[n]; M\}$ appear more compressed or more expanded as M is increased?

1.3   SLTI Systems

CT1.3.1 In a commercial AM (amplitude modulation) broadcasting system, the transmitted waveform $y(t)$ has the form

$$y(t) = A_c[1 + mx(t)]\cos(2\pi F_c t + \phi) \tag{24}$$

where $x(t)$ is the *message*, A_c is the *carrier amplitude*, m is the *modulation index*, F_c is the *carrier frequency*, and ϕ is an arbitrary phase angle. The message is normalized so that $|x(t)| < 1$ for all t and the modulation index is a number between 0 and 1. How would you describe the transformation from $x(t)$ to $y(t)$?
a) Linear? Justify your answer.
b) Time-invariant? Justify your answer.
c) Causal? Justify your answer.
d) Stable? Justify your answer.

CT1.3.2 In a SC-DSB (suppressed carrier double-sideband) communication system, the transmitted waveform $y(t)$ is given by

$$y(t) = Ax(t)\cos(2\pi F_o t + \phi) \tag{25}$$

where $x(t)$ is the *message*, and A, F_o, and ϕ are constants. What is the I-O relation between $x(t)$ and $y(t)$?
a) Linear? Justify your answer.
b) Time-invariant? Justify your answer.
c) Causal? Justify your answer.
d) Stable? Justify your answer.

CT1.3.3 In a commercial FM (frequency modulation) broadcasting system, the transmitted waveform $y(t)$ is given by

$$y(t) = A\cos\left(2\pi F_o t + m\int_{t_o}^{t} x(\lambda)\,d\lambda + \phi\right) \tag{26}$$

where $x(t)$ is the *message*, and A, F_o, and ϕ are constants; m is a positive number called the modulation index; t_o denotes the time that the FM transmission starts. What is the I-O relation from $x(t)$ to $y(t)$?
a) Linear? Justify your answer.
b) Time-invariant? Justify your answer.
c) Causal? Justify your answer.
d) Stable? Justify your answer.

CT1.3.4 Consider a system having the I-O relation

$$y(t) = \int_{-\infty}^{\infty} h(t, \lambda)x(\lambda)\,d\lambda \tag{27}$$

where $h(t, \lambda)$ is an arbitrary function of t and λ. What is the I-O relation between $x(t)$ and $y(t)$?
a) Linear? Justify your answer.
b) Time-invariant? Justify your answer.
c) Causal? Justify your answer.
d) Stable? Justify your answer. Hint:
$|y(t)| = |\int_{-\infty}^{\infty} h(t, \lambda)x(\lambda)d\lambda| \le \int_{-\infty}^{\infty} |h(t, \lambda)||x(\lambda)|\,d\lambda$

CT1.3.5 Consider the system of Problem CT1.3.4. Assume that $h(t, \lambda) = 0$ for $\lambda > t$. Show how this condition changes your answers to Problem CT1.3.4, if at all.

CT1.3.6 Consider the system of Problem CT1.3.4. Assume that $h(t, \lambda)$ depends only on the difference between t and λ: $h(t, \lambda) = h(t - \lambda)$. Show how this condition changes your answers to Problem CT1.3.4, if at all.

CT1.3.7 Consider the system of Problem CT1.3.4. Assume that $h(t, \lambda)$ satisfies

$$\int_{-\infty}^{\infty} |h(t, \lambda)| d\lambda < \infty \tag{28}$$

Show how this condition changes your answers to Problem CT1.3.4, if at all.

DT1.3.1 In a *time-division multiplexing* system, the DT sequences $x_1[n] = x_1(nT_s)$ and $x_2[n] = x_2(nT_s)$ are interleaved to produce the sequence $y[n]$ as shown below

$$\{\cdots, x_1[-2], x_2[-2], x_1[-1], x_2[-1], x_1[0], x_2[0], x_1[1], x_2[1], x_1[2], x_2[2], \cdots\}$$

$$\downarrow \quad \downarrow \quad \downarrow \quad \downarrow \quad \downarrow \quad \downarrow \quad \downarrow \quad \downarrow \quad \downarrow \quad \downarrow$$

$$y[n] = \{\cdots, y[-4], y[-3], y[-2], y[-1], y[0], y[1], y[2], y[3], y[4], y[5], \cdots\}$$

The resulting samples $y[n]$ occur in time at a rate $2/T_s$ samples per second. What are the transformations from $x_1[n]$ to $y[n]$ and from $x_2[n]$ to $y[n]$?
a) Linear? Justify your answer.
b) Time-invariant? Justify your answer.
c) Causal? Justify your answer.
d) Stable? Justify your answer.

DT1.3.2 In the DT version of an analog suppressed carrier double-sideband communication system, a message sequence $x[n]$ is transformed to

$$y[n] = Ax[n] \cos(2\pi f_o n + \phi) \tag{29}$$

where A, f_o, and ϕ are constants. What is the I-O relation between $x[n]$ and $y[n]$?
a) Linear? Justify your answer.
b) Time-invariant? Justify your answer.
c) Causal? Justify your answer.
d) Stable? Justify your answer.

DT1.3.3 Consider a system having the I-O relation

$$y[n] = \sum_{m=-\infty}^{\infty} h[n, m] x[m] \tag{30}$$

where $h[n, m]$ is an arbitrary function of n and m. What is the I-O relation between $x[n]$ and $y[n]$?
a) Linear? Justify your answer.
b) Time-invariant? Justify your answer.
c) Causal? Justify your answer.
d) Stable? Justify your answer. Hint:
$$|y[n]| = \left| \sum_{m=-\infty}^{\infty} h[n, m] x[m] \right| \leq \sum_{m=-\infty}^{\infty} |h[n, m]||x[m]|$$

DT1.3.4 Consider the system of Problem DT1.3.3. Assume that $h[n, m] = 0$ for $m > n$. Show how this condition changes your answers to Problem DT1.3.3, if at all.

DT1.3.5 Consider the system of Problem DT1.3.3. Assume that $h[n, m]$ depends only on the difference between n and m: $h[n, m] = h[n - m]$. Show how this condition changes your answers to Problem DT1.3.3, if at all.

DT1.3.6 Consider the system of Problem DT1.3.3. Assume that $h[n, m]$ satisfies

$$\sum_{m=-\infty}^{\infty} |h[n, m]| < \infty \tag{31}$$

Show how this condition changes your answers to Problem DT1.3.3, if at all.

1.4   Engineering Models

CT1.4.1 a) Show that an ideal amplifier is a linear system.
b) Show that an ideal amplifier is a time-variant system.
c) Show that an ideal amplifier is a causal system.
d) Show that an ideal amplifier is a stable system.

CT1.4.2 a) Show that an integrator is a linear system.
b) Show that an integrator is a time-invariant system.
c) Show that an integrator is a causal system.
d) Give an example to show that an integrator is an unstable system. Hint: Find and plot the output for the input

$$x(t) = \begin{cases} 1; & t \geq 0 \\ 0; & t < 0 \end{cases} \tag{32}$$

CT1.4.3 a) Show that a differentiator is a linear system.
b) Show that a differentiator is a time-invariant system.
c) Show that a differentiator is a causal system.
d) Give an example to show that a differentiator is an unstable system. Hint: Consider the input

$$x(t) = \begin{cases} 1; & t \geq \epsilon \\ t/\epsilon; & 0 \leq t < \epsilon \\ 0; & t < 0 \end{cases} \tag{33}$$

Make plots of $x(t)$ and $y(t)$. What happens as $\epsilon \to 0$?

CT1.4.4 a) Show that a delay element is a linear system.
b) Show that a delay element is a time-invariant system.
c) Show that a delay element is a causal system.
d) Show that a delay element is a stable system.

CT1.4.5 The *moving average* $y(t)$ of a waveform $x(t)$ is given by the equation

$$y(t) = \frac{1}{\tau} \int_{t-\tau}^{t} x(\lambda) \, d\lambda \tag{34}$$

where τ is the time over which the average is made.
a) Show how the moving average works by plotting a smooth waveform $x(\lambda)$ versus λ and shading the area under $x(\lambda)$ in the interval $t - \tau < \lambda < t$. Notice that the interval $t - \tau < \lambda < t$ moves as t increases.
b) Show that $y(t)$ can be put in the form

$$y(t) = \frac{1}{\tau} \int_{0}^{\tau} x(t - \alpha) \, d\alpha \tag{35}$$

CT1.4.6 Consider the moving average of Problem CT1.4.5.
    a) Show that the I-O relation is linear.
    b) Show that the I-O relation is time invariant.
    c) Show that the I-O relation is causal.
    d) Show that the I-O relation is stable.

DT1.4.1 a) Show that an ideal amplifier is a linear system.
    b) Show that an ideal amplifier is a time-invariant system.
    c) Show that an ideal amplifier is a causal system.
    d) Show that an ideal amplifier is a stable system.

DT1.4.2 a) Show that an accumulator is a linear system.
    b) Show that an accumulator is a time-invariant system.
    c) Show that an accumulator is a causal system.
    d) Give an example to show that an accumulator is an unstable system. Hint: Plot the output for the input

$$x[n] = \begin{cases} 1; & n \geq 0 \\ 0; & n < 0 \end{cases} \tag{36}$$

DT1.4.3 a) Show that a delay element is a linear system.
    b) Show that a delay element is a time-invariant system.
    c) Show that a delay element is a causal system.
    d) Show that a delay element is a stable system.

DT1.4.4 a) Show that a first difference is a linear system.
    b) Show that a first difference is a time-invariant system.
    c) Show that a first difference is a causal system.
    d) Show that a first difference is a stable system.

DT1.4.5 An N-point *moving average* $y[n]$ is computed by the equation

$$y[n] = \frac{x[n] + x[n-1] + x[n-2] + \cdots x[n-N+1]}{N}$$

$$= \frac{1}{N}\sum_{k=0}^{N-1} x[n-k] \tag{37}$$

    a) Show that we can write $y[n]$ as

$$y[n] = \frac{1}{N}\sum_{m=n-N+1}^{n} x[m] \tag{38}$$

    b) Show how the moving average works by plotting a sequence $x[m]$ versus m and identifying the elements of the sequence that are in the interval $n - N + 1 \leq m \leq n$. Notice that the interval $n - N + 1 \leq m \leq n$ moves as m increases.

DT1.4.6 Consider the moving average computer of Problem DT1.4.5.
    a) Show that a moving average computer is a linear system.
    b) Show that a moving average computer is a time-invariant system.
    c) Show that a moving average computer is a causal system.
    d) Show that a moving average computer is a stable system.

Miscellaneous Problems

1. A *quantizer* converts DT samples $x[n] = x(nT_s)$ to numbers taken from a discrete set $Y = \{Y_1, Y_2, \ldots, Y_Q\}$. Typically, Q is a power of 2: $Q = 2^b$, where b is a positive

Figure 1.16
A 3-bit uniform
quantizer.

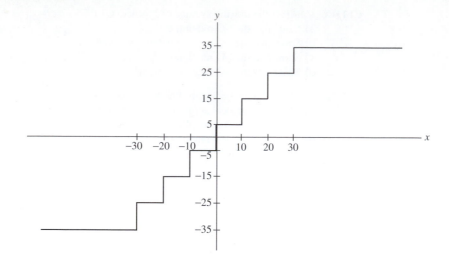

integer. The transformation from quantizer input values $x[n]$ to output values $y[n]$ for any n is given by a graph such as that shown in Figure 1.16. In the figure, $Q = 2^3 = 8$, and $Y = \{-35, -25, -15, -5, 5, 15, 25, 35\}$. These are the only possible values for the output $y[n]$. They are called *representation* (or *level* or *tred*) values. The values $X = \{-30, -20, -10, 0, 10, 20, 30\}$ on the horizontal axis are called *threshold* values. (If $x[n]$ equals a threshold value, $y[n]$ is defined to be the corresponding upper level value. For example, if $x[n] = 0$, then $y[n] = 5$.) The quantizer shown is called a uniform quantizer because consecutive level and threshold values increase by the same amount. It is called a 3-bit quantizer because each of its $8 = 2^3$ levels can be identified uniquely using a binary word $000, 001, 010, \ldots$, or 111 having length 3.

(a) Assume that the sequence $x[n] = \pi n$ is applied to the quantizer shown in the figure. Plot the quantizer input $x[n]$ and output $y[n]$.

(b) Is the transformation $x[n] \rightarrow y[n]$ linear? Justify your answer

(c) Is the transformation $x[n] \rightarrow y[n]$ shift invariant? Justify your answer.

(d) Is the transformation $x[n] \rightarrow y[n]$ causal? Justify your answer.

(e) Is the transformation $x[n] \rightarrow y[n]$ stable? Justify your answer.

 2. In this problem we consider the errors introduced when a real variable is expressed as a finite-length binary word using the uniform $Q = 2^b$ level quantizer of Figure 1.17. This quantizer is a generalization of the 3-bit quantizer of Figure 1.16 to b bits. It is called a b-bit quantizer because each of its $Q = 2^b$ levels can be identified uniquely using a binary word having length b.

The b-bit quantizer converts each DT input sample $x[n] = x(nT_s)$ to an output $y[n]$, whose value is a member of the set $Y = \{Y_1, Y_2, \ldots, Y_Q\}$, where $Y_i = i\Delta - \frac{1}{2}(Q+1)\Delta$; $1 \le i \le Q$ is the ith level value. The threshold values are $X = \{X_1, X_2, \ldots, X_{i-1}\}$, where $X_i = (i - \frac{1}{2}Q)\Delta$; $1 \le i \le Q - 1$. The quantization error is given by

$$e[n] = y[n] - x[n] \tag{39}$$

Assume that the quantizer input $x[n]$ is in the range $-A < x < +A$, shown on the horizontal axis.

Figure 1.17
A *b*-bit uniform quantizer.

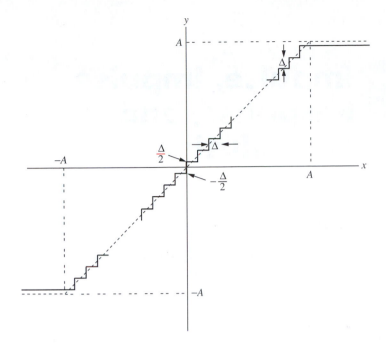

(a) Show from an inspection of the figure that for every n, $|e[n]| \le \frac{1}{2}\Delta$.

(b) Show from an inspection of the figure that $\frac{1}{2}\Delta = A2^{-b}$. Conclude from this equation that the ratio of the peak quantization error to the peak input value is

$$\frac{\Delta}{2A} = 2^{-b} \tag{40}$$

(c) Plot $\Delta/2A$ versus b for $1 \le b \le 10$. Comment on your result. Specifically, state how much $\Delta/2A$ decreases as b is increased by 1 bit.

(d) Decibel values are often used in engineering to express numbers having very large or very small magnitudes using more familiar numbers of moderate sizes. The decibel (dB) value of $\Delta/2A$ is defined as $(\Delta/2A)_{\text{dB}} = 20\log(\Delta/2A)$. For comparison to your plot in (c), plot $(\Delta/2A)_{\text{dB}}$ versus b for $1 \le b \le 10$. Comment on your result. Specifically, state how $(\Delta/2A)_{\text{dB}}$ decreases as b is increased by 1 bit.

2CT Impulse, Impulse Response, and Convolution

2CT.1 Introduction

In this chapter, we introduce the unit impulse, the unit impulse response, and convolution. These three concepts form the foundation for all that follows in this book. These concepts are important because:

- The unit impulse can be viewed as a building block for every waveform.
- An LTI system's response to an impulse completely defines the input-output properties of the system.
- The response of every LTI system to any input is given by a convolution between the input and the system's impulse response.

2CT.2 Unit Impulse

The unit impulse is a theoretical waveform that, along with its time-shifted versions, can serve as a building block for any waveform. Engineers might define a unit impulse as "any pulse that has an area of unity and is so short and so high that it is beyond our equipment's ability to display." A mathematical definition can start with a short rectangular pulse, $p_{\Delta\tau}(t)$, that has unit area and duration $\Delta\tau$ and is centered at the origin:

$$
p_{\Delta\tau}(t) = \begin{cases} 0; & t < -\Delta\tau/2 \\ 1/(2\Delta\tau); & t = -\Delta\tau/2 \\ 1/(\Delta\tau); & -\Delta\tau/2 < t < \Delta\tau/2 \\ 1/(2\Delta\tau); & t = \Delta\tau/2 \\ 0; & t > \Delta\tau/2 \end{cases} \tag{1}
$$

$p_{\Delta\tau}(t)$ is plotted in Figure 2CT.1.

To define a unit impulse, we take the limit $\Delta\tau \to 0$. As $\Delta\tau$ decreases, $p_{\Delta\tau}(t)$ becomes higher in amplitude and shorter in duration but retains its unit area. In the limit $\Delta\tau \to 0$, $p_{\Delta\tau}(t)$ is the unit impulse, symbolized as $\delta(t)$:

$$
\delta(t) = \lim_{\Delta\tau \to 0} p_{\Delta\tau}(t) \tag{2}
$$

Figure 2CT.1
The pulse $p_{\Delta\tau}(t)$.

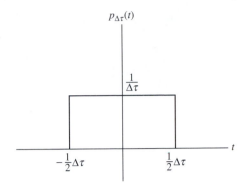

In the limit $\Delta\tau \to 0$, we can no longer draw $p_{\Delta\tau}(t)$ on paper or display it on any physical device. We depict it as illustrated in Figure 2CT.2. The arrow on the top of the impulse points to its unlimited amplitude. The number 1 to the side of the impulse indicates the area of the impulse. As illustrated in Figure 2CT.2, $\delta(t)$ is defined for all t. However, it is natural to say that the impulse $\delta(t)$ *occurs* when its argument t equals 0.

Figure 2CT.2
Unit impulse.

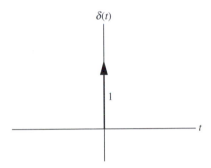

Equation (2) is one of many possible equivalent definitions. For example, $\delta(t)$ could be defined as the limit of a gaussian pulse (Figure 2CT.3)

$$\delta(t) = \lim_{\Delta\tau \to 0} g_{\Delta\tau}(t) \tag{3}$$

where

$$g_{\Delta\tau}(t) = \frac{1}{\sqrt{2\pi}\,\Delta\tau} e^{-0.5(t/\Delta\tau)^2} \tag{4}$$

Figure 2CT.3
Gaussian pulse.

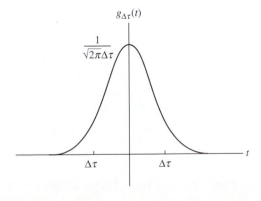

Figure 2CT.4
Sinc function.

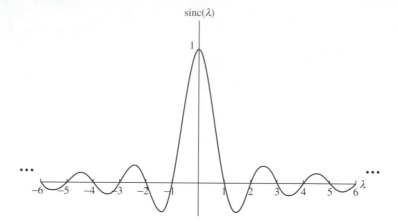

Figure 2CT.5
$(1/\Delta\tau)\,\mathrm{sinc}(t/\Delta\tau)$.

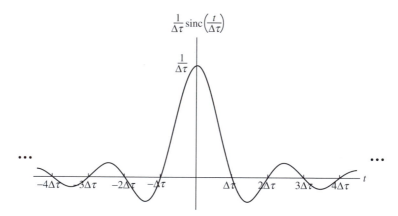

We can also define an impulse as the limit of a *sinc* function

$$\delta(t) = \lim_{\Delta\tau \to 0} \frac{1}{\Delta\tau} \mathrm{sinc}\left(\frac{t}{\Delta\tau}\right) \tag{5}$$

where

$$\mathrm{sinc}(\lambda) = \frac{\sin(\pi\lambda)}{\pi\lambda} \tag{6}$$

The sinc function Eq. (6) is well-known in engineering.[1] We will encounter it many times in upcoming chapters. It is plotted in Figure 2CT.4. Notice that $\mathrm{sinc}(\lambda) = 1$ for $\lambda = 0$, and $\mathrm{sinc}(\lambda) = 0$ for $\lambda = \pm1, \pm2, \ldots$. $\frac{1}{\Delta\tau}\mathrm{sinc}(t/\Delta\tau)$ is shown in Figure 2CT.5. Like $p_{\Delta\tau}(t)$ and $g_{\Delta\tau}(t)$, $\frac{1}{\Delta\tau}\mathrm{sinc}(t/\Delta\tau)$ becomes a unit impulse in the limit $\Delta\tau \to 0$.

The definition using $p_{\Delta\tau}(t)$ will be used in Section 2CT.4, where we show that any waveform can be represented as an integral of scaled and shifted impulse functions. The definition based on $g_{\Delta\tau}(t)$ is useful when we want to take derivatives of an impulse. Finally, the definition using the sinc function will arise naturally in upcoming derivations. We are free to use the definition that is easiest for the problem at hand.

---

[1] The sinc function Eq. (6) was introduced by P.M. Woodward [Woodward, P.M., 1964]. All text references are listed in the bibliography. Some books define the sinc function differently as $\mathrm{sinc}(\lambda) = \sin(\lambda)/\lambda$.

Figure 2CT.6
Delays and an adder.

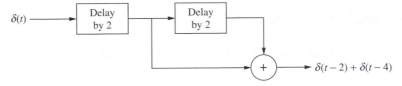

Figure 2CT.7
Shifted impulse
having area 8.2.

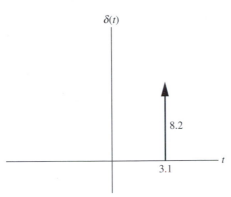

We can shift $\delta(t)$ by replacing its argument t with $t - \tau$ where τ is the amount of the shift. The argument of $\delta(t - \tau)$ equals 0 when $t = \tau$: It is for this value of t that the impulse occurs. Delay by τ can be accomplished by using a delay element. In Figure 2CT.6 we use two delays and an adder to obtain an output $\delta(t - 2) + \delta(t - 4)$ for an input $\delta(t)$.

We can obtain an impulse having some other area c by multiplying a unit impulse by c. The impulse $8.2\delta(t - 3.1)$ is illustrated in Figure 2CT.7. The value 8.2 is the area, not the amplitude, of the impulse. The area should be drawn to the side of the impulse, not at the top of the arrow.

Unit impulses possess a useful and important property called the *sampling property*. If you multiply an arbitrary waveform $x(t)$ by an impulse occurring at time τ, you obtain an impulse having area $x(\tau)$ occurring at time τ.

Sampling Property

$$x(t)\delta(t - \tau) = x(\tau)\delta(t - \tau) \tag{7}$$

We can view the impulse as "sampling" $x(t)$ at $t = \tau$. Equation (7) is illustrated in Figure 2CT.8. The sampling property applies to every $x(t)$ that is continuous at $t = \tau$.

Figure 2CT.8
Sampling property of
the unit impulse:
(a) $\delta(t - \tau)$; (b) $x(t)$;
(c) $x(t)\delta(t - \tau) =$
$x(\tau)\delta(t - \tau)$.

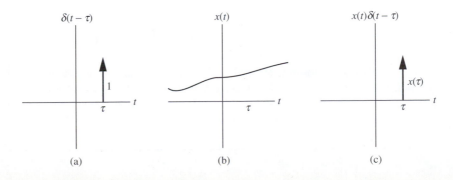

2CT.3   Impulse Response

The response of a CT LTI system to a unit impulse is called the system's *impulse response* and is denoted as $h(t)$. We will show in Section 2CT.5 that $h(t)$ completely characterizes the input-output (I-O) properties of an LTI system. For now, let's become acquainted with $h(t)$ by means of some examples.

EXAMPLE 2CT.1   The Integrator

One of the elementary LTI systems introduced in Chapter 1 is an *integrator*, defined by the I-O equation

$$y(t) = \int_{-\infty}^{t} x(\lambda)d\lambda \tag{8}$$

We can find the impulse response of an integrator by evaluating Eq. (8) for $x(\lambda) = \delta(\lambda)$. The region of integration in Eq. (8) is illustrated in Figure 2CT.9 where $x(\lambda) = \delta(\lambda)$.

Figure 2CT.9
Region of integration
for eq. (8) (shaded).

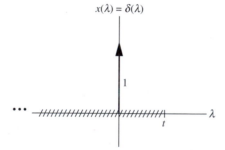

We can see from the figure that the result of the integration is

$$h(t) = \int_{-\infty}^{t} \delta(\lambda)d\lambda = \begin{cases} 0; & t < 0 \\ \frac{1}{2}; & t = 0 \\ 1; & t > 0 \end{cases} \tag{9}$$

The right side of Eq. (9) is easy to understand: For $t < 0$, the impulse is not included in the region of integration. Therefore $h(t) = 0$ for $t < 0$. For $t > 0$, the impulse is included in the region of integration. Therefore $h(t) = 1$ for $t > 0$. The value of $h(t)$ when $t = 0$ follows from the definition of the unit impulse because half of the area of $p_{\Delta\tau}(\lambda)$ and therefore of $\delta(\lambda)$ lies to the left of the origin and half lies to the right.

The result, Eq. (9), is important. It will be useful to define the function on the right side of Eq. (9) as

$$u(t) = \begin{cases} 0; & t < 0 \\ \frac{1}{2}; & t = 0 \\ 1; & t > 0 \end{cases} \qquad (10)$$

$u(t)$ is called the *unit step function*. It is plotted in Figure 2CT.10. Example 2CT.1 has shown us that, if we integrate a unit impulse, we get a unit step. It follows that a unit impulse is the derivative of a unit step.

$$\delta(t) = \frac{du(t)}{dt} \qquad (11)$$

Figure 2CT.10
Unit step function.

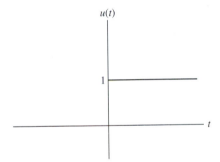

EXAMPLE 2CT.2   An Ideal Time Delay

Another elementary system introduced Chapter 1 is a time-delay element. The output of a time-delay element is simply the input shifted in time by some quantity $\tau \geq 0$. What is the impulse response of a time-delay element? Remember that the impulse response of a system is, by definition, the system's response to an impulse $\delta(t)$. A moment's thought tells us that the impulse response of a time-delay element is a time-delayed impulse:

$$h(t) = \delta(t - \tau) \qquad (12)$$

The term "time delay" implies that $\tau \geq 0$. More generally, we use the term "time shift" to denote the operation $x(t) \longrightarrow x(t - \tau)$ for positive or negative τ. The impulse response of a time shifter is given by Eq. (12).

EXAMPLE 2CT.3   A First-Order System

Consider the *first-order* system shown in Figure 2CT.11(a). An electrical engineering realization of this system, called an *RC low-pass filter*, is shown in Figure 2CT.11(b), where $x(t)$ and $y(t)$ are voltages and $RC = \tau_c$.

Notice that the output of the integrator in Figure 2CT.11(a) is $y(t)$. Therefore, the input to the integrator is $\frac{d}{dt}y(t)$. Looking at the block diagram, we see that

$$\frac{dy(t)}{dt} = \frac{1}{\tau_c}[x(t) - y(t)] \qquad (13a)$$

Equation (13a) is the I-O equation of the system. We can rearrange it into the standard form

$$\tau_c \frac{dy(t)}{dt} + y(t) = x(t) \tag{13b}$$

The equation is called a *first-order* differential equation because the highest derivative of y has order 1.

Figure 2CT.11
First-order system: (a)
block diagram; (b) *RC*
circuit realization.

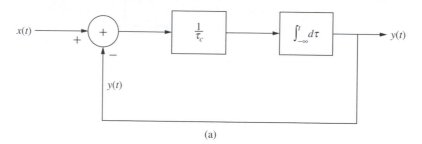

(a)

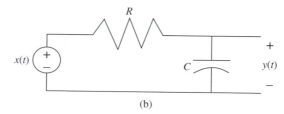

(b)

We find the impulse response by substituting $x = \delta(t)$ and solving for $y(t) \equiv h(t)$.

$$\tau_c \frac{dh(t)}{dt} + h(t) = \delta(t) \tag{14}$$

To solve this equation, we recognize that the block diagram represents a physical system that obeys cause and effect. The impulse occurs at $t = 0$. Since we are looking for the response to the impulse (and no other causes), there is no output before the impulse occurs. Therefore

$$h(t) = 0 \qquad \text{for } t < 0 \tag{15}$$

When the input impulse occurs, it causes an impulse

$$\frac{dy(t)}{dt} = \frac{\delta(t)}{\tau_c} \tag{16}$$

into the integrator. This creates an output

$$y(0^+) = h(0^+) = \int_{0^-}^{0^+} \frac{\delta(\tau)}{\tau_c} d\tau = \frac{1}{\tau_c} \tag{17}$$

immediately following the impulse. In Eq. (17), 0^- and 0^+ denote, respectively, the instants immediately before and immediately after $t = 0$, which in this case mean immediately before and immediately after the impulse. The output at $t = 0$ is $y(0) = h(0) = 1/(2\tau_c)$ because the impulse is just halfway integrated $t = 0$. After the impulse, the input goes back to 0 and Eq. (14) becomes

$$\tau_c \frac{dh(t)}{dt} + h(t) = 0 \qquad \text{for } t > 0 \tag{18}$$

We now need to find the function $h(t)$ that satisfies the preceding equation. Let's try

$$h(t) = Ke^{st} \tag{19}$$

where K and s are constants. When we substitute Eq. (19) into Eq. (18) we obtain

$$\tau_c s K e^{st} + K e^{st} = 0 \tag{20}$$

which simplifies to

$$\tau_c s + 1 = 0 \tag{21}$$

Therefore, our trial solution (19) works if $s = -1/\tau_c$. The value for K can be determined by the "initial condition" (17). If we set $t = 0^+$ in Eq. (19) and compare with Eq. (17), we find that $K = 1/\tau_c$. Therefore, we have found the impulse response of the first-order system:

$$h(t) = \begin{cases} 0; & t < 0 \\ \frac{1}{2\tau_c}; & t = 0 \\ \frac{1}{\tau_c}e^{-\frac{t}{\tau_c}}; & t > 0 \end{cases} \tag{22}$$

This impulse response is plotted in Figure 2CT.12. It is convenient to use a unit step function to write Eq. (22) compactly as

$$h(t) = \frac{1}{\tau_c}e^{-t/\tau_c}u(t) \tag{23}$$

The quantity τ_c is called the *time constant* of the first-order system. Referring to the circuit, Eq. (23) tells us that an input voltage impulse abruptly charges the capacitor to $h(0^+) = 1/\tau_c$ volts. For $t > 0$, the input returns to 0 volts and is equivalent to a short circuit. The charge on the capacitor then decays through the resistor with time constant $\tau_c = RC$.

Figure 2CT.12
Impulse response of a first-order system.

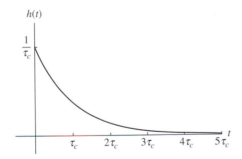

The first-order system of Figure 2CT.11 is important both in theory and practice. We shall return to it frequently. nth-order systems are described by nth-order differential equations. We will encounter some of these higher-order systems and equations as we proceed. A systematic development of nth-order systems is contained in Chapter 6CT.

EXAMPLE 2CT.4　Room Acoustics with Feedback

A final example is given by the room acoustics in Figures 1.1 and 1.3. The pertinent features are redrawn in Figure 2CT.13(a). We are interested in the response to an acoustical unit impulse (say, a single clap) input to the microphone. We take the output to be the sound out of the speaker. For simplicity, we assume that the system from microphone input to speaker output acts as an ideal amplifier. Let's first find the *open-loop* impulse response. The open-loop response can be obtained by hypothetically breaking the loop, as shown in Figure 2CT.13(b). (This cannot

Figure 2CT.13
Block diagram of
electroacoustic
system: (a) closed
loop; (b) hypothetical
open loop.

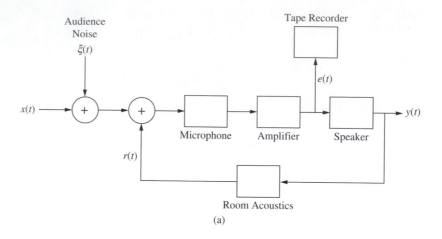

(a)

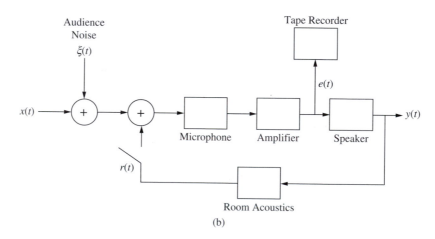

(b)

be done physically without the use of a timed switch.) The sound of the clap, modeled as $x(t) = \delta(t)$, goes into the microphone and appears as a sound $\alpha\delta(t)$ at the speaker output. α is a real constant accounting for the conversion by the microphone from acoustical to electrical energy, amplification, and the conversion from electrical energy to acoustic energy by the speaker. The acoustic signal $\alpha\delta(t)$ travels out from the speaker, bounces off the walls, and comes back to the microphone. Let's assume for simplicity that there is just one significant echo $\beta\alpha\delta(t - \tau)$, heard at the input to the microphone due to the original clap. τ is the time it takes for the sound to travel out from the speaker to the wall and return to the microphone as an echo. β is a real constant that accounts for the attenuation due to sound travel and any absorption at the wall where the sound reflection occurred. Since the loop is hypothetically open, as shown in the figure, the echo is not picked up by the microphone and does not come out of the speaker. Therefore the open-loop impulse response is just the sound at the speaker resulting from the original clap.

$$h_{\mathrm{OL}}(t) = \alpha\delta(t) \tag{24}$$

Now consider the loop closed, as we have in the actual physical situation of Figure 1.1. The closed loop system model is shown in Figure 2CT.13(a). The input to the microphone is the original clap sound $\delta(t)$. This produces the echo $\alpha\beta\delta(t - \tau)$. The echo, in turn, is picked up by the microphone, electronically amplified, and sent out of the speaker. At the speaker output, it is the acoustic signal $\alpha^2\beta\delta(t - \tau)$. This acoustic signal creates another echo at the microphone's input $(\alpha\beta)^2\delta(t - 2\tau)$. $(\alpha\beta)^2\delta(t - 2\tau)$ is, in turn, electronically amplified and broadcast becoming the speaker output $\alpha^3\beta^2\delta(t-2\tau)$. This second echo is, in turn, amplified, and so on. The acoustic signal at the output of the speaker is therefore

$$h(t) = \alpha\delta(t) + \alpha^2\beta\delta(t-\tau) + \alpha^3\beta^2\delta(t-2\tau) + \alpha^4\beta^3\delta(t-3\tau) + \cdots = \alpha\sum_{n=0}^{\infty}(\alpha\beta)^n\delta(t-n\tau) \qquad (25)$$

This is the system's closed-loop impulse response. A plot of $h(t)$ is shown in Figure 2CT.14

Figure 2CT.14
Impulse response of electroacoustic system.

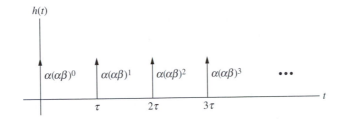

2CT.4

Waveform Represented as an Integral of Shifted Impulses

An arbitrary waveform $x(t)$ can be represented as an integral of shifted, weighted unit impulses. In this sense, the unit impulse is a building block for any waveform. We can derive this representation for a waveform by first approximating $x(t)$ with a sum (superposition) of adjacent pulses having duration $\Delta\tau$ and heights equal to the values of $x(t)$ at the center of each pulse (Figure 2CT.15). The pulses have the form $x(k\Delta\tau)p_{\Delta\tau}(t-k\Delta\tau)\Delta\tau$. The pulse $x(k\Delta\tau)p_{\Delta\tau}(t-k\Delta\tau)\Delta\tau$ is centered at $t=k\Delta\tau$, has amplitude $x(k\Delta\tau)$, and duration $\Delta\tau$. The waveform $x(t)$ is approximately given by the superposition

$$x(t) \approx \sum_{k=-\infty}^{\infty} x(k\Delta\tau)p_{\Delta\tau}(t-k\Delta\tau)\Delta\tau = \sum_{\tau=k\Delta\tau=-\infty}^{\infty} x(\tau)p_{\Delta\tau}(t-\tau)\Delta\tau \qquad (26)$$

Figure 2CT.15
Approximation to $x(t)$.

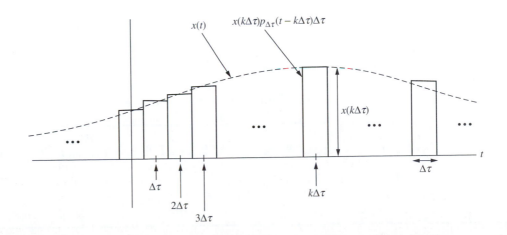

We now take the limit $\Delta\tau \to 0$. In this limit, the approximation (26) becomes an equality, τ approaches a real variable, and the sum becomes an integral of impulses occurring at time $t = \tau$ and having areas $x(\tau)d\tau$.

$$x(t) = \int_{-\infty}^{\infty} x(\tau)\delta(t - \tau)d\tau \qquad (27)$$

Another way to look at Eq. (27) is in terms of the *sifting property* of the unit impulse. The factor $\delta(t - \tau)$ can be thought of as "sifting" $x(t)$ out of all the terms in the integral.

It might appear that Eq. (27) is a circular way to represent a signal because the function $x(\cdot)$ is on both sides. Although the observation makes a good point, we see in the next section how Eq. (27) helps us understand the response of an LTI system to any input.

2CT.5   Convolution Integral

In this section we derive a formula for the response of any CT LTI system to any input. We start with Eq. (26), which is an approximate expression for the input $x(t)$.

If we apply just $p_{\Delta\tau}(t)$ to the LTI system, the response is an approximation $\widehat{h}(t)$ to the system's impulse response, $h(t)$.

$$p_{\Delta\tau}(t) \xrightarrow{h(t)} \widehat{h}(t) \qquad (28a)$$

The arrow is flow graph notation showing that the input on the left produces the output on the right. We have placed the actual impulse response $h(t)$ on top of the arrow to identify the system under consideration.

Consider next applying the pulse $x(k\Delta\tau)p_{\Delta\tau}(t - k\Delta\tau)\Delta\tau$ to the system. Notice that $x(k\Delta\tau)p_{\Delta\tau}(t - k\Delta\tau)\Delta\tau$ is a shifted and scaled version of $p_{\Delta\tau}(t)$. Therefore, if we apply $x(k\Delta\tau)p_{\Delta\tau}(t - k\Delta\tau)\Delta\tau$ to the system, the response is $x(k\Delta\tau)\widehat{h}(t - k\Delta\tau)\Delta\tau$ because the system is LTI. We can write this result as

$$x(k\Delta\tau)p_{\Delta\tau}(t - k\Delta\tau)\Delta\tau \xrightarrow{h(t)} x(k\Delta\tau)\widehat{h}(t - k\Delta\tau)\Delta\tau \qquad (28b)$$

By superposition, the response to the sum of pulses (26) is the sum of the responses to the individual pulses, i.e.,

$$\sum_{k=-\infty}^{\infty} x(k\Delta\tau)p_{\Delta\tau}(t - k\Delta\tau)\Delta\tau \xrightarrow{h(t)} \sum_{k=-\infty}^{\infty} x(k\Delta\tau)\widehat{h}(t - k\Delta\tau)\Delta\tau \qquad (28c)$$

We write the above as

$$\sum_{\tau=k\Delta\tau=-\infty}^{\infty} x(\tau)p_{\Delta\tau}(t - \tau)\Delta\tau \xrightarrow{h(t)} \sum_{\tau=k\Delta\tau=-\infty}^{\infty} x(\tau)\widehat{h}(t - \tau)\Delta\tau \qquad (28d)$$

We now take the limit $\Delta\tau \to 0$. In this limit, $p_{\Delta\tau}(t - \tau)$ becomes $\delta(t - \tau)$, τ approaches a real variable, $\widehat{h}(t - \tau)$ becomes $h(t - \tau)$, and the sums become integrals. Thus, in the limit $\Delta\tau \to 0$, Eq. (28d) yields the following basic result:

The Input-Output Relation for Every LTI System is a Convolution

$$x(t) \xrightarrow{h(t)} \int_{-\infty}^{\infty} x(\tau)h(t - \tau)\,d\tau \stackrel{\Delta}{=} x(t) * h(t) \qquad (29)$$

where we used Eqs. (26)–(27). The integral in (29) is called a *convolution* integral. $x(t) *$ $h(t)$ is a useful way to denote the convolution between $x(t)$ and $h(t)$. The symbol $*$ denotes the convolution operation.

Let's check Eq. (29) using the input $x(t) = \delta(t)$. The substitution of $x(t) = \delta(t)$ into Eq. (29) yields

$$y(t) = \int_{-\infty}^{\infty} \delta(\tau)h(t - \tau)d\tau = \delta(t) * h(t) = h(t) \tag{30}$$

(We used the sifting property to evaluate the integral.) The answer we obtained $h(t)$ is certainly correct because the response of any system to a unit impulse is indeed the system's impulse response!

It is not difficult to prove that the convolution operation ($*$) has the following properties:

- Commutativity:

$$x(t) * h(t) = h(t) * x(t) \tag{31}$$

- Superposition:

$$\{a_1x_1(t) + a_2x_2(t)\} * h(t) = a_1x_1(t) * h(t) + a_2x_2(t) * h(t) \tag{32}$$

- Associativity:

$$x(t) * \{h_1(t) * h_2(t)\} = \{x(t) * h_1(t)\} * h_2(t) \tag{33}$$

These properties are established in the problems. They apply for all $x(t), h(t), x_1(t),$ $x_2(t), a_1, a_2, h_1(t),$ and $h_2(t)$.

The operations performed by a convolution integral can be visualized with the aid of plots and a sliding tape, as illustrated in the following example.

EXAMPLE 2CT.5   Understanding Convolution

Consider the problem of finding the output of an LTI system having impulse response

$$h(t) = \begin{cases} 0; & t < 0 \\ 1 - \frac{t}{T}; & 0 \le t \le T \\ 0; & T < t \end{cases} \tag{34}$$

when the input is

$$x(t) = Au(t + a) \tag{35}$$

Figure 2CT.16(a) shows a plot of $x(t)$ and $h(t)$. To help us understand how to evaluate the convolution integral Eq. (29), we can imagine that $x(t)$ and $h(t)$ are recorded on analog tapes 1 and 2, respectively, shown to the right of the plots. The start of the tape containing $x(t)$ is aligned to the left of that containing $h(t)$ because the nonzero part of $x(t)$ begins at $t = -a$ and the nonzero part of $h(t)$ begins at $t = 0$.

Figure 2CT.16(b) shows graphs of the functions as they appear in the integral (29). Both functions $x(\tau)$ and $h(t - \tau)$ are plotted versus τ because it is the variable of integration. The corresponding analog tapes are again shown to the right of the graphs. Tape 2 has been reversed in direction and shifted by t to agree with the plot of $h(t - \tau)$. To compute Eq. (29) for a specific t, the corresponding values of $x(\tau)$ and $h(t - \tau)$ are multiplied and integrated. We can see these corresponding values by looking at the graphs or by looking at the tapes. As t increases, the function $h(t - \tau)$ and tape 2 slide to the right. By repeating the product and integration operation, $y(t)$ is computed for every value of t. The detailed calculations are shown on pages 30–31.

Figure 2CT.16
Graphs and sliding
tapes.

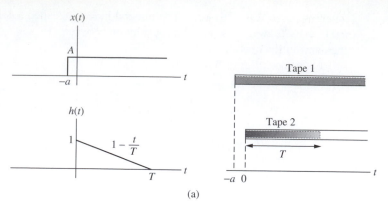

(a)

The two functions to be convolved,

$$x(t) = Au(t+a) \quad \text{and} \quad h(t) = \begin{cases} 0; & t < 0 \\ 1 - \frac{t}{T}; & 0 \le t \le T \\ 0; & T < t \end{cases}$$

as shown in Figure 2CT.16(a). The functions have been recorded on the tapes adjacent to the plots. Dark corresponds to high amplitude, white to 0 amplitude. Tape 1 starts at $t = -a$ and tape 2 starts at $t = 0$.

Figure 2CT.16
(*Cont.*)

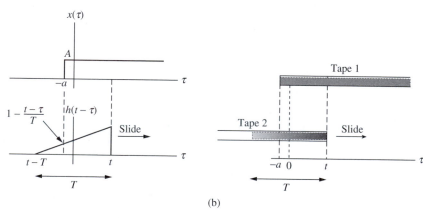

(b)

Figure 2CT.16(b) contains plots of the same functions as they appear in the convolution integral

$$y(t) = \int_{-\infty}^{\infty} h(t-\tau)x(\tau)\,d\tau$$

Here they are plotted versus the variable of integration τ. t is a parameter that can be positive or negative but that is assumed to be positive in the figure. Tape 2 has been reversed in direction and translated to the right by t. As t increases, the function $h(t-\tau)$ and tape 2 slide to the right.

To perform the convolution, fix t and multiply the corresponding values of $h(t-\tau)$ and $x(\tau)$ for every τ to obtain $h(t-\tau)x(\tau)$ as a function of τ. Then integrate $h(t-\tau)x(\tau)$ with respect to τ, as indicated by the convolution integral. You can imagine this operation being performed on either the plots or the tapes, whichever you prefer.

In this example, $h(t-\tau)$ ends when $\tau = t$. For $t < -a$ therefore, all of $h(t-\tau)$ appears before the start of $x(\tau)$ and the product $h(t-\tau)x(\tau)$ equals 0. Consequently

$$y(t) = \int_{-\infty}^{\infty} h(t-\tau)x(\tau)\,d\tau = \int_{-\infty}^{\infty} 0\,d\tau = 0 \quad \text{for } t < -a$$

For $-a \leq t < -a+T$ (as illustrated), $h(t-\tau)$ and $x(\tau)$ partially overlap. The product $h(t-\tau)x(\tau)$ equals 0 for $\tau < -a$ and $\tau > t$. The function $x(\tau) = A$ for $\tau > -a$. Therefore

$$y(t) = A \int_{-a}^{t} h(t-\tau)\, d\tau = A \int_{-a}^{t} \left(1 - \frac{t-\tau}{T}\right) d\tau$$

$$= A(t+a)\left(1 - \frac{t+a}{2T}\right) \qquad \text{for } -a \leq t \leq -a+T$$

For $t > -a+T$, $h(t-\tau)$ is entirely contained within the nonzero part of $x(\tau)$. The area of the product $h(t-\tau)x(\tau)$ does not change as t increases beyond $-a+T$. This area is just the area of a triangle having base T and height A and is therefore $AT/2$. Therefore

$$y(t) = \frac{1}{2}AT \qquad \text{for } t > -a+T$$

A plot of $y(t)$ is shown in Figure 2CT.16(c).

Figure 2CT.16
(Cont.)

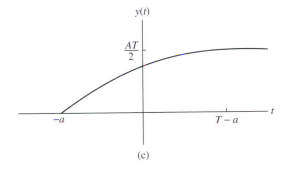

(c)

In the preceding example, we used analog tapes and plots to help us evaluate a convolution. In the following example, we discard the sliding tapes and work with just the plots. This method is called the graphical method of convolution.

EXAMPLE 2CT.6   Graphical Method of Convolution

Consider the problem of finding the output of an LTI system having the impulse response

$$h(t) = e^{-t/3}u(t) \tag{36}$$

for the input

$$x(t) = \begin{cases} 0; & t < 0 \\ e^t; & 0 \leq t \leq 1 \\ 0; & 1 < t \end{cases} \tag{37}$$

$x(t)$ and $h(t)$ are plotted in Figure 2CT.17.
The response is given by

$$y(t) = \int_{-\infty}^{\infty} x(\tau)h(t-\tau)\, d\tau \tag{38}$$

Figure 2CT.17
Input $x(t)$ and
impulse response
$h(t)$.

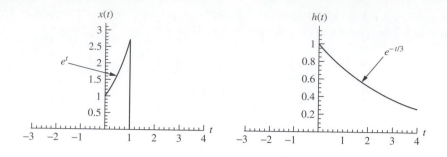

Figure 2CT.18
$x(\tau), h(t - \tau)$, and
$x(\tau)h(t - \tau)$: (a)
$t < 0$; (b) $0 \le t \le 1$;
(c) $t > 1$.

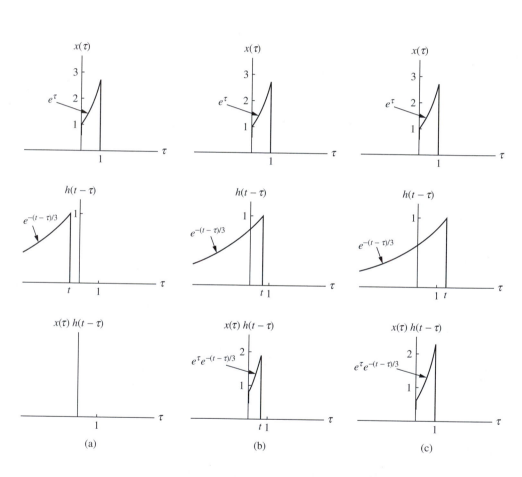

To evaluate this convolution we need to know the expression for $x(\tau)h(t - \tau)$. To find this expression, we can plot $x(\tau), h(t - \tau)$, and $x(\tau)h(t - \tau)$ versus τ, as shown in Figure 2CT.18. We include the formulas for $x(\tau), h(t - \tau)$, and $x(\tau)h(t - \tau)$ on the plots.

Figure 2CT.18(a) is a plot of $x(\tau), h(t - \tau)$ and the product $x(\tau)h(t - \tau)$ for $t < 0$. Notice that there is no overlap between the functions $x(\tau)$ and $h(t - \tau)$ for $t < 0$, and therefore $x(\tau)h(t - \tau) = 0$. Consequently (38) becomes

$$y(t) = \int_{-\infty}^{\infty} 0 \, d\tau = 0 \qquad \text{for } t < 0. \tag{39}$$

Figure 2CT.18(b) is a plot of $x(\tau), h(t-\tau)$ and the product $x(\tau)h(t-\tau)$ for $0 \le t \le 1$. Here $x(\tau)$ and $h(t-\tau)$ overlap in the range $0 \le \tau \le t$. Consequently $x(\tau)h(t-\tau)$ is nonzero in this range and Eq. (38) becomes

$$y(t) = \int_0^t e^{\tau} e^{-\frac{t-\tau}{3}} \, d\tau = \tfrac{3}{4}\left(e^t - e^{-\frac{t}{3}}\right) \qquad \text{for } 0 \le t \le 1 \tag{40}$$

Finally, for $t > 1$ (Figure 2CT.18[c]), $x(\tau)h(t-\tau)$ is nonzero in the range $0 \le \tau \le 1$ so that

$$y(t) = \int_0^1 e^{\tau} e^{-\frac{t-\tau}{3}} \, d\tau = \tfrac{3}{4}\left(e^{\frac{4}{3}} - 1\right)e^{-\frac{t}{3}} \qquad \text{for } t > 1 \tag{41}$$

In conclusion

$$y(t) = \begin{cases} 0; & \text{for } t < 0 \\ \tfrac{3}{4}\left(e^t - e^{-\frac{t}{3}}\right); & \text{for } 0 \le t \le 1 \\ \tfrac{3}{4}\left(e^{\frac{4}{3}} - 1\right)e^{-\frac{t}{3}}; & \text{for } t > 1. \end{cases} \tag{42}$$

The response $y(t)$ is plotted in Figure 2CT.19.

Figure 2CT.19
Output, $y(t)$.

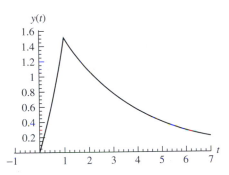

In conclusion

2CT.6   BIBO Stability

Recall from Chapter 1 that *a system is BIBO stable if every bounded input produces a bounded output.* We now state and prove a necessary and sufficient condition for a CT LTI system to be stable.

A CT LTI System is BIBO Stable If and Only If $h(t)$ is Absolutely Integrable

$$\int_{-\infty}^{\infty} |h(t)| \, dt < \infty \tag{43}$$

Proof: We first show that Eq. (43) is a *sufficient* condition. Suppose that we have an LTI system and we apply a bounded input: $|x(t)| \le B$. We can bound the output as follows:

$$|y(t)| = \left| \int_{-\infty}^{\infty} x(t-\tau)h(\tau) \, d\tau \right| \le \int_{-\infty}^{\infty} |x(t-\tau)||h(\tau)| \, d\tau \le \int_{-\infty}^{\infty} B|h(\tau)| \, d\tau$$

$$= B \int_{-\infty}^{\infty} |h(\tau)| \, d\tau \tag{44}$$

The first inequality states that the absolute value of an integral is upper bounded by the integral of absolute values of the integral's argument. The second inequality states that the input waveform is bounded. If we compare the left-and right-hand sides of Eq. (44) we see that $|y(t)|$ is finite if $\int_{-\infty}^{\infty} |h(\tau)|d\tau$ is finite. Therefore, the condition that $h(t)$ is absolutely integrable Eq. (43) is *sufficient* to ensure that a bounded input yields a bounded output.

We now show that Eq. (43) is a *necessary* condition. The condition is necessary if there is at least one bounded input that leads to an unbounded output when Eq. (43) is *not* true. The following waveform is one such input.

$$x(t) = \begin{cases} +1; & \text{if } h(-t) \geq 0 \\ -1; & \text{if } h(-t) < 0 \end{cases} \tag{45}$$

This input waveform consists of amplitudes ± 1 and therefore is bounded. For this input, we have

$$x(-t)h(t) = |h(t)| \tag{46}$$

for all t so that

$$y(0) = \int_{-\infty}^{\infty} x(-\tau)h(\tau)\,d\tau = \int_{-\infty}^{\infty} |h(\tau)|d\tau \tag{47}$$

We see from Eq. (47) that if Eq. (43) is not satisfied, then the input Eq. (45) leads to an output value $y(0)$ that is unbounded. Therefore we have proven that Eq. (43) is not only a sufficient condition for BIBO stability, but also a necessary one.

As an example, consider an integrator. The impulse response of an integrator is $h(t) = u(t)$. Here, condition (43) is not satisfied because the area under $|u(t)|$ is infinite. Thus, an integrator is not BIBO stable. We can check this result by applying the bounded input $x(t) = u(t)$ to the integrator. The output is

$$y(t) = \int_{-\infty}^{t} x(\lambda)d\lambda = \int_{-\infty}^{t} u(\lambda)d\lambda = \begin{cases} 0; & t < 0 \\ t; & t \geq 0 \end{cases} \tag{48}$$

The result is plotted in Figure 2CT.20. It is called a *unit ramp*.

$$r(t) = tu(t) \tag{49}$$

Figure 2CT.20
Unit ramp $r(t)$.

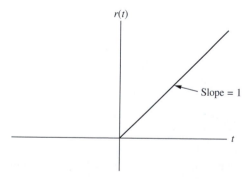

The unit ramp is an unbounded waveform: There is no finite number B for which $|r(t)| \leq B$ for all t. We see that the response of an integrator to the bounded input $x(t) = u(t)$ is the unbounded output $r(t)$. This checks our finding, based on Eq. (43), that an integrator is not BIBO stable.

2CT.7 Causal Systems

Recall from Chapter 1 that a system is causal if, for every input, the output $y(t_o)$ at any time t_o does not depend on input values $x(t_i)$ that occur at times *after* t_o, i.e., for $t_i > t_o$. We can find a necessary and sufficient condition for a CT LTI system to be causal by using Eqs. (29) and (31) to write

$$y(t_o) = \int_{-\infty}^{\infty} h(\tau)x(t_o - \tau)\,d\tau = \int_{-\infty}^{0_-} h(\tau)x(t_o - \tau)\,d\tau + \int_{0}^{\infty} h(\tau)x(t_o - \tau)\,d\tau \quad (50)$$

In the last integral, τ is greater than or equal to 0 because the range of integration runs from 0 to ∞. Therefore the last integral includes only values of the input $x(t_i) \equiv x(t_o - \tau)$ that occur at or before time t_o. In the next to last integral, τ is less than 0. Therefore this integral includes values of the input $x(t_i) \equiv x(t_o - \tau)$ that occur after time t_o. For a causal system, no input values occurring after time t_o effect the output at time t_o. Therefore, the system is causal if and only if the next to last integral is 0, and this is true for every input if and only if $h(\tau) = 0$ for $\tau < 0$.

An LTI System is Causal If and Only If $h(t) = 0$ For $t < 0$

2CT.8 Basic Connections of Two CT LTI Systems

Figure 2CT.21 shows ways to connect two LTI systems together to create a larger system. The unit impulse response of the larger system can be obtained in the usual way by applying a unit impulse to the input. The results are informative.

For example, Figure 2CT.21(a) shows a *cascade* (or *series*) connection of two LTI systems. If we apply a unit impulse to the input of Figure 2CT.21(a), we know that the output of the first system will be $h_1(t)$. This output is the input applied to the second system. The output of the second system is therefore given by the convolution of $h_1(t)$ with $h_2(t)$. The unit impulse response of the complete system is therefore

$$h(t) = h_1(t) * h_2(t) \quad (51)$$

By the commutation property (32) of convolution, we know that $h_1(t) * h_2(t) = h_2(t) * h_1(t)$. This proves that we can interchange the orders of cascaded LTI systems without changing the overall impulse response.

Figure 2CT.21(b) shows a *parallel* connection of two LTI systems. It is left as a problem to show that the overall unit impulse response of the system is

$$h(t) = h_1(t) + h_2(t) \quad (52)$$

Figure 2CT.21(c) shows an LTI system connected in a loop. We do not yet have the analytical tools to efficiently describe the unit impulse response of systems connected in loops. The following expression, however, can be obtained by tracing the signal around the loop when a unit impulse is applied. The expression is correct, but has limited use.

$$h(t) = \delta(t) + h_1(t) + h_1(t) * h_1(t) + h_1(t) * h_1(t) * h_1(t) + \cdots \quad (53)$$

Figure 2CT.21
Basic connections of
LTI systems:
(a) cascade;
(b) parallel; (c) loop.

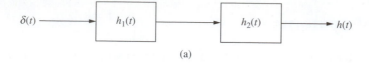

(a)

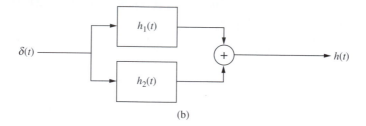

(b)

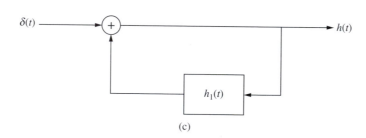

(c)

We develop more understanding and more useful expressions for the unit impulse response of loops as we progress in this book.

As an illustration of the preceding results, consider the system of Figure 2CT.22(a), which consists of a system having an impulse response $h_1(t)$ followed by a differentiator. The input is $x(t)$. By the commutativity property of LTI systems, it follows that the system of Figure 2CT.22(b) has the same output. By comparing Figures 2CT.22(a) and (b) we see that the response of an LTI system to the derivative of a signal $x(t)$ equals the derivative of the system's response to $x(t)$.

As a second illustration, consider the LTI system having the impulse response $h(t)$ shown Figure 2CT.23(a). We can sometimes find the output of this system more easily by first applying the derivative of the input to the system and integrating the output, as

Figure 2CT.22
Illustration of the
commutativity
property of LTI
systems: (a) system
with differentiator at
output; (b) equivalent
system with
differentiator at
input.

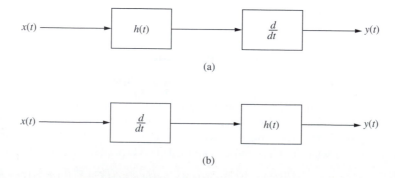

(a)

(b)

Figure 2CT.23
Calculating an output
by using an
equivalent system:
(a) original system;
(b) equivalent system.

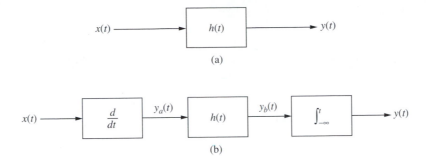

shown in Figure 2CT.23(b). The two systems are equivalent because of the commutation property of LTI systems and the fact that differentiation and integration are mutually inverse operators.

To see how the equivalent system concept can be useful, consider the problem of determining the step response of the first-order system of Figure 2CT.11. We see from Figure 2CT.23 that the step response is the integral of the impulse response Eq. (23):

$$y(t) = \int_{-\infty}^{t} \frac{1}{\tau_c} e^{-\lambda/\tau_c} u(\lambda) d\lambda = \begin{cases} 0; & t < 0 \\ 1 - e^{-t/\tau_c}; & t \geq 0 \end{cases} \tag{54a}$$

which is

$$y(t) = \left(1 - e^{-t/\tau_c}\right) u(t) \tag{54b}$$

The step response is plotted in Figure 2CT.24. We can see from Eq. (54b) that $y(\tau_c) = (1 - e^{-1}) \approx 0.632$, which is 63.2 percent of the "final," or "steady-state," value of 1 in one time constant. Similarly, $y(t)$ reaches approximately 86.5 percent, 95 percent, 98.2 percent, and 99.3 percent of its final value in 2, 3, 4, and 5 time constants, respectively. The time interval $5\tau_c$ is occasionally referred to as the system's transient response time because for most practical purposes, the final value has been reached.

For another example of the usefulness of the equivalent system idea, consider the problem of evaluating the convolution

$$\sqcap\left(\frac{t}{\tau}\right) * \sqcap\left(\frac{t}{\tau}\right) \tag{55}$$

Figure 2CT.24
Step response of
a first-order system:
$y(t)$ reaches
63.2 percent of its
final value in one
time constant and
99.3 percent of its
final value in 5 time
constants.

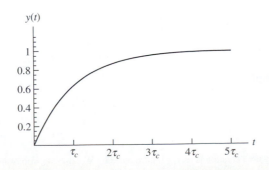

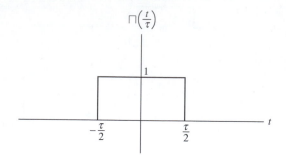

where $\sqcap(t/\tau)$ is a commonly used notation for a rectangular pulse[2] having amplitude 1. That is,

$$\sqcap\left(\frac{t}{\tau}\right) = \begin{cases} 0; & t < -0.5\tau \\ 0.5; & t = -0.5\tau \\ 1; & -0.5\tau < t < 0.5\tau \\ 0.5; & t = 0.5\tau \\ 0; & t > 0.5\tau \end{cases} \qquad (56)$$

The rectangular pulse is plotted in Figure 2CT.25. Notice that it can be written as the difference between two shifted step functions:

$$\sqcap\left(\frac{t}{\tau}\right) = u(t + 0.5\tau) - u(t - 0.5\tau) \qquad (57)$$

We can interpret the convolution (55) as the output of the system of Figure 2CT.23(a) where $x(t) = \sqcap(t/\tau)$ and $h(t) = \sqcap(t/\tau)$. Using the equivalent system of Figure 2CT.23(b), we have, by inspection,

$$y_a(t) = \delta\left(t + \frac{\tau}{2}\right) - \delta\left(t - \frac{\tau}{2}\right) \qquad (58)$$

and

$$y_b(t) = h\left(t + \frac{\tau}{2}\right) - h\left(t - \frac{\tau}{2}\right) = \sqcap\left(\frac{t + \frac{\tau}{2}}{\tau}\right) - \sqcap\left(\frac{t - \frac{\tau}{2}}{\tau}\right) \qquad (59)$$

The reader is encouraged to sketch $y_b(t)$ and show *by inspection* of the sketch that

$$y(t) = \int_{-\infty}^{t} y_b(\lambda)d\lambda = \begin{cases} 0; & t < -\tau \\ \tau + t; & -\tau \leq t < 0 \\ \tau - t; & 0 \leq t \leq \tau \\ 0; & t > \tau \end{cases} \qquad (60)$$

$y(t)$ can be written as

$$y(t) = \tau \wedge \left(\frac{t}{\tau}\right) \qquad (61)$$

---

[2] Another commonly used notation is $\text{rect}\left(\frac{t}{\tau}\right)$, introduced by Woodward [Woodward, P.M., 1964].

Figure 2CT.26
Triangular pulse
$\wedge\left(\frac{t}{\tau}\right)$.

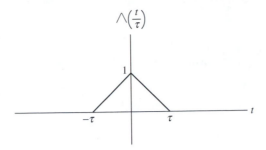

where $\wedge(t/\tau)$ is the *triangular pulse* defined by

$$\wedge\left(\frac{t}{\tau}\right) = \begin{cases} 0; & t < -\tau \\ 1 + \dfrac{t}{\tau}; & -\tau \le t < 0 \\ 1 - \dfrac{t}{\tau}; & 0 \le t \le \tau \\ 0; & t > \tau \end{cases} \tag{62}$$

The triangular pulse is shown in Figure 2CT.26.

Therefore, we have the result

$$\sqcap\left(\frac{t}{\tau}\right) * \sqcap\left(\frac{t}{\tau}\right) = \tau \wedge \left(\frac{t}{\tau}\right) \tag{63}$$

2CT.9   Singularity Functions and Their Applications

The unit impulse can be used to generate a family of signals called *singularity functions*. The singularity functions are obtained by repeatedly differentiating or integrating the unit impulse. In general, the ith singularity function is defined as

$$\delta_i(t) = \frac{d^i}{dt^i}\delta(t); \quad i = 1, 2, \ldots \tag{64}$$

$$\delta_0(t) = \delta(t); \tag{65}$$

$$\delta_{-i}(t) = \int_{-\infty}^{t} \cdots \int_{-\infty}^{\lambda_3}\int_{-\infty}^{\lambda_2}\delta(\lambda_1)d\lambda_1 d\lambda_2 \cdots d\lambda_i; \quad i = 1, 2, \ldots \tag{66}$$

The most frequently used singularity functions are:

■   The unit impulse: $\delta(t)$
■   The unit step

$$u(t) = \delta_{-1}(t) = \int_{-\infty}^{t}\delta(\lambda)d\lambda \tag{67}$$

■   The unit ramp

$$r(t) = \delta_{-2}(t) = \int_{-\infty}^{t} u(\lambda)\,d\lambda = tu(t) \tag{68}$$

We have already seen the unit impulse, unit step, and unit ramp.

Figure 2CT.27
Unit doublet as a limit: (a) derivative of $g_{\Delta\tau}(t)$; (b) the unit doublet $\delta_1(t)$ is defined as the limit of $g_{\Delta\tau}(t)$ as $\Delta\tau \to \infty$.

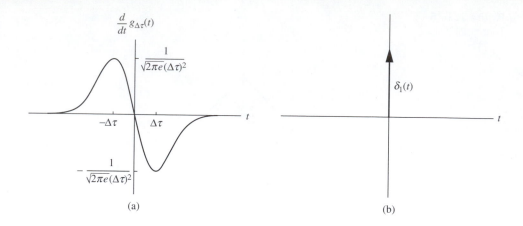

(a)   (b)

We can define the derivatives of the unit impulse (64) as a limit

$$\delta_i(t) = \frac{d^i}{dt^i}\delta(t) = \lim_{\Delta\tau\to\infty}\frac{d^i g_{\Delta\tau}(t)}{dt^i}; \quad i = 1, 2, \ldots \tag{69}$$

where $g_{\Delta\tau}(t)$ is the gaussian pulse (Figure 2CT.3). In particular, the first derivative of $\delta(t)$ is called the *unit doublet*.

$$\delta_1(t) = \frac{d}{dt}\delta(t) = \lim_{\Delta\tau\to\infty}\frac{dg_{\Delta\tau}(t)}{dt} \tag{70}$$

Plots of the derivative of $g_{\Delta\tau}(t)$, for finite $\Delta\tau$, and $\delta_1(t)$ are shown in Figures 2CT.27(a) and (b). We see from Figure 2CT.27(a) that the derivative of $g_{\Delta\tau}(t)$ resembles a positive pulse followed by a negative pulse. The pulse widths and amplitudes approach 0 and $\pm\infty$, respectively, as $\Delta\tau \to 0$. The doublet $\delta_1(t)$ has 0 area. There is no commonly accepted figure for a doublet. To identify the doublet (and distinguish it from an impulse), we write $\delta_1(t)$ next to the arrow, as shown in Figure 2CT.27(b).

Unit step and ramp functions are frequently used to describe waveforms that are made up of straight-line segments For example, we saw in Section 8 that a rectangular pulse can be expressed as the difference between the two shifted step functions

$$\sqcap\left(\frac{t}{\tau}\right) = u(t + 0.5\tau) - u(t - 0.5\tau) \tag{71}$$

Similarly, the *triangular pulse* $\wedge(t/\tau)$ shown in Figure 2CT.26 can be expressed as a sum of scaled and time-shifted ramp functions:

$$\wedge\left(\frac{t}{\tau}\right) = \frac{r(t + \tau) - 2r(t) + r(t - \tau)}{\tau} \tag{72}$$

An advantage of writing a waveform as a weighted sum of time-shifted step and ramp functions is to find the output of an LTI system. By using the properties of time invariance and linearity, we can often just write down an expression for the output.

EXAMPLE 2CT.7   Rectangular Pulse Response of a First-Order System

Let's find the response of the first-order system, or equivalently the RC low-pass filter, of Figure 2CT.11 to a rectangular pulse $A \sqcap (t/\tau)$. Recall that the step response was given in Eq. (54):

$$u(t) \longrightarrow h_{-1}(t) = (1 - e^{-t/\tau_c})u(t) \qquad (73)$$

where we have denoted the step response as $h_{-1}(t)$. Because the system is time invariant, its response to each time-shifted unit step in Eq. (71) is a time-shifted version of $h_{-1}(t)$. Because the system is also linear, its response to $A \sqcap (t/\tau)$ is the scaled sum of the two step responses, i.e.:

$$A \sqcap \left(\frac{t}{\tau}\right) \longrightarrow Ah_{-1}(t + 0.5\tau) - Ah_{-1}(t - 0.5\tau) \qquad (74a)$$

which is

$$A \sqcap \left(\frac{t}{\tau}\right) \longrightarrow A\left(1 - e^{-\frac{t+0.5\tau}{\tau_c}}\right) u(t + 0.5\tau) - A\left(1 - e^{-\frac{t-0.5\tau}{\tau_c}}\right) u(t - 0.5\tau) \qquad (74b)$$

With the help of a little algebra, we can write Eq. (74b) as

$$A \sqcap \left(\frac{t}{\tau}\right) \longrightarrow \begin{cases} 0; & t < -0.5\tau \\ A\left(1 - e^{-\frac{t+0.5\tau}{\tau_c}}\right); & -0.5\tau \leq t < 0.5\tau \\ A\left(1 - e^{-\frac{\tau}{\tau_c}}\right) e^{-\frac{t-0.5\tau}{\tau_c}}; & 0.5\tau \leq t \end{cases} \qquad (74c)$$

Figure 2CT.28 illustrates how the step responses add to produce the rectangular pulse response as indicated by Eq. (74b). The influence of the time constant τ_c on the pulse response is shown in Figure 2CT.29. Notice from the figure that the output becomes a closer approximation to the input pulse as the ratio τ_c/τ decreases.

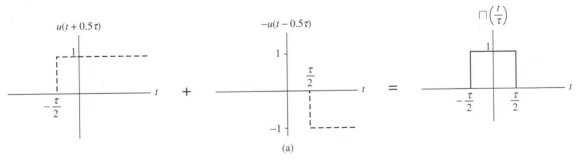

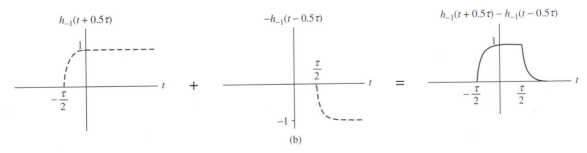

Figure 2CT.28
Response of a first-order system to a rectangular pulse: (a) the input pulse $A \sqcap (t/\tau)$ is the sum of two time-shifted step functions; (b) the response is the sum of the responses to each time-shifted step function. This figure assumes that $A = 1$ and $\tau_c = 0.1\tau$ where τ_c is the time constant of the first-order system.

Figure 2CT.29
Influence of the time constant on pulse response: The output becomes a closer approximation to the input as the system time constant τ_c decreases with respect to the pulse width τ : $\tau_c/\tau = 2.0$, 1.0, 0.4, 0.2.

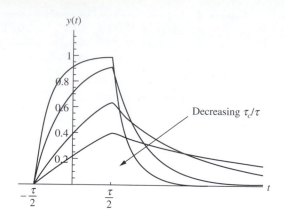

EXAMPLE 2CT.8   Baseband ASK

Information can be transmitted to a receiver by means of a sequence of pulses whose amplitudes carry the information. An example is given by the baseband *amplitude shift keying* (ASK) system in Figure 2CT.30. The system encodes binary digits from a source bit by bit into pulses having amplitudes that depend on the source output. Zeros are encoded into pulses with amplitudes $-A_o$ and ones are encoded into pulses having amplitudes $+A_o$. Each pulse has duration τ. The pulses are transmitted at a rate

$$R = \frac{1}{\tau} \tag{75}$$

pulses per second and are delivered to a receiver through a baseband communications channel. For illustration, let's assume that the communications channel is modeled by the RC low-pass filter of Figure 2CT.11. We studied the response of this filter to a single pulse in Example 2CT.7. The response of the channel to a series of such pulses, illustrated in Figure 2CT.30, is the superposition of the responses to the individual input pulses. For this figure, we choose $A_o = 1$. The unit of t is milliseconds: $R = 1000$ pulses per second (1 kpps) and $\tau_c = 100$ μs. We often want to transmit information as quickly as possible. Figure 2CT.31 shows what happens when we increase the pulse rate R. As R increases, the ratio τ_c/τ increases and each received pulse becomes increasingly attenuated and

Figure 2CT.30
A Baseband ASK communication system. The unit of t is milliseconds: $R = 1000$ pulses per second (1 kpps) and $\tau_c = 100$ microsecond. Pulse amplitude $A_o = 1$.

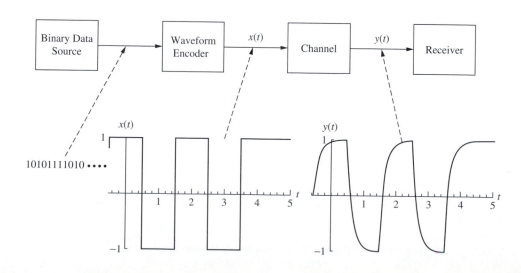

Figure 2CT.31
Effect of increasing
the transmitted pulse
rate $R = 1/\tau$ in ASK
communication: The
communications
channel is modeled as
an RC low-pass filter
with $\tau_c = 100\mu$s:
(a) $R = 1$ kpps;
(b) $R = 2$ kpps;
(c) $R = 4$ kpps;
(d) $R = 20$ kpps.

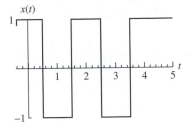
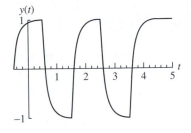

(a)

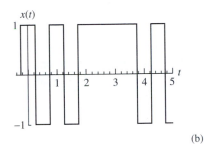
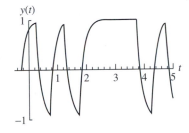

(b)

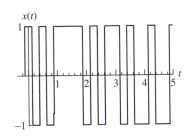
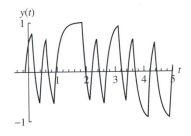

(c)

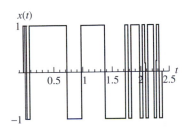
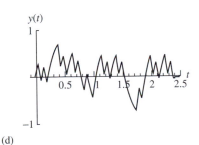

(d)

distorted, as was shown in Figure 2CT.29. Superposition applies for the sequence of pulses, and the distortion causes each received pulse to interfere with the next. The pulse-to-pulse interference is called *intersymbol interference*. As the figure illustrates, it becomes increasingly difficult to discern the information contained in the transmitted sequence as R increases. The example demonstrates that information cannot be transmitted at an arbitrarily high rate through a practical communications channel.[3]

---

[3] The fascinating topic of information transmission is addressed in information, coding, and communication theories.

2CT.10 Summary

This chapter introduced the topics of *unit impulse*, *unit impulse response*, and *convolution*. The unit impulse provides a building block for all other signals. We showed that every CT signal can be represented as an integral of scaled and shifted unit impulses and that every LTI CT system is characterized by its response to a unit impulse. We used the principle of superposition to obtain an expression for the output of any CT LTI system in terms of the input and the system's unit impulse response. This expression is called a *convolution integral*. We showed how the convolution integral can also be used to find necessary and sufficient conditions for a CT system to be stable and causal. We also found the overall impulse response of interconnected CT systems in terms of the impulse responses of the individual systems. Finally, we introduced the singularity functions that arise when a unit impulse is applied to a series connection of integrators or differentiators. We illustrated how to use singularity functions to construct rectangular pulses, triangular pulses, and other inputs and to find the response of an LTI system to these inputs. The analytical methods introduced in this chapter are very powerful: They form the foundation for all the chapters that follow. They are known as *time domain* methods because they deal with functions of time.

Summary of Definitions and Formulas

The Unit Impulse

$$\delta(t) = \lim_{\Delta\tau\to 0} p_{\Delta\tau}(t) = \lim_{\Delta\tau\to 0} g_{\Delta\tau}(t) = \lim_{\Delta\tau\to 0} \frac{1}{\Delta\tau}\text{sinc}\left(\frac{t}{\Delta\tau}\right)$$

where

$$p_{\Delta\tau}(t) = \frac{1}{\Delta\tau}\sqcap\left(\frac{1}{\Delta\tau}\right)$$

$$g_{\Delta\tau}(t) = \frac{1}{\sqrt{2\pi}\Delta\tau}e^{-0.5(t/\Delta\tau)^2}$$

$$\text{sinc}(\lambda) = \frac{\sin(\pi\lambda)}{\pi\lambda}$$

Properties of the Unit Impulse

$$x(t)\delta(t-\tau) = x(\tau)\delta(t-\tau)$$

$$x(t) = \int_{-\infty}^{\infty} x(\tau)\delta(t-\tau)\,d\tau$$

Singularity Functions

$$\delta_i(t) = \frac{d^i}{dt^i}\delta(t); \quad i = 1, 2, \ldots$$

$$\delta_0(t) = \delta(t)$$

$$\delta_{-i}(t) = \int_{-\infty}^{t}\cdots\int_{-\infty}^{\lambda_3}\int_{-\infty}^{\lambda_2}\delta(\lambda_1)d\lambda_1 d\lambda_2\cdots d\lambda_i; \quad i = 1, 2, \ldots$$

Unit Doublet, Unit Step, and Ramp Functions

$$\delta_1(t) = \frac{d}{dt}\delta(t)$$

$$u(t) = \delta_{-1}(t) = \int_{-\infty}^{t} \delta(\lambda)\, d\lambda = \begin{cases} 0; & t < 0 \\ \frac{1}{2}; & t = 0 \\ 1; & t > 0 \end{cases}$$

$$r(t) = \delta_{-2}(t) = \int_{-\infty}^{t} u(\lambda)\, d\lambda = tu(t)$$

Unit Impulse Response

$$\delta(t) \longrightarrow h(t)$$

$$x(t) \longrightarrow \int_{-\infty}^{\infty} x(\tau)h(t - \tau)\, d\tau \overset{\triangle}{=} x(t) * h(t)$$

For ideal time shift: $h(t) = \delta(t - \tau)$

For integrator: $h(t) = u(t)$

Properties of the Unit Impulse Response

Commutativity: $x(t) * h(t) = h(t) * x(t)$

Superposition: $\{a_1 x_1(t) + a_2 x_2(t)\} * h(t) = a_1 x_1(t) * h(t) + a_2 x_2(t) * h(t)$

Associativity: $x(t) * \{h_1(t) * h_2(t)\} = \{x(t) * h_1(t)\} * h_2(t)$

Stability

A CT LTI system is BIBO stable if and only if $\int_{-\infty}^{\infty} |h(t)|\, dt < \infty$.

Causality

An LTI system is causal if and only if $h(t) = 0$ for $t < 0$.

Rectangular Pulse and Triangular Pulse

$$\sqcap\left(\frac{t}{\tau}\right) = \begin{cases} 0; & t < -0.5\tau \\ 0.5; & t = -0.5\tau \\ 1; & -0.5\tau < t < 0.5\tau \\ 0.5; & t = 0.5\tau \\ 0; & t > 0.5\tau \end{cases}$$

$$\wedge\left(\frac{t}{\tau}\right) = \begin{cases} 0; & t < -\tau \\ 1 + \frac{t}{\tau}; & -\tau \le t < 0 \\ 1 - \frac{t}{\tau}; & 0 \le t \le \tau \\ 0; & t > \tau \end{cases}$$

$$\sqcap\left(\frac{t}{\tau}\right) * \sqcap\left(\frac{t}{\tau}\right) = \tau \wedge\left(\frac{t}{\tau}\right)$$

2CT.11   Problems

2CT.2   Unit Impulse

2CT.2.1 Plot:
  a) $x(t) = 5\delta(t)$.
  b) $x(t) = -9\delta(t - 5.3)$.
  c) $x(t) = \delta(t - 2.7)$.
  d) $x(t) = \delta(t + 1.2) - 2\delta(t) + \delta(t - 1.2)$.

2CT.2.2 a) Plot $p_{\Delta\tau}(t - t_o)$, where t_o is a constant and $\Delta\tau$ is finite.
  b) Plot an arbitrary function $x(t)$ that is continuous in the vicinity of t_o.
  c) Plot the product $x(t)p_{\Delta\tau}(t - t_o)$.
  d) Refer to your plots and describe what happens in the limit $\Delta\tau \to 0$. Is your result consistent with the sampling property of the unit impulse?

2CT.2.3 Simplify the expression $t^2\delta(t - 1) + \sin(t)\delta(t) + t^2\delta(t + 1)$, and plot the result.

2CT.2.4 a) Show that

$$\int_{-\infty}^{\infty} \delta(3t)\, dt = \tfrac{1}{3} \tag{76}$$

  b) Find the area of $\delta(at)$ where $a \neq 0$.
  c) Plot $\delta(aF)$ versus F for $a = 2\pi$.
  d) Plot $\delta(at)$ for $a = -5$.

2CT.3   Unit Impulse Response

2CT.3.1 a) The waveform $x(t) = e^{-t/\tau}u(t)$ is applied to a differentiator. Use the rules of differentiation to show that the output is

$$y(t) = \delta(t) - \frac{1}{\tau}e^{-\frac{t}{\tau}}u(t) \tag{77}$$

  b) Plot $x(t)$ and $y(t)$.

2CT.3.2 The *moving average* $y(t)$ of a waveform $x(t)$ is defined by the equation

$$y(t) = \frac{1}{\tau}\int_{t-\tau}^{t} x(\lambda)\, d\lambda \tag{78}$$

where τ is the time over which the average is made.
  a) Show that the impulse response associated with the system that computes a moving average is

$$h(t) = \frac{1}{\tau}[u(t) - u(t - \tau)] \tag{79}$$

  b) Plot $h(t)$.

2CT.3.3 Find and plot the response of the moving averager of Problem 2CT.3.2 to $x(t) = u(t)$.

2CT.3.4 Find the impulse response of the system shown in Figure 2CT.32.

2CT.3.5 a) Plot $p_{\Delta\tau}(t)$ and $\int_{-\infty}^{t} p_{\Delta\tau}(\lambda)\, d\lambda$ where $\Delta\tau$ is finite.
  b) Refer to your plots and describe what happens in the limit $\Delta\tau \to 0$. Is your result consistent with the equation $\int_{-\infty}^{t} \delta(\lambda)\, d\lambda = u(t)$?

Figure 2CT.32

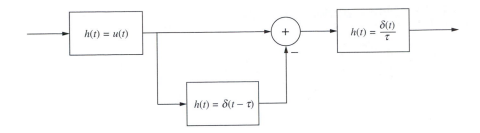

2CT.4   Waveform Represented as an Integral of Shifted Impulses

2CT.4.1 The sifting property of the unit impulse can be defined as

$$\int_{-\infty}^{\infty} g(t)\delta(t - t_o)\, dt = g(t_o) \tag{80}$$

where $g(t)$ is any function that is continuous at $t = t_o$. Use the sampling property of the unit impulse to derive the sifting property.

2CT.4.2 Use the sifting property of the unit impulse to simplify the following integrals:
a) $\int_{-\infty}^{\infty} \sqrt{\tau}\, \delta(25 - \tau)\, d\tau$
b) $\int_{-\infty}^{\infty} e^{-\tau} u(\tau)\delta(t - \tau)\, d\tau$

2CT.5   Convolution integral

2CT.5.1 We have shown that if a system is LTI, then the output is given by a convolution integral. Show that the converse is true; i.e., show that, if the output of system is given by a convolution integral, the system is LTI.

2CT.5.2 **a)** Consider the convolution $y(t) = x(t) * h(t)$, where

$$h(t) = \begin{cases} 0; & t < 0 \\ t; & 0 \le t \le T \\ 0; & T < t \end{cases} \tag{81}$$

and $x(t) = u(t)$.
b) Plot $x(t)$ and $h(t)$.
c) Use plots or the sliding tape method to help you find $y(t)$.
d) Plot $y(t)$.

2CT.5.3 An input

$$x(t) = \begin{cases} 0; & t < 0 \\ e^{-t}; & 0 \le t \le 1 \\ 0; & 1 < t \end{cases} \tag{82}$$

is applied to an unstable LTI system having impulse response

$$h(t) = e^{+t/2} u(t) \tag{83}$$

a) Plot $x(t)$ and $h(t)$.
b) Use plots of $x(\tau)$, $h(t - \tau)$, and $x(\tau)h(t - \tau)$ to help you find $y(t) = x(t) * h(t)$.
c) Plot $y(t)$.

2CT.5.4 Repeat Problem 2CT.5.3 for

$$x(t) = \begin{cases} 0; & t < 0 \\ e^{\alpha t}; & 0 \le t \le 1 \\ 0; & 1 < t \end{cases} \tag{84}$$

and

$$h(t) = e^{\beta t} u(t) \tag{85}$$

where $\alpha \ne \beta$. Use the values $\alpha = -1$ and $\beta = -\frac{1}{2}$ for your plot.

2CT.5.5 Repeat Problem 2CT.5.3 for

$$x(t) = \begin{cases} 0; & t < 0 \\ e^{\alpha t}; & 0 \le t \le 1 \\ 0; & 1 < t \end{cases} \tag{86}$$

and

$$h(t) = e^{\alpha t} u(t) \tag{87}$$

Use the value $\alpha = -1$ for your plot.

2CT.5.6 A unit step function is applied to a system having the impulse response $h(t)$. Show by substituting $x(t) = u(t)$ into Eq. (29) that the response is given by

$$y(t) = u(t) * h(t) = \int_{-\infty}^{t} h(\lambda) \, d\lambda \tag{88}$$

2CT.5.7 An input $x(t)$ is applied to a system having impulse response $h(t) = u(t)$. Show that the response is given by

$$y(t) = x(t) * u(t) = \int_{-\infty}^{t} x(\lambda) \, d\lambda \tag{89}$$

2CT.5.8 A unit step function $x(t) = u(t)$ is applied to a system having impulse response $h(t) = (1 - e^{-t})u(t)$.
a) Find the output by using the graphical method of convolution.
b) Check your answer to part (a) using a different method.

2CT.5.9 Change the variable of integration from τ to $\lambda = t - \tau$ to show that

$$y(t) = \int_{-\infty}^{\infty} x(\tau)h(t - \tau) \, d\tau = \int_{-\infty}^{\infty} h(\lambda)x(t - \lambda) \, d\lambda \tag{90}$$

This result proves that the convolution operator is commutative: $x(t) * h(t) = h(t) * x(t)$.

2CT.5.10 Show for every waveform $x(t)$:
a) $\delta(t) * x(t) = x(t)$
b) $\delta(t - t_o) * x(t) = x(t - t_o)$
c) $e^{j2\pi Ft} * x(t) = e^{j2\pi Ft} X(F)$, where $X(F)$ depends on F but not on t. What is the formula for $X(F)$?

2CT.5.11 Use the graphical method of convolution to find the output of the moving average of Problem 2CT.3.2 for $x(t) = u(t)$. Plot your answer.

2CT.5.12 Find $\delta(t - 1.1) * \delta(t - 2.2) * \delta(t - 3.3)$.

2CT.5.13 When the input $x(t)$ is applied to an LTI system, the response is $y(t)$. Does it follow that the response of the same system to the input $x(2t)$ is $y(2t)$? Justify your answer.

2CT.6   BIBO Stability

2CT.6.1 State which of the following systems are BIBO stable and which are not. Justify your answers.

a) $h(t) = u(t)$

b) $h(t) = \delta(t)$

c) $h(t) = \dfrac{1}{\tau} e^{-t/\tau} u(t)$

d) $h(t) = u(t) - u(t - 2)$

e) $h(t) = \cos(100t)\, u(t)$

2CT.6.2 The energy in a waveform $h(t)$, is defined as

$$E_h = \int_{-\infty}^{\infty} |h(t)|^2 dt \qquad (91)$$

a) Show that the impulse response of every stable system has finite energy. Hint: Use the mathematical inequality

$$\int_{-\infty}^{\infty} |h(t)|^2 dt \leq \left(\int_{-\infty}^{\infty} |h(t)| dt \right)^2 \qquad (92)$$

b) Does it follow that if $h(t)$ has finite energy, then the system is BIBO stable? Hint: Consider $h(t) = \dfrac{1}{t} u(t - 1)$.

2CT.7   Causal Systems

2CT.7.1 State which of the following systems are causal and which are not causal. Justify your answers.

a) $h(t) = \delta(t - 3)$

b) $h(t) = \delta(t + 0.01)$

c) $h(t) = u(t - 5) - u(t + 5)$

d) $y(t) = x(-t)$

2CT.7.2 State which of the following systems are causal and which are not causal. Justify your answers.

a) $h(t) = u(-t + 1)$

b) $h(t) = \cos(100t)\, u(t)$

c) $h(t) = \text{sinc}(t)$

d) $y(t) = x^2(3t)$

2CT.8   Basic Connections of CT LTI Systems

2CT.8.1 In Section 2CT.8 of the text we evaluated the convolution $\sqcap(t/\tau) * \sqcap(t/\tau)$ with the help of the system of Figure 2CT.23(b). Fill in the details of the derivation by plotting $y_b(t)$ of Eq. (59) and using your plot to verify Eq. (60).

2CT.8.2 The *step response* of a system is defined as the response of that system to a unit step input.

a) It can be shown that the unit step response of the RC circuit of Figure 2CT.11 is $y(t) = \left(1 - e^{-t/\tau_c}\right) u(t)$. Show that the derivative of $y(t)$ equals the impulse response of this circuit as given in Eq. (23).

b) Prove that the impulse response of an LTI system is always given by the derivative of its step response.

c) Show that the step response of an LTI system is always given by the integral of its impulse response.

2CT.8.3 A ramp is applied at time t_o to an unstable LTI system, whose impulse response is also a ramp. Use any method to find $r(t - t_o) * r(t)$. Plot your answer for $t_o = 1$.

2CT.8.4 a) Use Eq. (53) to find the impulse response of the closed-loop system of Figure 2CT.21(c) for $h_1(t) = \delta(t-1)$.
b) Show how to obtain the same result by inspection of the system.

2CT.8.5 Is it possible to connect stable systems in such a way that the overall system is unstable? If your answer is yes, give an example.

2CT.9   Singularity Functions

2CT.9.1 a) Use the graphical method of convolution to show that

$$r(t) = u(t) * u(t) \tag{93}$$

b) Interpret your result using an integrator having input $u(t)$.

2CT.9.2 Find the impulse response of a differentiator.

2CT.9.3 Use the graphical method of convolution to show that

$$\sqcap\left(\frac{t}{\tau}\right) * \sqcap\left(\frac{t}{\tau}\right) = \tau \wedge \left(\frac{t}{\tau}\right) \tag{94}$$

2CT.9.4 Plot:
a) $\wedge(2t)$.
b) $\wedge(t/3)$.

2CT.9.5 An LTI system has a unit impulse response $h(t)$, unit step response $h_{-1}(t)$, and unit ramp response $h_{-2}(t)$. Show that
a)

$$h_{-1}(t) = \int_{-\infty}^{t} h(\lambda)\, d\lambda \tag{95}$$

b)

$$h(t) = \frac{d}{dt} h_{-1}(t) \tag{96}$$

c)

$$h_{-2}(t) = \int_{-\infty}^{t} h_{-1}(t)\, d\lambda \tag{97}$$

2CT.9.6 An engineer wants to build a system that has the impulse response $h(t) = 5 \sqcap\left(\frac{t-1}{2}\right)$.
a) Plot $h(t)$.
b) Show how to construct the system using a delay, an integrator, gains, and adders.
c) Find an expression for the output of this system when the input is an arbitrary waveform $x(t)$.

2CT.9.7 Show how to express $x(t)$ of Figure 2CT.33 as a sum of seven scaled and time-shifted triangular pulses. Hint: The dotted lines provide help.

2CT.9.8 Suppose we know that the system's ramp response is $h_{-2}(t)$:

$$r(t) \xrightarrow{h(t)} h_{-2}(t) \tag{98}$$

Show that

$$\wedge(t) \xrightarrow{h(t)} h_{-2}(t+1) - 2h_{-2}(t) + h_{-2}(t-1) \tag{99}$$

2CT.9.9 A system's ramp response is $h_{-2}(t)$. Find the system's response to the input $\wedge(t/6)$ in terms of $h_{-2}(\cdot)$.

Figure 2CT.33

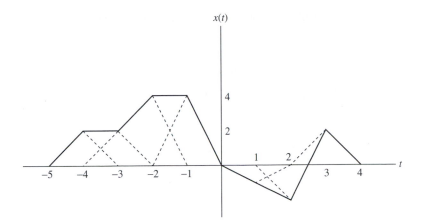

Miscellaneous Problems

2CT.1. A time-varying horizontal force $F(t)$ is applied to a mass that lies on a horizontal surface.
   a) Show that the equation relating the resulting velocity v of the mass to the force F is
      $m\,dv(t)/dt + kv(t) = F(t)$, where m is the mass and k is the coefficient of friction.
   b) Draw the block diagram of the system where $F(t)$ is the input and $v(t)$ is the output.
   c) Find $v(t)$ for $F(t) = F_o u(t)$. What velocity does the mass approach as t increases?

2CT.2. Consider the RC circuit of Figure 2CT.11. The current entering the capacitor is given by Ohm's law as $i(t) = [y(t) - x(t)]/R$. The voltage difference across the capacitor is $y(t) = q(t)/C$, where $q(t) = \int_{-\infty}^{t} i(\lambda)\,d\lambda$, where $q(t)$ is the charge on the capacitor. Use these two equations to derive I-O Eq. (13).

2CT.3. Find the impulse response of the RC circuit of Figure 2CT.11 when the output is taken across the terminals of the resistor.

2CT.4. Repeat Problem 2CT.3 to find the step response of the RC circuit when the output is taken across the terminals of the resistor.

2CT.5. Find the equation relating output $y(t)$ to input $x(t)$ for an LTI system having the impulse response $h(t) = r(t) = tu(t)$.

2CT.6. a) In the feedback system of Figure 2CT.34, A, G, and β are constants and $0 < \beta < 1$. Assume that $\xi(t) = 0$ and find the equation relating input $x(t)$ to output $y(t)$. Hint: Start by writing down the equation for the output of the adder $v(t)$. Then find another equation you can use to eliminate $v(t)$.

Figure 2CT.34

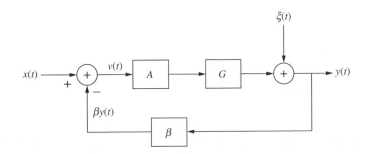

b) Show that when $AG\beta \gg 1$, the I-O equation of part (a) reduces to the approximate result

$$y(t) \approx \frac{1}{\beta}x(t) \tag{100}$$

c) What inequality must A satisfy for (100) to be true?

2CT.7. a) Find the expression for the output of the feedback system of Figure 2CT.34, where the "signal" $x(t)$ and "noise" $\xi(t)$ are both nonzero.

b) Explore what happens to your answer of part (a) for $AG\beta \gg 1$.

2CT.8. The techniques of this chapter can be extended to two dimensions. The extension provides a way to describe imaging systems that are linear and space invariant. For simplicity, we consider only grayscale (i.e., an achromatic or "black and white") images in this problem. A two-dimensional unit impulse can be defined as

$$\delta(x, y) = \delta(x)\delta(y) \tag{101}$$

where x and y are the spatial coordinates of the x, y plane. We can define $\delta(x)$ as $\delta(x) = \lim_{\Delta L \to 0} p_{\Delta L}(x)$. A similar definition applies to $\delta(y)$. We can imagine $\delta(x, y)$ as a point of light having unit "intensity" located at the center of an otherwise dark x, y plane. The response of a linear space-invariant imaging system to $\delta(x, y)$ is the *point spread function* $h(x, y)$. The output image intensity resulting from an arbitrary input image intensity $I_i(x, y)$ is given by a two-dimensional convolution integral.

$$I_o(x, y) = \int_{-\infty}^{\infty} \int_{-\infty}^{\infty} I_i(\alpha, \beta) h(x - \alpha, y - \beta) d\alpha d\beta \tag{102}$$

a) Plot the product $p_{\Delta L}(x) p_{\Delta L}(y)$ above the x, y plane and describe what happens as $\Delta L \to 0$.

b) Find $I_o(x, y)$ for $I_i(x, y) = \delta(x - x_o, y - y_o)$ where x_o and y_o are constants. Express your answer in terms of $h(\cdot, \cdot)$.

c) Use shading to plot the *intensity edge* $I_i(x, y) = u(x)$ as an image in the x, y plane. Let the value 0 correspond to dark and the value 1 correspond to bright.

d) Evaluate the convolution integral to find $I_o(x, y)$ when $I_i(x, y)$ is an intensity edge and $h(x, y) = \delta(x, y)$.

2CT.9. Consider the linear space-invariant imaging system of Problem 2CT.8. Blur due to camera motion in the x direction during the image exposure time can be described by the point spread function $h(x, y) = \sqcap(x/\Delta L)\delta(y)$.

a) Use shading to plot $h(x, y)$ as an image. Your plot should look like a streak of light, showing that motion caused the point source $\delta(x, y)$ to be imaged as a streak.

b) Evaluate the convolution integral to find $I_o(x, y)$ when $I_i(x, y)$ is an intensity edge $I_i(x, y) = u(x)$ and $h(x, y) = \sqcap(x/\Delta L)\delta(y)$. Use shading to plot $I_o(x, y)$ as an image.

Impulse, Impulse Response, and Convolution

2DT.1 Introduction

In this chapter, we introduce the discrete-time unit impulse, the discrete-time unit impulse response, and convolution sum. These three concepts form the foundation for all that follows in this book. These concepts are important because:

- The unit impulse can be viewed as a building block for every sequence.
- An LTI system's response to an impulse completely defines the input-output properties of the system.
- The response of every LTI system to any input is given by a convolution between the input and the system's impulse response.

2DT.2 Discrete-Time Unit Impulse

The *discrete-time unit impulse* (abbreviated as *unit impulse*) is a special sequence that, along with its shifted versions, can serve as a building block for any sequence. The unit impulse is defined as

$$\delta[n] = \begin{cases} 0 & n \neq 0 \\ 1 & n = 0 \end{cases} \tag{1}$$

Figure 2DT.1 shows a popular way to plot $\delta[n]$. Here, $\delta[n]$ is plotted as a dot supported by a stem having length 1. Sometimes, $\delta[n]$ is plotted using just the dot, a circle, an X, or some other mark. The stem is optional.

The unit impulse is defined for all n. However, it is natural to say that it *occurs* at $n = 0$.

We can shift $\delta[n]$ by replacing n with $n - m$. For example, if we replace n by $n - 3$, we have

$$\delta[n-3] = \begin{cases} 1 & \text{for } n - 3 = 0 \\ 0 & \text{for } n - 3 \neq 0 \end{cases} \tag{2}$$

which is

$$\delta[n-3] = \begin{cases} 1 & \text{for } n = 3 \\ 0 & \text{for } n \neq 3 \end{cases} \tag{3}$$

$\delta[n-3]$ is plotted in Figure 2DT.2.

53

Figure 2DT.1
Unit impulse:
(a) plotted as a dot
with a stem;
(b) plotted as an ×
without a stem.

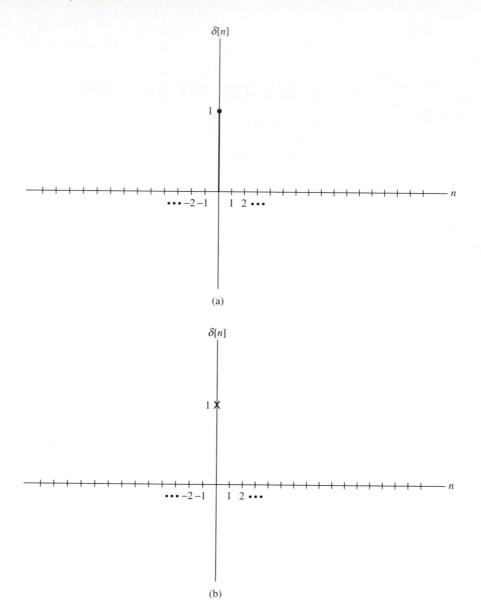

(a)

(b)

More generally, we have

$$\delta[n - m] = \begin{cases} 1 & \text{for } n = m \\ 0 & \text{for } n \neq m \end{cases} \tag{4}$$

where m is an integer. Delay by m can be accomplished using m unit delay elements in series. In Figure 2DT.3, we use four unit delays and an adder to obtain the output $\delta[n - 2] + \delta[n - 4]$ for input $\delta[n]$.

Figure 2DT.2
Shifted unit impulse.

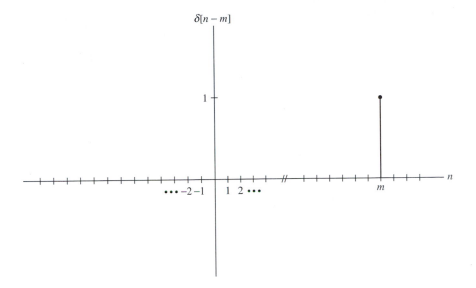

Figure 2DT.3
Four unit delay
elements and an
adder.

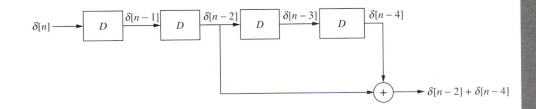

2DT.3   The Unit Impulse Response

The response of a DT LTI system to a unit impulse is called the system's *unit impulse response* and is denoted by $h[n]$. We will show in Section 2DT.4 that $h[n]$ characterizes the input-output (I-O) properties of a DT LTI system. For now, let's become acquainted with $h[n]$ by means of some examples.

EXAMPLE 2DT.1   Five-Point Moving-Average Computer

Consider the system shown in Figure 2DT.4. This system computes the average of the present and previous four input values. Four unit memories are connected in cascade (series) to accomplish this. A cascade of unit memories is called a *shift register*. When n increases by one unit, the value stored in each memory cell is fed into the adjacent cell that follows. We see from the figure that the output of the system for an input sequence x is given by:

$$y[n] = \tfrac{1}{5}\{x[n] + x[n-1] + x[n-2] + x[n-3] + x[n-4]\} \qquad (5)$$

This is a *moving* average because the average is updated as n increases. We can find the unit impulse response of the system from the preceding equation by substituting $x[n] = \delta[n]$ and setting $y[n] \equiv h[n]$. The result is

$$h[n] = \tfrac{1}{5}\{\delta[n] + \delta[n-1] + \delta[n-2] + \delta[n-3] + \delta[n-4]\} \qquad (6)$$

$h[n]$ is plotted in Figure 2DT.5

Figure 2DT.4
Five-point
moving-average
computer.

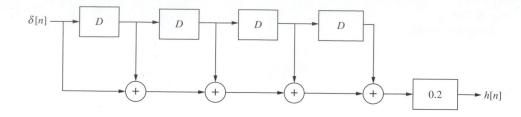

Figure 2DT.5
Unit impulse
response of five-point
moving-average
computer.

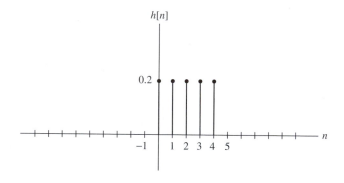

Notice that because $h[n]$ is defined as the system's response to the input $\delta[n]$, there are no other causes for the output. By definition of $h[n]$, the memory units are initially empty (contain 0 values).

The preceding example described a 5-point moving average computer. More generally, an N-term moving average is given by the equation

$$y[n] = \frac{1}{N} \sum_{k=0}^{N-1} x[n-k] = \frac{1}{N} \sum_{m=n-N+1}^{n} x[m] \qquad (7)$$

In the next example, we consider a system whose output is the sum of the present and all past inputs.

EXAMPLE 2DT.2   An Accumulator

Consider a *running sum*, or *accumulator*, defined by the equation

$$y[n] = \sum_{k=-\infty}^{n} x[k] \qquad (8)$$

An accumulator is a familiar item to anyone who shops for groceries. The cash register in the checkout line displays the running sum of the prices of the items coming down a conveyor belt. In Eq. (8), $x[n]$ is the price of the nth item, and $y[n]$ is the running total cost of all items up to and including the nth item (not including tax). As is obvious from this example, another way to write this sum is

$$y[n] = y[n-1] + x[n] \qquad (9)$$

which says that the total shown on the cash register is increased by the price of the next item. The block diagram for Eq. (9) is shown in Figure 2DT.6. Notice that it requires only one unit memory—the cash register in our example. A block diagram for directly realizing Eq. (8) requires an infinite number of memory units!

Figure 2DT.6
Feedback diagram for
accumulator.

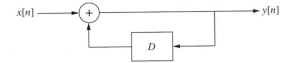

We can find the unit impulse response of an accumulator by substituting $x[n] = \delta[n]$ into Eq. (8). This substitution
yields

$$h[n] = \sum_{k=-\infty}^{n} \delta[k] = \begin{cases} 0 & n < 0 \\ 1 & n \geq 0 \end{cases} \tag{10}$$

The result can be confirmed by noticing that if $n < 0$, then the term $\delta[0] = 1$ does not appear in the sum. The term
$\delta[0] = 1$ is included in the sum if $n \geq 0$. The result (10) can also be confirmed by experience: If you place (only)
a \$1 item one the conveyor belt at $n = 0$, the running sum price of #1 appears on the display at $n = 0$ and remains
thereafter.

The result, Eq. (10) is important. It will be useful to define the sequence on the right
side of Eq. (10) as

$$u[n] = \begin{cases} 0 & n < 0 \\ 1 & n \geq 0 \end{cases} \tag{11}$$

$u[n]$ is called the *unit step* sequence. It is plotted in Figure 2DT.7. Example 2DT.2 has
shown us that the unit step sequence is the running sum of the unit impulse. If we substitute
$u[n]$ and $\delta[n]$ for $y[n]$ and $x[n]$, respectively, in Eq. (9), and rearrange, we obtain

$$\delta[n] = u[n] - u[n-1] \tag{12}$$

which shows that the unit impulse is the difference of consecutive unit steps. We can write
Eq. (12) as

$$\delta[n] = \Delta u[n] \tag{13}$$

where Δ is the *first difference* operator, defined for any sequence $x[n]$ as

$$\Delta x[n] = x[n] - x[n-1] \tag{14}$$

Figure 2DT.7
Unit step sequence.

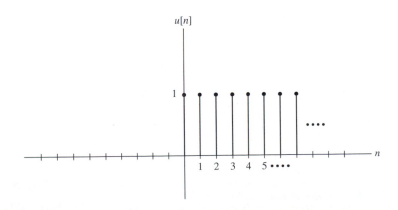

EXAMPLE 2DT.3   A First-Order System

The final example is given by the system of Figure 2DT.8. The system shown is called a first-order system because the equation (called a difference equation) describing it has one unit of delay in the output variable $y[n]$. The difference equation is

$$y[n] = x[n] + ay[n-1] \tag{15}$$

Here again, we find the unit impulse response by substituting $x[n] = \delta[n]$ and solving for $y[n] \equiv h[n]$.

$$h[n] = \delta[n] + ah[n-1] \tag{16}$$

We can find $h[n]$ by referring to the block diagram. We assume that this block diagram represents a causal system[1] and seek a causal solution to Eq. (16). We are solving for the impulse response, which by definition is the output due to $\delta[n]$. Because the system is causal and $\delta[n]$ occurs at $n = 0$, there is no response until $n = 0$; i.e.,

$$h[n] = 0 \qquad \text{for } n < 0 \tag{17}$$

Now we can solve Eq. (16) by recursion, starting with $n = 0$

$$h[0] = 1 + a \cdot 0 = 1 \tag{18}$$

$$h[1] = 0 + a = a \tag{19}$$

$$h[2] = 0 + a \cdot a = a^2 \tag{20}$$

$$h[3] = 0 + a \cdot a^2 = a^3 \tag{21}$$

$$\bullet \bullet \bullet$$

$$h[n] = a^n \qquad \text{for } n \geq 0 \tag{22}$$

In summary, the unit impulse response of the system in Figure 2DT.8 is

$$h[n] = \begin{cases} 0 & n < 0 \\ a^n & n \geq 0 \end{cases} \tag{23}$$

We can use a unit step to write this result more compactly as

$$h[n] = a^n u[n] \tag{24}$$

Figure 2DT.8
First-order system.

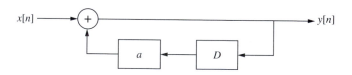

The preceding first-order system is important both in theory and practice. We shall return to it frequently. nth-order systems are described by nth-order difference equations.

---

[1] The assumption that a block diagram represent a causal system is routine when the block diagram is composed of adders, gains, and delays, all of which represent physical systems. However, there is also a noncausal solution to Eq. (16). (See Problem 2DT.3.10.)

We encounter some of these higher-order systems and equations as we proceed. A systematic development of nth-order systems is contained in Chapter 6DT.

<h1>2DT.4   Data Sequence Represented as a Sum of Shifted Impulses</h1>

An arbitrary sequence $x[n]$ can be represented as a sum of shifted, weighted unit impulses. In this sense, the unit impulse is a building block for any sequence. As an example, consider the sequence plotted in Figure 2DT.9.

Looking at the sequence of Figure 2DT.9, we see that

$$x[n] = 2\delta[n] + 5\delta[n-3] - 0.5\delta[n-4]$$
$$= x[0]\delta[n] + x[3]\delta[n-3] + x[4]\delta[n-4] \tag{25}$$

Figure 2DT.9
Sequence $x[n]$: This sequence is composed of three impulses, $x[n] = 2\delta[n] + 5\delta[n-3] - 0.5\delta[n-4]$, which can be written as $x[n] = x[0]\delta[n] + x[3]\delta[n-3] + x[4]\delta[n-4]$.

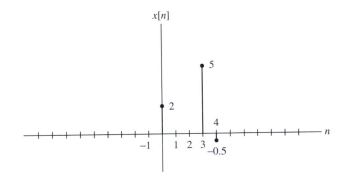

For the general input sequence of Figure 2DT.10 we have

$$x[n] = \sum_{m=-\infty}^{\infty} x[m]\delta[n-m] \tag{26}$$

The term $x[m]\delta[n-m]$ is the impulse occurring at index $n = m$ and having amplitude $x[m]$.

Figure 2DT.10
Arbitrary sequence $x[n]$: $x[n]$ is composed of an arbitrarily large number of impulses, $x[n] = \cdots + x[-1]\delta[n+1] + x[0]\delta[n] + x[1]\delta[n-1] + \cdots = \sum_{m=-\infty}^{\infty} x[m]\delta[n-m]$.

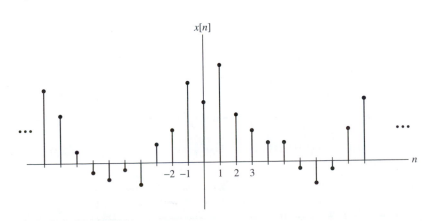

Equation (26) represents an arbitrary sequence as a sum of scaled, shifted impulses. Another way to look at Eq. (26) is in terms of the *sifting property* of the unit impulse. The factor $\delta[n - m]$ is thought of as "sifting" $x[n]$ out of all the terms in the sum.

It might appear that Eq. (26) is a circular way to represent a signal because the sequence $x[\cdot]$ is on both sides. Although the observation makes a good point, we see in the next section how Eq. (26) helps us understand the response of an LTI system to any input.

2DT.5   Convolution Sum

In this section, we derive a formula for the response of any DT LTI system to any input. Let us first derive the response of any LTI system to the specific input sequence $\cdots 0, 0, 0, 2, 0, 0, 5, -0.5, 0, 0, 0 \cdots$, where the arrow denotes the $n = 0$ point. We can write this sequence as

$$x[n] = 2\delta[n] + 5\delta[n - 3] - 0.5\delta[n - 4] = x[0]\delta[n] + x[3]\delta[n - 3] + x[4]\delta[n - 4] \quad (27)$$

Because the system is linear, the response to the sum of the impulses in Eq. (27) is the sum of the individual responses to each impulse acting alone. For each impulse acting alone we have

$$2\delta[n] \xrightarrow{h[n]} 2h[n] \quad (28)$$

$$5\delta[n - 3] \xrightarrow{h[n]} 5h[n - 3] \quad (29)$$

$$-0.5\delta[n - 4] \xrightarrow{h[n]} -0.5h[n - 4] \quad (30)$$

where we have used the properties of linearity and time invariance. The arrow is flow graph notation and the unit impulse response indicated above the arrow indicates the system under consideration. The input is on the left, and the output is on the right. These three I-O expressions can be written respectively as

$$x[0]\delta[n] \xrightarrow{h[n]} x[0]h[n] \quad (31)$$

$$x[3]\delta[n - 3] \xrightarrow{h[n]} x[3]h[n - 3] \quad (32)$$

$$x[4]\delta[n - 4] \xrightarrow{h[n]} x[4]h[n - 4] \quad (33)$$

By superposition, we have the response to $x[n]$

$$x[n] \xrightarrow{h[n]} y[n] = 2h[n] + 5h[n - 3] - 0.5h[n - 4] \quad (34)$$

which is

$$x[n] \xrightarrow{h[n]} y[n] = x[0]h[n] + x[3]h[n - 3] + x[4]h[n - 4] \quad (35)$$

Now we can generalize by recalling that any input is given by the sum (26). In that summation, the $x[m]$'s are just constants. Linearity and time invariance tell us that the response to the sum is given by the sum of the individual responses to each term in the sum as in (27)–(34). Therefore, we have the following basic result:

The Input-Output Relation for Every LTI System is a Convolution

$$x[n] \xrightarrow{h[n]} y[n] = \sum_{m=-\infty}^{\infty} x[m]h[n - m] \triangleq x[n] * h[n] \quad (36)$$

The sum in (36) is called a *convolution sum*. $x[n] * h[n]$ is a useful way to denote the convolution between $x[n]$ and $h[n]$. The symbol $*$ denotes the *convolution operation*.

Let's check Eq. (36) using the input $x[n] = \delta[n]$. The substitution of $x[n] = \delta[n]$ into Eq. (36) yields:

$$y[n] = \sum_{m=-\infty}^{\infty} \delta[m]h[n-m] = \delta[n] * h[n] = h[n] \tag{37}$$

We used the sifting property to evaluate the sum. The answer we obtained $h[n]$ is certainly correct because the response of any system to a unit impulse is indeed the system's unit impulse response!

It is not difficult to prove that the convolution operation $(*)$ has the following properties:

■ Commutativity:

$$x[n] * h[n] = h[n] * x[n] \tag{38}$$

■ Superposition:

$$\{a_1 x_1[n] + a_2 x_2[n]\} * h[n] = a_1 x_1[n] * h[n] + a_2 x_2[n] * h[n] \tag{39}$$

■ Associativity:

$$x[n] * \{h_1[n] * h_2[n]\} = \{x[n] * h_1[n]\} * h_2[n] \tag{40}$$

These properties are established in the problems. They apply for all $x[n], h[n], x_1[n], x_2[n],$ $a_1, a_2, h_1[n],$ and $h_2[n]$.

The operations performed by a convolution sum can be visualized with the aid of plots and a sliding tape, as illustrated in the following example.

EXAMPLE 2DT.4   Understanding Convolution

Consider the problem of finding the output of an LTI system having the impulse response

$$h[n] = \begin{cases} 0 & n < 0 \\ 7 - n & 0 \le n \le 6 \\ 0 & 6 < n \end{cases} \tag{41}$$

when the input is

$$x[n] = \delta[n] + 5\delta[n-1] + 2\delta[n-2] \tag{42}$$

Figure 2DT.4(a) shows plots of $x[n]$ and $h[n]$. To help us understand how to evaluate the convolution sum (36), we can imagine that $x[n]$ and $h[n]$ are recorded on tapes 1 and 2, respectively, shown to the right of the plots.

Figure 2DT.4(b) shows plots of the sequences as they appear in the sum Eq. (36). Both sequences, $x[m]$ and $h[n-m]$, are plotted versus m because it is the index of summation. The corresponding tapes are again shown to the right of the graphs. Tape 2 has been reversed in direction and shifted by n to agree with the plot of $h[n-m]$. To compute Eq. (36) for a specific n, the corresponding values of $x[m]$ and $h[n-m]$ are multiplied and summed. We can see these corresponding values by looking at the graphs or by looking at the tapes. As n increases, the sequence $h[n-m]$ and Tape 2 slide to the right. By repeating the product and sum operation, $y[n]$ is computed for every value of n. The detailed calculations are shown in the text of the figure.

Figure 2DT.11
Plots and sliding
tapes.

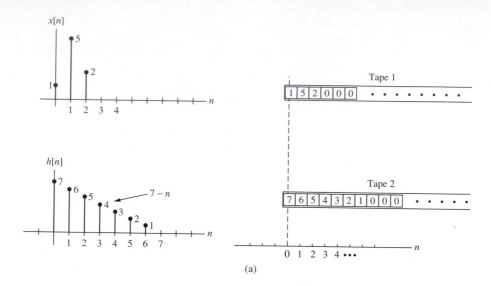

(a)

The two sequences to be convolved

$$x[n] = \delta[n] + 5\delta[n-1] + 2\delta[n-2] \quad \text{and} \quad h[n] = \begin{cases} 0 & n < 0 \\ 7-n & 0 \le n \le 6 \\ 0 & 6 < n \end{cases}$$

are shown in Figure 2DT.11(a). The sequences have been recorded on the tapes adjacent to the plots.

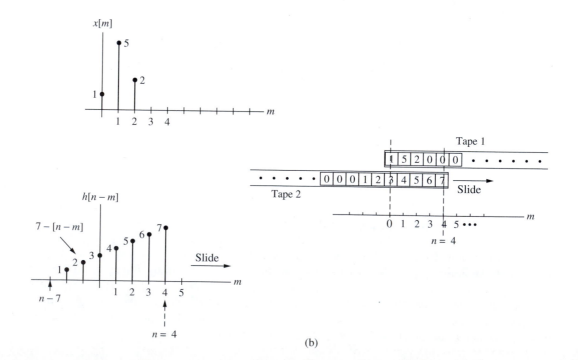

(b)

Figure 2DT.11(b) contains plots of the same sequences as they appear in the convolution sum

$$y[n] = \sum_{m=-\infty}^{\infty} h[n-m]x[m]$$

Here they are plotted versus the summation index m. n is a parameter that can be positive or negative. It is assumed that $n = 4$ in the figure. Tape 2 has been reversed in direction and translated to the right by n. As n increases, $h[n-m]$ and Tape 2 slide to the right.

To perform the convolution, fix the value of n and multiply the corresponding values of $h[n-m]$ and $x[m]$ for every m. Then sum the products $h[n-m]x[m]$ over m, as indicated by the convolution sum. You can imagine this operation being performed on either the plots or the tapes, whichever you prefer. For example, consider $n = 4$. First position the plot of $h[n-m]$ or the tape as illustrated in the figure for $n = 4$. Then multiply the values of $h[n-m]$ and $x[m]$ that are vertically aligned, i.e., $3 \times 1, 4 \times 5$, and 5×2. Finally, sum the products. This gives the result

$$y[4] = 3 \times 1 + 4 \times 5 + 5 \times 2 = 33$$

Other values of $y[n]$ are obtained similarly.

$$y[0] = 7 \times 1 = 7$$
$$y[1] = 6 \times 1 + 7 \times 5 = 41$$
$$y[2] = 5 \times 1 + 6 \times 5 + 7 \times 2 = 49$$
$$y[3] = 4 \times 1 + 5 \times 5 + 6 \times 2 = 41$$

etc.

For $n < 0$ and for $n > 9$, the product $h[n-m]x[m]$ equals 0 for all m. Consequently

$$y[n] = \sum_{m=-\infty}^{\infty} h[n-m]x[m] = \sum_{m=-\infty}^{\infty} 0 = 0 \qquad \text{for } n < 0 \text{ and for } n > 9$$

The sequence $y[n]$ is given by

$$\{y[n]\} = \{\cdots, 0, 0, 7, 41, 49, 41, 33, 25, 17, 9, 2, 0, 0, \cdots\}$$
$$\uparrow$$

where $\uparrow$ indicates the $n = 0$ element. This result is plotted in Figure 2DT.11(c).

Figure 2DT.11
(*Cont.*)

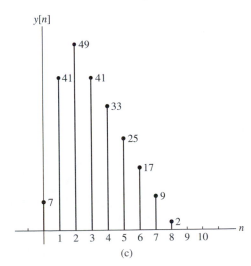

(c)

In the preceding example, we used tapes and plots to help us evaluate a convolution. In the following example, we discard the sliding tapes and work with just the plots. This method is called the *graphical method of convolution*.

EXAMPLE 2DT.5   The Graphical Method of Convolution

Consider the problem of finding the output of an LTI system having the impulse response

$$h[n] = (0.9)^n u[n] \tag{43}$$

for the input

$$x[n] = \begin{cases} 0 & n < 0 \\ (1.2)^n & 0 \le n \le 10 \\ 0 & 10 < n \end{cases} \tag{44}$$

$x[n]$ and $h[n]$ are plotted in Figure 2DT.12.

The response is given by

$$y[n] = \sum_{m=-\infty}^{\infty} x[m]h[n-m] \tag{45}$$

To evaluate this convolution we need to know the expression for $x[m]h[n-m]$. To find this expression, we can plot $x[m], h[n-m]$, and $x[m]h[n-m]$ versus m, as shown in Figure 2DT.13. We include the formulas for $x[m], h[n-m]$ and $x[m]h[n-m]$ on the plots.

Figure 2DT.13(a) is a representative plot of $x[m]$, $h[n-m]$, and the product $x[m]h[n-m]$ for $n < 0$. In this plot $n = -5$. Notice that there is no overlap between the sequences $x[m]$ and $h[n-m]$ for $n < 0$, and therefore $x[m]h[n-m] = 0$. Consequently Eq. (45) becomes

$$y[n] = \sum_{m=-\infty}^{\infty} 0 = 0 \qquad \text{for } n < 0 \tag{46}$$

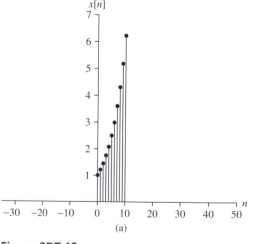

(a)

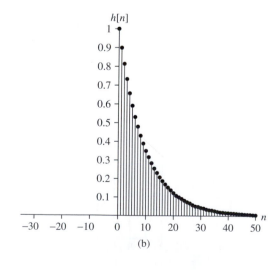

(b)

Figure 2DT.12

Input $x[n]$ and impulse response $h[n]$.

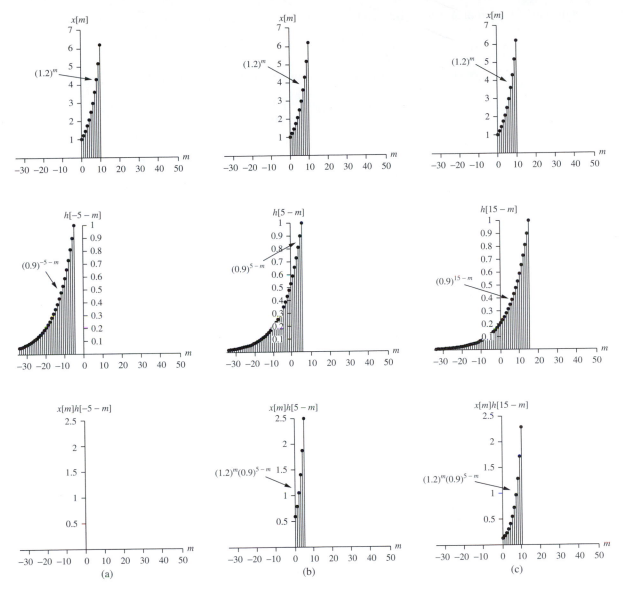

Figure 2DT.13
$x[m]$, $h[n-m]$, and $x[m]h[n-m]$: (a) $n = -5$, (b) $n = 5$, (c) $n = 15$.

Figure 2DT.13(b) is a representative plot of $x[m]$, $h[n-m]$ and the product $x[m]h[n-m]$ for $0 \le n \le 10$. In this plot $n = 5$. Here $x[m]$ and $h[n-m]$ overlap in the range $0 \le m \le n$. Consequently, $x[m]h[n-m]$ is nonzero in this range and Eq. (45) becomes

$$y[n] = \sum_{m=0}^{n} (1.2)^m (0.9)^{(n-m)} = (0.9)^n \sum_{m=0}^{n} \left(\frac{1.2}{0.9}\right)^m \qquad \text{for } 0 \le n \le 10 \qquad (47)$$

which, with the aid of Eq. (1) of Appendix B, becomes

$$y[n] = (0.9)^n \frac{1-(1.2/0.9)^{n+1}}{1-(1.2/0.9)} = \frac{1}{0.3}\{(1.2)^{n+1} - (0.9)^{n+1}\} \qquad \text{for } 0 \le n \le 10 \tag{48}$$

Finally, for $n > 10$ [Figure 2DT.13(c)], $x[m]h[n-m]$ is nonzero in the range $0 \le m \le 10$ so that

$$y[n] = \sum_{m=0}^{10}(1.2)^m(0.9)^{(n-m)} = (0.9)^n \sum_{m=0}^{10}\left(\frac{1.2}{0.9}\right)^m$$

$$= (0.9)^n \frac{1-(1.2/0.9)^{11}}{1-(1.2/0.9)} = 3\left[\left(\frac{4}{3}\right)^{11} - 1\right](0.9)^n \qquad \text{for } n > 10 \tag{49}$$

In conclusion,

$$y[n] = \begin{cases} 0 & \text{for } n < 0 \\ \frac{1}{0.3}\{(1.2)^{n+1} - (0.9)^{n+1}\} & \text{for } 0 \le n \le 10 \\ 3\left\{\left(\frac{4}{3}\right)^{11} - 1\right\}(0.9)^n & \text{for } n > 10. \end{cases} \tag{50}$$

The response, $y[n]$, is plotted in Figure 2DT.14.

Figure 2DT.14
Output $y[n]$.

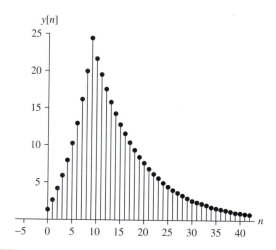

2DT.6 BIBO Stability

Recall from Chapter 1 that *a system is BIBO stable if every bounded input produces a bounded output*. We now state and prove a necessary and sufficient condition for a DT LTI system to be stable.

A DT LTI System is BIBO Stable If and Only If $h[n]$ is Absolutely Summable

$$\sum_{k=-\infty}^{\infty} |h[k]| < \infty \tag{51}$$

We first show that Eq. (51) is a sufficient condition. Suppose that we have an LIT system and we apply a bounded input: $|x[n]| \leq B$ for all n. We can bound the output as follows:

$$|y[n]| = \left| \sum_{k=-\infty}^{\infty} x[n-k]h[k] \right| \leq \sum_{k=-\infty}^{\infty} |x[n-k]||h[k]| \leq \sum_{k=-\infty}^{\infty} B|h[k]| = B \sum_{k=-\infty}^{\infty} |h[k]| \tag{52}$$

The first inequality states that the absolute value of a sum is upper bounded by the sum of absolute values. The second inequality states that the input sequence is bounded. If we compare the left- and right-hand sides of Eq. (52), we see that $|y[n]|$ is finite if $\sum_{k=-\infty}^{\infty} |h[k]|$ is finite. Therefore, the condition that $h[n]$ is absolutely summable Eq. (51) is *sufficient* to ensure that a bounded input yields a bounded output.

We now show that Eq. (51) is a *necessary* condition. It is necessary if there is at least one bounded input that leads to an unbounded output when Eq. (51) is *not* true. The following input is one such input.

$$x[k] = \begin{cases} +1 & h[-k] \geq 0 \\ -1 & h[-k] < 0 \end{cases} \tag{53}$$

This input sequence consists of impulses with amplitudes ± 1 and therefore is bounded. For this input, we have

$$x[-k]h[k] = |h[k]| \tag{54}$$

for all k so that

$$y[0] = \sum_{k=-\infty}^{\infty} x[-k]h[k] = \sum_{k=-\infty}^{\infty} |h[k]| \tag{55}$$

We see that if Eq. (51) is not satisfied, then the input Eq. (53) leads to an output that is unbounded. Therefore we have proven that Eq. (51) is not only a sufficient condition for BIBO stability, but also a necessary one.

As an example, consider an accumulator. The unit impulse response of an accumulator is $h[n] = u[n]$. Here, condition Eq. (51) is not satisfied because the sum of $|u[n]|$ over all n is infinite. Thus an accumulator is not BIBO stable. We can check this finding by applying the bounded input $x[n] = u[n]$. The output is

$$y[n] = \sum_{k=-\infty}^{n} x[k] = \sum_{k=-\infty}^{n} u[k] = (n+1)u[n+1] = r[n+1] \tag{56}$$

where $r[n]$ is the unit ramp sequence, defined as

$$r[n] = nu[n] \tag{57}$$

The unit ramp is plotted in Figure 2DT.15.

The unit ramp is an unbounded sequence: There is no finite number B for which $|r[n]| \leq B$ for all n. In terms of our cash register analogy of Section 2DT.3, if you repeatedly place \$1 items on the conveyor belt, the dollar total increases linearly without limit. This checks our finding, based on Eq. (51), that an accumulator is not BIBO stable.

Figure 2DT.15
Unit ramp $r[n]$.

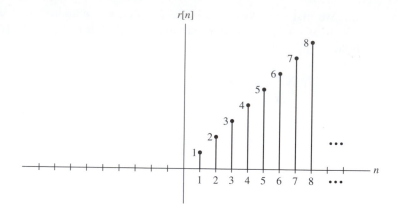

2DT.7 Causal Systems

Recall from Chapter 1 that a system is causal if, for every input, the output $y[n]$, at any n, is not effected by input data values $x[m]$ where $m > n$. We can find a necessary and sufficient condition for a DT LTI system to be causal if we use Eq. (38) to write Eq. (36) as

$$y[n] = \cdots + h[-2]x[n+2] + h[-1]x[n+1] + h[0]x[n-0]$$
$$+ h[1]x[n-1] + h[2]x[n-2] + \cdots \tag{58}$$

We see from Eq. (58) that $y[n]$ does not depend on $x[n+1]$, $x[n+2]$, $\cdots$, if and only if $h[n] = 0$ for $n < 0$. Thus:

An LTI System is Causal If and Only If $h[n] = 0$ for $n < 0$

2DT.8 Basic Connections of DT LTI Systems

Figure 2DT.16 shows ways to connect two LTI systems to create a larger system. The overall unit impulse response of the system can be obtained in the usual way by applying a unit impulse to the input and calculating the output. The results are informative. For example, Figure 2DT.16(a) shows a *cascade* connection of two LTI systems. If we apply a unit impulse to the input in Figure 2DT.16(a) we know that the output of the first system is $h_1[n]$. This output is applied as input to the second system. The output of the second system is given by the convolution of $h_1[n]$ with $h_2[n]$. Therefore, the unit impulse response of the complete system is

$$h[n] = h_1[n] * h_2[n] \tag{59}$$

By the commutativity property of convolution, we know that $h_1[n] * h_2[n] = h_2[n] * h_1[n]$. This means that we can interchange the orders of cascaded LTI systems without changing the overall unit impulse response.

Figure 2DT.16
Basic connections of
LTI systems.

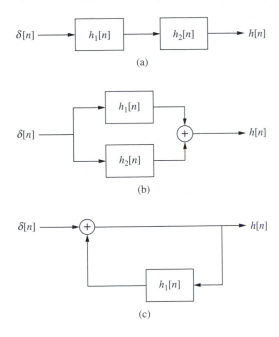

Figure 2DT.16(b) shows a *parallel* connection of two LTI systems. It is left as a problem to show that the overall unit impulse response of the system is

$$h[n] = h_1[n] + h_2[n] \tag{60}$$

Figure 2DT.16(c) shows an LTI system connected in a loop. We do not yet have the analytical tools for efficiently describing the unit impulse response of systems connected in loops. The following expression, however, can be obtained by tracing the signal around the loop when a unit impulse is applied. The expression is correct, but has limited use.

$$h[n] = \delta[n] + h_1[n] + h_1[n] * h_1[n] + h_1[n] * h_1[n] * h_1[n] + \cdots \tag{61}$$

We develop more understanding and more efficient expressions for the unit impulse response of loops as we progress in this book.

As an application of these results, consider the system of Figure 2DT.17(a), which consists of a system having an impulse response $h[n]$ followed by a difference operator. The input is $x[n]$. By the commutativity property of LTI systems, it follows that the system of Figure 2DT.17(b) has the same output. Comparing Figures 2DT.17(a) and (b), we see that the response of an LTI system to the first difference of a sequence $x[n]$ equals the first difference of the system's response to $x[n]$.

Figure 2DT.17
Application of the
commutativity
property of LTI
systems: (a) system
with difference
operator at output;
(b) equivalent system.

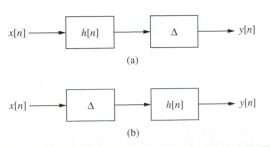

Figure 2DT.18
Calculating an output
by using an
equivalent system:
(a) original system;
(b) equivalent system.

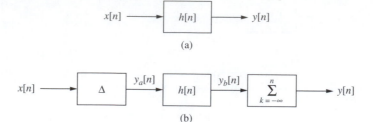

(a)

(b)

As a second application, consider the LTI system having the unit impulse response $h[n]$, shown Figure 2DT.18(a). We can sometimes find the output of this system more easily by first applying the first difference of the input to the system and summing the output, as shown in Figure 2DT.18(b). The two systems are equivalent because of the commutation property of LTI systems and the fact that difference and summation are mutually inverse operators.

EXAMPLE 2DT.6   Step Response of First-Order System

Consider the problem of determining the step response of the first-order system of Figure 2DT.8. We see from Figure 2DT.18 that the step response is the running sum of the impulse response Eq. (24):

$$y[n] = \sum_{k=-\infty}^{n} a^k u[k] = \begin{cases} 0 & n < 0 \\ \sum_{k=0}^{n} a^k & n \geq 0 \end{cases} \tag{62}$$

The following closed form for the sum is given by identity 1 of Appendix B:

$$\sum_{k=0}^{n} a^k = \frac{1 - a^{n+1}}{1 - a} \tag{63}$$

where $n \geq 0$. The substitution Eq. (63) into Eq. (62) yields

$$y[n] = \sum_{k=-\infty}^{n} a^k u[k] = \begin{cases} 0 & n < 0 \\ \dfrac{1 - a^{n+1}}{1 - a} & n \geq 0 \end{cases} \tag{64a}$$

We can write Eq. (64a) as

$$y[n] = \frac{1 - a^{n+1}}{1 - a} u[n] \tag{64b}$$

Notice from Eq. (64b) that

$$\lim_{n \to \infty} y[n] = \frac{1}{1 - a} \qquad \text{for } |a| < 1 \tag{65}$$

The step response Eq. (64b) is illustrated in Figures 2DT.19(a1), (b1), and (c1) for $a > 0$ and (a2), (b2), and (c2) for $a < 0$. It is interesting to notice that the response is oscillatory for $a < 0$. This oscillatory response of a first-order DT system is truly remarkable to anyone familiar with only CT systems: The response of a first-order CT system is *never* oscillatory. It is follows directly from the sum in Eq. (64a) that

$$y[n] = \begin{cases} u[n] & \text{for } a = 1 \\ \frac{1}{2}\{1 + (-1)^n\}u[n] & \text{for } a = -1 \end{cases} \tag{66}$$

$|y[n]|$ increases without bound as n increases for $|a| > 1$. The unbounded response for $|a| > 1$ is illustrated in Figures 2DT.19(c1) and (c2).

Figure 2DT.19
Step response of a
first-order system:
(a1) $a = 0.6$,
(a2) $a = -0.6$,
(b1) $a = 0.9$;
(b2) $a = -0.9$;
(c1) $a = 1.2$,
(c2) $a = -1.2$.

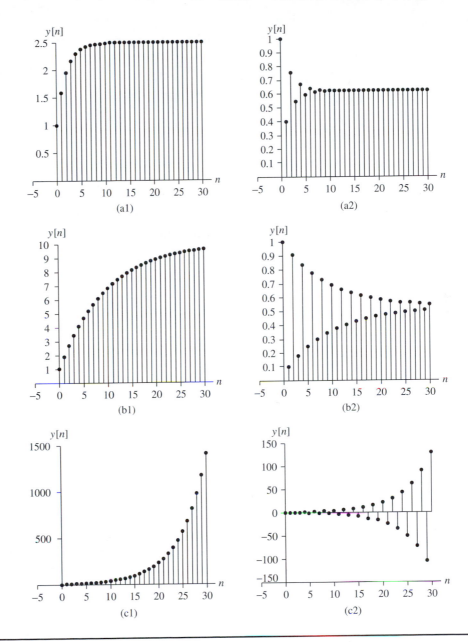

The preceding example shows that the step response of the stable first-order system of
Figure 2DT.8 approaches $1/(1 - a)$ as $n \to \infty$. It is often advantageous to use a first-order
system whose step response approaches unity as $n \to \infty$. We can achieve this result by
cascading the system of Figure 2DT.8 with a gain $1 - a$, as shown in Figure 2DT.20. The
impulse response and step response of the modified system are, respectively

$$h[n] = (1 - a)a^n u[n] \qquad (67)$$

Figure 2DT.20
Modified first-order
system.

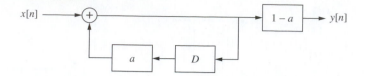

and

$$y[n] = (1 - a^{n+1})u[n] \qquad (68)$$

The step response of the modified system is illustrated in Figure 2DT.21.

For another application of the equivalent system concept (Figure 2DT.18), consider the problem of evaluating the convolution

$$\sqcap [n; K] * \sqcap [n; K] \qquad (69)$$

where $\sqcap [n; K]$ is *a rectangular pulse* sequence, defined as

$$\sqcap [n; K] = \begin{cases} 0 & n < -K \\ 1 & -K \leq n \leq K \\ 0 & K < n \end{cases} \qquad (70)$$

for $K = 0, 1, 2, \ldots$.

The rectangular pulse is plotted in Figure 2DT.22 for $K = 3$. Notice that $\sqcap [n; K]$ is composed of an odd number $2K + 1$ of impulses (7 in the figure). Notice also that it can be written as the difference between two shifted step sequences.

$$\sqcap [n; K] = u[n + K] - u[n - K - 1] \qquad (71)$$

We can interpret the convolution Eq. (69) as being the output of the system of Figure 2DT.18(a), where $x[n] = \sqcap [n; K]$ and $h[n] = \sqcap [n; K]$. Using the equivalent system of Figure 2DT.18(b), we have by inspection

$$y_a[n] = \delta[n + K] - \delta[n - K - 1] \qquad (72)$$

and

$$y_b[n] = h[n + K] - h[n - K - 1] = \sqcap [n + K; K] - \sqcap [n - K - 1; K] \qquad (73)$$

The reader is encouraged to sketch $y_b[n]$ and show from an inspection of this sketch that

$$y[n] = \sum_{k=-\infty}^{n} y_b[n] = \begin{cases} 0 & n < -2K \\ 2K + 1 + n & -2K \leq n < 0 \\ 2K + 1 - n & 0 \leq n \leq 2K \\ 0 & n > 2K \end{cases} \qquad (74)$$

$y[n]$ can be written as

$$y[n] = (2K + 1) \wedge [n; 2K] \qquad (75)$$

where $\wedge [n; 2K]$ is the *triangular pulse* defined by

$$\wedge [n; M] = \begin{cases} 0 & n \leq -M - 1 \\ 1 - \frac{|n|}{M+1} & -M \leq n \leq M \\ 0 & M + 1 \leq n \end{cases} \qquad (76)$$

where $M = 0, 1, 2, \ldots$

Figure 2DT.21

Step response of a modified first-order system: (a1) $a = 0.6$; (a2) $a = -0.6$; (b1) $a = 0.9$; (b2) $a = -0.9$; (c1) $a = 1.2$; (c2) $a = -1.2$.

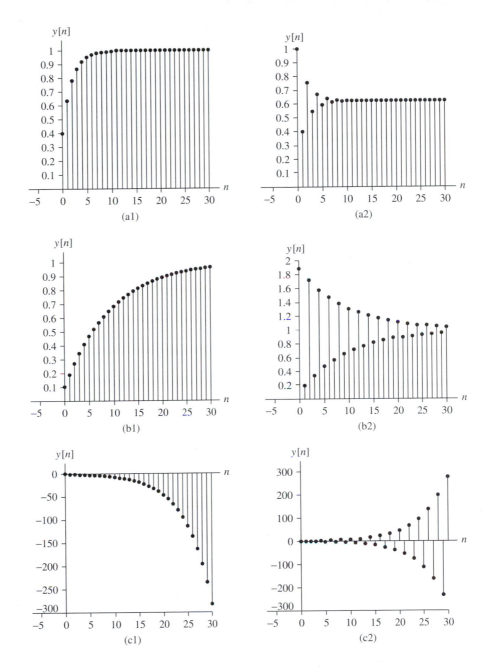

The triangular pulse is shown in Figure 2DT.23 for $M = 6$. Notice that $\wedge [n; M]$ is composed of an odd number $2M + 1$ of impulses (13 for $M = 6$).

In conclusion, we have obtained the result

$$\sqcap [n; K] * \sqcap [n; K] = (2K + 1) \wedge [n; 2K] \qquad (77)$$

Figure 2DT.22
Rectangular pulse
⊓ [n; 3].

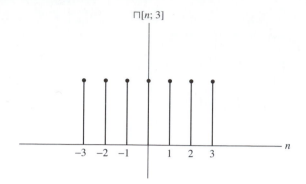

Figure 2DT.23
Triangular pulse
∧ [n; 6].

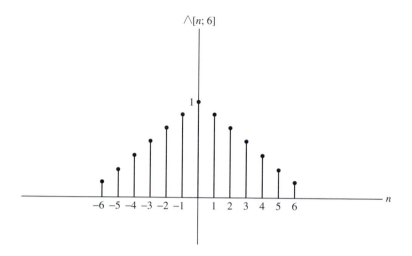

<table>
</table>

2DT.9 **Singularity Sequences and Their Applications**

The discrete-time unit impulse can be used to generate a family of signals we call *singularity sequences*.[2] The singularity sequences are obtained by repeatedly taking the first difference of, or accumulating, the discrete-time unit impulse. In general, the ith singularity sequence is defined as

$$\delta_i[n] = \Delta^i \delta[n] \quad i = 1, 2, \ldots \tag{78}$$

$$\delta_0[n] = \delta[n] \tag{79}$$

$$\delta_{-i}[n] = \sum_{n_i=-\infty}^{n} \cdots \sum_{n_2=-\infty}^{n_3} \sum_{n_1=-\infty}^{n_2} \delta[n_1] \quad i = 1, 2, \ldots \tag{80}$$

[2] There is nothing "singular" about the DT singularity sequences. The terminology is motivated by an obvious parallelism to the CT singularity functions described in Chapter 2CT.

The most frequently used singularity sequences are:

- The unit impulse: $\delta[n]$
- The unit step:

$$u[n] = \delta_{-1}[n] = \sum_{k=-\infty}^{n} \delta[k]$$

- The shifted unit ramp:

$$\delta_{-2}[n] = \sum_{k=-\infty}^{n} u[k] = (n+1)\,u[n+1] = r[n+1] \tag{82a}$$

where

$$r[n] = nu[n] \tag{82b}$$

We have already seen the unit impulse, the unit step, and the unit ramp.

Another singularity sequence of interest is the discrete-time doublet, which is defined as

$$\delta_1[n] = \Delta\delta[n] = \delta[n] - \delta[n-1] \tag{83}$$

Unit step and ramp sequences are frequently used to describe sequences that are made up of "straight-line" segments For example, we saw in Section 2DT.8 that a rectangular pulse can be expressed as the difference between the two shifted step sequences.

$$\sqcap [n; K] = u[n+K] - u[n-K-1] \tag{84}$$

Similarly, the triangular pulse $\wedge [n; M]$, illustrated in Figure 2DT.23, for $M = 6$ can be expressed as a sum of scaled and shifted ramp sequences.

$$\wedge [n; M] = \frac{r[n+M+1] - 2r[n] + r[n-M-1]}{M+1} \tag{85}$$

An advantage of writing a sequence as a weighted sum of shifted step and ramp sequences is to find the output of a LTI system. By using the properties of time invariance and linearity, we can often just write an expression for the output.

EXAMPLE 2DT.7   Rectangular Pulse Response of First-Order System

Consider the response of the first-order system of Figure 2DT.20 to a rectangular pulse $A \sqcap [n; K]$. The step response of this system is given by Eq. (68)

$$u[n] \longrightarrow h_{-1}[n] = (1 - a^{n+1})\,u[n] \tag{86}$$

where we have denoted the step response as $h_{-1}[n]$. Because the system is time invariant, its response to each shifted unit step in Eq. (84) is a shifted version of $h_{-1}[n]$. Because the system is also linear, its response to $\sqcap [n; K]$ is the sum of the two step responses, i.e.,

$$\sqcap [n; K] \longrightarrow h_{-1}[n+K] - h_{-1}[n-K-1] \tag{87}$$

which is

$$\sqcap [n; K] \longrightarrow (1 - a^{n+K+1})\,u[n+K] - (1 - a^{n-K})\,u[n-K-1] \tag{88}$$

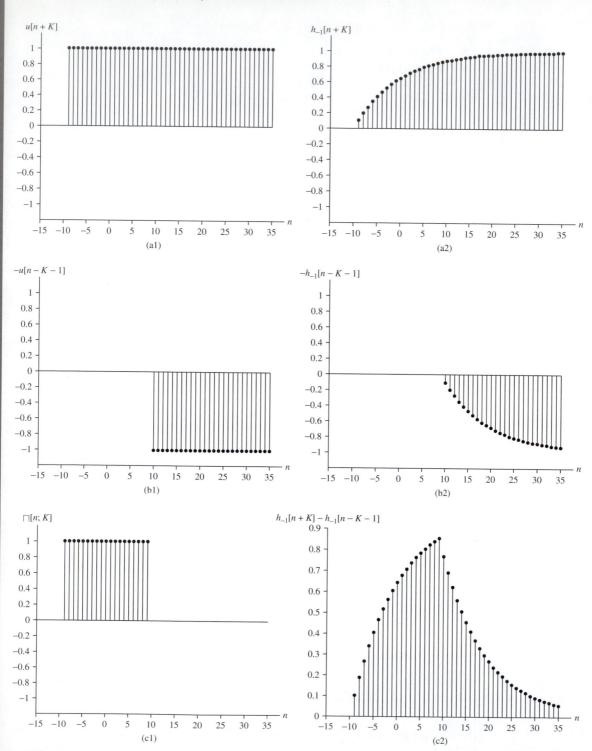

Figure 2DT.24
First-order system response to a rectangular pulse. The pulse ⊓[n; K], (c1), is the sum of the shifted step sequences (a1) and (b1). The pulse response, (c2), is the sum of the shifted step responses (a2) and (b2). The figure assumes $K = 9$ and $a = 0.9$.

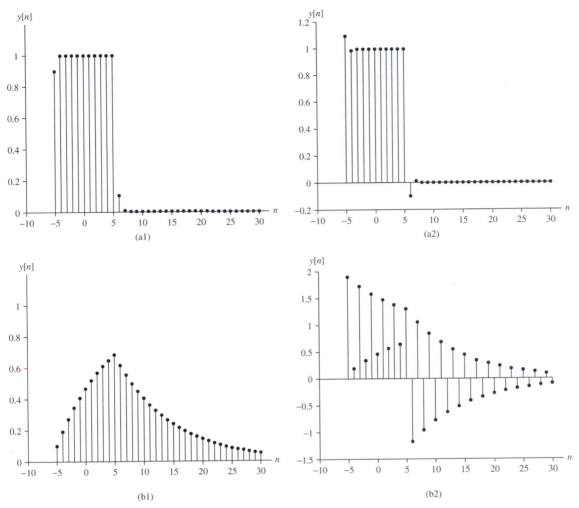

Figure 2DT.25

Influence of parameter a on rectangular pulse response: (a1) $a = 0.1$, (a2) $a = -0.1$, (b1) $a = 0.9$, (b2) $a = -0.9$.

With the help of a little algebra, we can write Eq. (88) as

$$\sqcap [n; K] \longrightarrow \begin{cases} 0 & n < -K \\ 1 - a^{n+K+1} & -K \leq n \leq K \\ (1 - a^{2K+1})a^{n-K} & K < n \end{cases} \tag{89}$$

Figure 2DT.24 illustrates how the step responses add to produce the rectangular pulse response, as indicated by Eq. (89). The influence of the first order system's parameter a on the input rectangular pulse response is shown in Figure 2DT.25, where $K = 5$. In parts: a1 and b1 of the figure, $a = 0.1$ and 0.9 respectively. Notice that the output is approximately equal to the input rectangular pulse for $a = 0.1$ but is a substantially degraded rectangular pulse for $a = 0.9$. A similar result is shown in parts a2 and b2 where $a = -0.1$ and -0.9, respectively. For $a = -0.1$, we see that the response is somewhat oscillatory but is still a good approximation to a rectangular pulse. On the other hand, the response is highly oscillatory and bears no resemblance to the rectangular pulse input for $a = -0.9$. These plots illustrate the ability of the system to pass the rectangular input pulse for $|a| \approx 0$ and its increasing inability to pass the input pulse as $|a|$ increases. The student is encouraged to relate the behavior shown in Figure 2DT.25 to the analytical results given in Eq. (89).

2DT.10 **Summary**

This chapter introduced the topics of *unit impulse*, *unit impulse response*, and *convolution*. The discrete-time unit impulse provides a building block for all other sequences. We showed that every DT signal is a sum of scaled and shifted unit impulses and that every LTI DT system is completely characterized by its response to a unit impulse. We used the principle of superposition to obtain an expression for the output of any DT LTI system in terms of the input and the system's unit impulse response. This expression is called a *convolution sum*. We showed how the convolution sum can also be used to find necessary and sufficient conditions for a DT system to be stable and causal. We also found the overall impulse response of interconnected DT systems in terms of the impulse responses of the individual systems. Finally, we introduced the discrete-time singularity sequences that arise when a unit impulse is applied to a series connection of accumulators or first difference operators. We illustrated how to use singularity sequences to construct rectangular pulses, triangular pulses, and other inputs and obtained greater understanding the response of an LTI system to these inputs. The analytical methods introduced in this chapter are very powerful: They form the foundation for all the chapters that follow. The analytical methods described here are known as *sequence domain* (also called *time domain*) methods because they deal directly with DT sequences.

Summary of Definitions and Formulas

The Unit Impulse

$$\delta[n] = \begin{cases} 0 & n \neq 0 \\ 1 & n = 0 \end{cases}$$

Properties of the Unit Impulse

Sifting property:

$$x[n] = \sum_{m=-\infty}^{\infty} x[m]\delta[n-m]$$

Singularity Sequences

$$\delta_i[n] = \Delta^i \delta[n] \qquad i = 1, 2, \ldots$$

$$\delta_0[n] = \delta[n]$$

$$\delta_{-i}[n] = \sum_{n_i=-\infty}^{n} \cdots \sum_{n_2=-\infty}^{n_3} \sum_{n_1=-\infty}^{n_2} \delta[n_1] \qquad i = 1, 2, \ldots$$

where Δ is the first difference operator

$$\Delta x[n] = x[n] - x[n-1]$$

Discrete-time Doublet, Unit Step, and Ramp Sequences

$$\delta_1[n] = \Delta\delta[n] = \delta[n] - \delta[n-1]$$

$$u[n] = \delta_{-1}[n] = \sum_{k=-\infty}^{n} \delta[k] = \begin{cases} 0 & n < 0 \\ 1 & n \geq 0 \end{cases}$$

$$\delta_{-2}[n] = \sum_{k=-\infty}^{n} u[k] = (n+1)\,u[n+1] = r[n+1]$$

where $r[n]$ is the unit ramp

$$r[n] = nu[n]$$

Unit Impulse Response

$$\delta[n] \longrightarrow h[n]$$

$$x[n] \longrightarrow y[n] = \sum_{m=-\infty}^{\infty} x[m]h[n-m] \overset{\Delta}{=} x[n] * h[n]$$

For ideal shift: $h[n] = \delta[n-m]$

For accumulator: $h[n] = u[n]$

Properties of the Unit Impulse Response

Commutativity: $x[n] * h[n] = h[n] * x[n]$

Superposition: $\{a_1 x_1[n] + a_2 x_2[n]\} * h[n] = a_1 x_1[n] * h[n] + a_2 x_2[n] * h[n]$

Associativity: $x[n] * \{h_1[n] * h_2[n]\} = \{x[n] * h_1[n]\} * h_2[n]$

Stability

A DT LTI system is BIBO stable if and only if $\sum_{k=-\infty}^{\infty} |h[k]| < \infty$.

Causality

An LTI system is causal if and only if $h[n] = 0$ for $n < 0$.

Rectangular Pulse And Triangular Pulse

$$\sqcap[n; K] = \begin{cases} 0 & n < -K \\ 1 & -K \leq n \leq K \\ 0 & K \leq n \end{cases} \tag{90}$$

$$\wedge[n; M] = \begin{cases} 0 & n \leq -M-1 \\ 1 - \dfrac{|n|}{M+1} & -M \leq n \leq M \\ 0 & M+1 \leq n \end{cases} \tag{91}$$

$$\sqcap[n; K] * \sqcap[n; K] = (2K+1)\wedge[n; 2K] \tag{92}$$

2DT.11 Problems

2DT.2   Unit Impulse

2DT.2.1 Plot:
    a) $x[n] = 7.2\delta[n]$.
    b) $x[n] = 3\delta[n+1] + 4.2\delta[n] + 2\delta[n-5]$.
    c) $x[n] = \sum_{m=0}^{5} \delta[n-m]$.
    d) $x[n] = \sum_{m=0}^{\infty} \delta[n-3m]$.
    e) $x[n] = \sum_{k=-\infty}^{\infty} \delta[k-m]$.

2DT.2.2 Simplify the expression $n^2\delta[n-1] + \sin(n)\delta[n] + n^2\delta[n+1]$ and plot the result.

2DT.3   Unit Impulse Response

2DT.3.1 The sequence $x[n] = a^n u[n]$ is applied to a first differencer. Find the response $y[n] = \Delta x[n]$.

2DT.3.2 Find the equation for a 3-term moving average. Find and plot the unit pulse response $h[n]$, and draw the system.

2DT.3.3 Draw a system using a 2-element shift register that computes $y[n] = -x[n] + 2x[n-1] - x[n-2]$.
Find the unit impulse response $h[n]$.

2DT.3.4 An engineer wants to build a system that has output $h[n] = 3.5\delta[n-2] - \delta[n] + 7\delta[n-5]$ when the input is $\delta[n]$. Show how to do this using a shift register, gains, and adders. Then find an expression for the output of this system when the input is an arbitrary sequence $x[n]$.

2DT.3.5 **a)** Derive Eq. (9) mathematically from Eq. (8).
    b) Derive Eq. (8) mathematically from Eq. (9).

2DT.3.6 Verify Eq. (12) by plotting $u[n]$ and $u[n-1]$ and taking their difference.

2DT.3.7 Draw a block diagram of the system that computes the first difference Eq. (14) of a sequence.

2DT.3.8 The unit impulse response of the first-order system of Figure 2DT.8 depends dramatically on the value of a. Sketch $h[n]$ for the system for the following values of a:
    a) 0.9
    b) 1
    c) -1
    d) -0.9
    e) 2

2DT.3.9 Express the data sequence shown in Figure 2DT.26 as a sum of weighted unit impulses.

2DT.3.10 Find the solution to Eq. (16) if the system is noncausal. Hint: Rewrite Eq. (16) as

$$h[n-1] = -a^{-1}\delta[n] + a^{-1}h[n] \qquad (93)$$

and assume that $h[n] = 0$ for $n = 0, 1, 2, \ldots$.

Figure 2DT.26

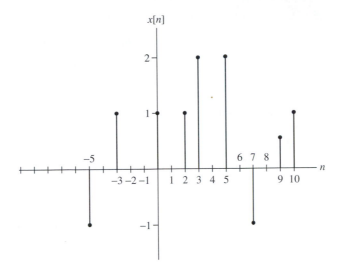

2DT.4   Data Sequence Represented as a Sum of Shifted Impulses

2DT.4.1 The sifting property of the unit impulse can be defined as

$$\sum_{m=-\infty}^{\infty} g[m]\delta[m - n_o] = g[n_o] \tag{94}$$

where $g[n]$ is any sequence. Use the sifting property of the unit impulse to simplify the following sums:
a) $\sum_{n=-\infty}^{\infty} \sqrt{n}\delta[25 - n]$
b) $\sum_{m=-\infty}^{\infty} a^m u[m]\delta[n - m]$

2DT.4.2 Show that $x[n]\delta[n - m] = x[m]\delta[n - m]$ for every sequence $x[n]$.

2DT.5   Convolution Sum

2DT.5.1 We have shown that if a DT system is LTI, the output is given by a convolution sum. Show that the converse is true; i.e., show that, if the output of a system is given by a convolution sum, the system is LTI.

2DT.5.2 A sequence $x[n]$ is applied to an accumulator. Show that the output is given by $y[n] = x[n] * u[n]$.

2DT.5.3 Consider the convolution $u[n] * u[n]$.
a) Use the graphical method to help you find $y[n] = u[n] * u[n]$.
b) Plot your result.

2DT.5.4 Assume that $x[n] = \delta[n] - \delta[n - 10]$ is applied to the system of Figure 2DT.8. Assume that $a = 0.9$.
Find and plot $y[n]$.

2DT.5.5 Assume that $x[n] = \delta[n] + \delta[n - 1] + \delta[n - 2] + \delta[n - 3] + \delta[n - 4]$ is applied to a 5-point moving-average computer. Find and plot the output, $y[n]$.

2DT.5.6 Consider the convolution $y[n] = h[n] * x[n]$ where

$$h[n] = (1.2)^n u[n] \tag{95}$$

and

$$x[n] = (0.9)^n u[n] \tag{96}$$

a) Plot $x[n]$ and $h[n]$.
b) Use the graphical method to help you find $y[n]$. Hint: See Eq. (1) of Appendix B.
c) Plot $y[n]$.

2DT.5.7 Consider the convolution $y[n] = h[n] * x[n]$ where

$$h[n] = a^n u[n] \tag{97}$$

and

$$x[n] = b^n u[n]. \tag{98}$$

a) Show that $y[n] = (a^{n+1} - b^{n+1})/(a - b) \, u[n]$ for $a \neq b$. Hint: See Eq. (1) of Appendix B.
b) Show that $y[n] = (n + 1)a^n u[n]$ for $a = b$.

2DT.5.8 Show by change of variable from m to $k = n - m$ that

$$y[n] = \sum_{m=-\infty}^{\infty} x[m]h[n - m] = \sum_{k=-\infty}^{\infty} h[k]x[n - k] = h[n] * x[n] \tag{99}$$

It follows from this result that convolution is commutative.

$$x[n] * h[n] = h[n] * x[n] \tag{100}$$

2DT.5.9 Show that for every sequence $x[n]$:
a) $\delta[n] * x[n] = x[n]$.
b) $\delta[n - n_o] * x[n] = x[n - n_o]$, where n_o is an integer.
c) $e^{j2\pi fn} * x[n] = e^{j2\pi fn} X(e^{j2\pi f})$, where $X(e^{j2\pi f})$ depends on f but not on n. What is the formula for $X(e^{j2\pi f})$?

2DT.5.10 Find $\delta[n - 1] * \delta[n - 2] * \delta[n - 3]$.

2DT.5.11 When the input $x[n]$ is applied to an LTI system, the response is $y[n]$. Does it follow that the response of the same system to the input $x[2n]$ is $y[2n]$? Justify your answer.

2DT.6   BIBO Stability

2DT.6.1 State which of the following systems are stable and which are unstable. Justify your answers.
a) $h[n] = \delta[n]$
b) $h[n] = a^n u[n]$, where $a = -0.7$
c) $h[n] = a^n u[n]$, where $a = 1.001$
d) $h[n] = u[n] - u[n - 200]$
e) $h[n] = \cos(100n) \, u[n]$

2DT.6.2 Show that the first-order system of Figure 2DT.8 is BIBO stable if and only if $|a| < 1$.

2DT.6.3 The energy in a sequence $h[n]$ is defined as

$$E_h = \sum_{n=-\infty}^{\infty} |h[n]|^2 \tag{101}$$

a) Show that the impulse response of every stable system has finite energy. Hint: Use the following mathematical inequality.

$$\sum_{n=-\infty}^{\infty} |h[n]|^2 \le \left(\sum_{n=-\infty}^{\infty} |h[n]|\right)^2 \tag{102}$$

b) Does it follow that, if $h[n]$ has finite energy, then the system is stable? Hint: Consider $h[n] = (1/n)u[n-1]$.

2DT.7 Causal Systems

2DT.7.1 State which of the following systems are causal and which are not causal. Justify your answers.
a) $h[n] = \delta[n-3]$
b) $h[n] = \delta[n+1]$
c) $h[n] = u[n-5] - u[n+5]$
d) $y[n] = x[-n]$

2DT.7.2 State which of the following systems are causal and which are not causal. Justify your answers.
a) $h[n] = u[-n+1]$
b) $h[n] = \cos(100n)\,u[n]$
c) $y[n] = x^2[3n]$

2DT.8 Basic Connections of DT LTI Systems

2DT.8.1 In Section 2DT.8 of the text we evaluated the convolution $\sqcap[n; K] * \sqcap[n; K]$ with the help of the system of Figure 2DT.18(b). Fill in the details of the derivation by plotting $y_b[n]$ of Eq. (73) and using your plot to verify Eq. (74).

2DT.8.2 The *step response* of a system is defined as the response of the system to a unit step input.
a) It was stated in the text that the unit step response of the first-order system of Figure 2DT.20 is $y[n] = (1 - a^{n+1})u[n]$. Show that the first difference $\Delta y[n]$ equals the impulse response, $h[n]$, of this system as given in Eq. (67).
b) Prove that the impulse response of an LTI system is always given by the first difference of its step response.
c) Show that the step response of an LTI system is always given by summing its impulse response $h[m]$ from $m = -\infty$ to $m = n$.

2DT.8.3 An engineer wants to build a system that has output $h[n] = 5u[n] - 5u[n-4]$ when the input is $\delta[n]$. Draw the block diagram of the system and find an expression for the output when the input is an arbitrary sequence $x[n]$.

2DT.8.4 Is it possible to connect stable systems in such a way that the overall system is unstable? If your answer is yes, give an example.

MATLAB® Problems

MATLAB® problems are in this separate section to help you get started. In later chapters, MATLAB problems will be placed in the same sections as the other problems. All MATLAB problems in this book are identified by the MATLAB icon.

This text does not include a tutorial on MATLAB. If you are new to MATLAB, you should read or peruse the "Getting started with MATLAB" tutorial that comes with MATLAB.

To help you get started with a particular MATLAB problem, you can use the code suggested for the problem. You can also improve on the suggested code or write your own. Never use the suggested code blindly. First understand it. Use the help window or the help function in MATLAB for descriptions of MATLAB functions and commands. In the help window, you can search for the function you need help on. You can also type `help` followed by the name of a function following the MATLAB prompt >> in the command window to learn about the function. For example, type "`help zeros`" to learn about the function zeros.

2DT.1

Sequences should be plotted using the MATLAB stem function. Avoid using the plot function because it interpolates between plotted data points using straight lines. This interpolation can make the plotted data look like a function of continuous time instead of a sequence, as demonstrated in part (a) of this problem.

(a) Make a plot of a discrete-time impulse using the plot function.

```
impulse=[zeros(1,25) 1 zeros(1,25)]
plot(impulse)
```

(b) Compare your plot of part (a) with the plot given by stem.

```
impulse=[zeros(1,25) 1 zeros(1,25)]
stem(impulse)
```

(c) At what value of n does the impulse in your plots occur? Revise your code of part (b) to generate an impulse at $n = 0$.

```
impulse=[zeros(1,25) 1 zeros(1,25)]
n=[-25:25]
stem(n,impulse)
```

2DT.2

(a) As shown in the text, the response of an accumulator to an impulse is a unit step. Refer to MATLAB Problem 2DT.1(c) and use the `cumsum` function to generate a unit step. Plot the unit step using stem.

```
impulse=[zeros(1,25) 1 zeros(1,25)]
n=[-25:25]
step=cumsum(impulse)
stem(n,step)
```

(b) Use the `cumsum` function to demonstrate that the step response of an accumulator is a shifted ramp.

(c) Use the `diff` function to demonstrate that the first difference of a ramp is a step.

(d) Use the `diff` function to demonstrate that the first difference of a step is an impulse.

2DT.3

Plot $u[n + 5] + u[n] + u[n - 5]$ for $-25 \leq n \leq 25$.

2DT.4

In MATLAB, exponentiation has higher priority than multiplication, and the symbols for element-by-element exponentiation and multiplication are `^` and `.*`, respectively. Therefore the sequence $(0.9)^n u[n]$ can be encoded as `0.9.^n.*step` or, equivalently, `(0.9.^n).*step` where `step` was generated in Problem 2DT.2. Plot $a^n u[n]$ for:

   (a) $a = 0.9$.
   (b) $a = -0.9$.
   (c) $a = 1.1$.
   (d) $a = -1.1$.

2DT.5

   (a) The MATLAB function `filter` can be used to describe the first-order LTI system of Figure 2DT.8 for $a = 0.9$ if we set `a=[1,-.9];b=[1]`. Plot the step response of this filter for $-25 \leq n \leq 25$. Here is a possible code:

```
m=10; N=51;
n=[-m:N-m-1];
u=[zeros(1,m) ones(1,N-m)];
a=[1,-.9];b=[1];
y=filter(b,a,u);
stem(n,y)
```

   (b) Find and plot the impulse response of the filter of part (a).
   (c) Demonstrate that if you accumulate the impulse response of part (b), the result is the filter's step response. Hint: Use the `cumsum` function.
   (d) Experiment with the impulse and step response of the first-order filter by selecting different values of a.

2DT.6

Use the MATLAB function `conv` to demonstrate the commutation property of LTI systems using two DT systems of your choice.

2DT.7

Use the MATLAB function `filter` to find the output of two first-order filters connected in series. Assume that each filter has the same parameter $a = 0.9$.

2DT.8

Use the MATLAB conv function and plot $\sqcap (n; K_1) * \sqcap (n; K_2)$, where:

   (a) $K_1 = K_2 = 5$.
   (b) $K_1 = 5$, $K_2 = 10$.

2DT.9

Use MATLAB to compute and plot the response of the first-order system of Figure 2DT.20 to $\sqcap (n; K)$ using the `conv` function and a variety of values for K and a.

2DT.10

Use MATLAB to compute and plot the response of the first-order system of Figure 2DT.20 to $\sqcap (n; K)$ using the function `filter` and a variety of values for K and a.

Miscellaneous Problems

2DT.1. The techniques of this chapter can be extended to two dimensions. The extension applies to image processing. For simplicity, we consider only grayscale (i.e., achromatic or "black and white") images. A two-dimensional discrete-space unit impulse is defined as

$$\delta[n, m] = \begin{cases} 1 & \text{for } n = m = 0 \\ 0 & \text{otherwise} \end{cases} \tag{103}$$

where n and n are the discrete coordinates of a two-dimensional array. We can imagine $\delta[n, m]$ as a point of light having unit "intensity" located at the center of the otherwise dark array. The response of a discrete-space linear space-invariant image processor to $\delta[n, m]$ is the discrete-space *point spread sequence* $h[n, m]$. The output image intensity resulting from an arbitrary input image intensity $I_i[n, m]$ is given by a two-dimensional convolution sum.

$$I_o[n, m] = \sum_{i=-\infty}^{\infty} \sum_{k=-\infty}^{\infty} I_i[i, k]h[n - i, m - k] \tag{104}$$

a) Find $I_o[n, m]$ for $I_i[i, k] = \delta[i - n_o, k - m_o]$, where n_o and m_o are integers. Express your answer in terms of $h[\cdot, \cdot]$.

b) Assume that

$$h[n, m] = \left(\frac{1}{2K + 1}\right)^2 \sqcap (n; K) \sqcap (m; K) \tag{105}$$

What operation does the processor perform on the input image? State your answer both in words and with a mathematical expression.

2DT.2. Write and run a MATLAB program that performs the processing described in Problem 2DT.1 on the MATLAB image "durer" or some other image. Use values of K in the range $0 \leq K \leq 25$, and comment on the resulting image. ($K = 0$ yields the original image.)

```
k=input('Enter averaging array size variable K>=0:')
n=2*k+1

%compute processing filter array
H=ones(n,n)./(n*n);

load durer
%image data is in the variable X

%perform 2D convolution
C=conv2(X,H);

%Find minimum and minimum intensity values in X
I1=min(min(X));
I2=max(max(X));

%display the convolved image using same dynamic range as
input image
imagesc(C,[I1 I2])
axis image
colormap(gray)
```

2DT.3. Image enhancement can be performed using the two-dimensional point spread sequence.

$$h'[n, m] = \frac{\alpha\delta[n, m] - (1 - \alpha)h[n, m]}{2\alpha - 1} \qquad (106)$$

where $h[n, m]$ is the point spread sequence given by Eq. (105), and α is a variable in the range $0.5 < \alpha \leq 1$. The technique is called *unsharp masking*.

a) Show that if the input image $I_i[i, k]$ equals a constant c for all i, k, then the output image also equals c. Explain why this property of $h'[n, m]$ is desirable for an image processor.

b) Describe in words what $h'[n, m]$ does to an input image; i.e., how does the unsharp masking technique work? Hint: Try to answer this question on your own, but if you need help, you can try to find the answer using a Web search for unsharp masking.

c) Modify the MATLAB program in Problem 2DT.2 to compute and display the durer or other image using $h'[n, m]$. Use the same commands as in the suggested program to display the processed image. Comment on your results.

3CT Frequency Response Characteristics

3CT.1 Introduction

We saw in Chapter 2CT that LTI systems are characterized by their responses to an impulse. LTI systems can also be characterized by their responses to sinusoidal inputs. This characterization is called a system's *frequency response characteristic*, which is a very natural and convenient way to describe an LTI system. Like the impulse response, the frequency response characteristic provides a complete description of the I-O properties of an LTI system. If you know a system's frequency response characteristic, you can find the output for any input.

The frequency response characteristic is based on an important and unique property of sinusoidal waveforms: If you apply a sinusoidal waveform to an LTI system, the output is also a sinusoidal waveform. The frequency of the sinusoidal output is always the same as that of the input, but the output amplitude and phase are generally different from those of the input. The system's frequency response characteristic describes how the amplitudes and phases of the output sinusoid compare to those of the input sinusoid as a function of frequency.

3CT.2 Sinusoidal Waveforms

We can write any real sinusoidal waveform as

$$x(t) = A\cos(2\pi Ft + \phi) \tag{1}$$

where A is the *amplitude*, F is the *frequency* in hertz (Hz), and ϕ is the phase angle in radians. The frequency in radians per second is $\Omega = 2\pi F$. Figure 3CT.1 shows a plot of Eq. (1).

Real sinusoids are related to complex sinusoids by Euler's formula:

$$Ae^{j(2\pi Ft + \phi)} = A\cos(2\pi Ft + \phi) + jA\sin(2\pi Ft + \phi) \tag{2}$$

The left-hand side of Eq. (2) is a *complex sinusoid*, $Ae^{j(2\pi Ft + \phi)} = (Ae^{j\phi})e^{j2\pi Ft}$. It can be plotted in the complex plane as shown in Figure 3CT.2. In circuit theory $X \triangleq Ae^{j\phi}$ is called a *phasor*, and $Xe^{j2\pi Ft}$ is called a *rotating phasor*. F is the number of revolutions per second. The real sinusoid $A\cos(2\pi Ft + \phi)$ is the projection of $Xe^{j2\pi Ft}$ on the real axis.

$$A\cos(2\pi Ft + \phi) = \mathcal{R}\{Ae^{j(2\pi Ft + \phi)}\} = \mathcal{R}\{Xe^{j2\pi Ft}\} \tag{3}$$

Figure 3CT.1
The sinusoidal
waveform: $x(t) = A\cos(2\pi F t + \phi)$.

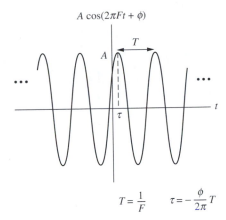

$$T = \frac{1}{F} \qquad \tau = -\frac{\phi}{2\pi}T$$

Figure 3CT.2
The complex
sinusoid: $Ae^{j(2\pi F t + \phi)}$.

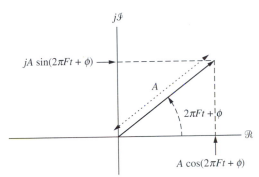

where $\mathcal{R}\{\}$ means "real part of." The phasor, or complex amplitude X, tells us the amplitude and the phase of the real sinusoid: $A = |X|$ and $\phi = \angle X$. One complete revolution and one period of $A\cos(2\pi F t + \phi)$ take $T = 1/F$ s.

3CT.3   Response of LTI Systems to Sinusoidal Input

You know from personal experience that echoes can make it difficult to understand what someone is saying. Echoes are scaled, delayed replicas of a waveform, and, when they are added to the waveform, the shape of the waveform sum is different from the original waveform. There is one bounded waveform, however, whose shape is *not* changed by echoes. This is the sinusoidal waveform. A pure sinusoidal sound tone having frequency F is still a pure sinusoidal tone having frequency F regardless of the strength and number of echoes. The property "sinusoid in, sinusoid out" applies for every stable LTI system.[1]

**3CT.3.1
LTI System
Response to a
Complex Sinusoid**

In this section we determine the response of an LTI system to a complex sinusoidal input. Consider the general I-O equation for an LTI system

$$y(t) = \int_{-\infty}^{\infty} h(\lambda)x(t-\lambda)\,d\lambda \qquad (4)$$

---

[1] It also applies to some unstable systems, including ideal filters.

and assume that

$$x(t) = Xe^{j2\pi Ft} \tag{5}$$

When we substitute Eq. (5) into Eq. (4), we obtain

$$y(t) = \int_{-\infty}^{\infty} h(\lambda)Xe^{j2\pi F(t-\lambda)}d\lambda = \left\{ \int_{-\infty}^{\infty} h(\lambda)e^{-j2\pi F\lambda}d\lambda \right\} Xe^{j2\pi Ft} \tag{6}$$

which is

$$y(t) = H(F)Xe^{j2\pi Ft} \tag{7}$$

The function $H(F)$ is called the *frequency response characteristic* of the system. It is given by the term in the braces in Eq. (6), which we can rewrite:

Frequency Response Characteristic

$$H(F) = \int_{-\infty}^{\infty} h(t)e^{-j2\pi Ft}dt \tag{8}$$

The integral in Eq. (8) is the *Fourier transform*[2] of $h(t)$. This integral transforms the time domain function $h(t)$ to the frequency domain function $H(F)$. $H(F)$ exists for every BIBO stable LTI system and for some LTI systems that are not BIBO stable. We see from Eq. (7) that, when the complex sinusoid $Xe^{j2\pi Ft}$ is applied to an LTI system, the response is another complex sinusoid having the same frequency.

$$y(t) = H(F)Xe^{j2\pi Ft} \tag{9}$$

Equation (9) is the basic result of this section. We can summarize this key finding using flow graph notation:

The Response of a Stable LTI System to a Complex Sinusoid Having Frequency F is a Complex Sinusoid Having Frequency F

$$Xe^{j2\pi Ft} \xrightarrow{\ h(t)\ } Ye^{j2\pi Ft} \tag{10a}$$

where

$$Y = H(F)X \tag{10b}$$

This result applies to systems having both real and complex impulse responses. It assumes that the input is a complex sinusoid for all t.[3]

We can extend the result to real sinusoidal waveforms that are applied to real systems. When a system is real, its impulse response $h(t)$ is a member of the set of real functions. To accomplish the extension, we need to learn how the form of $H(F)$ depends on the assumption that $h(t)$ is real. We do this in the next section.

---

[2] A Fourier transform can be approximately computed using a computer algorithm called the FFT. This topic is considered in Chapter 5DT.

[3] You can conclude that the input is sinusoidal for all t because no restriction is placed on the range of t in Eq. (5). The response of an LTI system to a input that is 0 for $t < t_0$ and complex sinusoidal for $t \geq t_0$ is considered in Problem 3CT.3.3.

**3CT.3.2
Symmetry
Properties of
$H(F)$**

When $h(t)$ is real, $H(F)$ has *conjugate symmetry*

$$H(-F) = H^*(F) \tag{11}$$

You can easily verify Eq. (11) by inspection of Eq. (8) if you replace F with $-F$: The effect of a change in the sign of F is equivalent to a change in the sign of j. The conjugate symmetry of $H(F)$ implies that the *amplitude* and *phase* characteristics $|H(F)|$ and $\angle H(F)$ of every real LTI system satisfy

$$|H(-F)| = |H(F)| \tag{12}$$

and

$$\angle H(-F) = -\angle H(F) \tag{13}$$

Thus, real systems have even amplitude characteristics and odd phase characteristics.[4]

Now that we have established the symmetry properties of $H(F)$, we are ready to consider the response of a real system to a real sinusoid.

**3CT.3.3
LTI System
Response to Real
Sinusoid**

We have shown that a complex sinusoidal input $Xe^{j2\pi Ft}$ yields a complex sinusoidal output $H(F)Xe^{j2\pi Ft}$

$$Xe^{j2\pi Ft} \xrightarrow{h(t)} H(F)Xe^{j2\pi Ft} \tag{14}$$

for every stable LTI system. We can use this result and superposition to find the response of a real LTI system to a real sinusoidal input. By Euler's formula, a real sinusoid is the sum of two complex sinusoids

$$A\cos(2\pi Ft + \phi) = Xe^{j2\pi Ft} + X^*e^{-j2\pi Ft} \tag{15}$$

where

$$X = \tfrac{1}{2}Ae^{j\phi} \tag{16}$$

If we replace F by $-F$ and X by X^* in Eq. (14), we obtain

$$X^*e^{-j2\pi Ft} \xrightarrow{h(t)} H(-F)X^*e^{-j2\pi Ft} \tag{17}$$

By superposition:

$$Xe^{j2\pi Ft} + X^*e^{-j2\pi Ft} \xrightarrow{h(t)} H(F)Xe^{j2\pi Ft} + H(-F)X^*e^{-j2\pi Ft} \tag{18}$$

which, with the aid Euler's formula and the conjugate symmetry property of $H(F)$, becomes

$$A\cos(2\pi Ft + \phi) \xrightarrow{h(t)} A|H(F)|\cos(2\pi Ft + \phi + \angle H(F)) \qquad \text{for real stable LTI systems.} \tag{19}$$

In conclusion, we see that the response of a real LTI system to a sinusoidal input is a sinusoid having the same frequency. The system's frequency response characteristic describes how

---

[4] In general, a function $g(\psi)$ is defined as *even* if $g(-\psi) = g(\psi)$. Similarly, $g(\psi)$ is defined as *odd* if $g(-\psi) = -g(\psi)$.

the output amplitude and phase are related to the input amplitude and phase. Real and complex sinusoids are the only bounded signals that reproduce themselves when applied to LTI systems.

Let's now consider some examples of a system's frequency response characteristic $H(F)$.

EXAMPLE 3CT.1   First-Order System

Consider the RC circuit of Figure 3CT.3. Recall that this circuit can be modeled as the first-order system of Figure 2CT.11.

Figure 3CT.3
RC low-pass filter.

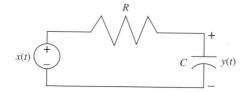

We showed in Eq. (2CT.23) that the impulse response of this circuit is $h(t) = 1/\tau_c e^{-t/\tau_c} u(t)$, where $\tau_c = RC$. We write this result as $h(t) = \alpha e^{-\alpha t} u(t)$ where $\alpha = 1/\tau_c$. The Fourier transform of $\alpha e^{-\alpha t} u(t)$ is

$$H(F) = \alpha \int_{-\infty}^{\infty} e^{-\alpha t} u(t) e^{-j2\pi F t} dt = \alpha \int_{0}^{\infty} e^{-(\alpha + j2\pi F)t} dt$$

$$= \alpha \frac{-e^{-j2\pi F t} e^{-\alpha t}}{\alpha + j2\pi F} \Big|_{0}^{\infty} = \frac{\alpha}{\alpha + j2\pi F} \qquad \text{for } \alpha > 0 \qquad (20)$$

The Fourier transform (20) exists only for $\alpha > 0$. For the circuit, the condition $\alpha > 0$ is automatically satisfied because $\alpha = 1/RC$, where R and C are always positive. By substituting $\alpha = 1/\tau_c$ into (20), we obtain

$$H(F) = \frac{1}{1 + \tau_c j2\pi F} \qquad (21)$$

Equation (21) is the result we sought.

Let's compare Eq. (21) with the result obtained using the method of circuit theory. This alternative method starts with the equation relating input voltage to output voltage.

$$\tau_c \frac{dy(t)}{dt} + y(t) = x(t) \qquad (22)$$

and applies the basic property (10). It applies a rotating phasor input

$$x(t) = X e^{j2\pi F t} \qquad (23)$$

and assumes a solution of the form

$$y(t) = Y e^{j2\pi F t} \qquad (24)$$

The substitution of Eq. (23) and Eq. (24) into Eq. (22) yields

$$\tau_c j2\pi F Y e^{j2\pi F t} + Y e^{j2\pi F t} = X e^{j2\pi F t} \qquad (25)$$

which simplifies to

$$\tau_c j2\pi F Y + Y = X \qquad (26)$$

It follows that

$$Y = \frac{1}{1 + \tau_c\, j2\pi F}\, X \tag{27}$$

and

$$H(F) = \frac{Y}{X} = \frac{1}{1 + \tau_c\, j2\pi F} \tag{28}$$

which is identical to Eq. (21). Circuit theory enthusiasts recognize $H(F)$ as a *voltage transfer function* and recognize that it can be written as

$$H(F) = \frac{\frac{1}{j2\pi FC}}{R + \frac{1}{j2\pi FC}} \tag{29}$$

which is the formula for an AC voltage divider based on impedances.

It follows from Eq. (19) that an input $A\cos(2\pi Ft + \phi)$ produces the output

$$y(t) = |H(F)|A\cos(2\pi Ft + \phi + \angle H(F)) \tag{30}$$

where

$$|H(F)| = \frac{1}{\sqrt{1 + (\tau_c 2\pi F)^2}} \tag{31}$$

and

$$\angle H(F) = -\arctan(\tau_c 2\pi F) \tag{32}$$

Plots of $|H(F)|$ and $\angle H(F)$ are given in Figure 3CT.4. This circuit is a *low-pass filter*. For low frequencies, i.e., $F \ll 1/2\pi\tau_c$, the absolute value of $1/j2\pi FC$ is large compared to R, and the output voltage is approximately equal to the input voltage: $H(F) \approx 1$. For high frequencies, i.e., $F \gg 1/2\pi\tau_c$, the absolute value of $1/j2\pi FC$ is small compared to R, and the output voltage is near zero: $H(F) \approx 0$. The frequency $F = 1/2\pi\tau_c$ is called the *half-power frequency* of the filter: $H(1/2\pi\tau_c) = (1/\sqrt{2})\angle - 45°$.

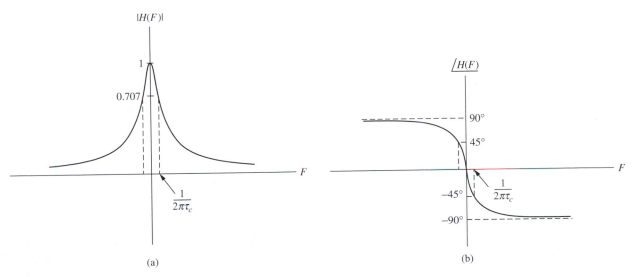

(a)   (b)

Figure 3CT.4
RC Low-pass filter frequency response characteristics: (a) amplitude; (b) phase.

We see from Example 3CT.1 that the results of systems and signals theory and AC circuit theory are in perfect agreement. This example demonstrates that you do not have to evaluate the Fourier transform integral (8) to find $H(F)$. You can also find $H(F)$ by applying $Xe^{j2\pi Ft}$ as the input and solving the I-O equation for the output $Ye^{j2\pi Ft}$. The frequency response characteristic $H(F)$ is the ratio Y/X. This method assumes that $H(F)$ exists but does not require you to first find $h(t)$. In fact, once you have found $H(F)$, you can find $h(t)$ by means of the inverse Fourier transform discussed in Section 3CT.5.

EXAMPLE 3CT.2   Moving Average

The *moving average* $y(t)$ of a waveform $x(t)$ is defined by the equation

$$y(t) = \frac{1}{\tau} \int_{t-\tau}^{t} x(\lambda)\, d\lambda \tag{33}$$

where τ is the time over which the average is made. The impulse response associated with the system that computes a moving average was found in Problem 2CT.3.2.

$$h(t) = \frac{1}{\tau}[u(t) - u(t-\tau)] = \frac{1}{\tau} \sqcap \left(\frac{t-0.5\tau}{\tau}\right) \tag{34}$$

The substitution of Eq. (34) into Eq. (8) yields

$$H(F) = \int_{-\infty}^{\infty} \frac{1}{\tau}[u(\lambda) - u(\lambda - \tau)]e^{-j2\pi F\lambda}d\lambda =$$

$$\frac{1}{\tau}\int_{0}^{\tau} e^{-j2\pi F\lambda}d\lambda = \frac{1}{-j2\pi F\tau}e^{-j2\pi F\lambda}\Big|_{0}^{\tau} = \frac{1}{-j2\pi F\tau}\left[e^{-j2\pi F\tau} - 1\right] =$$

$$= \frac{1}{j2\pi F\tau}\left[e^{+j\pi F\tau} - e^{-j\pi F\tau}\right]e^{-j\pi F\tau} = \frac{\sin(\pi F\tau)}{\pi F\tau}e^{-j\pi F\tau} \tag{35}$$

Equation (35) can be written as

$$H(F) = \text{sinc}(F\tau)e^{-j\pi F\tau} = \text{sinc}(F\tau) \times (1\angle - \pi F\tau) \tag{36}$$

where the *sinc function* was defined in Eq. (2CT.6). We have chosen not to express $H(F)$ in terms of absolute value and phase because that would spoil its simplicity. A plot of $\text{sinc}(F\tau)$ and the linear phase $-\pi F\tau$ is shown in Figure 3CT.5. The factors $\text{sinc}(F\tau)$ and $1\angle - \pi F\tau$ can each be interpreted in a simple way. $\text{Sinc}(F\tau)$ is the Fourier transform of a rectangular pulse, $(1/\tau) \sqcap (t/\tau)$. Except for a time shift, this rectangular pulse performs the moving average. The linear phase $-\pi F\tau$ corresponds to a time delay of $\tau/2$ that makes the process of averaging causal.

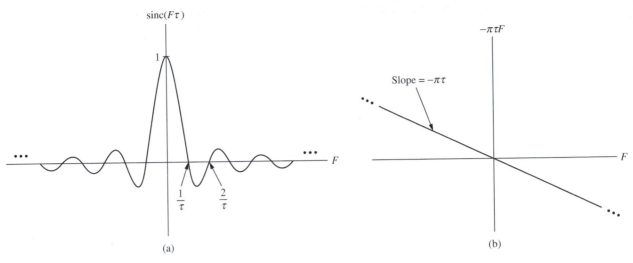

Figure 3CT.5
Frequency response characteristic: $H(F) = \text{sinc}(F\tau) \times (1\angle{-\pi F\tau})$: (a) sinc $(F\tau)$; (b) $-\pi F\tau$.

EXAMPLE 3CT.3   Cell Phone Call with Echo

Let's use the Fourier transform to find the audio frequency response characteristic from one cell phone to another when the audio contains an echo: When you speak into the cell phone, the person at the other end hears your voice plus the echo. A simplified model of this system is shown in Figure 3CT.6. We have modeled the sending and receiving cell phones as ideal amplifiers having unity gain. The transmission channel is modeled as having one direct path and one path for the echo.

Figure 3CT.6
Model for audio communication in the presence of echo.

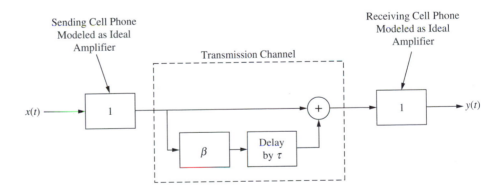

We see from the figure that the I-O relation of the model is

$$x(t) \xrightarrow{h(t)} x(t) + \beta x(t - \tau) \tag{37}$$

where β is a scale factor and τ is the time delay of the echo. It follows that

$$\delta(t) \xrightarrow{h(t)} h(t) = \delta(t) + \beta\delta(t - \tau) \tag{38}$$

We can find $H(F)$ from Eq. (8) easily using the sifting property of the impulse. The result is

$$H(F) = 1 + \beta e^{-j2\pi F\tau} \tag{39}$$

The amplitude characteristic is

$$|H(F)| = \sqrt{1 + \beta^2 + 2\beta \cos(2\pi F\tau)} \tag{40}$$

and the phase characteristic is

$$\angle H(F) = \arctan\left\{\frac{-\beta \sin(2\pi F\tau)}{1 + \beta \cos(2\pi F\tau)}\right\} \tag{41}$$

If you apply a real sinusoidal input $A\cos(2\pi Ft + \phi)$, the output is $|H(F)|A\cos(2\pi Ft + \phi + \angle H(F))$.

The frequency response characteristic of the cell phone audio link is plotted in Figure 3CT.7 for $\beta = 0.7$.

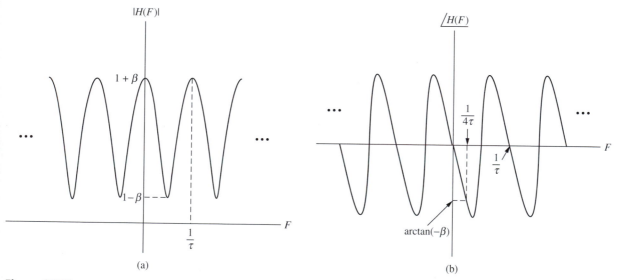

Figure 3CT.7
Cell phone communication audio frequency response characteristic for $\beta = 0.7$: (a) amplitude; (b) phase.

The shape of the amplitude characteristic can be explained by the constructive or destructive interference occurring due to the sum of the input sinusoid $A\cos(2\pi Ft + \phi)$ and its echo $\beta A\cos(2\pi F(t - \tau) + \phi)$. For frequencies near $k/\tau, k = 0, 1, 2, \ldots$, the input sinusoid and its echo are approximately in phase (i.e., they have approximately the same phase shift except for multiples of 2π) and therefore constructive interference occurs. For frequencies near $(k+0.5)/\tau$, $k = 0, 1, 2, \ldots$, the input sinusoid and its echo are approximately 180° out of phase and destructive interference occurs. Whenever you obtain an answer to an engineering problem, you should consider if the answer makes sense, as we did here. The cell phone communications model used in this example is refined in Section 3CT.7.

EXAMPLE 3CT.4   Room Acoustics with Feedback

Let's return to the system in Figure 1.1. The voice of the performer on stage is picked up by a microphone and broadcast through a speaker. The sound from the speaker comes back as an echo into the microphone to be amplified again, and the feedback of echoes repeats ad infinitum. The impulse response of the closed-loop system from microphone input to speaker output is, from Eq. (2CT.25).

$$h(t) = \alpha \sum_{n=0}^{\infty} (\alpha\beta)^n \delta(t - n\tau) \tag{42}$$

To find $H(F)$, we substitute Eq. (42) into Eq. (8) and interchange the orders of integration and summation. This step yields

$$H(F) = \int_{-\infty}^{\infty} h(t)e^{-j2\pi Ft}\,dt = \alpha \sum_{n=0}^{\infty} (\alpha\beta)^n \int_{-\infty}^{\infty} \delta(t - n\tau)e^{-j2\pi Ft}\,dt \tag{43}$$

which, with the aid of the sifting property and identity (B.4) of Appendix B, is

$$H(F) = \alpha \sum_{n=0}^{\infty} \left(\alpha\beta e^{-j2\pi F\tau} \right)^n = \frac{\alpha}{1 - \alpha\beta e^{-j2\pi F\tau}} \qquad \text{for } |\alpha\beta| < 1. \tag{44}$$

The amplitude and phase characteristics

$$|H(F)| = \alpha \sqrt{\frac{1}{1 + \alpha^2\beta^2 - 2\alpha\beta\cos(2\pi F\tau)}} \tag{45}$$

and

$$\angle H(F) = -\arctan\left\{ \frac{\sin(2\pi F\tau)}{1 - \alpha\beta\cos(2\pi F\tau)} \right\} \tag{46}$$

are plotted in Figure 3CT.8 for $0 < \alpha\beta < 1$.

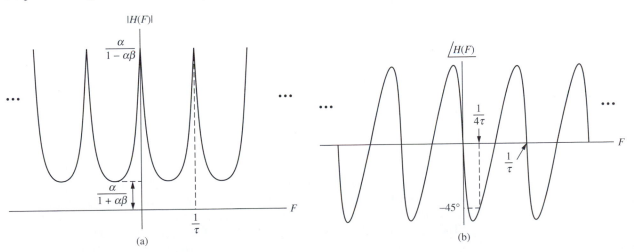

(a)   (b)

Figure 3CT.8
Frequency response characteristic of room with feedback: The peak value of $|H(F)|$ increases without bound as $\alpha\beta \to 1$; $H(F)$ does not exist for $\alpha\beta \geq 1$.

We can see from Figure 3CT.8 that the frequency response is not flat in spite of our assumption that the system from microphone input to speaker output has a constant gain for all frequencies! The frequency response is not flat because constructive and destructive interference occurs. The strength of the total audio input (original voice plus echo) to the microphone depends on frequency. We can see in Figure 3CT.8 that the variation in gain $|H(F)|$ from one frequency to another (called amplitude distortion) depends on the product $\alpha\beta$. There is no amplitude distortion for $\beta = 0$, which is the condition for no feedback. The amplitude distortion increases as $\alpha\beta$ increases toward unity. For $\alpha\beta \geq 1$, $H(F)$ no longer exists. It is at the point $\alpha\beta = 1$ that the system becomes unstable and one hears a loud shriek generated by the feedback loop.

The analysis in Example 3CT.4 captures some of the properties of a real-life setup like that shown in Figure 1.1. It is unrealistic, however, in the sense that the amplitude characteristic $|H(F)|$ does not display an overall decrease in gain as frequency increases. This unrealistic property comes about primarily because we have assumed that the microphone, amplifier, and speaker have frequency responses that are constant for all frequencies. This aspect of the problem is considered in Section 3CT.7.

3CT.4   Multiple Sinusoidal Components

Because sinusoidal waveforms propagate throughout LTI systems, as shown in Eqs. (10) and (19), we are motivated to use them as building blocks for nonsinusoidal waveforms. In this section we begin a fundamental development in signals and system theory that eventually brings us to the Fourier representations of signals of Chapters 4CT–5CT: The development is based on the observation that, for every LTI system, properties (10) and (19) can be extended easily to sums of sinusoids. To see how, consider the input

$$x(t) = A_1 \cos(2\pi F_1 t + \phi_1) + A_2 \cos(2\pi F_2 t + \phi_2) \tag{47}$$

The output of an LTI system follows immediately from Eq. (19) and superposition.

$$y(t) = |H(F_1)|A_1 \cos(2\pi F_1 t + \phi_1 + \angle H(F_1)) + |H(F_2)|A_2 \cos(2\pi F_2 t + \phi_2 + \angle H(F_2)) \tag{48}$$

The output amplitudes and phases are determined by the input and by the frequency response characteristic of the system.

Two graphic aids are used to visualize the action of the system on the input signal. The first is shown in Figure 3CT.9 and uses a *one-sided* plot of the frequency response characteristic $H(F)$. In the figure, the amplitudes and phases of the input sinusoids are plotted above the frequencies of the sinusoids to which they refer. These plots are called *one-sided amplitude and phase spectrums*. Similar plots are made for the output sinusoidal components. The output amplitude spectrum is given by the product of the input amplitude spectrum and the amplitude characteristic of the filter. The output phase spectrum is given by the sum of the input phase spectrum and the phase characteristic of the filter at frequencies F_1 and F_2. [See Eq. (48).] Both the input and the output spectrums are sometimes called *line spectrums* because they consist of vertical lines.

The other graphic aid uses a *two-sided* plot of the frequency response characteristic $H(F)$. This representation is shown in Figure 3CT.10. We use Euler's formula to express the input (47) as a sum of four complex sinusoids,

$$x(t) = X_1 e^{j2\pi F_1 t} + X_1^* e^{-j2\pi F_1 t} + X_2 e^{j2\pi F_2 t} + X_2^* e^{-j2\pi F_2 t} \tag{49}$$

Figure 3CT.9
One-sided frequency domain plots:
(a) input spectrum;
(b) filter frequency response characteristic;
(c) output spectrum.

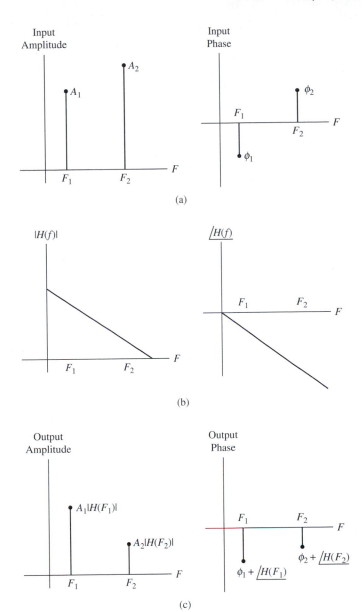

where $X_1 = \frac{1}{2}A_1 e^{j\phi_1}$ and $X_2 = \frac{1}{2}A_2 e^{j\phi_2}$. It then follows from Eq. (10) and superposition that

$$y(t) = H(F_1)X_1 e^{j2\pi F_1 t} + H(-F_1)X_1^* e^{-j2\pi F_1 t} + H(F_2)X_2 e^{j2\pi F_2 t} + H(-F_2)X_2^* e^{-j2\pi F_2 t}$$

(50)

The two-sided input and output spectrums are shown in Figures 3CT.10(a) and (c), respectively. The plots are called two-sided amplitude and phase spectrums. The output amplitude spectrum is given by the product of the input amplitude spectrum and the amplitude characteristic of the filter [Figure 3CT.10(b). The output phase spectrum is given by the sum of the input phase spectrum and the phase characteristic of the filter at frequencies $\pm F_1$ and

Figure 3CT.10
Two-sided frequency
domain plots:
(a) input spectrum;
(b) filter frequency
response
characteristic;
(c) output spectrum.

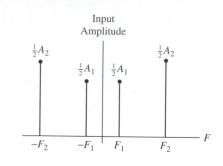

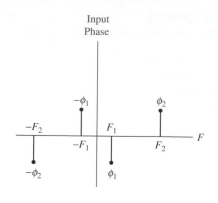

(a)

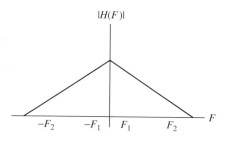

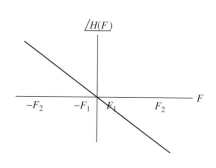

(b)

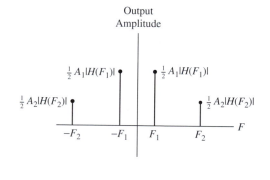

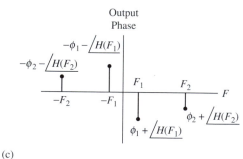

(c)

$\pm F_2$. Notice that the figure assumes the amplitude and phase symmetry properties of $H(F)$ that apply for a real system.

The generalization of these results to an input composed of many sinusoidal components is straightforward. Consider an input composed of the sum of a constant (DC) term A_0 and I sinusoidal components:

$$x(t) = A_0 + \sum_{i=1}^{I} A_i \cos(2\pi F_i t + \phi_i) \tag{51}$$

It follows from Eq. (19) and superposition that the output is

$$y(t) = A_0 H(0) + \sum_{i=1}^{I} A_i |H(F_i)| \cos(2\pi F_i t + \phi_i + \angle H(F_i)) \qquad (52)$$

The complex version of the preceding is

$$x(t) = \sum_{i=-I}^{I} X_i e^{j2\pi F_i t} \qquad (53)$$

$$y(t) = \sum_{i=-I}^{I} X_i H(F_i) e^{j2\pi F_i t} \qquad (54)$$

where

$$X_i = \begin{cases} A_0 & i = 0 \\ \frac{1}{2} A_i e^{j\phi_i} & i = 1, 2, \ldots, I \\ X_{-i}^* & i = -1, -2, \ldots, -I \end{cases} \qquad (55)$$

and

$$F_i = \begin{cases} 0 & i = 0 \\ -F_{-i} & i = -1, -2, \ldots, -I \end{cases} \qquad (56)$$

Flow graph notation shows the I-O relationship as

$$\sum_{i=-I}^{I} X_i e^{j2\pi F_i t} \xrightarrow{h(t)} \sum_{i=-I}^{I} X_i H(F_i) e^{j2\pi F_i t} \qquad (57)$$

We will show in Chapter 4 that most waveforms found in engineering practice can be represented as sums or integrals of complex sinusoidal components. This representation is very advantageous because we can more easily understand the effect of an LTI system on an input signal if we look at the input signal's spectrum and the system's frequency response characteristics than if we look at the time domain input waveform and the system's impulse response. The time domain operation is a convolution. The frequency domain operation involves the much simpler operation of complex multiplication, $X_i H(F_i)$, as demonstrated. This is the reason we find a graphic equalizer on audio amplifiers like the one depicted in Figure 1.1, instead of a plot of the equalizer's impulse response $h(t)$. The equalizer band adjustments provide the values of $|H(F_i)|$. No phase adjustments are included in an audio amplifier because human hearing is relatively insensitive to phase.

3CT.5 $h(t)$ and $H(F)$ are a Fourier Transform Pair

We have seen that the Fourier transform of a system's impulse response

$$H(F) = \int_{-\infty}^{\infty} h(t) e^{-j2\pi F t} dt \qquad (58)$$

provides us with useful understanding of how the system responds to both real and complex sinusoidal inputs. But this is not all. It turns out that $H(F)$ is also a *complete description* of the I-O properties of a stable LTI system. This means that starting from $H(F)$ we can work backward to find $h(t)$. The formula for the *inverse Fourier transform* is

$$h(t) = \int_{-\infty}^{\infty} H(F)e^{j2\pi Ft}dF \tag{59}$$

We can prove that this formula is correct by substituting it into the integral in Eq. (8). This substitution needs a change of variable $F \rightarrow F'$ to distinguish the variable of integration in Eq. (59) from the argument F in Eq. (8). The result of the substitution is

$$\int_{-\infty}^{\infty} h(t)e^{-j2\pi Ft}dt = \int_{-\infty}^{\infty} \left\{ \int_{-\infty}^{\infty} H(F')e^{j2\pi F't}dF' \right\} e^{-j2\pi Ft}dt$$

$$= \int_{-\infty}^{\infty} H(F') \left\{ \int_{-\infty}^{\infty} e^{j2\pi(F'-F)t}dt \right\} dF' \tag{60}$$

We can evaluate the integral involving the exponential as a limit

$$\int_{-\infty}^{\infty} e^{j2\pi(F'-F)t}dt = \lim_{T\to\infty} \int_{-T/2}^{T/2} e^{j2\pi(F'-F)t}dt = \lim_{T\to\infty} T\,\text{sinc}(T(F'-F)) = \delta(F'-F) \tag{61}$$

where we have used Eq. (2CT.5). The substitution of $\delta(F'-F)$ into Eq. (60) yields

$$\int_{-\infty}^{\infty} h(t)e^{-j2\pi Ft}dt = \int_{-\infty}^{\infty} H(F')\delta(F'-F)dF' = H(F) \tag{62}$$

which proves that Eq. (59) is correct. $h(t)$ and $H(F)$ are a *Fourier transform pair*, and their one-to-one relationship is symbolized as

$$h(t) \overset{FT}{\longleftrightarrow} H(F) \tag{63}$$

The operator notation

$$\mathcal{FT}\{h(t)\} = \int_{-\infty}^{\infty} h(t)e^{-j2\pi Ft}dt \tag{64}$$

and

$$\mathcal{FT}^{-1}\{H(F)\} = \int_{-\infty}^{\infty} H(F)e^{j2\pi Ft}dF \tag{65}$$

are often used to denote the Fourier transform and its inverse. Tables of Fourier transform pairs and Fourier transform properties are developed in Chapter 4CT.

3CT.6 Frequency Response Characteristics of Ideal Systems

Ideal systems are useful for describing what we might want a practical system to do. They provide goals for practical filter design, as described in Chapter 7. The frequency response characteristics of some important ideal systems are shown in Figure 3CT.11.

All the systems shown can be called filters. However, the term *ideal filter* is usually reserved for the systems of Figures 3CT.11(a)–(d). These systems are called ideal filters because they perfectly pass the part of the input signal spectrum lying within their passbands

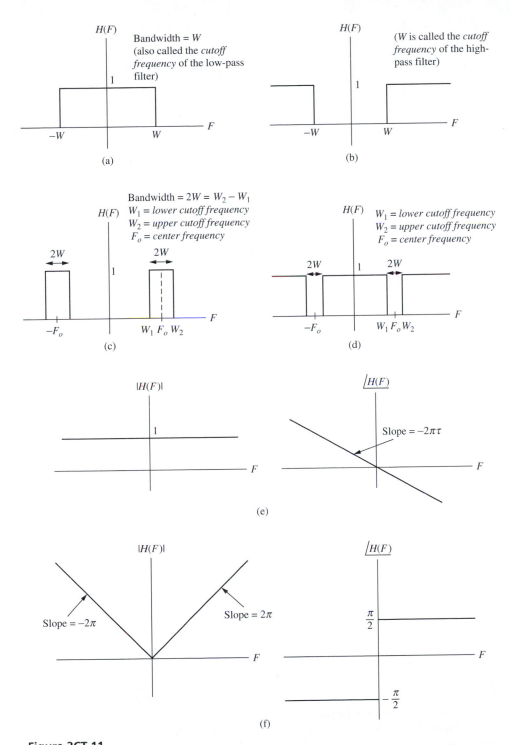

Figure 3CT.11

Frequency response characteristics of ideal systems: (a) low-pass filter; (b) high-pass filter; (c) band-pass filter (the center frequency F_o is halfway between W_1 and W_2); (d) stop-band filter (also called a *band rejection* or *notch* filter; the center frequency F_o is half way between W_1 and W_2) ; (e) delay; (f) differentiator; (g) integrator.

Figure 3CT.11
(*Cont.*)

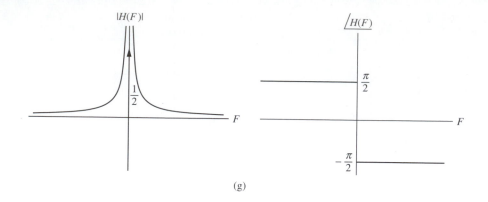

(g)

(where $H(F) = 1$) and completely reject the part of the input signal spectrum lying within their rejection bands (where $H(F) = 0$). The figure defines the relevant terminology used in connection with these filters. There is no need to plot amplitude and phase characteristics for these filters because the frequency response characteristics are purely real.

EXAMPLE 3CT.5   Two Sinusoids Applied to an Ideal Low-Pass Filter

Assume that an input

$$x(t) = 10\cos(2\pi 25t + 22°) + 250\cos(2\pi 100t) \tag{66}$$

is applied to the ideal low-pass filter of Figure 3CT.11(a). The output is given by

$$y(t) = |H(25)|10\cos(2\pi 25t + 22°) + |H(100)|250\cos(2\pi 100t) \tag{67}$$

where we have used the fact that the angle of H equals 0.

The output depends on the value of the filter's *cutoff frequency* W. If $W < 25$, there is no output because the input frequencies $F = 25$ and $F = 100$ are outside the filter passband and are therefore rejected by the filter. If $25 < W < 100$, the output is $10\cos(2\pi 25t + 22°)$ because this is the only component passed by the filter. If $100 < W$, then both components are passed by the filter and the output is $10\cos(2\pi 25t + 22°) + 250\cos(2\pi 100t)$.

The impulse responses of the ideal filters can be obtained from Eq. (59). For example, the impulse response of the ideal low-pass filter is

$$h(t) = \int_{-W}^{W} e^{j2\pi Ft}\,dF = \frac{1}{j2\pi t}e^{j2\pi Ft}\Big|_{-W}^{W} = \frac{1}{j2\pi t}\left(e^{j2\pi Wt} - e^{-j2\pi Wt}\right) = \frac{\sin(2\pi Wt)}{\pi t} \tag{68}$$

which is

$$h(t) = 2W\,\text{sinc}(2Wt) \tag{69}$$

Figure 3CT.12 shows a plot of the impulse response of the ideal low-pass filter. We can see from this plot that an ideal low-pass filter is a noncausal system.

An important special case of Eq. (69) occurs in the limit $W \to \infty$. In this limit, the frequency response characteristic equals unity: $H(F) = 1$ for all F. We can see from Figure 3CT.12 that $h(t)$ becomes narrower and narrower and higher and higher as W increases. It is easy to show that the area of $2W\,\text{sinc}(2Wt)$ is unity (See Problem 3CT.6.1). Therefore, in the limit $W \to \infty$, $h(t) = \delta(t)$. This provides a derivation of the Fourier

Figure 3CT.12
Impulse response of
the ideal low-pass
filter.

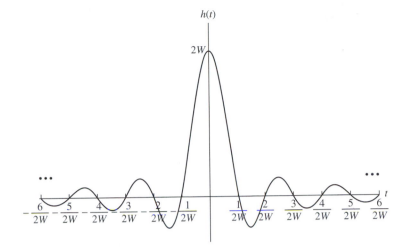

Table 3CT.1
Ideal Filters

Name	$H(F)$	$h(t)$
Ideal low-pass filter	$\sqcap\left(\dfrac{F}{2W}\right)$	$2W\,\text{sinc}(2Wt)$
Ideal band-pass filter	$\sqcap\left(\dfrac{F-F_o}{2W}\right)+\sqcap\left(\dfrac{F+F_o}{2W}\right)$	$4W\,\text{sinc}(2Wt)\cos(2\pi F_o t)$
Ideal high-pass filter	$1-\sqcap\left(\dfrac{F}{2W}\right)$	$\delta(t)-2W\,\text{sinc}(2Wt)$
Ideal stop-band filter	$1-\sqcap\left(\dfrac{F-F_o}{2W}\right)-\sqcap\left(\dfrac{F+F_o}{2W}\right)$	$\delta(t)-4W\,\text{sinc}(2Wt)\cos(2\pi F_o t)$

transform pair

$$\delta(t)\overset{FT}{\longleftrightarrow}1 \qquad (70)$$

The conclusion is that, if a filter has the impulse response $h(t)=\delta(t)$, then that filter passes all frequency components of its input without change, i.e., $H(F)=1$ for all F.

The impulse response waveforms of all the ideal filters are listed in Table 3CT.1. All the impulse responses listed in the table can be derived in a manner similar to Eq. (68). In the following chapter, we show how to derive them more easily by using properties of the Fourier transform. It is remarkable that all the ideal filters have nonzero impulse responses for $t<0$. This means that all ideal filters are noncausal systems.[5]

> **All Ideal Filters are Noncausal**

Table 3CT.2 lists the impulse responses of the remaining ideal systems of Figure 3CT.11. Let's derive its entries. Figure 3CT.11(e) shows the frequency response characteristic of an ideal time shift element. The time domain output is a time-shifted replica of the input:

$$y(t)=x(t-\tau) \qquad (71)$$

---

[5] Another remarkable property of the ideal filters listed in Table 3CT.1 is that they are not BIBO stable. This result follows fom the fact that they each have an impulse response that is not absolutely integrable.

Table 3CT.2
Ideal Systems

Name	$H(F)$	$h(t)$
Time shift	$e^{-j2\pi F\tau}$	$\delta(t-\tau)$
Differentiator	$j2\pi F$	$\dfrac{d}{dt}\delta(t) = \delta_1(t)$
Integrator	$\frac{1}{2}\delta(F) + \dfrac{1}{j2\pi F}$	$u(t)$

where τ is the delay (τ is nonnegative for a causal system). The impulse response of the time shift element is

$$h(t) = \delta(t-\tau) \tag{72}$$

The frequency response characteristic is easily found by substituting Eq. (72) into Eq. (8) and applying the sifting property of an impulse. The result is

$$H(F) = e^{-j2\pi F\tau} \tag{73}$$

which is listed in Table 3CT.2.

Figure 3CT.11(f) shows the frequency response characteristic of an ideal differentiator. The time domain output is the derivative of the input

$$y(t) = \frac{d}{dt}x(t) \tag{74}$$

The impulse response is a unit doublet

$$h(t) = \frac{d}{dt}\delta(t) = \delta_1(t) \tag{75}$$

We can find the frequency response characteristic easily by substituting $Xe^{j2\pi Ft}$ into Eq. (74). The derivative is $j2\pi F X e^{j2\pi Ft}$. Therefore

$$Xe^{j2\pi Ft} \longrightarrow j2\pi F X e^{j2\pi Ft} \tag{76}$$

and it follows from Eq. (10) that the frequency response characteristic is

$$H(F) = j2\pi F \tag{77}$$

which is listed in Table 3CT.2.

Figure 3CT.11(g) shows the frequency response characteristic of an ideal integrator. We know from our previous work that an integrator has an impulse response $h(t) = u(t)$. The derivation of the frequency response characteristic $H(F)$ requires considerable care. The derivation is given in Appendix 3CT.A. The result is

$$H(F) = \frac{1}{2}\delta(F) + \frac{1}{j2\pi F} \tag{78}$$

as listed in Table 3CT.2.

3CT.7   Basic Connections of LTI Systems

Figure 3CT.13 depicts frequency domain results for connecting two LTI systems to create a third, larger system. We can obtain the overall frequency response characteristics of these systems by applying an input $Xe^{j2\pi Ft}$, as shown in the figure.

For example, Figure 3CT.13(a) shows a *cascade* connection of two LTI systems. To obtain the overall frequency response characteristic, we apply $Xe^{j2\pi Ft}$ to the first system. We know from Eq. (10) that the output of the first system is $H_1(F)Xe^{j2\pi Ft}$. This complex sinusoid is the input to the second system, and the output is $H_2(F)H_1(F)Xe^{j2\pi Ft}$ (again by Eq. (10a)). Therefore, the overall frequency response characteristic is

$$H(F) = H_1(F)H_2(F) \tag{79}$$

Because $H_1(F)H_2(F) = H_2(F)H_1(F)$, we can interchange the orders of cascaded LTI systems. This conclusion confirms the commutation property (2CT.31). By comparing Eq. (79) with Eq. (2CT.51) we also see that the time domain operation of convolution $h_1(t) * h_2(t)$, corresponds to the frequency domain operation of multiplication $H_1(F)$ $H_2(F)$. Clearly, this is an important result: Multiplication is much simpler than convolution to visualize and evaluate!

Figure 3CT.13(b) shows a *parallel* connection of two LTI systems. We leave it as a problem to show that

$$H(F) = H_1(F) + H_2(F) \tag{80}$$

Figure 3CT.13(c) shows an LTI systems connected in a loop. To obtain the frequency response characteristic of the system, we apply an input $Xe^{j2\pi Ft}$ and assume that the complex sinusoidal form propagates throughout the system. In the figure, the output of the system is denoted as $Ye^{j2\pi Ft}$, and the output of the adder is denoted as $Ve^{j2\pi Ft}$. We apply

Figure 3CT.13
Basic connections of
CT LTI systems in the
time domain with
input $x(t) = Xe^{j2\pi Ft}$:
(a) cascade;
(b) parallel; (c) loop.

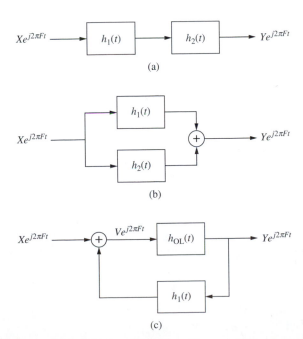

the basic property (10), and find by inspection of the figure that

$$Ye^{j2\pi Ft} = H_{OL}(F)Ve^{j2\pi Ft} \tag{81}$$

and

$$Ve^{j2\pi Ft} = Xe^{j2\pi Ft} + YH_1(F)e^{j2\pi Ft} \tag{82}$$

We can solve the above equations for $H(F) = Y/X$. The result is

$$H(F) = \frac{H_{OL}(F)}{1 - H_{OL}(F)H_1(F)} \tag{83}$$

Equation (83) is an important result describing the frequency response characteristic for systems connected in a loop. Our method and the result Eq. (83), assume that $H(F)$ exists. Notice that $H(F) = H_{OL}(F)$ where there is no feedback, i.e., when $H_1(F) = 0$. $H_{OL}(F)$ is the "open loop" frequency response characteristic of the sytem. A special case of Eq. (83) occurs for $H_{OL}(F) = 1$:

$$H(F) = \frac{1}{1 - H_1(F)} \tag{84}$$

This special case reduces the system to that shown in Figure 2CT.21c. We can see that Eq. (84) agrees with Eq. (2CT.53) because Eq. (84) can be expanded (identity (B.4) of Appendix B) to give

$$H(F) = 1 + H_1(F) + H_1^2(F) + H_1^3(F) + \cdots \tag{85}$$

for $|H(F)| < 1$.

These results are summarized in Figure 3CT.14.

Let's now use the results to obtain more realistic answers to previous examples.

Figure 3CT.14
Basic connections of CT LTI systems in the frequency domain:
(a) cascade;
(b) parallel; (c) loop.

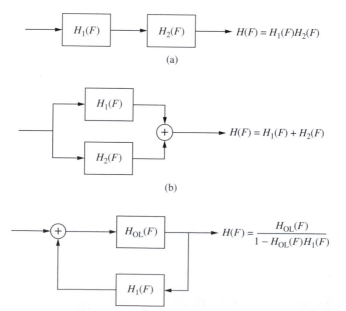

EXAMPLE 3CT.6   Improved Model of Cell Phone Call with Echo

We can improve our solution to the cell phone communication problem in Example 3CT.3. The model of this system, shown in Figure 3CT.6, assumed that the frequency response of a cell phone is flat. We can illustrate that assumption by setting $\beta = 0$ in Figure 3CT.6. Then, Eq. (37) reduces to.

$$y(t) = x(t) \tag{86}$$

where $x(t)$ is the audio waveform input of the sending cell phone and $y(t)$ is the audio output of the receiving cell phone. Equation (86) implies that the impulse response of a cell phone without an echo is $h_{\mathrm{CP}}(t) = \delta(t)$, which by Eq. (70), implies that $H_{\mathrm{CP}}(F) = 1$. Thus, the amplitude characteristic implied by Eq. (86) is flat for all F. Is Eq. (86) really true for a practical cell phone? The answer is no, but it is still a reasonable model *provided that $x(t)$ is a voice signal*, i.e., a signal for which the cell phone was designed. Equation (86) is not true if $x(t)$ is any signal having a frequency component beyond the audio frequency range: These are not communicated by a real-world cell phone.

We can refine our model of Figure 3CT.6 by incorporating a more realistic frequency response for the cell phone. The simplest way to incorporate finite bandwidth into the model is by means of an ideal band-pass filter $H_{\mathrm{CP}}(F) = H_{\mathrm{BP}}(F)$. The band-pass filter rejects frequencies $F < W_1$ and $F > W_2$ outside the audio frequency range of the cell phone. The improved model for cell phone communication is shown in Figure 3CT.15. The dotted lines show that the improved model is a cascade connection of three systems: the sending and receiving cell phones modeled as ideal band-pass filters and the transmission channel. The overall frequency response is the product of the frequency responses of each system.

$$\begin{aligned} H(F) &= H_{\mathrm{CP}}(F)H_{\mathrm{CH}}(F)H_{\mathrm{CP}}(F) \\ &= H_{\mathrm{BP}}(F)H_{\mathrm{CH}}(F)H_{\mathrm{BP}}(F) = H_{\mathrm{CH}}(F)H_{\mathrm{BP}}(F) \end{aligned} \tag{87}$$

where the last step follows from the fact that the value of $H_{\mathrm{BP}}(F)$ is 1 in the pass band and 0 in the stop band: $H_{\mathrm{BP}}^2(F) = H_{\mathrm{BP}}(F)$.

Further refinements can be made. For example, we could use the frequency response of an actual cell phone for $H_{\mathrm{CP}}(F)$ instead of an ideal band-pass filter $H_{\mathrm{BP}}(F)$. We could include the effects of multiple echoes and the overall transmission delay that occur between audio input and output.

Figure 3CT.15
Improved model of cell phone communication channel.

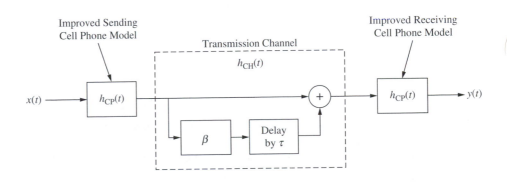

EXAMPLE 3CT.7 Room Acoustics with Feedback

Let us return to the system in Figure 1.1. We found that the frequency response characteristic of the closed-loop system from microphone input to speaker output is given by Eq. (44). This result is based on the assumption that the microphone, amplifier, and speaker all have frequency response characteristics that are constant for all F. Let's improve our model by including more realistic frequency responses for these components. Real-life microphones, audio amplifiers, and speakers have amplitude response characteristics that fall off sharply for frequencies below about 30 Hz, are approximately flat for 30 Hz $< F <$ 20 kHz, and fall off sharply for frequencies greater than about 20 kHz. To mathematically incorporate the microphone $H_M(F)$, audio amplifier $H_A(F)$, and speaker frequency response characteristics $H_S(F)$ into our closed loop problem, we refer to Eq. (83), which provides us with the expression for the block diagram of Figure 3CT.14c. We associate $H_{OL}(F) = H_M(F)H_A(F)H_S(F)$ with the open-loop system response from microphone input to speaker output, and $H_1(F)$ with the room acoustics from speaker output to microphone input. We model the room acoustics from speaker output to microphone input as a single echo for which $h_1(t) = \beta\delta(t - \tau)$ and $H_1(F) = \beta e^{-j2\pi F\tau}$. The substitution of this frequency response characteristics into Eq. (83) yields

$$H(F) = \frac{H_{OL}(F)}{1 - H_{OL}(F)\beta e^{-j2\pi F\tau}} \tag{88}$$

Notice that Eq. (88) reduces to Eq. (44) for the special case that $H_{OL}(F) = \alpha$. For simplicity, let's now model $H_{OL}(F)$ as $H_{OL}(F) = \alpha H_{BP}(F)$ where $H_{BP}(F)$ is an ideal band-pass filter having lower and upper cutoff frequencies $W_1 = 30$ Hz and $W_2 = 20$ kHz. Then $H(F)$ becomes

$$H(F) = \frac{\alpha H_{BP}(F)}{1 - \alpha H_{BP}(F)\beta e^{-j2\pi F\tau}} = \frac{\alpha}{1 - \alpha\beta e^{-j2\pi F\tau}} H_{BP}(F) \tag{89}$$

The last equality in (89) follow from the fact that $H_{BP}(F)$ is either 0 or 1.

3CT.8 Power Waveforms, Power Spectrums, and Parseval's Theorem

The sinusoid $A\cos(2\pi Ft + \phi)$ of Eq. (1) and the multiple sinusoidal component waveform of Eq. (51) are members of a class of waveforms called *power waveforms*. The *average power* in a real or complex waveform $x(t)$ is defined as

$$P = <|x(t)|^2> \tag{90}$$

in which $|x(t)|^2$ is the *instantaneous power* in $x(t)$ and

$$<(\cdot)> = \lim_{T_0 \to \infty} \frac{1}{2T_0} \int_{-T_0}^{T_0} (\cdot)\, dt \tag{91}$$

denotes the *time average value* of its argument $(\cdot)$. P is the time average of the instantaneous power carried by $x(t)$. The terms "power" and "power waveform" are motivated by electrical engineering. Physically, P would equal the average power dissipated in the form of heat if $x(t)$ were a real voltage applied across a 1 Ω resistor.

A waveform $x(t)$ is called a *power waveform* if P is a nonzero finite number. Real-life waveforms, like noise and voice, are often modeled as power waveforms. A pulse waveform like $A \sqcap (t/\tau_c)$ is not a power waveform because the average power it carries is 0.

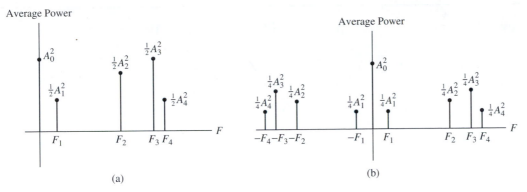

Figure 3CT.16
Power spectrums: (a) one sided; (b) two sided.

Many power waveforms are periodic. A waveform $x(t)$ is *periodic* with period T if

$$x(t) = x(t + T) \qquad \text{for all } t \tag{92}$$

where T is the smallest nonzero time interval for which Eq. (92) applies. For periodic waveforms, we can we compute the time average $<(\cdot)>$ over a period T.

$$<(\cdot)> = \frac{1}{T} \int_{-T/2}^{T/2} (\cdot)\, dt \tag{93}$$

For a periodic waveform, Eqs. (91) and (93) give the same results. We leave it as a problem to show that the real sinusoid $x(t) = A \cos(2\pi F_1 t + \phi)$ has average power $P = \frac{1}{2}A^2$ for $F_1 \neq 0$, $P = A^2 \cos^2 \phi$ for $F_1 = 0$, and the complex sinusoid $Xe^{j2\pi Ft}$ has average power $|X|^2$ for every F (see Problem 3CT.8.2).

It is often useful to display the average power of the sinusoidal components in a waveform having the form Eq. (51) versus frequency, as shown in Figure 3CT.16. Figure 3CT.16(a) shows a *one-sided power spectrum* for $I = 4$. The height of each line $P_i = \frac{1}{2}A_i^2$, $i = 1, 2, \ldots, I$, tells us the average power of the sinusoid having the frequency F_i, indicated by the position of the line. The line at $F = 0$ shows the power in the DC component $P_0 = A_0^2$.

Figure 3CT.16(b) shows a *two-sided power spectrum* of $x(t)$. A two-sided power spectrum consists of lines, each having amplitude $A_i^2/4$ positioned at frequencies $\pm F_i$, as shown in Figure 3CT.16(b). (See Eq. (56).) The height of each line is equal to the average power carried by the complex sinusoid having the frequency indicated by the position of the line. We can obtain a two-sided power spectrum by breaking up the real sinusoids in Eq. (51) into pairs of complex sinusoids as in Eq. (53). The average power carried by $X_i e^{j2\pi F_i t}$ is $< |X_i e^{j2\pi F_i t}|^2 > = |X_i|^2 = A_i^2/4$ for $i = \pm 1, \pm 2, \ldots, \pm I$ and $P_0 = |X_0|^2 = A_0^2$. (See Eq. (55)).

A fundamental property of both one-sided and two-sided power spectrums is that the total average power carried by a waveform is equal to the sum of the heights of the lines in the waveform's power spectrum. This property is called *Parseval's theorem* for multiple sinusoidal component waveforms.

Parseval's Theorem for Multiple Sinusoidal Component Waveforms

$$\text{Complex or real } x(t) : \; <|x(t)|^2> = \sum_{i=-I}^{I} |X_i|^2 \qquad (94a)$$

$$\text{Real } x(t) : \; <x^2(t)> = \sum_{i=-I}^{I} |X_i|^2 = A_0^2 + \frac{1}{2}\sum_{i=1}^{I} A_i^2 \qquad (94b)$$

We can prove Eq. (94a) by substituting Eq. (53) into Eq. (90) and writing the absolute value squared as a double sum

$$<|x(t)|^2> = \; <\left|\sum_{i=-I}^{I} X_i e^{j2\pi F_i t}\right|^2> = \; <\sum_{i=-I}^{I} X_i e^{j2\pi F_i t}\sum_{k=-I}^{I} X_k^* e^{-j2\pi F_k t}>$$

$$(95)$$

$$= \; <\sum_{i=-I}^{I}\sum_{k=-I}^{I} X_i e^{j2\pi F_i t} X_k^* e^{-j2\pi F_k t}> = \sum_{i=-I}^{I}\sum_{k=-I}^{I} X_i X_k^* <e^{j2\pi F_i t} e^{-j2\pi F_k t}>$$

where the last step follows by the definition of the average value operator. The key step in the proof comes next. The average value term in Eq. (95) simplifies to

$$<e^{j2\pi F_i t} e^{-j2\pi F_k t}> = \; <e^{j2\pi(F_i-F_k)t}> = \begin{cases} 1 & \text{for } i = k \\ 0 & \text{for } i \neq k \end{cases} \qquad (96)$$

The upper part of this equality follows immediately because $e^{j2\pi(F_i-F_k)t} = 1$ for $i = k$. The lower part follows from Euler's formula because

$$<e^{j2\pi(F_i-F_k)t}> = \; <\cos(2\pi(F_i - F_k)t)> + j<\sin(2\pi(F_i - F_k)t)> \qquad (97)$$

and the average value of any sinusoid is 0 (the area under one cycle is 0). If we substitute Eq. (96) back into Eq. (95), we obtain the first equality in Parseval's theorem. Parseval's theorem for real valued waveforms Eq. (94b), follows directly from Eq. (55). It can also be derived directly from Eq. (51) and the property that

$$<\cos(2\pi F_i t + \phi_i)\cos(2\pi F_k t + \phi_k)> = \begin{cases} 0.5 & \text{for } i = k \\ 0 & \text{for } i \neq k \end{cases} \qquad (98)$$

for all ϕ_i and ϕ_k (Problem 3CT.8.7).

The term *orthogonality* is used to describe the important lower equalities in Eqs. (96) and (98). In general, two power waveforms $x_1(t)$ and $x_2(t)$, are called *orthogonal* if

$$<x_1(t)x_2^*(t)> = 0 \qquad (99)$$

Sinusoids having different frequencies are orthogonal. Parseval's theorem is based on this orthogonality.

The power spectrum of the output of an LTI system is related in a simple way to the power spectrum of the input. Equation (52) shows us the output $y(t)$ of an LTI system having multiple sinusoidal input components. The average power in the output sinusoid having frequency F_i is $P_i = \frac{1}{2}|H(F_i)|^2 A_i^2$, for $i = 1, 2, \ldots, I$ and $P_0 = |H(0)|^2 A_0^2$. Therefore, the one-sided power spectrum of $y(t)$ equals the product of the one-sided power spectrum of the input and the square of the filter amplitude characteristic. Similarly, (54) implies that the average power in the output complex sinusoid having frequency F_i is $P_i = |H(F_i)|^2|X_i|^2 = \frac{1}{4}|H(F_i)|^2 A_i^2$ for $i = \pm 1, \pm 2, \ldots, \pm I$ and $P_0 = |H(0)|^2|X_0|^2 = |H(0)|^2 A_0^2$. Therefore the two-sided power spectrum of the filter output equals the product of the two-sided power spectrum of the input times the square of the filter amplitude characteristic.

3CT.9   Summary

In this chapter, we described systems from the frequency domain viewpoint, which is based on complex sinusoidal signals. We showed that, if you apply a complex sinusoid $Xe^{j2\pi Ft}$ to an LTI system having impulse response $h(t)$, the output is $H(F)Xe^{j2\pi Ft}$, where $H(F)$ is the Fourier transform of $h(t)$. Therefore the system's frequency response characteristic $H(F)$ describes how the input and output sinusoids are related. We stated that $H(F)$ exists for all BIBO stable LTI systems and some LTI systems that are not BIBO stable. We described the frequency response characteristics and impulse responses for ideal filters, ideal delays, ideal differentiators, and ideal integrators. We found the overall frequency response of systems placed in cascade, in parallel, and in a loop. We introduced one- and two-sided amplitude and phase spectrums to show graphically how signal amplitude and phase are distributed with respect to frequency. The amplitude and phase spectrums of the output of an LTI system are obtained by multiplying the spectrum of the input by the system's frequency response characteristic.

Sinusoids and waveforms composed of multiple sinusoids are members of a class of waveforms called power waveforms. We introduced signal power spectrums to show graphically how average signal power is distributed with respect to frequency. The power spectrum of the output of an LTI system is obtained by multiplying the power spectrum of the input by the square of the system's amplitude response characteristic. Parseval's theorem states that the total average power carried by a waveform is equal to the sum of the heights of the lines in the waveform's power spectrum.

This chapter has shown that a system's frequency response characteristic is a useful analytical aid for finding the response of an LTI system to sinusoids and waveforms composed of multiple sinusoids. In the following chapter we show that the frequency response characteristic is similarly useful for finding the response of an LTI system to waveforms such as $\sqcap(t/\tau)$, $\wedge(t/\tau)$, and $e^{-t}u(t)$.

Summary of Definitions and Formulas

Complex and Real Sinusoids

$$Ae^{j(2\pi Ft+\phi)} = A\cos(2\pi Ft + \phi) + jA\sin(2\pi Ft + \phi)$$

$$A\cos(2\pi Ft + \phi) = \mathcal{R}\{Ae^{j(2\pi Ft+\phi)}\}$$

Response of Stable LTI System to Complex Sinusoid

The input is applied at $t = -\infty$.

$$Xe^{j2\pi Ft} \xrightarrow{h(t)} H(F)Xe^{j2\pi Ft}$$

where

$$H(F) = \int_{-\infty}^{\infty} h(\lambda)e^{-j2\pi F\lambda}d\lambda \quad \text{and} \quad h(t) = \int_{-\infty}^{\infty} H(F)e^{j2\pi Ft}dF$$

are a Fourier transform pair.

Properties of $H(F)$ for Real $h(t)$

$$H(-F) = H^*(F)$$

$$|H(-F)| = |H(F)|$$

$$\angle H(-F) = -\angle H(F)$$

Response of Stable LTI System to Real Sinusoid

The input is applied at $t = -\infty$.

$$A\cos(2\pi Ft + \phi) \xrightarrow{h(t)} |H(F)|A\cos(2\pi Ft + \phi + \angle H(F)) \qquad \text{for real } h(t)$$

Response of Stable LTI System to Multiple Sinusoids

$$\sum_{i=-I}^{I} X_i e^{j2\pi F_i t} \xrightarrow{h(t)} \sum_{i=-I}^{I} X_i H(F_i) e^{j2\pi F_i t}$$

$$\sum_{i=0}^{I} A_i \cos(2\pi F_i t + \phi_i) \xrightarrow{h(t)} |H(F_i)|A_i\cos(2\pi F_i t + \phi_i + \angle H(F_i)) \qquad \text{for real } h(t)$$

Ideal Systems

See Tables 3CT.1 and 3CT.2.

Periodic Waveforms

$$x(t) = x(t + T) \qquad \text{for all } t$$

Power Waveforms

$$P = <|x(t)|^2> \text{ is nonzero and finite.}$$

For an arbitrary waveform:

$$<(\cdot)> = \lim_{T_o \to \infty} \frac{1}{2T_o} \int_{-T_o}^{T_o} (\cdot)\, dt$$

For a periodic waveform (period T):

$$<(\cdot)> = \frac{1}{T} \int_{-T/2}^{T/2} (\cdot)\, dt$$

Parseval's Theorem

Real or complex $x(t)$:

$$<|x(t)|^2> = \sum_{i=-I}^{I} |X_i|^2$$

Real $x(t)$:

$$<x^2(t)> = \sum_{i=-I}^{I} |X_i|^2 = A_0^2 + \tfrac{1}{2}\sum_{i=1}^{I} A_i^2$$

3CT.10 Problems

3CT.3 Response of LTI Systems to Sinusoidal Input

3CT.3.1 Verify Eq. (11).

3CT.3.2 Show that Eq. (11) implies Eq. (12) and Eq. (13).

3CT.3.3 The basic result (10) was obtained from the assumption that $x(t)$ is a complex sinusoid for all t. For comparison, this problem examines the response of an LTI system to an input $x(t)$ that is 0 for $t < t_0$ and complex sinusoidal for $t \geq t_0$.

$$x(t) = Xe^{j2\pi Ft}u(t - t_0) \tag{100}$$

a) Show, by substitution of Eq. (100) into Eq. (4) that

$$y(t) = Xe^{j2\pi Ft}\left\{ H(F) - \int_{t-t_0}^{\infty} h(\lambda)e^{-j2\pi F\lambda}\,d\lambda \right\} \tag{101}$$

How does this result compare to Eq. (10a)?

b) Assume that

$$h(t) = \alpha e^{-\alpha t}u(t) \tag{102}$$

and show that for $x(t)$ given by Eq. (100)

$$y(t) = \left[H(F)Xe^{j2\pi Ft} - H(F)Xe^{-\alpha(t-t_0)+j2\pi Ft_0} \right]u(t - t_0) \tag{103}$$

where

$$H(F) = \frac{\alpha}{\alpha + j2\pi F} \tag{104}$$

c) Assume that $x(t) = A\cos(2\pi Ft + \phi)u(t - t_0)$ and $h(t) = \alpha e^{-\alpha t}u(t)$. Use the result of part (b) and superposition to show that

$$y(t) = [A|H(F)|\cos(2\pi Ft + \phi + \angle H(F)) - A|H(F)|\cos(2\pi Ft_0 + \phi$$
$$+ \angle H(F))e^{-\alpha(t-t_0)}]u(t - t_0) \tag{105}$$

Discuss this result. In particular, show that, for $t > t_0 + (5/\alpha)$, the output has approximately reached the sinusoidal steady-state response, $A|H(F)|\cos(2\pi Ft + \phi + \angle H(F))$.

3CT.3.4 Consider an LTI system described by the second-order differential equation

$$\frac{d^2y(t)}{dt^2} + 2\frac{dy(t)}{dt} + y(t) = 3\frac{dx(t)}{dt} + x(t) \tag{106}$$

Show that the frequency response characteristic of this system is

$$H(F) = \frac{3(j2\pi F) + 1}{(j2\pi F)^2 + 2(j2\pi F) + 1} \tag{107}$$

Hint: Apply the complex sinusoidal input $x(t) = Xe^{j2\pi Ft}$ to the differential equation and assume a solution of the form $y(t) = Ye^{j2\pi Ft}$.

3CT.3.5 a) Use the method of Problem 3CT.3.4 to find $H(F)$ for an LTI system described by the n-th order differential equation

$$a_n \frac{d^n}{dt^n} y(t) + a_{n-1} \frac{d^{n-1}}{dt^{n-1}} y(t) + \cdots + a_1 \frac{d}{dt} y(t) + a_0 y(t) = b_m \frac{d^m}{dt^m} x(t)$$

$$+ b_{m-1} \frac{d^{m-1}}{dt^{m-1}} x(t) + \cdots + b_1 \frac{d}{dt} x(t) + b_0 x(t) \qquad (108)$$

b) State any restrictions or qualifications that should be placed on your result of part (a).

3CT.3.6 a) Show how to work backward from Eq. (29) to find the differential I-O equation of the circuit of Figure 3CT.3.

b) Find the differential I-O equation of a system whose frequency response characteristic is $H(F) = 1/[(1 + j2\pi F)(2 + j2\pi F)]$

3CT.3.7 a) Show that, if a system has the impulse response $h(t) = \delta(t) + \beta\delta(t - \tau)$, the I-O equation of the system is $y(t) = x(t) + \beta x(t - \tau)$.

b) Find $H(F)$ by applying $Xe^{j2\pi Ft}$ to the I-O equation $y(t) = x(t) + \beta x(t - \tau)$.

3CT.3.8 Find $H(F)$ in Problem 3CT.3.7 by substituting $x(t) = A\cos(2\pi Ft + \phi)$ directly into the I-O equation $y(t) = x(t) + \beta x(t - \tau)$ and using trigonometric identities. Hint: Set $y(t) = |H(F)|A\cos(2\pi Ft + \phi + \angle H(F))$; then solve for $|H(F)|$ and $\angle H(F)$.

3CT.3.9 Find the output of the system that computes the moving average (33) for an input $A\cos(2\pi Ft + \phi)$. At what frequencies is the output 0? Give a simple explanation why the output is 0 at these frequencies.

3CT.3.10 Starting from Eq. (44), derive Eqs. (45) and (46).

3CT.4   Multiple Sinusoidal Components

3CT.4.1 The moving average of a waveform $x(t) = 1 + 3\cos(2\pi 500t + 30°) + 5\sin(2\pi 1,000t + 30°)$ is computed where $\tau = 1$ ms. (See Example 3CT.2.) Find the output waveform.

3CT.4.2 Repeat Problem 3CT.4.1 for the first-order system of Example 3CT.1, where $\tau_c = 1$ ms.

3CT.4.3 Repeat Problem 3CT.4.1 for an ideal differentiator.

3CT.4.4 A plot of the sum of two sinusoids having slightly different frequencies looks like a single sinusoid whose amplitude changes slowly and periodically with time. Consider the sum

$$x(t) = \cos\{2\pi(F_x + \Delta F_x)t\} + \cos\{2\pi(F_x - \Delta F_x)t\} \qquad (109)$$

where $\Delta F_x \ll F_x$.

a) Sketch $x(t)$ for $F_x = 10$ and $\Delta F_x = 1$. If you prefer, use a computer to plot of $x(t)$. Hint: If you decide to sketch $x(t)$, you can use the identity $\cos(A + B) + \cos(A - B) = 2\cos(A)\cos(B)$ to help you understand what the plot should look like.

b) If $x(t)$ were an audio signal with F_x in the approximate range 20 Hz to 20 kHz and with ΔF_x on the order of a few hertz, then $x(t)$ would sound like a sinusoidal tone whose amplitude changes periodically. The periodic changes in amplitude are called *beats*. Find a general formula for the period of the beats in $x(t)$. Express your answer in terms of ΔF_x.

3CT.4.5 a) Find a closed form for $x(t) = 1 + 2\cos(2\pi F_1 t) + 2\cos(2\pi 2F_1 t) + 2\cos(2\pi 3F_1 t) + \cdots + 2\cos(2\pi N F_1 t)$.

b) Sketch or use a computer to plot $x(t)$ for $N = 10$. Comment on your plot.

3CT.4.6 Write a computer program to plot

$$x(t) = \sum_{k=1,3,\cdots}^{K} \frac{4(-1)^{\frac{k-1}{2}}}{\pi k} \cos(2\pi k t) \qquad (110)$$

for $-1 < t < 1$. Run your program for $K = 1, 5, 7$, and 55 and comment on the results.

3CT.5 $h(t)$ and $H(F)$ are a Fourier Transform Pair

3CT.5.1 Find $h(t)$ if $H(F) = \delta(F)$. What does this system do?

3CT.5.2 Find $H(F)$ if $h(t) = \delta(t)$. What does this system do?

3CT.5.3 Find $h(t)$ if $H(F) = \frac{1}{2}\delta(F - 100) + \frac{1}{2}\delta(F + 100)$.

3CT.5.4 Find $H(F)$ if $h(t) = \delta(t) + 3e^{-6t}u(t) + 5e^{-4t}u(t)$

3CT.6   Frequency Response Characteristics of Ideal Systems

3CT.6.1 We saw in Eq. (69) that the impulse response of an ideal low-pass filter is given by

$$h(t) = 2W \operatorname{sinc}(2Wt) \qquad (111)$$

Show that $h(t)$ has area 1. Hint: Write down the integral for the area of $h(t)$ and compare it to the definition of the Fourier transform of $h(t)$. Ask yourself how these integrals are related.

3CT.6.2 Use the method of Problem 3CT.6.1 to find the areas of the impulse responses of:
  a) An ideal high-pass filter.
  b) An ideal band-pass filter.
  c) An ideal notch filter.
  d) An ideal differentiator.

3CT.6.3 Use L'Hôspital's rule from calculus to show that $\sin(\pi t)/\pi t = 1$ at $t = 0$. Then confirm this result using some other method.

3CT.6.4 **a)** Show that the step response of an ideal low-pass filter having frequency response characteristic $\sqcap (F/2W)$ is given by

$$h_{-1}(t) = \int_{-\infty}^{2Wt} \operatorname{sinc}(v)dv \qquad (112)$$

  b) Show that $h_{-1}(t)$ can be put in the form

$$h_{-1}(t) = \frac{1}{\pi} \int_{-\infty}^{2\pi Wt} \frac{\sin(\xi)}{\xi} d\xi = \frac{1}{2\pi} + \frac{1}{\pi} \operatorname{Si}(2\pi Wt) \qquad (113)$$

  where $\operatorname{Si}(x)$ is the *sine integral*

$$\operatorname{Si}(x) = \int_{0}^{x} \frac{\sin(\lambda)}{\lambda} d\lambda \qquad (114)$$

  c) Write a computer program to plot $\operatorname{Si}(x)$.

3CT.6.5 Find and sketch a bounded waveform $x(t)$ that yields unbounded output when applied to an ideal low-pass filter. Hint: You can find such a waveform by looking at the proof that an LTI system is BIBO stable if and only if $h(t)$ is absolutely integrable. See Section 2CT.6.

3CT.7   Basic Connections of CT LTI Systems

3CT.7.1 Two LTI systems having frequency response characteristics $H_1(F)$ and $H_2(F)$ are connected in parallel. Show that the frequency response characteristic of the overall system is given by $H(F) = H_1(F) + H_2(F)$.

3CT.7.2 In the text we described a series connection of two filters having impulse responses $h_1(t)$ and $h_2(t)$. The impulse response of this series connection is $h(t) = h_1(t) * h_2(t)$. We showed that the frequency response characteristic corresponding $h(t) = h_1(t) * h_2(t)$ is $H(F) = H_1(F)H_2(F)$. This means that $h_1(t) * h_2(t)$ and $H_1(F)H_2(F)$ are a Fourier transform pair.

a) Does it follow that $a(t) * b(t)$ and $A(F)B(F)$ are a Fourier transform pair if $a(t)$ and $b(t)$ are arbitrary waveforms, where $A(F) = \mathcal{FT}\{a(t)\}$ and $B(F) = \mathcal{FT}\{b(t)\}$? Explain.

b) Does it follow that $x(t) * h(t)$ and $X(F)H(F)$ are a Fourier transform pair if $x(t)$ and $h(t)$ are arbitrary waveforms, where $X(F) = \mathcal{FT}\{x(t)\}$ and $H(F) = \mathcal{FT}\{h(t)\}$? Consider this result carefully. Does it provide a new way to look at I-O properties of an LTI system? State how.

3CT.8 Power Waveforms, Power Spectrums, and Parseval's Theorem

3CT.8.1 Plot the following waveforms and state which are power waveforms. Find the average power in the power waveforms.

a) $\sum_{k=-\infty}^{\infty} \Pi(t - 2k)$

b) $\sum_{k=-\infty}^{\infty} \delta(t - 2k)$

c) $\sum_{k=-\infty}^{\infty} e^{-(t-k)} u(t - k)$

d) $e^{-t} u(t)$

e) $\wedge(t)$

f) $\cos(2\pi F_a t) \cos(2\pi F_b t)$, where $F_a \neq F_b$

3CT.8.2 a) Show that the real sinusoid $x(t) = A \cos(2\pi F_1 t + \phi)$ has average power $P = \frac{1}{2} A^2$ for $F_1 \neq 0$ and $P = A^2 \cos^2 \phi$ for $F_1 = 0$.

b) Show that the complex sinusoid $X e^{j2\pi Ft}$ has average power $|X|^2$ for every F.

3CT.8.3 The waveform $x(t) = A \cos(2\pi F_a t + \phi_a) + B \cos(2\pi F_b t + \phi_b)$, where $F_a \neq F_b$, may or may not be periodic depending on F_a and F_b. Find the condition on F_a and F_b that makes $x(t)$ periodic, and give the formula for the period.

3CT.8.4 Consider the waveform

$$x(t) = 2 + 3\cos(2\pi 100t) + 3\sin(2\pi 100t) + 5\cos(2\pi 250t) \quad (115)$$

a) Plot the two-sided power spectrum.

b) What is the total average power carried by the DC (i.e., constant) component?

c) What is the total average power carried by the 100-Hz component $3\cos(2\pi 100t) + 3\sin(2\pi 100t)$?

d) What is the average power carried by the 250-Hz component $5\cos(2\pi 250t)$?

e) Show that the average powers of all the components add to give the total average power carried by $x(t)$.

3CT.8.5 In a one-sided power spectrum, the average powers of real sinusoidal frequency components are plotted versus frequency for $F \geq 0$. Plot the one-sided power spectrum for the waveform of Problem 3CT.8.4 and show that the sum of all the average powers of each frequency component add to give the total average power.

3CT.8.6 Two real-valued power waveforms $x_1(t)$ and $x_2(t)$ are called *orthogonal* if

$$< x_1(t)x_2(t) > = 0 \quad (116)$$

This equation is a special case of Eq. (101). Assume that orthogonal power waveforms $x_1(t)$ and $x_2(t)$ are added to produce a third signal

$$x(t) = x_1(t) + x_2(t) \tag{117}$$

a) Show that the average power contained in $x(t)$ equals the sum of the average powers contained in $x_1(t)$ and $x_2(t)$.

b) Suppose that we are told that $x(t) = x_1(t) + x_2(t)$ and that the average power contained in $x(t)$ equals the sum of the average powers contained in $x_1(t)$ and $x_2(t)$. Must it necessarily follow that $x_1(t)$ and $x_2(t)$ are orthogonal?

3CT.8.7 Refer to the definition of orthogonal waveforms given in Problem 3CT.8.6.

a) Show that real sinusoids having different frequencies are orthogonal.

b) Show that real sinusoids having the same frequency are orthogonal if they differ in phase by $90°$.

c) Use your result from part (a) to derive Parseval's theorem (94b) for waveforms composed of real multiple sinusoidal components. That is, show that

$$< x(t)^2 > = A_0^2 + \tfrac{1}{2} \sum_{i=1}^{I} A_i^2 \tag{118}$$

3CT.8.8 Two complex waveforms $x_1(t)$ and $x_2(t)$ are defined as *orthogonal* if

$$< x_1(t)x_2^*(t) > = 0 \tag{119}$$

Assume that orthogonal power waveforms $x_1(t)$ and $x_2(t)$ are added to produce a third signal

$$x(t) = x_1(t) + x_2(t) \tag{120}$$

a) Show that the average power $< |x(t)|^2 >$, contained in $x(t)$ equals the sum of the average powers contained in $x_1(t)$ and $x_2(t)$.

b) Suppose that we are told that $x(t) = x_1(t) + x_2(t)$ and that the average power contained in $x(t)$ equals the sum of the average powers contained in $x_1(t)$ and $x_2(t)$. Must it necessarily follow that $x_1(t)$ and $x_2(t)$ are orthogonal?

3CT.8.9 Show that, in the multiple sinusoidal component waveform (53), we can obtain the coefficients $X_k, -I \leq k \leq I$, using the formula

$$X_k = < x(t)e^{-j2\pi F_k t} > \tag{121}$$

Miscellaneous Problems

3CT.1. An amplifier has input voltage $x(t) = A_{in} \cos(\Omega t + \phi_{in})$ and output voltage $y(t) = A_{out} \cos(\Omega t + \phi_{out})$. The physical average power supplied to the amplifier by its input is $P_{in} = \tfrac{1}{2} A_{in}^2 / R_{in}$, where R_{in} is the input resistance of the amplifier. The physical average power delivered by the amplifier is $P_{out} = \tfrac{1}{2} A_{out}^2 / R_{load}$, where R_{load} the load resistance. Physical power has the unit watts (W).

(a) Find the relation between the amplifier's decibel (dB) amplitude gain, defined as

$$G_{dB \text{ amplitude}} = 20 \log \frac{A_{out}}{A_{in}} \tag{122}$$

and its decibel power gain, defined as

$$G_{dB \text{ power}} = 10 \log \frac{P_{out}}{P_{in}} \tag{123}$$

(b) Show that the dB power and voltage gains are equal when $R_{load} = R_{in}$.

(c) Show that if the decibel voltage gain $G_{\text{dB amplitude}}$ is increased or decreased by q dB, where q is some number, then the decibel power gain $G_{\text{dB power}}$ is also increased or decreased by q dB, regardless of the values for R_{load} and R_{in}.

3CT.2. Refer to miscellaneous Problems 2CT.8 and 2CT.9. The two-dimensional frequency response characteristic of an imaging system having point spread function $h(x, y)$ is given by the two-dimensional Fourier transform

$$H(F_x, F_y) = \int_{-\infty}^{\infty} \int_{-\infty}^{\infty} h(x, y) e^{-j2\pi(F_x x + F_y y)} \, dx \, dy \qquad (124)$$

(a) Find $H(F_x, F_y)$ for motion blur in the x direction where
$h(x, y) = \sqcap(x/\Delta L)\delta(y)$.

(b) Show that in general if $h(x, y)$ is separable $h(x, y) = h_1(x)h_2(y)$, then $H(F_x, F_y)$ is also separable $H(F_x, F_y) = H_1(F_x)H_2(F_y)$.

3CT.3. Refer to Problem 3CT.2. The point spread functions of many imaging systems have circular symmetry. When $h(x, y)$ is circularly symmetric, it can be written as

$$h(x, y) = h(r) \qquad (125)$$

where $r = \sqrt{x^2 + y^2}$ and $h(r)$ is some function of r. A famous result says that the two-dimensional frequency response characteristic of an imaging system having a circularly symmetric point spread function is itself circularly symmetric and is given by the Hankel transform of $h(r)$.

$$H(F_x, F_y) = H(F_r) = 2\pi \int_0^\infty h(r) J_0(2\pi r F_r) r \, dr \qquad (126)$$

where $F_r = \sqrt{F_x^2 + F_y^2}$. In the preceding equation

$$J_0(\chi) = \frac{1}{2\pi} \int_0^{2\pi} e^{-j\chi \cos(\theta - \phi)} d\theta \qquad (127)$$

is the 0th-order Bessel function of the first kind, where ϕ is an arbitrary constant.

Perform the coordinate conversion $(x, y) = (r \cos\theta, r \sin\theta)$ in Eq. (124) to obtain Eq. (127). Hint: This problem looks a lot harder than it is. Use the identity $A \cos\theta + B \sin\theta = \sqrt{A^2 + B^2} \cos(\theta - \phi)$, where $\phi = \tan^{-1}(B/A)$.

Appendix 3CT.A Frequency Response Characteristic of an Ideal Integrator

The frequency response characteristic of an ideal integrator is given by the Fourier transform of $h(t) = u(t)$, where $u(t)$ is the unit step function. Our task, therefore, is to find the Fourier transform $U(F)$ of $u(t)$. The direct substitution of $u(t)$ into the Fourier integral yields an indeterminate result. We therefore express $u(t)$ as a limit

$$u(t) = \lim_{\alpha \to 0} h_\alpha(t) \qquad (A1)$$

where

$$h_\alpha(t) \overset{\Delta}{=} e^{-\alpha t} u(t) \qquad (A2)$$

We define

$$U(F) = \lim_{\alpha \to 0} H_\alpha(F) \qquad (A3)$$

where $H_\alpha(F)$ is the Fourier transform of $h_\alpha(t)$.

Figure 3CT.17
$H_{\alpha r}(F)$ and $H_{\alpha i}(F)$:
(a) $H_{\alpha r}(F)$; (b) $H_{\alpha i}(F)$.

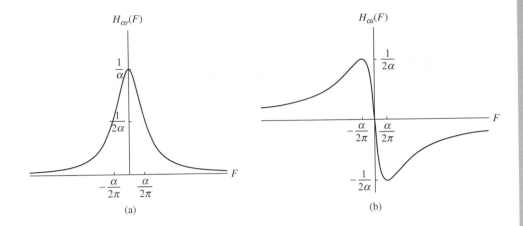

(a)   (b)

It follows in a manner similar to Example 3CT.1 that

$$H_\alpha(F) = \frac{1}{\alpha + j2\pi F} \qquad \text{for} \quad \alpha > 0 \tag{A4}$$

To take the limit $\alpha \to 0$, we write $H_\alpha(F)$ as

$$H_\alpha(F) = H_{\alpha r}(F) + jH_{\alpha i}(F) \tag{A5}$$

where $H_{\alpha r}(F)$ and $H_{\alpha i}(F)$, respectively, denote the real and the imaginary parts of $H_\alpha(F)$. A little algebra yields

$$H_{\alpha r}(F) = \frac{\alpha}{\alpha^2 + (2\pi F)^2} \tag{A6}$$

and

$$H_{\alpha i}(F) = -\frac{2\pi F}{\alpha^2 + (2\pi F)^2} \tag{A7}$$

$H_{\alpha r}(F)$ and $H_{\alpha i}(F)$ are plotted in Figure 3CT.17.

The area under $H_{\alpha r}(f)$ is $\frac{1}{2}$. We can establish this result as follows:

$$\int_{-\infty}^{\infty} H_{\alpha r}(F)dF = \int_{-\infty}^{\infty} \mathcal{R}\{H_\alpha(F)\}dF = \mathcal{R}\left\{\int_{-\infty}^{\infty} H_\alpha(F)dF\right\}$$

$$= \mathcal{R}\left\{\left.\int_{-\infty}^{\infty} H_\alpha(F)e^{j2\pi Ft}dF\right|_{t=0}\right\} = \mathcal{R}\{h_\alpha(0)\} = \frac{1}{2} \tag{A8}$$

where $\mathcal{R}$ is the real part operator. As α decreases toward 0, $H_{\alpha r}(F)$ retains the area of $\frac{1}{2}$ while becoming narrower and higher. Thus

$$\lim_{\alpha \to 0} H_{\alpha r}(F) = \frac{1}{2}\delta(F) \tag{A9}$$

The limit of $H_{\alpha i}(F)$ as α decreases toward 0 follows directly from Eq. (A7) and is

$$\lim_{\alpha \to 0} H_{\alpha i}(F) = -\frac{1}{2\pi F} \tag{A10}$$

In conclusion, the frequency response characteristic of an ideal integrator is

$$U(F) = \lim_{\alpha \to 0} H_\alpha(F) = \frac{1}{2}\delta(F) + \frac{1}{j2\pi F} \tag{A11}$$

3DT Frequency Response Characteristics

3DT.1 Introduction

We saw in Chapter 2DT that LTI systems are characterized by their responses to an impulse. LTI systems can also be characterized by their responses to sinusoidal inputs. This characterization is called a system's *frequency response characteristic,* which is a very natural and convenient way to describe an LTI system. Like the impulse response, the frequency response characteristic provides a complete description of the I-O properties of an LTI system. If you know a system's frequency response characteristic, you can find the output for any input.

The frequency response characteristic is based on an important and unique property of sinusoidal sequences: If you apply a sinusoidal sequence to an LTI system, the output is also a sinusoidal sequence. The frequency of the sinusoidal output is always the same as that of the input, but the output amplitude and phase are generally different from those of the input. The system's frequency response characteristic describes how the amplitudes and phases of the output sinusoid compare to those of the input sinusoid as a function of frequency.

3DT.2 Sinusoidal Sequences

If we sample a CT sinusoidal signal every T_s seconds, as shown in Figure 3DT.1(a), we obtain the unique sinusoidal sequence shown in Figure 3DT.1(b). But assume that we *start* with the sinusoidal sequence of Figure 3DT.1(b). Can we identify the original CT sinusoid? The answer is no, not necessarily. A phenomenon called *aliasing* makes unique identification impossible.

3DT.2.1 Aliasing

In the movie industry, aliasing is called the *strobe effect*, which creates weird visual effects. You see one of them when you watch the spoke wheels of a stagecoach in a western movie. The wheels actually revolve faster as the stagecoach speeds up. But in the movie, the wheels appear to rotate faster only initially. Then, when the stagecoach reaches a certain speed, the wheels appear to rotate backward. This happens because each frame of the movie is a DT image of the CT object. Our visual system interprets the sequence of DT images in a way that seems simplest, which in this case is a backward rotating wheel!

We can understand aliasing by considering a rotating phasor

$$e^{j2\pi Ft} = \cos(2\pi Ft) + j\sin(2\pi Ft) \tag{1}$$

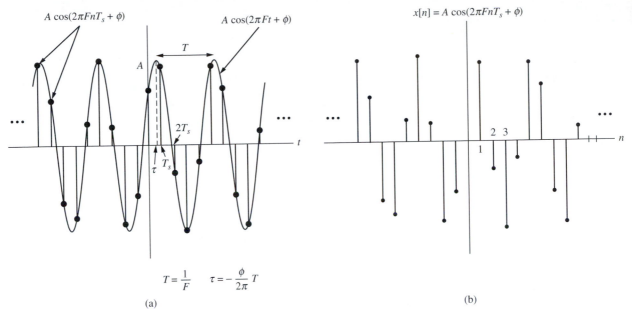

Figure 3DT.1
$A\cos(2\pi F t + \phi)$ and its samples $x(nT_s) = A\cos(2\pi F nT_s + \phi)$: (a) $A\cos(2\pi F t + \phi)$ and its samples $x(nT_s)$; (b) the sequence $x[n] = x(nT_s)$.

drawn in the complex plane: The phasor rotates counterclockwise at a rate of F revolutions per second. Imagine that the rotating phasor is captured on movie film every T_s seconds. The complex value of the phasor in the nth frame is obtained by setting $t = nT_s = n/F_s$.

$$e^{j2\pi F t}\Big|_{t=n/F_s} = e^{j2\pi F n/F_s} = \cos\left(2\pi\frac{F}{F_s}n\right) + j\sin\left(2\pi\frac{F}{F_s}n\right) \tag{2}$$

where F_s is the number of frames per second. Figure 3DT.2(a) illustrates four frames of $e^{j2\pi F n/F_s}$, where the rotation rate is $20°$ per frame ($F/F_s = 20/360 = 1/18$). The frame numbers are written next to the phasors. Your visual system would interpret the movie as a slowly counterclockwise rotating phasor. Figure 3DT.2(b) shows four frames for a rotation of $160°$ per frame ($F/F_s = 160/360 = 4/9$). Your visual system would register this as rapid counterclockwise rotation. For any rotation rate between $0°$ and $180°$ per frame (that is, for $0 < F/F_s < 0.5$), the phasor looks like it is rotating counterclockwise.

Aliasing occurs for $F \geq F_s/2$. Figure 3DT.2(c) shows four frames of $e^{j2\pi F n/F_s}$ at a rotation rate of $180°$ per frame: $F/F_s = 0.5$. Here the phasor points repeatedly from right to left $180°$ apart. It would be impossible to tell from the movie if the rotation is counterclockwise or clockwise. The fastest rotation you can see in the movie occurs when $F/F_s = 0.5$, but its direction of rotation is ambiguous.

As F increases from $F_s/2$ to F_s, the actual counterclockwise rotation appears as to be *clockwise*. Figure 3DT.2(d) shows four frames for a rotation of $200°$ per frame ($F/F_s = 200/360 = 5/9$). Your visual system would register this as a rapid *clockwise* rotation. Figure 3DT.2(e) shows four consecutive frames for a rotation of $340°$ per frame ($F/F_s = 340/360 = 17/18$). This movie would look like a slow clockwise phasor rotation. For $F = F_s$ the phasor rotates exactly one revolution per frame so that it is in the same position each frame (Figure 3DT.2(f)). The clockwise rotation appears fastest when F is slightly larger than $F_s/2$ and slows gradually to a stop when $F = F_s$.

Figure 3DT.2
Complex sinusoid as captured in four consecutive movie frames (coordinate system omitted for simplicity; rotation rates in degrees per frame): (a) 20; (b) 160; (c) 180; (d) 200; (e) 340; (f) 360.

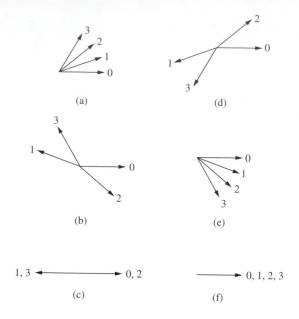

As F increases beyond F_s, the DT rotation looks the same as it did when F was in the range $0 < F \le F_s$ because you cannot see the extra revolution(s) taking place between each frame.

The properties of a sampled rotating phasor naturally show up in plots of the projections of the phasor onto the real axis. The phasor plots of Figures 3DT.2(a)–(f) correspond to time domain plots of Figures 3DT.3(a)–(f). The waveforms and the four samples are the projections of the corresponding rotating phasors on the real axis of the complex plane.

Alternating samples of Figure 3DT.3(c) correspond to the right-left motion of the arrows in Figure 3DT.2(c). Constant amplitude samples in Figure 3DT.3(f) correspond to the stationary appearance of the phasor in Figure 3DT.2(f). A close examination reveals that the samples in Figures 3DT.3(a) and (e) are identical, as are those in Figures 3DT.3(b) and (d). Figure 3DT.4 shows the identical sample values on the same plot as the two CT waveforms from which they might have been taken.

Mathematical understanding of aliasing is provided by the equation

$$e^{j2\pi(F+kF_s)n/F_s} = e^{j2\pi Fn/F_s}e^{j2\pi kn} = e^{j2\pi Fn/F_s} \tag{3}$$

(where k is an integer and $e^{j2\pi kn} = 1$). Equation (3) tells us that the value of $e^{j2\pi(F+kF_s)n/F_s}$ is exactly the same as the value of $e^{j2\pi Fn/F_s}$ for all F. This means that $e^{j2\pi Fn/F_s}$ is a periodic function of F having period F_s. It tells us that there is *no way* to tell the actual rotation rate F of the rotating phasor from its samples. If we are given only the samples, then all we can say is that the rotation rate is $F + kF_s$, where k is some integer, positive, negative, or 0.

However, if we are told that actual rotation rate of the rotating phasor is in the range

$$-0.5F_s < F < 0.5F_s \tag{4}$$

Figure 3DT.3
Real sinusoid and samples corresponding to Figure 3DT.2 (sampling rates in degrees per sample): (a) 20°; (b) 160°; (c) 180°; (d) 200°; (e) 340°; (f) 360°.

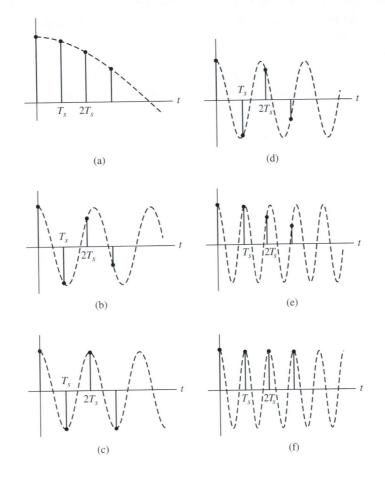

then the aliasing problem does not arise. The condition (4) can be restated as follows:

$$\text{Aliasing does not occur if } F_s > 2|F| \qquad (5)$$

If the sampling rate F_s satisfies Eq. (5) and we make use of that information in interpreting the samples, then we can identify the value of F. For the single real or complex sinusoid

Figure 3DT.4
Two sinusoids with samples shown on same plot (sampling rates in degrees per sample): (a) 20° and 340°; (b) 160°; and 200°.

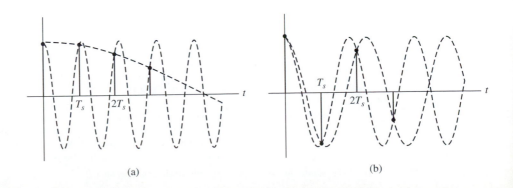

having frequency F that we have been discussing here, the frequency $2|F|$ is called the *Nyquist rate*.[1]

3DT.2.2 Normalized Frequency

In Section 3DT.2, we obtained DT sinusoids $e^{j2\pi FT_s n}$, $\cos(2\pi FT_s n)$, and $\sin(2\pi FT_s n)$ by sampling CT sinusoids. In future work, it will be more convenient to write these DT sinusoids as $e^{jn2\pi f}$, $\cos(n2\pi f)$, and $\sin(n2\pi f)$, where the *normalized frequency* f is defined as

$$f = FT_s = F/F_s \qquad (6)$$

By definition, f is the ratio of the frequency F of the CT sinusoid being sampled to the sampling frequency F_s and has the unit cycles per sample. The sinusoids $e^{jn2\pi f}$, $\cos(n2\pi f)$, and $\sin(n2\pi f)$ are all periodic functions of f having period 1. We can rephrase the conclusion of the preceding section by stating that aliasing does not occur if

$$-\tfrac{1}{2} < f < \tfrac{1}{2} \qquad (7a)$$

We can also define normalized frequency in radians per sample as $\omega = 2\pi f$. In terms of ω, aliasing does not occur if

$$-\pi < \omega < \pi \qquad (7b)$$

3DT.3 Response of LTI System to Sinusoidal Input

3DT.3.1 LTI System Response to a Complex Sinusoid

In the steady state, sinusoidal sequences reproduce themselves throughout LTI systems with only a change in amplitude and phase. The property "sinusoid in, sinusoid out" applies for *every* stable LTI system.[2] We can prove this property using the general I-O equation for an LTI system.

$$y[n] = \sum_{k=-\infty}^{\infty} x[n-k]h[k] \qquad (8)$$

If we substitute

$$x[n] = Xe^{j2\pi fn} \qquad (9)$$

into Eq. (8) we obtain

$$y[n] = X\sum_{k=-\infty}^{\infty} e^{j(n-k)2\pi f}h[k] = \left\{\sum_{k=-\infty}^{\infty} e^{-jk2\pi f}h[k]\right\}Xe^{jn2\pi f} \qquad (10)$$

which is

$$y[n] = H(e^{j2\pi f})Xe^{jn2\pi f} \qquad (11)$$

[1] This section has considered sampling a sinusoidal waveform. In Chapter 4CT we consider sampling a real waveform composed of multiple CT sinusoids. We show there that aliasing does not occur if and only if F_s exceeds twice the highest frequency in the multiple sinusoidal waveform. For a waveform having multiple sinusoidal components, the Nyquist rate is $2F_{max}$ where F_{max} is the highest frequency in the waveform.

[2] It also applies to some unstable systems, including ideal filters.

The function $H(e^{j2\pi f})$ is called the *frequency response characteristic* of the system. It is given by the term in the braces in Eq. (10) which we can rewrite:

Frequency Response Characteristic

$$H(e^{j2\pi f}) = \sum_{k=-\infty}^{\infty} h[k]e^{-jk2\pi f} \tag{12}$$

The sum on the right-hand side of Eq. (12) is the *discrete-time Fourier transform*[3] (DTFT) of $h[n]$. This sum transforms the sequence $h[n]$ to the frequency domain function $H(e^{j2\pi f})$. $H(e^{j2\pi f})$ exists for every BIBO stable LTI system and for some LTI systems that are not BIBO stable. We see from Eq. (11) that, when the complex sinusoid $Xe^{j2\pi fn}$ is applied to any LTI system, the response is another complex sinusoid having the same frequency.

$$y[n] = H(e^{j2\pi f})Xe^{j2\pi fn} \tag{13}$$

Equation (13) is the basic result of this section. We can summarize this key finding using flow graph notation.

The Response of a Stable LTI System to a Complex Sinusoid Having Normalized Frequency f is a Complex Sinusoid Having Normalized Frequency f

$$Xe^{j2\pi fn} \xrightarrow{h[n]} Ye^{j2\pi fn} \tag{14a}$$

where

$$Y = H(e^{j2\pi f})X \tag{14b}$$

This result applies to systems having both real and complex impulse responses. It assumes that the input is a complex sinusoid for all n.[4]

We can extend the result to real sinusoidal sequences that are applied to real systems. When a system is real, its impulse response $h[n]$ is a member of the set of real sequences. To accomplish the extension, we need to learn how the form of $H(e^{j2\pi f})$ depends on the assumption that $h[n]$ is real. We do this in the next section.

3DT.3.2 Symmetry Properties of $H(e^{j2\pi f})$

When $h[n]$ is real, $H(e^{j2\pi f})$ has *conjugate symmetry*.

$$H(e^{-j2\pi f}) = H^*(e^{j2\pi f}) \tag{15}$$

You can easily verify Eq. (15) by replacing f with $-f$ in Eq. (12). The effect of a change in the sign of f is equivalent to a change in the sign of j. The conjugate symmetry of

---

[3] A discrete-time Fourier transform can be approximately computed using a computer algorithm called the *fast Fourier transform* (FFT). This topic is considered in depth in Chapter 5DT.

[4] You can conclude that the input is sinusoidal for all n because no restriction is placed on the range of n in Eq. (9). The response of an LTI system to a input that is 0 for $n < n_0$ and complex sinusoidal for $n \geq n_0$ is considered in Problem 3.6.

$H(e^{j2\pi f})$ implies that the *amplitude* and *phase* characteristics $|H(e^{j2\pi f})|$ and $\angle H(e^{j2\pi f})$ of every real LTI system satisfy

$$|H(e^{-j2\pi f})| = |H(e^{j2\pi f})| \qquad (16)$$

and

$$\angle \mathrm{H}(e^{-j2\pi f}) = -\angle \mathrm{H}(e^{j2\pi f}) \qquad (17)$$

Thus, real systems have even amplitude characteristics and odd phase characteristics.[5]

A second important symmetry property of $H(e^{j2\pi f})$ is *translational symmetry*. Specifically, $H(e^{j2\pi f})$ is a periodic function of f with period 1.

$$H(e^{j2\pi(f+1)}) = H(e^{j2\pi f}) \qquad (18)$$

This periodicity follows immediately from the fact that the argument $e^{j2\pi f}$ of $H(e^{j2\pi f})$ is itself a periodic function of f: $e^{j2\pi(f+1)} = e^{j2\pi f}$. The periodicity property of $H(e^{j2\pi f})$ means that we do not need to plot or specify the frequency response characteristic of a DT system outside the range $-\frac{1}{2} \leq f < \frac{1}{2}$. If you are given or see a plot of $H(e^{j2\pi f})$ for $-\frac{1}{2} \leq f < \frac{1}{2}$ (or for any other period), remember that $H(e^{j2\pi f})$ just repeats outside the plot range.

Now that we have established the symmetry properties of $H(e^{j2\pi f})$, we are ready to consider the response of a real system to a real sinusoid.

3DT.3.3
LTI System
Response to a
Real Sinusoid

We have shown that a complex sinusoidal input sequence $Xe^{j2\pi fn}$ yields a complex sinusoidal output $H(e^{j2\pi f})Xe^{j2\pi fn}$

$$Xe^{j2\pi fn} \xrightarrow{h[n]} H(e^{j2\pi f})Xe^{j2\pi fn} \qquad (19)$$

for every stable LTI system. We can use this result and superposition to find the response of a real LTI system to a real sinusoidal input. By Euler's formula, a real sinusoid is the sum of two complex sinusoids

$$A\cos(2\pi fn + \phi_x) = Xe^{j2\pi fn} + X^*e^{-j2\pi fn} \qquad (20)$$

where

$$X = \tfrac{1}{2}Ae^{j\phi_x} \qquad (21)$$

We can replace f by $-f$ and X by X^* in Eq. (19) to obtain

$$X^*e^{-j2\pi fn} \xrightarrow{h[n]} H(e^{-j2\pi f})X^*e^{-j2\pi fn} \qquad (22)$$

---

[5] In general, a function $g(\psi)$ is defined as *even* if $g(-\psi) = g(\psi)$. Similarly, $g(\psi)$ is defined as *odd* if $g(-\psi) = -g(\psi)$.

By superposition:

$$Xe^{j2\pi fn} + X^*e^{-j2\pi fn} \xrightarrow{h[n]} H(e^{j2\pi f})Xe^{j2\pi fn} + H(e^{-j2\pi f})X^*e^{-j2\pi fn} \qquad (23)$$

which, with the aid of Euler's formula and the conjugate symmetry property of $H(e^{j2\pi f})$, becomes

$$A\cos(2\pi fn + \phi_x) \xrightarrow{h[n]} A|H(e^{j2\pi f})|\cos(2\pi fn + \phi_x + \angle H(e^{j2\pi f})) \qquad (24)$$

In conclusion, we see that the response of a real LTI system to a sinusoidal input is a sinusoid having the same frequency. The system's frequency response characteristic describes how the output amplitude and phase are related to the input amplitude and phase. Real and complex sinusoids are the only bounded signals that reproduce themselves when applied to LTI systems.

Let's now consider some examples of a system's frequency response characteristic $H(e^{j2\pi f})$.

EXAMPLE 3DT.1   Five-Point Moving Average

Let's find the frequency response characteristic of the 5-point averager described in Example 3DT.1 of Chapter 2DT. This system has been redrawn in Figure 3DT.5.

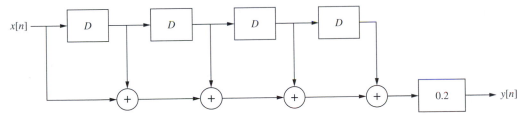

Figure 3DT.5
Five-point moving average computer.

The unit impulse response of the system is

$$h[n] = \tfrac{1}{5}\{\delta[n] + \delta[n-1] + \delta[n-2] + \delta[n-3] + \delta[n-4]\} \qquad (25)$$

The substitution of Eq. (25) into Eq. (12) yields

$$H(e^{j2\pi f}) = \sum_{k=-\infty}^{\infty} e^{-j2\pi fk}\tfrac{1}{5}\{\delta[k] + \delta[k-1] + \delta[k-2] + \delta[k-3] + \delta[k-4]\}$$
$$= \tfrac{1}{5}\{1 + e^{-j2\pi f} + e^{-j4\pi f} + e^{-j6\pi f} + e^{-j8\pi f}\} \qquad (26)$$

One period of the amplitude and phase characteristics is plotted in Figure 3DT.6. These plots were computed using software that forced the phase characteristic to remain in the range $-180° < \angle H(e^{-j2\pi f}) \le 180°$. Problem 3DT.3.7

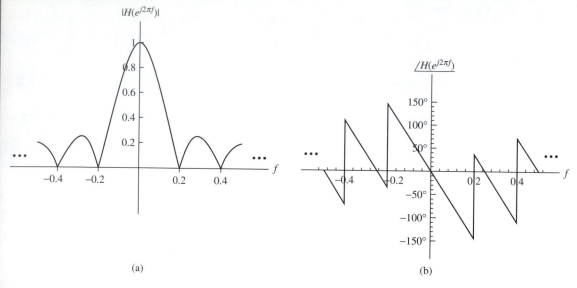

(a)   (b)

Figure 3DT.6
Frequency response characteristic of 5-point averager: (a) amplitude; (b) phase.

shows that there is a closed form for Eq. (26), and it considers these amplitude and phase characteristics in greater detail.

We obtained $H(e^{j2\pi f})$, Eq. (26), by taking the DTFT of $h[n]$. Before leaving this example, let's see if we can also solve for $H(e^{j2\pi f})$ another way, using the basic result Eq. (14a). We start with the equation relating input voltage to output voltage

$$y[n] = \tfrac{1}{5}\{x[n] + x[n-1] + x[n-2] + x[n-3] + x[n-4]\} \tag{27}$$

and apply the basic result Eq. (14a). We set $x[n] = Xe^{j2\pi fn}$ and obtain

$$y[n] = \tfrac{1}{5}\{Xe^{j2\pi fn} + Xe^{j2\pi f(n-1)} + Xe^{j2\pi f(n-2)} + Xe^{j2\pi f(n-3)} + Xe^{j2\pi f(n-4)}\}$$

$$= \tfrac{1}{5}\{1 + e^{-j2\pi f} + e^{-j4\pi f} + e^{-j6\pi f} + e^{-j8\pi f}\}Xe^{j2\pi fn} \tag{28}$$

Equation (28) can be written as

$$y[n] = Ye^{j2\pi fn} \tag{29}$$

where

$$Y = H(e^{j2\pi f})X \tag{30}$$

and

$$H(e^{j2\pi f}) = \tfrac{1}{5}\{1 + e^{-j2\pi f} + e^{-j4\pi f} + e^{-j6\pi f} + e^{-j8\pi f}\} \tag{31}$$

This result agrees with Eq. (26), as it must. Figure 3DT.7 shows the signals that appear throughout the system.

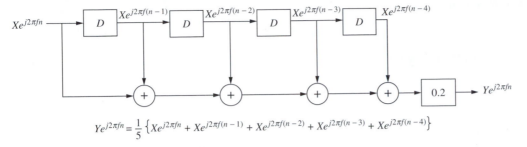

$$Ye^{j2\pi fn} = \frac{1}{5}\left\{Xe^{j2\pi fn} + Xe^{j2\pi f(n-1)} + Xe^{j2\pi f(n-2)} + Xe^{j2\pi f(n-3)} + Xe^{j2\pi f(n-4)}\right\}$$

Figure 3DT.7
Five-point averager in the complex sinusoidal steady state.

Circuit enthusiasts will notice that the DT system works in a way similar to an AC circuit: DT complex sinusoids having the same frequency as the input appear throughout the system. The associated frequency domain drawing of the system is shown in Figure 3DT.8.

Figure 3DT.8
Five-point averager
drawn in the
frequency domain.

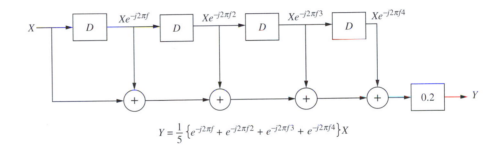

$$Y = \frac{1}{5}\left\{e^{-j2\pi f} + e^{-j2\pi f2} + e^{-j2\pi f3} + e^{-j2\pi f4}\right\}X$$

Example 3DT.1 demonstrates that you do not have to evaluate a discrete-time Fourier transform to find $H(e^{j2\pi f})$. You can also find $H(e^{j2\pi f})$ by applying $x[n] = Xe^{j2\pi fn}$ as the input and solving the I-O equation for the output $Ye^{j2\pi fn}$. The frequency response characteristic $H(e^{j2\pi f})$ is the ratio Y/X. This method assumes that $H(e^{j2\pi f})$ exists but does not require you first to find $h[n]$. In fact, once you have found $H(e^{j2\pi f})$, you can find $h[n]$ by means of the inverse DTFT discussed in Section 3DT.5.

EXAMPLE 3DT.2   First-Order DT Filter

For a second example of the DTFT, consider the first-order system of Figure 2DT.20. This system is redrawn in Figure 3DT.9.

Figure 3DT.9
First-order system.

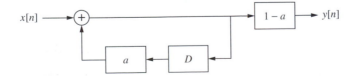

The impulse response of the system is

$$h[n] = (1 - a)a^n u[n] \tag{32}$$

The substitution of Eq. (32) into Eq. (12) yields

$$H(e^{j2\pi f}) = (1 - a) \sum_{k=-\infty}^{\infty} e^{-j2\pi fk} a^k u[k] = (1 - a) \sum_{k=0}^{\infty} (ae^{-j2\pi f})^k \tag{33}$$

The right-hand side of Eq. (33) is an infinite geometric series. The series converges, and a closed form for it exists if $|a| < 1$ (identity (B.4)). The result is

$$H(e^{j2\pi f}) = \frac{1 - a}{1 - ae^{-j2\pi f}} \qquad \text{for } |a| < 1 \tag{34}$$

If $|a| \geq 1$, then the geometric series does not converge. This means that $H(e^{j2\pi f})$ does not exist for $|a| \geq 1$. Figure 3DT.10 illustrates the amplitude and phase characteristics for $a = \pm 0.9, \pm 0.6$, and ± 0.1. The amplitude characteristic equals 1 when $f = 0$ and $(1 - a)/(1 + a)$ when $f = \pm 0.5$. For $a > 0$, input sinusoids having high frequencies $f(f \approx 0.5)$ are attenuated by the filter. For $a < 0$, high frequencies inputs are amplified by the filter.

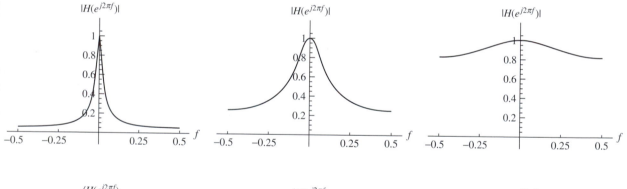

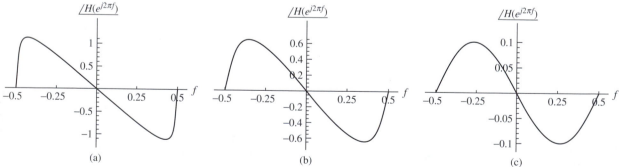

(a)   (b)   (c)

Figure 3DT.10
Amplitude and phase characteristics of first-order system ($|H(e^{j2\pi f})| = (1 - a)/[\sqrt{1 + a^2 - 2a \cos(2\pi f)}]$, $\angle H(e^{j2\pi f}) = \tan^{-1}\{[a \sin(2\pi f)]/[1 - a \cos(2\pi f)]\}$ with unit radians): (a) $a = 0.9$; (b) $a = 0.6$; (c) $a = 0.1$; (d) $a = -0.9$; (e) $a = -0.6$; (f) $a = -0.1$.

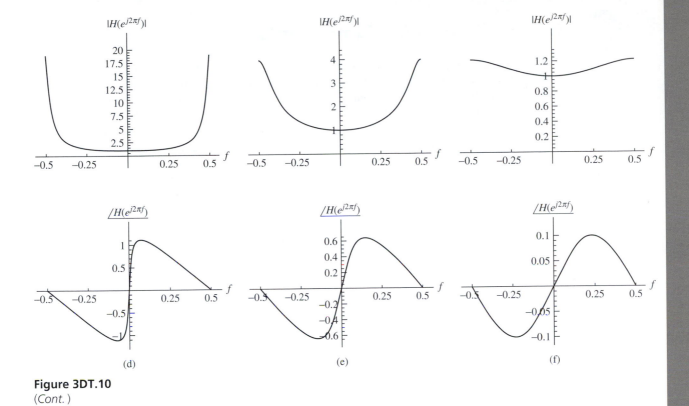

(d)   (e)   (f)

Figure 3DT.10
(*Cont.*)

Figure 3DT.11
$|H(e^{j2\pi f})|$ of Figure
3DT.10 plotted above
a unit circle:
$|H(e^{j2\pi f})| =$
$1/\sqrt{1+a^2-2a\cos(2\pi f)},$
with $a = 0.9$.

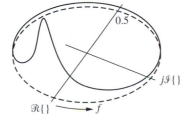

As illustrated in the preceding examples, we typically plot frequency response characteristics versus normalized frequency f. It is occasionally helpful to plot $H(e^{j2\pi f})$ above a unit circle. Notice that the argument of $H(e^{j2\pi f})$, $e^{j2\pi f}$, is a complex variable having magnitude 1 and angle $2\pi f$. As f varies, therefore, $e^{j2\pi f}$ generates a unit circle in the complex plane. We can plot $|H(e^{j2\pi f})|$ and $\angle H(e^{j2\pi f})$ in three-dimensional space above this unit circle. An example is shown in Figure 3DT.11, which is a replot of the amplitude characteristics of Figure 3DT.10(a).

3DT.4   Multiple Sinusoidal Components

Because sinusoidal sequences propagate throughout LTI systems, as shown in Eqs. (14a) and (24), we are motivated to use them as building blocks for nonsinusoidal sequences. Here, in this section, we begin a fundamental development in signals and system theory that eventually brings us to the discrete Fourier representations of signals of Chapters 4DT–5DT: The development is based on the observation that for every LTI system, properties (14a) and (24) can be extended easily to sums of sinusoids. To see how, consider the input

$$x[n] = A_1 \cos(2\pi f_1 n + \phi_1) + A_2 \cos(2\pi f_2 n + \phi_2) \tag{35}$$

It follows immediately from Eq. (24) and superposition that

$$y[n] = |H(e^{j2\pi f_1})| A_1 \cos(2\pi f_1 n + \phi_1 + \angle H(e^{j2\pi f_1}))$$
$$+ |H(e^{j2\pi f_2})| A_2 \cos(2\pi f_2 n + \phi_2 + \angle H(e^{j2\pi f_2})) \tag{36}$$

The output amplitudes and phases are determined by the input and by the frequency response characteristic of the system.

Two graphical aids are used to visualize the action of the system on the input signal. The first is shown in Figure 3DT.12 and uses a *one-sided* plot of the frequency response characteristic $H(e^{j2\pi f})$. In the figure, the amplitudes and phases of the input sinusoids are plotted above the frequencies of the sinusoids to which they refer. These plots are called one-sided amplitude and phase spectrums. Similar plots are made for the output sinusoidal components. Notice that the output amplitude spectrum is given by the product of the input amplitude spectrum and the amplitude characteristic of the filter. The output phase spectrum is given by the sum of the input phase spectrum and the phase characteristic of the filter at normalized frequencies f_1 and f_2. Both the input and the output spectrums are sometimes called *line spectrums* because they consist of vertical lines.

The other graphical aid uses a *two-sided* frequency response characteristic $H(e^{j2\pi f})$. We use Euler's formula to express the input (35) as a sum of four complex sinusoids,

$$x[n] = X_1 e^{j2\pi f_1 n} + X_1^* e^{-j2\pi f_1 n} + X_2 e^{j2\pi f_2 n} + X_2^* e^{-j2\pi f_2 n} \tag{37}$$

where $X_1 = \frac{1}{2} A_1 e^{j\phi_1}$ and $X_2 = \frac{1}{2} A_2 e^{j\phi_2}$. It then follows from Eq. (14a) and superposition that

$$y[n] = H(e^{j2\pi f_1}) X_1 e^{j2\pi f_1 n} + H(e^{-j2\pi f_1}) X_1^* e^{-j2\pi f_1 n}$$
$$+ H(e^{j2\pi f_2}) X_2 e^{j2\pi f_2 n} + H(e^{-j2\pi f_2}) X_2^* e^{-j2\pi f_2 n} \tag{38}$$

The two-sided input and output spectrums are shown in Figures 3DT.13(a) and (c), respectively. The plots are called two-sided amplitude and phase spectrums. The output amplitude spectrum is given by the product of the input amplitude spectrum and the amplitude characteristic of the filter (Figure 3DT.13(b)). The output phase spectrum is given by the sum of the input phase spectrum and the phase characteristic of the filter at normalized frequencies $\pm f_1$ and $\pm f_2$. Notice that the figure assumes the amplitude and phase symmetry properties of $H(e^{j2\pi f})$ that apply for a real system.

Figure 3DT.12
One-sided frequency domain plots:
(a) input spectrum;
(b) filter frequency response characteristic;
(c) output spectrum.

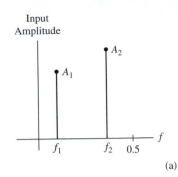
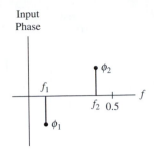

(a)

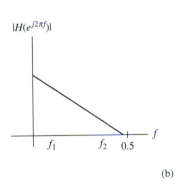
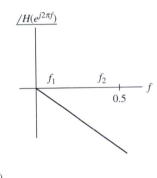

(b)

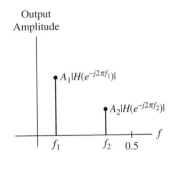
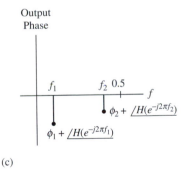

(c)

The generalization of these results to an input composed of many sinusoidal components is straightforward. Consider an input composed of the sum of a constant (DC) term A_0 and I sinusoidal components:

$$x[n] = A_0 + \sum_{i=1}^{I} A_i \cos(2\pi f_i n + \phi_{xi}) \tag{39}$$

It follows from Eq. (24) and superposition that the output is

$$y[n] = A_0 H(e^{j0}) + \sum_{i=1}^{I} A_i |H(e^{j2\pi f_i})| \cos(2\pi f_i n + \phi_{xi} + \angle H(e^{j2\pi f_i})) \tag{40}$$

Figure 3DT.13
Two-sided frequency
domain plots:
(a) input amplitude
and phase spectrum;
(b) filter frequency
response
characteristic;
(c) output amplitude
and phase spectrum.

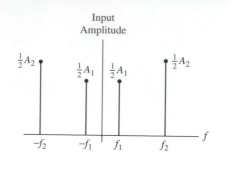
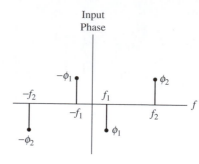

(a)

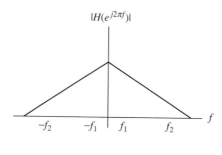
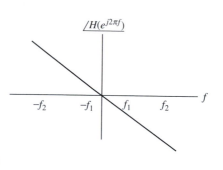

(b)

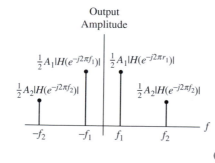
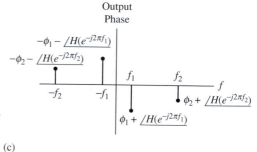

(c)

The complex version of the preceding is

$$x[n] = \sum_{i=-I}^{I} X_i e^{j2\pi f_i n} \tag{41}$$

$$y[n] = \sum_{i=-I}^{I} X_i H(e^{j2\pi f_i}) e^{j2\pi f_i n} \tag{42}$$

where

$$X_i = \begin{cases} A_0 & i = 0 \\ \frac{1}{2} A_i e^{j\phi_i} & i = 1, 2, \ldots, I \\ X_{-i}^* & i = -1, -2, \ldots, -I \end{cases} \tag{43}$$

and

$$f_i = \begin{cases} 0 & i = 0 \\ -f_{-i} & i = -1, -2, \ldots, -I \end{cases} \qquad (44)$$

Flow graph notation shows the I-O relationship as

$$\sum_{i=-I}^{I} X_i e^{j2\pi f_i n} \xrightarrow{h[n]} \sum_{i=-I}^{I} X_i H(e^{j2\pi f_i}) e^{j2\pi f_i n} \qquad (45)$$

We will show in Chapter 4 that most sequences found in engineering practice can be expressed as sums or integrals of complex sinusoidal components. This representation is very advantageous because we can more easily understand the effect of an LTI system on an input signal if we look at the input signal's spectrum and the system's frequency response characteristics than if we look at the input sequence and the system's impulse response. The sequence domain operation is a convolution. The frequency domain operation involves the much simpler operation of complex multiplication $X_i H(e^{j2\pi f_i})$, as just demonstrated.

3DT.5 $h[n]$ and $H(e^{j2\pi f})$ are a Discrete-Time Fourier Transform (DTFT) Pair

We have seen that the discrete time Fourier transform of an LTI system's unit impulse response

$$H(e^{j2\pi f}) = \sum_{k=-\infty}^{\infty} h[k] e^{-j2\pi f k} \qquad (46)$$

provides us with useful understanding of how the system responds to both real and complex sinusoidal inputs. But this is not all. It turns out that $H(e^{j2\pi f})$ is a *complete description* of the I-O properties of a stable system. This means that starting from $H(e^{j2\pi f})$ we can find $h[n]$. The formula for the *inverse DTFT* is

$$h[n] = \int_1 H(e^{j2\pi f}) e^{jn2\pi f} df \qquad (47a)$$

where $\int_1$ denotes an integration over any period of $H(e^{j2\pi f})$. We will routinely take the integration from $f = -0.5$ to $f = 0.5$ and write Eq. (47a) as

$$h[n] = \int_{-0.5}^{0.5} H(e^{j2\pi f}) e^{jn2\pi f} df \qquad (47b)$$

We can prove that this formula is correct by substituting the expression for $H(e^{j2\pi f})$ from Eq. (46) into the right-hand side of Eq. (47). The result of the substitution is

$$\int_1 H(e^{j2\pi f}) e^{jn2\pi f} df = \int_1 \sum_{k=-\infty}^{\infty} e^{-j2\pi f k} h[k] e^{jn2\pi f} df$$

$$= \sum_{k=-\infty}^{\infty} h[k] \int_1 e^{j(n-k)2\pi f} df \qquad (48)$$

For $k = n$, $e^{j(n-k)2\pi f}$ equals unity and the value of the integral equals one. For $k \neq n$, we can write

$$e^{j(n-k)2\pi f} = \cos\{(n-k)2\pi f\} + j\sin\{(n-k)2\pi f\} \tag{49}$$

The integral of $e^{j(n-k)2\pi f}$ from $f = -0.5$ to $f = 0.5$ equals the sum of the areas under each term on the right-hand side of Eq. (49). The result is 0 for $k \neq n$ because the integration is over $(n - k)$ complete cycles of each sinusoid. Therefore

$$\int_1 e^{j(n-k)2\pi f} df = \begin{cases} 1 & \text{for } k = n \\ 0 & \text{for } k \neq n \end{cases} \tag{50}$$

Using this result in the right-hand side of Eq. (48) proves that

$$\int_1 H(e^{j2\pi f})e^{jn2\pi f} df = h[n] \tag{51}$$

which is what we wanted to show. $h[n]$ and $H(e^{j2\pi f})$ are a *discrete-time Fourier transform pair*, and their one-to-one relationship is symbolized as

$$h[n] \overset{\text{DTFT}}{\longleftrightarrow} H(e^{j2\pi f}) \tag{52}$$

The operator notations

$$\mathcal{DTFT}\{h[n]\} = \sum_{k=-\infty}^{\infty} h[k]e^{-j2\pi fk} \tag{53}$$

and

$$\mathcal{DTFT}^{-1}\{H(e^{j2\pi f})\} = \int_1 H(e^{j2\pi f})e^{jn2\pi f} df \tag{54}$$

are often useful to denote the DTFT and its inverse. Tables of DTFT pairs and properties are developed in Chapter 4DT.

3DT.6 Frequency Response Characteristics of Ideal Systems

Ideal systems are useful for describing what we might want a practical system to do. They provide goals for practical filter design, as will be described in Chapter 7. The frequency response characteristics of some important ideal systems are shown in Figure 3DT.14.

All of the systems shown can be called filters. However, the term *ideal filter* is usually reserved for the systems of Figures 3DT.14(a)–(d). These systems are called ideal filters because they perfectly pass the part of the input signal spectrum lying within their pass-bands (where $H(e^{j2\pi f}) = 1$) and completely reject the part of the input signal spectrum lying within their rejection bands (where $H(e^{j2\pi f}) = 0$). The figure defines the relevant terminology used in connection with these filters. There is no need to plot amplitude and phase characteristics for these filters because the frequency response characteristics are purely real.

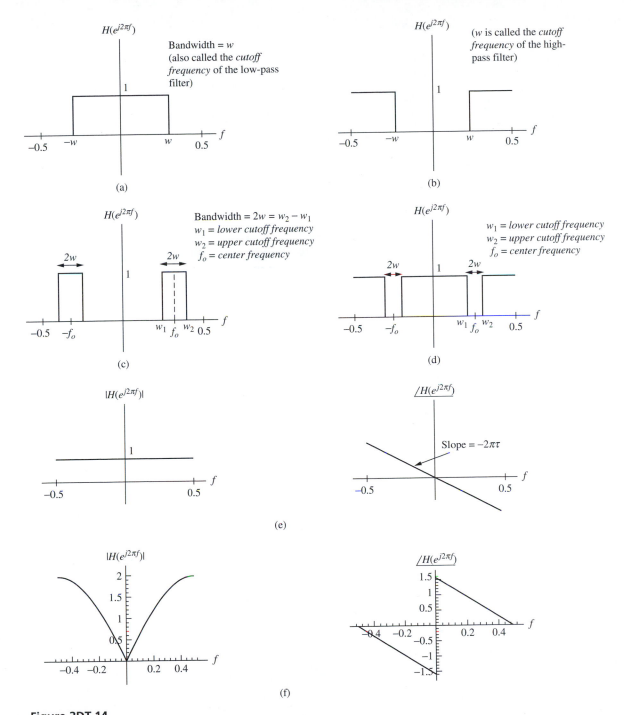

Figure 3DT.14
Frequency response characteristics of ideal systems: (a) low-pass filter; (b) high-pass filter; (c) band-pass filter (the center frequency f_o is halfway between w_1 and w_2); (d) stop-band filter (also called a *band rejection* or *notch* filter; the center frequency f_o is halfway between w_1 and w_2); (e) unit delay; (f) first difference; (g) accumulator.

Figure 3DT.14
(Cont.)

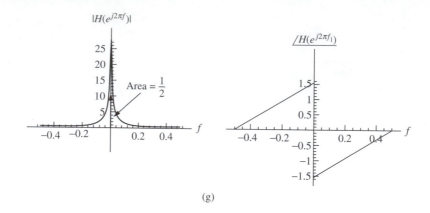

(g)

EXAMPLE 3DT.3   Two Sinusoids Applied to an Ideal Low-Pass Filter

Assume that an input

$$x[n] = 10\cos(2\pi 0.1n + 22°) + 255\cos(2\pi 0.25n) \tag{55}$$

is applied to the ideal low-pass filter of Figure 3DT.14(a). The output is given by

$$y[n] = |H(e^{j2\pi 0.1})|10\cos(2\pi 0.1n + 22°) + |H(e^{j2\pi 0.25})|255\cos(2\pi 0.25n) \tag{56}$$

where we have used the fact that the angle of H equals 0.

The preceding output depends on the value of the filter's *cutoff frequency* w. If $w < 0.1$, there is no output because the input frequency components are outside the filter passband and are therefore rejected by the filter. If $0.1 < w < 0.25$, the output is $10\cos(2\pi 0.1n + 22°)$ because this is the only component passed by the filter. If $0.25 < w$, then both components are passed by the filter and the output is $10\cos(2\pi 0.1n + 22°) + 255\cos(2\pi 0.25n)$.

The impulse responses of the ideal filters can be obtained from Eq. (47b). For example, the unit impulse response of the ideal low-pass filter is

$$h[n] = \int_{-w}^{w} e^{j2\pi fn}df = \frac{1}{j2\pi n}e^{j2\pi fn}\Big|_{-w}^{w} = \frac{1}{j2\pi n}(e^{j2\pi wn} - e^{-j2\pi wn}) = \frac{\sin(2\pi wn)}{\pi n} \tag{57}$$

for $n \neq 0$, and

$$h[0] = \int_{-w}^{w} df = 2w \tag{58}$$

These results can be written as

$$h[n] = 2w\,\text{sinc}(2wn) \tag{59}$$

where

$$\text{sinc}(\lambda) = \frac{\sin(\pi\lambda)}{\pi\lambda} \tag{60}$$

It can be shown using L'Hôspital's rule from calculus that sinc(0) = 1. A plot of sinc(λ) appears in Figure 2CT.4.

Figure 3DT.15 shows a plot of the unit impulse response of the ideal low-pass filter for $w = 0.25$. We can see from this plot that an ideal low-pass filter is a noncausal system.

Figure 3DT.15
Unit impulse response
of an ideal low-pass
filter for $w = 0.25$.

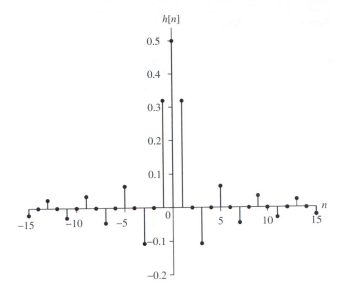

An important special case of Eq. (59) is obtained by taking the limit $w \to \frac{1}{2}$. In this limit, the frequency response characteristic equals unity: $H(e^{j2\pi f}) = 1$ for all f. We can see from Eq. (59) that $h[n]$ equals $\text{sinc}(n) \equiv \delta[n]$, in the limit $w \to \frac{1}{2}$. This provides a derivation of the Fourier transform pair

$$\delta[n] \stackrel{\text{DTFT}}{\longleftrightarrow} 1 \tag{61}$$

The conclusion is that if a filter has impulse response $h[n] = \delta[n]$, then that filter passes all frequency components of its input without change, i.e. $H(e^{j2\pi f}) = 1$ for all f.

The unit impulse response sequences of all the ideal filters (Figure 3DT.14(a)–(d)) are listed in Table 3DT.1. In the table

$$\sqcap\left(\frac{f}{2w}\right) = \begin{cases} 0 & f < -w \\ 0.5 & f = -w \\ 1 & -w < f < w \\ 0.5 & f = w \\ 0 & f > w \end{cases} \tag{62}$$

denotes a rectangular pulse having amplitude 1 and width $2w$ in the variable f.

Table 3DT.1
Ideal Filters

Name	$H(e^{j2\pi f}); -\frac{1}{2} \le f < \frac{1}{2}$	$h[n]$
Ideal low-pass filter	$\sqcap\left(\dfrac{f}{2w}\right)$	$2w\,\text{sinc}(2wn)$
Ideal band-pass filter	$\sqcap\left(\dfrac{f-f_o}{2w}\right) + \sqcap\left(\dfrac{f+f_o}{2w}\right)$	$4w\,\text{sinc}(2wn)\cos(2\pi f_o n)$
Ideal high-pass filter	$1 - \sqcap\left(\dfrac{f}{2w}\right)$	$\delta[n] - 2w\,\text{sinc}(2wn)$
Ideal stop-band filter	$1 - \sqcap\left(\dfrac{f-f_o}{2w}\right) - \sqcap\left(\dfrac{f+f_o}{2w}\right)$	$\delta[n] - 4w\,\text{sinc}(2wn)\cos(2\pi f_o n)$

Table 3DT.2

Ideal Systems

Name	$H(e^{j2\pi f}); -\frac{1}{2} \leq f < \frac{1}{2}$	$h[n]$
Unit delay	$e^{-j2\pi f}$	$\delta[n-1]$
First difference	$1 - e^{-j2\pi f}$	$\delta[n] - \delta[n-1]$
Accumulator	$\frac{1}{2}\delta(f) + \dfrac{1}{1 - e^{-j2\pi f}}$	$u[n]$

All the impulse responses listed in Table 3DT.1 can be derived in a manner similar to Eq. (57). In the following chapter, we show how to derive them more easily by using properties of the Fourier transform. It is remarkable that all the ideal filters have nonzero impulse responses for $n < 0$. This means that all the ideal filters are noncausal systems.[6]

> ## All Ideal Filters are Noncausal

Table 3DT.2 lists the unit impulse responses of the remaining ideal systems of Figure 3DT.14. Let's derive its entries.

The output of a unit delay is an indexed-shifted replica of the input

$$y[n] = x[n-1] \tag{63}$$

The unit impulse response is

$$h[n] = \delta[n-1] \tag{64}$$

The frequency response characteristic is easily found from Eq. (46). The result is

$$H(e^{j2\pi f}) = e^{-j2\pi f} \tag{65}$$

which is listed in Table 3DT.2.

The output of the first differencer is

$$y[n] = \Delta x[n] = x[n] - x[n-1] \tag{66}$$

The unit impulse response is

$$h[n] = \delta[n] - \delta[n-1] \tag{67}$$

We can find the frequency response characteristic of the first differencer easily from Eq. (46). The result is

$$H(e^{j2\pi f}) = 1 - e^{-j2\pi f} \tag{68}$$

which is listed in Table 3DT.2.

---

[6] Another remarkable property of the ideal filters listed in Table 3DT.1 is that they are not BIBO stable. This result follows fom the fact that each has an impulse response that is not absolutely summable.

Recall that an accumulator has a unit impulse response $h[n] = u[n]$. We can view $u[n]$ as the limit of $h_a[n] \triangleq a^n u[n]$ as $a \to 1$. To find the frequency response characteristic $H(e^{j2\pi f})$ of an accumulator, we take the $a \to 1$ of $H_a(e^{j2\pi f}) = \mathcal{DTFT}\{h_a[n]\}$. The derivation requires considerable care and is given in Appendix 3DT. The result is

$$H(e^{j2\pi f}) = \tfrac{1}{2}\delta(f) + \frac{1}{1 - e^{-j2\pi f}} \qquad \text{for} \quad -\tfrac{1}{2} \le f < \tfrac{1}{2} \tag{69}$$

which is listed in Table 3DT.2. In the preceding equation, $\delta(f)$ denotes a unit impulse in the real variable f. Unit impulses with arguments taken from the set of real numbers can be viewed as infinitely high, infinitely narrow pulses having area unity.[7] A commonly used definition is

$$\delta(f) = \lim_{w \to 0} \frac{1}{2w} \, \sqcap \left(\frac{f}{2w} \right) \tag{70}$$

The bold upward arrow at the origin in Figure 3DT.14(g) depicts the impulse $\tfrac{1}{2}\delta(f)$ in Eq. (60). As always, $H(e^{j2\pi f})$ is periodic with period 1.

3DT.7 Basic Connections of LTI Systems

Figure 3DT.16 depicts frequency domain results for connecting two LTI systems to create a third, larger system. Important properties of these systems can be derived by applying an input $Xe^{j2\pi fn}$, as shown in the figure.

For example, Figure 3DT.16(a) shows a *cascade* connection of two LTI systems. To obtain the overall frequency response characteristic, we apply $Xe^{j2\pi fn}$ to the first system. We

---

[7] The unit impulse $\delta(\cdot)$ was introduced in Section 2CT.2. Students who have not read Section 2CT.2 can learn about them from Problems 3DT.5.3 and 3DT.5.4.

Figure 3DT.16
Basic connections of LTI systems in the sequence domain with input $x[n] = Xe^{j2\pi fn}$:
(a) cascade;
(b) parallel; (c) loop.

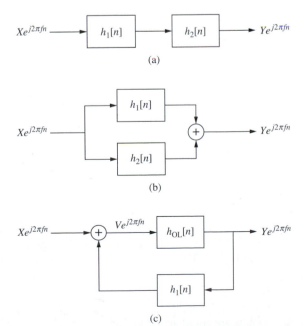

know from Eq. (14) that the output of the first system is $H_1(e^{j2\pi f})Xe^{j2\pi fn}$. This complex sinusoid is applied as input to the second system, and the output is $H_2(e^{j2\pi f})H_1(e^{j2\pi f})Xe^{j2\pi fn}$ (again by Eq. (14)). Therefore, the overall frequency response characteristic is

$$H(e^{j2\pi f}) = H_1(e^{j2\pi f})H_2(e^{j2\pi f}) \tag{71}$$

Because $H_1(e^{j2\pi f})H_2(e^{j2\pi f}) = H_2(e^{j2\pi f})H_1(e^{j2\pi f})$, we can interchange the orders of cascaded LIT systems. This result confirms the commutation property Eq. (2DT.38). By comparing Eq. (71) with Eq. (2DT.59), we also see that the sequence operation of convolution $h_1[n] * h_2[n]$, corresponds to the frequency domain operation of multiplication, $H_1(e^{j2\pi f})H_2(e^{j2\pi f})$. Clearly, this is an important result: Multiplication is much simpler than convolution to visualize and evaluate!

Figure 3DT.16(b) shows a *parallel* connection of two LTI systems. We leave it as a problem for you to show that

$$H(e^{j2\pi f}) = H_1(e^{j2\pi f}) + H_2(e^{j2\pi f}) \tag{72}$$

Figure 3DT.16(c) shows LTI systems connected in a loop. To obtain the frequency response characteristic of the overall system, we apply an input $Xe^{j2\pi fn}$ to the system and assume that the complex sinusoidal form propagates throughout the system. We denote the output of the system as $Ye^{j2\pi fn}$, and we denote the output of the adder as $Ve^{j2\pi fn}$. We apply the basic property (14) and find from the figure that

$$Ye^{j2\pi fn} = H_{OL}(e^{j2\pi f})Ve^{j2\pi fn} \tag{73}$$

and

$$Ve^{j2\pi fn} = Xe^{j2\pi fn} + YH_1(e^{j2\pi f})e^{j2\pi fn} \tag{74}$$

We can solve the preceding equations for $H(e^{j2\pi f}) = Y/X$. The result is

$$H(e^{j2\pi f}) = \frac{H_{OL}(e^{j2\pi f})}{1 - H_{OL}(e^{j2\pi f})H_1(e^{j2\pi f})} \tag{75}$$

Equation (75) is an important result describing the frequency response characteristic for systems connected in a loop. Our method and the result, (75), assume that $H(e^{j2\pi f})$ exists. Notice that $H(e^{j2\pi f}) = H_{OL}(e^{j2\pi f})$ where there no feedback, i.e., when $H_1(e^{j2\pi f}) = 0$. $H_{OL}(e^{j2\pi f})$ is the "open loop" frequency response characteristic of the system. A special case of Eq. (75) occurs for $H_{OL}(e^{j2\pi f}) = 1$.

$$H(e^{j2\pi f}) = \frac{1}{1 - H_1(e^{j2\pi f})} \tag{76}$$

This special case reduces the system to that shown in Figure 2DT.13(c). We can see that Eq. (76) agrees with Eq. (2DT.51) because Eq. (76) can be expanded (Appendix B) to give

$$H(e^{j2\pi f}) = 1 + H_1(e^{j2\pi f}) + H_1^2(e^{j2\pi f}) + H_1^3(e^{j2\pi f}) + \cdots \tag{77}$$

The above results are summarized in Figure 3DT.17.

Figure 3DT.17
Basic connections of
DT LTI systems in the
frequency domain
(a) cascade;
(b) parallel; (c) loop.

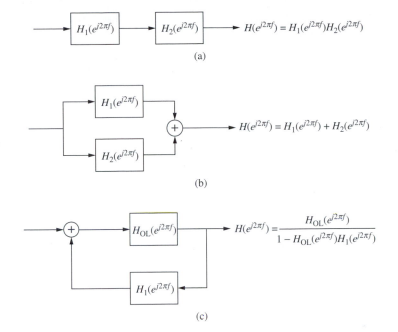

$$H(e^{j2\pi f}) = H_1(e^{j2\pi f})H_2(e^{j2\pi f})$$

(a)

$$H(e^{j2\pi f}) = H_1(e^{j2\pi f}) + H_2(e^{j2\pi f})$$

(b)

$$H(e^{j2\pi f}) = \frac{H_{OL}(e^{j2\pi f})}{1 - H_{OL}(e^{j2\pi f})H_1(e^{j2\pi f})}$$

(c)

3DT.8 **Power Sequences, Power Spectrums, and Parseval's Theorem**

The sinusoid $A\cos(2\pi f n + \phi)$ and the multiple sinusoidal component sequence described in this chapter are members of a class of sequences called *power sequences*. The *average power* in a real or complex sequence $x[n]$ is defined as

$$P = < |x[n]|^2 > \tag{78}$$

The quantity $|x[n]|^2$ is the power in the sample $x[n]$ and

$$<(\cdot)> \; = \; \lim_{N_o \to \infty} \frac{1}{2N_o + 1} \sum_{n=-N_o}^{N_o} (\cdot) \tag{79}$$

denotes the *average value* of the argument $(\cdot)$. Real-life sequences, like noise and voice sequences, are often modeled as power sequences.

All periodic sequences are power sequences. A sequence $x[n]$ is *periodic* with period N if

$$x[n] = x[n + N] \qquad \text{for all } n \tag{80}$$

where N is the smallest positive integer for which Eq. (80) applies. For periodic sequences, we can we compute the time average $<(\cdot)>$ over a period N:

$$<(\cdot)> \; = \; \frac{1}{N} \sum_{n=0}^{N-1} (\cdot) \tag{81}$$

For a periodic sequence, Eqs. (79) and (81) give the same results.

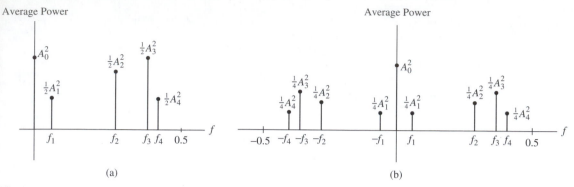

Figure 3DT.18
Power spectrums: (a) one sided; (b) two sided.

The sinusoidal sequences $A\cos(2\pi f n+\phi)$ and $Xe^{j2\pi fn}$ are periodic if an only if f is a rational number, i.e., $f = m/N$, where m and N are integers (Problem 3DT.8.1). The sinusoidal sequences are always power sequences. The real sinusoid $x[n] = A\cos(2\pi f n + \phi)$ has average power $P = \frac{1}{2}A^2$ when f is not an integer multiple of $\frac{1}{2}$, and average power $P = A^2\cos^2(\phi)$ when f is an integer multiple of $\frac{1}{2}$. The complex sinusoid $Xe^{j2\pi fn}$ has average power $|X|^2$ for every f. These results are established in Problem 3DT.8.2.

It is useful to display the average power in the sinusoidal components in a sequence having the form (39) versus frequency, as shown in Figure 3DT.18 for $I = 4$. Figure 3DT.18(a) shows a *one-sided power spectrum*. The height of each line, $P_i = \frac{1}{2}A_i^2$, $i = 1, 2, \ldots, I$, tells us the average power of the sinusoid having the frequency f_i, indicated by the position of the line. The line at $f = 0$ shows the power in the DC component $P_0 = A_0^2$.

Figure 3DT.18(b) shows the *two-sided power spectrum* of $x[n]$. A two-sided power spectrum of $x[n]$ consists lines, each having amplitude $A_i^2/4$ positioned at frequencies $\pm f_i$, as shown in Figure 3DT.18(b). (See Eq. (44).) The height of each line is equal to the average power carried by the complex sinusoid having the frequency indicated by the position of the line. We can obtain a two-sided power spectrum by breaking up the real sinusoids in Eq. (39) into pairs of complex sinusoids as in Eq. (41). The average power carried by $X_i e^{j2\pi f_i n}$ is $<|X_i e^{j2\pi f_i n}|^2> = |X_i|^2 = A_i^2/4$ for $i = \pm1, \pm2, \ldots, \pm I$ and $P_0 = |X_0|^2 = A_0^2$. (See Eq. (43).)

A fundamental property of both one-sided and two-sided power spectrums is that the total average power carried by a sequence is equal to the sum of the heights of the lines in the sequence's power spectrum. This property is called *Parseval's theorem* for multiple sinusoidal component sequences.

Parseval's Theorem for Multiple Sinusoidal Component Sequences

Complex or real $x[n]$:

$$< |x[n]|^2 > = \sum_{i=-I}^{I} |X_i|^2 \tag{82a}$$

Real $x[n]$:

$$<x^2[n]> = \sum_{i=-I}^{I} |X_i|^2 = A_0^2 + \frac{1}{2}\sum_{i=1}^{I} A_i^2 \tag{82b}$$

We can prove Parseval's theorem using Eq. (41) if we write the absolute value squared as a double sum

$$<|x[n]|^2> = \, <\left| \sum_{i=-I}^{I} X_i e^{j2\pi f_i n} \right|^2> = \, <\sum_{i=-I}^{I} X_i e^{j2\pi f_i n} \sum_{k=-I}^{I} X_k^* e^{-j2\pi f_k n}>$$

$$= \, <\sum_{i=-I}^{I} \sum_{k=-I}^{I} X_i X_k^* e^{j2\pi f_i n} e^{-j2\pi f_k n}> = \, \sum_{i=-I}^{I} \sum_{k=-I}^{I} X_i X_k^* <e^{j2\pi f_i n} e^{-j2\pi f_k n}> \tag{83}$$

where the last step follows by the definition of the average value operator. The key element of the proof comes next. The average value term in Eq. (83) simplifies to

$$<e^{j2\pi f_i n} e^{-j2\pi f_k n}> = \, <e^{j2\pi (f_i - f_k)n}> = \begin{cases} 1 & \text{for } i = k \\ 0 & \text{for } i \ne k \end{cases} \tag{84}$$

The upper part of this equality follows immediately because $e^{j2\pi (f_i - f_k)n} = 1$ for $i = k$. The lower part follows from Euler's formula because

$$<e^{j2\pi (f_i - f_k)n}> = \, <\cos(2\pi (f_i - f_k)n)> + j<\sin(2\pi (f_i - f_k)n)> \tag{85}$$

and the average value of any sinusoid is 0. If we substitute Eq. (84) into Eq. (83), we obtain the first equality in Parseval's theorem. The second equality follows directly from Eq. (43).

The term *orthogonality* is used to describe the important lower equality in Eq. (84). In general, two power sequences, $x_1[n]$ and $x_2[n]$, are called *orthogonal* if

$$<x[n]x_2^*[n]> = 0 \tag{86}$$

Sinusoids having different frequencies are orthogonal. Parseval's theorem is based on this orthogonality.

The power spectrum of the output of an LTI system is related in a simple way to the power spectrum of the input. Equation (40) shows us the output $y[n]$ of an LTI system having multiple sinusoidal input components. The average power in the output sinusoid having frequency f_i is $P_i = \frac{1}{2}|H(e^{j2\pi f_i})|^2 A_i^2$, for $i = 1, 2, \ldots, I$ and $P_0 = |H(1)|^2 A_0^2$. Therefore, the one-sided power spectrum of $y[n]$ equals the product of the one-sided power spectrum of the input and the square of the filter amplitude characteristic. Similarly, Eq. (42) implies that the average power in the output complex sinusoid having frequency f_i is $P_i = |H(e^{j2\pi f_i})|^2|X_i|^2 = \frac{1}{4}|H(e^{j2\pi f_i})|^2 A_i^2$ for $i = \pm1, \pm2, \ldots, \pm I$ and $P_0 = |H(1)|^2|X_0|^2 = |H(1)|^2 A_0^2$. Therefore the two-sided power spectrum of the filter output equals the product of the two-sided power spectrum of the input times the square of the filter amplitude characteristic.

3DT.9   Summary

In this chapter, we described systems from the frequency domain viewpoint. The frequency domain viewpoint is based on complex sinusoidal signals. We showed that, if you apply a complex sinusoid $Xe^{j2\pi f n}$ to an LTI system having impulse response $h[n]$, then the output is $H(e^{j2\pi f})Xe^{j2\pi f n}$, where $H(e^{j2\pi f})$ is the discrete-time Fourier transform of $h[n]$. Therefore the system's frequency response characteristic $H(e^{j2\pi f})$ describes how the input and output sinusoids are related. We stated that $H(e^{j2\pi f})$ exists for all BIBO stable LTI systems and some LTI systems that are not BIBO stable. We described the frequency response characteristics and impulse responses for ideal filters, delays, differencers, and accumulators. We found the overall frequency response of systems placed in cascade, in parallel, and in a loop. We introduced one- and two-sided amplitude and phase spectrums to show graphically how signal amplitude and phase are distributed with respect to frequency. The amplitude and phase spectrum of the output of an LTI system are obtained by multiplying the spectrum of the input by the system's frequency response characteristic.

Sinusoids and sequences composed of multiple sinusoids are members of a class of sequences called power sequences. We introduced signal power spectrums to show graphically how average signal power is distributed with respect to frequency. The power spectrum of the output of an LTI system is obtained by multiplying the power spectrum of the input by the square of the system's amplitude response characteristic. Parseval's theorem states that the total average power carried by a sequence is equal to the sum of the heights of the lines in the sequence's power spectrum.

This chapter has shown that a system's frequency response characteristic is a useful analytical aid for finding the response of an LTI system to sinusoids and sequences composed of multiple sinusoids. In the following chapter we show that the frequency response characteristic is similarly useful for finding the response of an LTI system to sequences such as $\sqcap[n; K]$, $\wedge[n; M]$, and $a^n u[n]$.

Summary of Definitions and Formulas

Sinusoidal Sequence as Sampled Sinusoidal Waveform

If we sample $Ae^{j(2\pi Ft+\phi)}$ at rate $F_s = 1/T_s$, we obtain $Ae^{j(2\pi fn+\phi)}$, where $f = F/F_s$. The real variable f is called the *normalized frequency*. It is the ratio of the frequency F of the sinusoid being sampled to the sampling frequency F_s.

By Euler's formula, $A\cos(2\pi fn + \phi) = \mathcal{R}\{Ae^{j(2\pi fn+\phi)}\}$.

Aliasing

If a real or complex CT sinusoid having frequency F is sampled at rate $F_s = 1/T_s$, aliasing does not occur if

$$F_s > 2|F|$$

The frequency $2|F|$ is called the *Nyquist rate*.

Response of Stable LTI System to Complex Sinusoid

The input is applied at $n = -\infty$.

$$Xe^{jn2\pi f} \xrightarrow{h[n]} H(e^{j2\pi f})Xe^{jn2\pi f}$$

where

$$H(e^{j2\pi f}) = \sum_{k=-\infty}^{\infty} h[k]e^{-jk2\pi f} \qquad \text{and} \qquad h[n] = \int_1 H(e^{j2\pi f})e^{jn2\pi f}\,df$$

are a discrete-time Fourier transform pair.

Properties of $H(e^{j2\pi f})$ for Real $h[n]$

$$H(e^{-j2\pi f}) = H^*(e^{j2\pi f})$$
$$|H(e^{-j2\pi f})| = |H(e^{j2\pi f})|$$
$$\angle H(e^{-j2\pi f}) = -\angle H(e^{j2\pi f})$$

Response of Stable LTI System to Real Sinusoid

The input is applied at $n = -\infty$.

$$A\cos(2\pi fn + \phi_x) \xrightarrow{h[n]} A|H(e^{j2\pi f})|\cos(2\pi fn + \phi_x + \angle\mathrm{H}(e^{j2\pi f})) \qquad \text{for real } h[n]$$

Response of Stable LTI System to Multiple Sinusoids

$$\sum_{i=-I}^{I} X_i e^{j2\pi f_i n} \xrightarrow{h[n]} \sum_{i=-I}^{I} X_i H(e^{j2\pi f_i}) e^{j2\pi f_i n}$$

$$\sum_{i=0}^{I} A_i \cos(2\pi f_i t + \phi_i) \xrightarrow{h[n]} |H(e^{j2\pi f_i})| A_i \cos(2\pi f_i t + \phi_i + \angle H(e^{j2\pi f_i})) \qquad \text{for real } h[n]$$

Ideal Systems

Tables 3DT.1 and 3DT.2.

Periodic Sequences

$$x[n] = x[n + N] \qquad \text{for all } n$$

Power Sequences

$$P = \, < |x[n]|^2 > \text{ is nonzero and finite.}$$

For an arbitrary sequence:

$$<(\cdot)> = \lim_{N_o \to \infty} \frac{1}{2N_o + 1} \sum_{n=-N_o}^{N_o} (\cdot)$$

For a periodic sequence (period N):

$$<(\cdot)> = \frac{1}{N} \sum_{n=0}^{N-1} (\cdot)$$

Parseval's Theorem

Real or complex $x[n]$:

$$<|x[n]|^2> = \sum_{i=-I}^{I} |X_i|^2$$

Real $x[n]$:

$$<x^2[n]> = \sum_{i=-I}^{I} |X_i|^2 = A_0^2 + \tfrac{1}{2}\sum_{i=1}^{I} A_i^2$$

3DT.10   Problems

3DT.2   Sinusoidal Sequences

3DT.2.1 A source produces a pure sinusoidal waveform having an unknown frequency in the range 0 Hz to 1 kHz. The source is sampled at a rate F_s. Find the minimum value of F_s for which it is possible to determine the unknown frequency of the source from the samples.

3DT.2.2 Suppose that a sampling frequency of 1 kHz is used in the system of Problem 3DT.2.1. Again, the source produces a pure sinusoidal waveform having an unknown frequency in the range 0 Hz to 1 kHz. Could it be possible to unambiguously determine the frequency of the source from the samples without any additional information? If the answer depends on the frequency actually produced by the source, state the range of source frequencies that can be determined and those that cannot be determined.

3DT.2.3 Repeat Problem 3DT.2.2 for a sampling frequency of 1.9 kHz.

3DT.2.4 **a)** What is the minimum sampling rate F_s for which $A \cos(2\pi F_a t + \phi)$ can be sampled without aliasing? Prove your answer.
b) What is the minimum sampling rate F_s for which $A_1 \cos(2\pi F_a t + \phi_1) + A_2 \cos(2\pi F_b t + \phi_2)$ can be sampled without aliasing? (Assume that $F_b > F_a$.) Prove your answer.
c) Generalize your analysis of part (b) to a waveform composed of the sum of many sinusoids.

3DT.3   Response of LTI System to Sinusoidal Input

3DT.3.1 Are there any restrictions on the property that $H(e^{j2\pi(f+1)}) = H(e^{j2\pi f})$? Prove your answer.

3DT.3.2 Use the definition $H(e^{j2\pi f}) = \sum_{k=-\infty}^{\infty} h[k]e^{-jk2\pi f}$ to prove that for any DT LTI system, $H(e^{-j2\pi f}) = H^*(e^{j2\pi f})$ if and only if $h[n]$ is real.

3DT.3.3 Show that if $H(e^{-j2\pi f}) = H^*(e^{j2\pi f})$, then $|H(e^{-j2\pi f})| = |H(e^{j2\pi f})|$ and $\angle H(e^{-j2\pi f}) = -\angle H(e^{j2\pi f})$.

3DT.3.4 A system has the frequency response characteristic $H(e^{j2\pi f}) = j2\pi f$ for $-0.5 \le f < 0.5$.
a) Plot the amplitude and phase characteristics of the system for $-1 \le f < 1$.
b) Is the system real? Prove your answer.

3DT.3.5 A real system has the frequency response characteristic 1 for $0 \le f < 0.25$ and 0 for $0.25 \le f < 0.5$. Plot $H(e^{j2\pi f})$ for $-1 \le f < 1$.

 3DT.3.6 The basic result (8) was obtained from the assumption that $x[n]$ is a complex sinusoid for all n. For comparison, this problem examines the response of an LTI system to an input that $x[n]$ is 0 for $n < n_0$ and complex sinusoidal for $n \ge n_0$

$$x[n] = Xe^{j2\pi fn}u[n - n_0] \tag{87}$$

a) Show, by substitution of Eq. (87) into (8) that

$$y[n] = Xe^{j2\pi fn}\left\{H(e^{j2\pi f}) - \sum_{m=n-n_0}^{\infty} h[m]e^{-j2\pi fm}\right\}$$

How does this result compare to Eq. (14)?
b) Assume that

$$h[n] = a^n u[n]$$

where $|a| < 1$ and show that, for $x[n]$ given by Eq. (87),

$$y[n] = \left[H(e^{j2\pi f})Xe^{j2\pi fn} - H(e^{j2\pi f})Xa^{(n-n_0)}e^{j2\pi fn_0}\right]u[n - n_0]$$

where

$$H(e^{j2\pi f}) = \frac{1}{1 - ae^{-j2\pi f}}$$

c) Assume that $x[n] = A\cos(2\pi fn + \phi)u[n - n_0]$ and $h[n] = a^n u[n]$. Use the result of part (b) and superposition to show that

$$y[n] = \big\{ A|H(e^{j2\pi f})| \cos\left(2\pi rn + \phi + \angle H(e^{j2\pi f})\right)$$
$$- A|H(e^{j2\pi f})| \cos\left(2\pi rn_0 + \phi + \angle H(e^{j2\pi f})\right) a^{n-n_0} \big\} u[n - n_0]$$

Discuss this result. In particular, state approximately how large n must be for the transient term to be negligible.

3DT.3.7 Greater understanding of the amplitude and phase characteristics of the 5-point averager of Example 3DT.1 can be obtained by writing $H(e^{j2\pi f})$ of Eq. (31) in the form

$$H(e^{j2\pi f}) = \frac{\sin(5\pi f)}{5\sin(\pi f)} e^{-j4\pi f} \qquad (88)$$

The derivation of Eq. (88) is based on identities in Appendix B.
a) Read Appendix B. Then show how to obtain Eq. (88).
b) Use L'Hôspital's rule from calculus to show that $\sin(5\pi f)/5\sin(\pi f)\big|_{f=0} = 1$. Then confirm this result using some other method.
c) Use hand calculations to plot $\sin(5\pi f)/5\sin(\pi f)$ for $-\frac{1}{2} \le f < \frac{1}{2}$.
d) Plot the angle of $e^{-j4\pi f}$ for $-\frac{1}{2} \le f < \frac{1}{2}$.
e) Describe how Figure 3DT.6 is related to your plots from parts (c) and (d)

3DT.3.8 Find the output of the moving average computer of Example 3DT.1 for an input $A\cos(2\pi fn + \phi)$. At what normalized frequencies is the output 0? Give a simple explanation why the output is 0 at these frequencies.

3DT.3.9 a) Find the difference equation of the system of Example 3DT.2.
b) Confirm Eq. (34) by substituting $Xe^{j2\pi fn}$ into the system's difference equation.
c) Does the condition $|a| < 1$ appear anywhere in your derivation? Explain why.

3DT.3.10 a) Find the frequency response characteristic $H(e^{j2\pi f})$ of an LTI system described by the second-order difference equation

$$cy[n-2] + by[n-1] + y[n] = x[n]$$

Hint: Apply the complex sinusoidal input

$$x[n] = Xe^{j2\pi fn}$$

and assume a solution of the form

$$y[n] = Ye^{j2\pi fn}$$

b) State any restrictions that might apply to your answer for part (a).

3DT.3.11 Find the frequency response characteristic $H(e^{j2\pi f})$ of an LTI system described by the lth-order difference equation

$$a_l y[n-l] + a_{l-1}y[n-(l-1)] + \cdots + a_1 y[n-1] + a_0 y[n] =$$
$$b_m x[n-m] + b_{m-1}x[n-(m-1)] + \cdots + b_1 x[n] + b_0 x[n] \qquad (89)$$

See the hint for Problem 3DT.3.10.

3DT.3.12 Find the I-O difference equation of a system having the frequency response characteristic $H(e^{j2\pi f}) = e^{-j2\pi f}/(1 + e^{-j2\pi f})$.

3DT.3.13 Find the I-O difference equation of a system having the frequency response characteristic

$$H(e^{j2\pi f}) = \frac{1}{1 + ae^{-j2\pi f}} + \frac{1}{1 + be^{-j2\pi f}}$$

3DT.3.14 a) Find $H(e^{j2\pi f})$ for the system described by the equation

$$2\Delta y[n] + y[n] = 3\Delta x[n] + x[n]$$

b) Are you sure that $H(e^{j2\pi f})$ exists? If necessary, check your answer or part (b) using some other method.

3DT.3.15 MATLAB uses the symbols i and j to denote the imaginary unit $\sqrt{-1}$. If a and b are real numbers (or vectors having the same length), then the command z=a+jb creates a complex number (or vector of complex numbers), z. The functions abs, angle, imag, real, and conj all have the obvious meanings: abs(z) is the vector of absolute values of the components of z, and so on. Consult MATLAB help for more information.

a) Use the plot command to plot the amplitude and phase characteristics for the 5-point averager of Example 3DT.1 for $-0.5 \leq f < 0.5$.

b) Use the plot command to plot the amplitude and phase characteristics for the first-order filter of Example 3DT.2 for $-0.5 \leq f < 0.5$. Experiment with different values for the parameter a. Hint: Remember that $H(e^{j2\pi f})$ exists if and only if $|a| < 1$.

3DT.4 Multiple Sinusoidal Components

3DT.4.1 The sequence $x[n] = 1 + 3\cos(2\pi 0.1n + 30°) + 5\sin(2\pi 0.2n + 30°)$ is applied to a 5-term averager. Find the output sequence.

3DT.4.2 Repeat Problem 3DT.4.1 for the first-order system of Example 3DT.2 where $a = 0.1$.

3DT.4.3 Repeat Problem 3DT.4.1 for the system $y[n] = \Delta x[n]$.

3DT.4.4 Find the output of the system of Problem 3DT.3.4 if $x[n] = 2.2\cos(2\pi 0.15n + 15°) + 3\sin(2\pi 0.32n + 15°)$.

3DT.4.5 Find the output of the system of Problem 3DT.3.5 for an input $x[n] = 2.2\cos(2\pi 0.15n + 15°) + 3\sin(2\pi 0.32n + 15°)$.

3DT.5 $h[n]$ and $H(e^{j2\pi f})$ are a Discrete-Time Fourier Transform (DTFT) Pair

3DT.5.1 Find $h[n]$ if $H(e^{j2\pi f}) = 1$. What does this system do?

3DT.5.2 Find $h[n]$ if $H(e^{j2\pi f}) = 1 + [0.5/(1 - 0.2e^{-j2\pi f})] + [0.5/(1 + 0.2e^{-j2\pi f})]$

3DT.5.3 In Eq. (70) of the text, we defined the unit impulse as $\delta(f) = \lim_{w\to 0} 1/2w \sqcap (f/2w)$.

a) Explain the fundamental difference between the symbols $\delta(f)$ and $\delta[n]$.

b) Show that the area of $1/2w \sqcap (f/2w)$ is unity, independent of the value of w.

c) Plot $1/2w \sqcap (f/2w)$ with $-0.5 \leq f < 0.5$ for $w = 1$ and $w = 0.1$.

d) Show that $h[n] = \text{sinc}(2wn)$ and $H(e^{j2\pi f}) = 1/2w \sqcap (f/2w)$; $-0.5 \leq f < 0.5$ are a DTFT pair.

e) Take the limit $w \to 0$ and use your answer to part (c) to show that
$$1 \overset{DTFT}{\longleftrightarrow} \sum_{k=-\infty}^{\infty} \delta(f - k).$$

3DT.5.4 a) Let g(f) be any smooth function of f. Show that

$$g(f)\delta(f - f_o) = g(f_o)\delta(f - f_o) \qquad (90)$$

where f_o is a constant. The preceding relation is called the *sampling property* of the unit impulse. Hint: Sketch the product $g(f)1/2w \sqcap ((f - f_0)/2w)$ for small w. Consider the limit $w \to 0$. Assume that f and f_o are in the range -0.5 to 0.5.

b) Use your answer to part (a) to show that

$$\int_{-1/2}^{1/2} g(f)\delta(f - f_o)\, df = g(f_o) \tag{91}$$

where f_o is in the range -0.5 to 0.5. The preceding relation is called the *sifting property* of the unit impulse $\delta(f)$.

c) Repeat parts (a) and (b) where $\delta(f)$ is defined as $\delta(f) = \lim_{w \to 0} 1/w\ \text{sinc}(f/w)$ to demonstrate that other equivalent definitions of $\delta(f)$ are possible.

d) Use (91) to find $h[n]$ if $H(e^{j2\pi f}) = \frac{1}{2}\delta(f - 0.1) + \frac{1}{2}\delta(f + 0.1);\ -0.5 \le f < 0.5.$

3DT.6   Frequency Response Characteristics of Ideal Systems

3DT.6.1 We saw in Eq. (59) that the impulse response of an ideal low-pass filter is given by

$$h[n] = 2w\ \text{sinc}(2wn) \tag{92}$$

Show that $\sum_{n=-\infty}^{\infty} h[n] = 1$. Hint: Compare this sum to the definition of the DTFT of $h[n]$. Ask yourself how these sums are related.

3DT.6.2 Use the method of Problem 3DT.6.1 to find $\sum_{n=-\infty}^{\infty} h[n]$ for:

a) An ideal high-pass filter.

b) An ideal band-pass filter.

c) An ideal notch filter.

d) An ideal differencer.

3DT.6.3 a) Show that the step response of an ideal low-pass filter having the frequency response characteristic $\sqcap(f/2w)$ is given by

$$h_{-1}[n] = \tfrac{1}{2}(1 + 2w) + \text{sgn}(n)\sum_{k=1}^{n} 2w\ \text{sinc}(2wk)$$

where

$$\text{sgn}(\lambda) = \begin{cases} 1 & \lambda > 0 \\ 0 & \lambda = 0 \\ -1 & \lambda < 0 \end{cases}$$

b) Write a computer program to plot $h_{-1}[n]$ for various values of w in the range $0 < w < 0.5$.

3DT.6.4 Find and sketch a bounded sequence $x[n]$ that yields unbounded output when applied to an ideal low-pass filter. Hint: You can find such a waveform by looking at the proof that an LTI system is BIBO stable if and only if $h[n]$ is absolutely summable. See Section 2DT.6.

3DT.7   Basic Connections of LTI Systems

3DT.7.1 Go through the detailed steps described in the text to find the frequency response characteristic of systems connected in series.

3DT.7.2 Go through the detailed steps described in the text to find the frequency response characteristic of systems connected in a loop.

3DT.7.3 Two LTI systems having frequency response characteristics $H_1(e^{j2\pi f})$ and $H_2(e^{j2\pi f})$ are connected in parallel. Show that the frequency response characteristic of the overall system is given by $H(e^{j2\pi f}) = H_1(e^{j2\pi f}) + H_2(e^{j2\pi f})$.

Figure 3DT.19

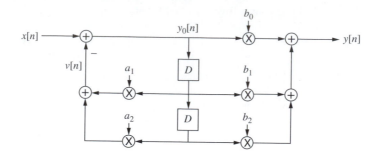

3DT.7.4 Consider the system of Figure 3DT.19

a) Draw Figure 3DT.19 on homework paper and write the formula $x[n] = Xe^{j2\pi f n}$ as the input to the figure. Then write down the formulas for the outputs of every system on your figure. Assume that the complex sinusoidal form propagates through the system but that the complex amplitudes differ.

b) Write down two equations you can solve to obtain $y_0[n] = Y_0 e^{j2\pi f n}$.

c) Solve your equations from part (b) to obtain the frequency response characteristic for Y_0/X.

d) Find the frequency response characteristic for Y/Y_0.

e) Use your answers from parts (c) and (d) to find the frequency response characteristic Y/X of the system.

f) Use your answer from part (e) to find the difference equation relating input and output sequences of the system.

3DT.7.5 Consider the system of Figure 3DT.20

a) Generalize the method of Problem 3DT.7.4 to find the frequency response characteristic Y/X of the system.

b) Use your answer from part (a) to find the difference equation relating the input and output sequences of the system.

Figure 3DT.20

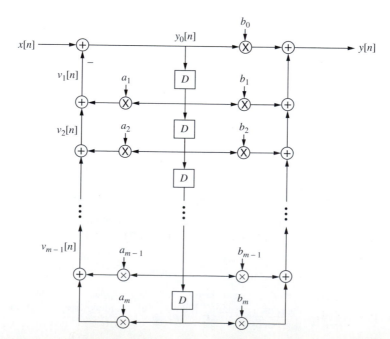

3DT.7.6 In the text we described a series connection of two filters having impulse responses $h_1[n]$ and $h_2[n]$. The impulse response of this series connection is $h[n] = h_1[n] * h_2[n]$. We showed that the frequency response characteristic corresponding to $h[n] = h_1[n] * h_2[n]$ is $H(e^{j2\pi f}) = H_1(e^{j2\pi f})H_2(e^{j2\pi f})$. This means that $h_1[n] * h_2[n]$ and $H_1(e^{j2\pi f})H_2(e^{j2\pi f})$ are a DTFT pair.

a) Does it follow that $a[n] * b[n]$ and $A(e^{j2\pi f})B(e^{j2\pi f})$ are a Fourier transform pair if $a[n]$ and $b[n]$ are arbitrary sequences, where $A(e^{j2\pi f}) = \mathcal{DTFT}\{a[n]\}$ and $B(e^{j2\pi f}) = \mathcal{DTFT}\{b[n]\}$? Explain.

b) Does it follow that $x[n] * h[n]$ and $X(e^{j2\pi f})H(e^{j2\pi f})$ are a DTFT pair if $x[n]$ and $h[n]$ are arbitrary sequences, where $X(e^{j2\pi f}) = \mathcal{DTFT}\{x[n]\}$ and $H(e^{j2\pi f}) = \mathcal{DTFT}\{h[n]\}$? Consider this result carefully. Does it provide a new way to look at I-O properties of an LTI system? State how.

3DT.8   Power Sequences, Power Spectrums, and Parseval's Theorem

3DT.8.1 Show that $A\cos(2\pi f n + \phi)$ and $Xe^{j2\pi f n}$ are periodic if an only if f is a rational number, i.e., $f = m/N$, where m and N are integers. Hint: Apply the definition $x[n] = x[n + N]$.

3DT.8.2 **a)** Show that the complex sinusoid, $Xe^{j2\pi f n}$ has average power $|X|^2$ for every f.

b) Show that the real sinusoid $x[n] = A\cos(2\pi f n + \phi)$ has average power $P = \frac{1}{2}A^2$ when f is not an integer multiple of $\frac{1}{2}$ and average power $P = A^2\cos^2(\phi)$ when f is an integer multiple of $\frac{1}{2}$. Hint:

$$<\cos^2(2\pi f n + \phi)> = \tfrac{1}{2} + \tfrac{1}{2}<\cos(2\pi 2 f n + 2\phi)>$$

$$<\cos(2\pi 2 f n + 2\phi)> = \mathcal{R}\{<e^{j(2\pi 2 f n + 2\phi)}>\} = \mathcal{R}\{e^{2\phi}<e^{j(2\pi 2 f n)}>\}$$

$$<e^{j(2\pi 2 f n)}> = \lim_{K\to\infty}\frac{1}{2K+1}\sum_{n=-K}^{K}e^{j(2\pi 2 f n)} = \lim_{K\to\infty}\frac{\sin((2K+1)\pi 2 f)}{(2K+1)\sin(\pi 2 f)}$$

where the last equality follows from identity (B.3) of Appendix B.

3DT.8.3 State which of the following sequences are power signals. Find the average power in the power signals.

a) $\sum_{k=-\infty}^{\infty} \sqcap[n - 5k; 1]$
b) $\sum_{k=-\infty}^{\infty} \delta[n - 2k]$
c) $\sum_{k=-\infty}^{\infty} a^{n-k} u[n - k]$
d) $a^n u[n]$
e) $\wedge[n; M]$
f) $\cos(2\pi f_a n)\cos(2\pi f_b n)$, where $0 < f_a < 0.5, 0 < f_b < 0.5$ and $f_a \neq f_b$

3DT.8.4 Consider the sequence

$$x[n] = 2 + 3\cos(2\pi 0.1 n) + 3\sin(2\pi 0.1 n) + 5\cos(2\pi 0.25 n)$$

a) Plot the two-sided power spectrum of the sequence.
b) What is the total average power carried by the DC (i.e., constant) component?
c) What is the total average power carried by the component $3\cos(2\pi 0.1 n) + 3\sin(2\pi 0.1 n)$?
d) What is the average power carried by the component $5\cos(2\pi 0.25 n)$?
e) Show that the average powers of all the components add to give the total average power carried by $x[n]$.

3DT.8.5 In a one-sided power spectrum, the average powers of real sinusoidal frequency components are plotted versus frequency for $0 \leq f < 0.5$.

Plot the one-sided power spectrum for the sequence of Problem 3DT.9.1, and show that the sum of all the average powers of each frequency component add to give the total average power.

3DT.8.6 Two real power sequences $x_1[n]$ and $x_2[n]$ are called *orthogonal* if

$$<x_1[n]x_2[n]> \; = 0 \tag{93}$$

Assume that real orthogonal sequences $x_1[n]$ and $x_2[n]$ are added to produce a third sequence

$$x[n] = x_1[n] + x_2[n] \tag{94}$$

a) Show that the average power contained in $x[n]$ equals the sum of the average powers contained in $x_1[n]$ and $x_2[n]$.

b) Suppose that we are told that $x[n] = x_1[n] + x_2[n]$ and that the average power contained in $x[n]$ equals the sum of the average powers contained in $x_1[n]$ and $x_2[n]$. Must it necessarily follow that $x_1[n]$ and $x_2[n]$ are orthogonal?

3DT.8.7 Refer to the definition of real orthogonal sequences given in Problem 3DT.8.6.

a) Show that sinusoidal sequences having different normalized frequencies f_1 and f_2 are orthogonal, provided that $f_1 \neq f_2 + k$ where k is an integer.

b) Show that sinusoidal sequences having the same normalized frequency f are orthogonal if they differ in phase by $90°$ and if $f \neq 0, \pm\frac{1}{2}, \pm\frac{2}{2}, \pm\frac{3}{2}, \ldots$.

Miscellaneous Problems

3DT.1. Aliasing can occur in space as well as time. Aliasing in sampled images produces patterns known as moiré patterns. The following MATLAB program provides a dramatic demonstration of moiré patterns.

```
I=zeros(301,301);
f=100;
for n=1:300;
  for m=1:301;
    ang=f*atan2(m-150,n-150);
    I(n,m)=100*sin(ang)+120;
  end
end
  imagesc(I);
  colormap(gray);
```

a) Run this program and describe in your own words what you are see. Refer to the concept of aliasing in your description.

b) Describe in your own words what you would see if the array of samples I were replaced by a continuous space image $I_c(x, y) = 120 + 100 \sin(\theta)$, where $\theta = F * \tan^{-1}(y/x)$?

c) Point out some locations in the image of part (a) where aliasing occurs. (In these locations, the image of part (a) appears to have low spatial frequencies, whereas the image function $I_c(x, y)$ of part (c) has high spatial frequencies.)

3DT.2. Refer to miscellaneous Problem 2DT.1. The two-dimensional frequency response characteristic of an image processor having discrete-space point spread $h[n, m]$ is given by the two-dimensional discrete-space Fourier transform

$$H(e^{j2\pi f_a n}, e^{j2\pi f_b m}) = \sum_{n=-\infty}^{\infty} \sum_{m=-\infty}^{\infty} h[n, m] e^{-j2\pi f_a n} e^{-j2\pi f_b m} \tag{95}$$

a) Find $H(e^{j2\pi f_a n}, e^{j2\pi f_b m})$ for

$$h[n, m] = \left(\frac{1}{2K+1}\right)^2 \sqcap(n; K) \sqcap(m; K) \qquad (96)$$

b) Show that in general if $h[n, m]$ is *separable* $h[n, m] = h_1[n]h_2[m]$, then $H(F_x, F_y)$ is also separable $H(e^{j2\pi f_a n}, e^{j2\pi f_b m}) = H_1(e^{j2\pi f_a n})H_2(e^{j2\pi f_b m})$.

Appendix 3DT.A Frequency Response Characteristic of an Accumulator

The frequency response characteristic of an accumulator is given by the discrete-time Fourier transform of $h[n] = u[n]$. Our task, therefore, is to find the Fourier transform $U(e^{j2\pi f})$ of $u[n]$. The direct substitution of $u[n]$ into the DTFT sum yields an indeterminate result. We therefore express $u[n]$ as a limit

$$u(n) = \lim_{a \to 1} h_a[n] \qquad (A1)$$

where

$$h_a[n] \overset{\Delta}{=} a^n u[n] \qquad (A2)$$

We define

$$U(e^{j2\pi f}) = \lim_{a \to 1} H_a(e^{j2\pi f}) \qquad (A3)$$

where $H_a(e^{j2\pi f})$ is the DTFT of $h_a[n]$.

The following DTFT of $a^n u[n]$ can be found easily, either directly from Eq. (12), or by scaling the result of Example 3DT.2.

$$H_a(e^{j2\pi f}) = \frac{1}{1 - ae^{-j2\pi f}} \qquad (A4)$$

where $|a| < 1$. To take the limit $a \to 1$, we write $H(e^{j2\pi f})$ as

$$H_a(e^{j2\pi f}) = H_{ar}(f) + jH_{ai}(f) \qquad (A5)$$

where $H_{ar}(f)$ and $H_{ai}(f)$ respectively denote the real and the imaginary parts of $H_a(e^{j2\pi f})$. A little algebra yields

$$H_{ar}(f) = \frac{1 - a\cos 2\pi f}{1 - 2a\cos 2\pi f + a^2} \qquad (A6)$$

and

$$H_{ai}(f) = -\frac{a\sin 2\pi f}{1 - 2a\cos 2\pi f + a^2} \qquad (A7)$$

$H_{ar}(f)$ and $H_{ai}(f)$ are plotted in Figure 3DT.21 for $a = 0.7$.

Both functions have period 1. The area under one period of $H_{ar}(f)$ is unity. We can establish this result as follows:

$$\int_{-\frac{1}{2}}^{1/2} H_{ar}(f)df = \int_{-1/2}^{1/2} \mathcal{R}\{H_a(f)\}df = \mathcal{R}\left\{\int_{-1/2}^{1/2} H_a(f)df\right\}$$

$$= \mathcal{R}\left\{\int_{-1/2}^{1/2} H_a(f)e^{j2\pi f n}df\Big|_{n=0}\right\} = \mathcal{R}\{h_a[0]\} = 1 \qquad (A8)$$

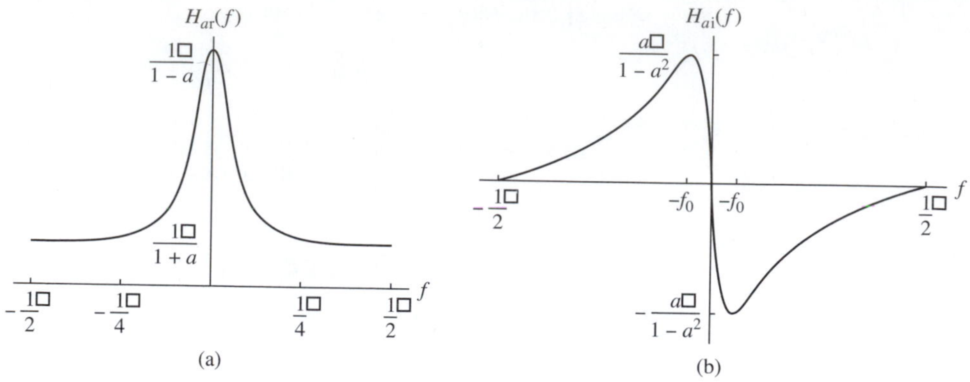

Figure 3DT.21
$H_{ar}(f)$ and $H_{ai}(f)$ for $a = 0.7$: (a) $H_{ar}(f)$; (b) $H_{ai}(f)$.

where $\mathcal{R}$ is the real part operator. As a increases toward unity, $H_{ar}(2\pi f)$ retains the area of 1. The part of $H_{ar}(2\pi f)$ near the origin approaches an impulse in the real variable f having area $\frac{1}{2}$ and the part away from the origin approaches a constant $\frac{1}{2}$.

$$\lim_{a\to 1} H_{ar}(f) = \tfrac{1}{2}\delta(f) + \tfrac{1}{2} \tag{A9}$$

for $-0.5 \le f < 0.5$ and where $\delta(f)$ is a unit impulse in the real variable f. The peak values of $H_{ai}(f)$ occur at $\pm f_o$, where $f_o = 1/2\pi \cos^{-1}(2a/(1+a^2))$. The limit of $H_{ai}(f)$ as a increases towards unity follows directly from Eq. (A7)

$$\lim_{a\to 1} H_{ai}(f) = \frac{\sin 2\pi f}{2(1-\cos 2\pi f)} \tag{A10}$$

In conclusion, the frequency response characteristic of an accumulator is

$$U(e^{j2\pi f}) = \lim_{a\to 1} H_a(e^{j2\pi f}) = \tfrac{1}{2}\delta(f) + \tfrac{1}{2} - j\frac{\sin 2\pi f}{2(1-\cos 2\pi f)} \tag{A11}$$

or $-0.5 \le f < 0.5$. The above expression simplifies to

$$U(e^{j2\pi f}) = \tfrac{1}{2}\delta(f) + \frac{1}{1-e^{-j2\pi f}} \tag{A12}$$

Spectrums

4CT.1 Introduction

Signals, like LTI systems, can be described in either the time domain or the frequency domain. The frequency domain representation of a waveform is called its *spectrum*. We obtain a signal's spectrum by Fourier transforming the signal. We recover a signal from its spectrum by inverse Fourier transforming the spectrum.

Relationship Between a Signal and Its Spectrum

Spectrum:

$$X(F) = \int_{-\infty}^{\infty} x(t)e^{-j2\pi Ft}dt \tag{1}$$

Signal:

$$x(t) = \int_{-\infty}^{\infty} X(F)e^{j2\pi Ft}dF \tag{2}$$

$X(F)$ is sometimes called a *density spectrum* because if $x(t)$ has the unit volt (V), then $X(F)$ has the unit volts per hertz (V/Hz). Every waveform that can be completely drawn on paper or displayed on a monitor has a spectrum.

Spectrums play a central and simplifying role in the theory of signals and systems. When a waveform $x(t)$ is applied to an LTI system, the spectrum of the output waveform $y(t)$ is given by the product $Y(F) = X(F)H(F)$, where $H(F)$ is the system's frequency response characteristic. The relationship between $Y(F)$ and $X(F)$ is central to signal processing. It provides a new and powerful way to find the output of an LTI system. As we shall see, we can often find $y(t)$ by an inspection of $X(F)H(F)$.

4CT.2 Spectrums and LTI Systems

The concept of a spectrum follows directly from material we have presented so far. This section presents two derivations of Eq. (1). Each derivation discloses the property $Y(F) = H(F)X(F)$ that is central to LTI signal processing.

In the first derivation we simply substitute

$$h(t) = \int_{-\infty}^{\infty} H(F)e^{j2\pi Ft}dF \tag{3}$$

from Chapter 3CT into

$$y(t) = \int_{-\infty}^{\infty} h(t-\sigma)x(\sigma)d\sigma. \tag{4}$$

from Chapter 2CT. The substitution gives us

$$y(t) = \int_{-\infty}^{\infty} \left\{ \int_{-\infty}^{\infty} H(F)e^{j2\pi F(t-\sigma)}dF \right\} x(\sigma)d\sigma$$

$$= \int_{-\infty}^{\infty} H(F) \left\{ \int_{-\infty}^{\infty} x(\sigma)e^{-j2\pi F\sigma}d\sigma \right\} e^{j2\pi Ft}dF \tag{5}$$

The term in braces is the spectrum of $x(t)$:

$$X(F) = \int_{-\infty}^{\infty} x(t)e^{-j2\pi Ft}dt = \mathcal{FT}\{x(t)\} \tag{6}$$

When we substitute Eqs. (6) into (5), we obtain the important result:

$$y(t) = \int_{-\infty}^{\infty} X(F)H(F)e^{j2\pi Ft}dF \tag{7}$$

The integral on the right-hand side of Eq. (7) is an inverse Fourier transform. This means that $X(F)H(F)$ is the Fourier transform of $y(t)$.

$$Y(F) = X(F)H(F) \tag{8}$$

Thus, the output spectrum equals the product of the input spectrum and the system's frequency response characteristic. We can obtain $y(t)$ by inverse Fourier transforming $Y(F)$, as in Eq. (7). The time and frequency domain input-output relationships (4) and (8) are summarized in the following two flow graphs.

Time domain:

$$x(t) \xrightarrow{h(t)} y(t) = x(t) * h(t) \tag{9a}$$

Frequency domain:

$$X(F) \xrightarrow{H(F)} Y(F) = X(F)H(F) \tag{9b}$$

Figure 4CT.1 shows the relationships between waveforms and spectrums given by Eqs. (6)–(8) in pictorial form. In the first stage of processing, the Fourier transforms the waveform $x(t)$ into its spectrum $X(F)$, as in Eq. (6). The second stage consists of the multiplication $X(F)H(F)$, indicated by Eq. (8). In the final stage, $y(t)$ is obtained by inverse Fourier transforming the product $Y(F) = X(F)H(F)$, as in Eq. (7). The three-step processing shown is equivalent to a convolution $y(t) = x(t)*h(t)$. This block diagram in Figure 4CT.1 cannot be realized in hardware without modification. However, it provides motivation and a starting point for the practical signal processing techniques developed in Chapter 5DT.

The second derivation shows that the filtering concepts connected with multiple sinusoids apply directly to filtering any Fourier transformable waveform $x(t)$. The derivation

Figure 4CT.1
Relationships
between waveforms
and spectrums: a
pictorial summary of
Eqs. (6)–(8).

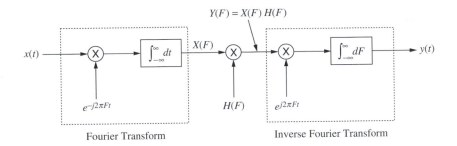

starts with Eq. (3CT.57).

$$\underbrace{\sum_{i=-I}^{I} X_i e^{j2\pi F_i t}}_{x(t)} \xrightarrow{h(t)} \underbrace{\sum_{i=-I}^{I} X_i H(F_i) e^{j2\pi F_i t}}_{y(t)} \qquad (3\text{CT}.57)$$

Recall that the preceding I-O relation is a special case of the superposition principle. It states that the response of an LTI system to a sum of sinusoids is the sum of the responses to the individual sinusoids. Recall also that X_i tells us the amplitude and phase of the input sinusoid $X_i e^{j2\pi F_i t}$ and $X_i H(F_i)$ tells us the amplitude and phase of the corresponding output sinusoid $X_i H(F_i) e^{j2\pi F_i t}$. If we set $F_i = i\Delta F$ and $X_i = X(i\Delta F)\Delta F$ in Eq. (3CT.57) and take the limit $\Delta F \to 0, i\Delta F \to F$ and $I\Delta F \to \infty$, then Eq. (3CT.57) becomes

$$\underbrace{\int_{-\infty}^{\infty} X(F)e^{j2\pi F t} dF}_{x(t)} \xrightarrow{h(t)} \underbrace{\int_{-\infty}^{\infty} H(F)X(F)e^{j2\pi F t} dF}_{y(t)} \qquad (10)$$

The inverse Fourier transforms in Eq. (10) are limiting forms of Eq. (3CT.57). We can interpret the integral on the left-hand side of Eq. (10) as a sum of infinitesimal sinusoids, each having the form $X(F)e^{j2\pi F t} dF$. The input spectrum $X(F)$ describes the amplitudes and phases of these sinusoidal components. Similarly, we can interpret the right-hand side of Eq. (10) as a sum of infinitesimal sinusoids, each having the form $H(F)X(F)e^{j2\pi F t} dF$. The output spectrum is $Y(F) = H(F)X(F)$. The term $H(F)$ in Eq. (10) acts the same way as the term $H(F_i)$ in Eq. (3CT.57) to filter the sinusoidal components in the input.

EXAMPLE 4CT.1   Distinguishing Added Signals

Figure 4CT.2(a) shows a signal $x_1(t)$ and its spectrum $X_1(F)$. By looking at the spectrum, we see that $x_1(t)$ is composed of a band of complex sinusoidal components centered at ± 11 Hz. Similarly, Figure 4CT.2(b) shows a signal, $x_2(t)$, and its spectrum $X_2(F)$. $x_2(t)$ is composed of complex sinusoidal components centered at ± 19 Hz. Figure 4CT.2(c) shows the sum $x(t) = x_1(t) + x_2(t)$ and its spectrum $X(F)$. The spectrum $X(F)$ equals $X_1(F) + X_2(F)$ because the sinusoidal components in $x(t)$ are the sum of the sinusoidal components in $x_1(t)$ and $x_2(t)$. Notice that $X_1(F)$ and $X_2(F)$ are clearly distinguishable by inspection of $X(F)$, even though $x_1(t)$ and $x_2(t)$ are not clearly distinguishable by inspection of $x(t)$. This figure shows us that it can be much easier to understand a waveform by looking at its spectrum than by looking at the waveform itself. This understanding is made possible by the Fourier transform's ability to distinguish sinusoidal components having different frequencies.

Suppose you wanted to process $x(t)$ in such a way as to recover the signals $x_1(t)$ and $x_2(t)$. It would be difficult to see how to accomplish this separation if you consider only the plot of $x(t)$. However, when you look at the spectrum

Figure 4CT.2
Two signals and their
spectrums: (a) $x_1(t)$
and its spectrum;
(b) $x_2(t)$ and its
spectrum;
(c) $x(t) = x_1(t) + x_2(t)$
and its spectrum.

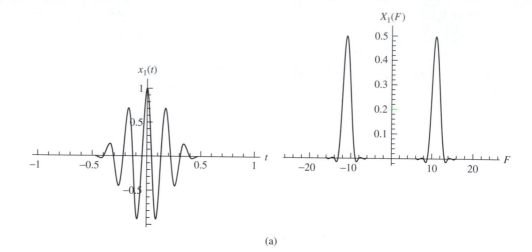

(a)

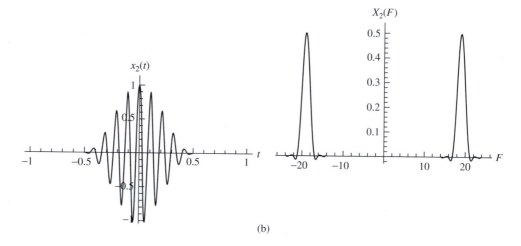

(b)

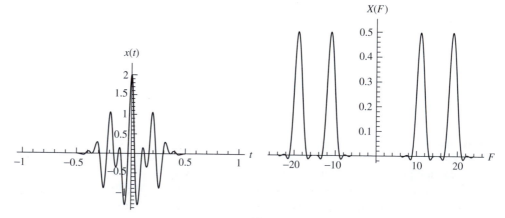

(c)

Figure 4CT.3
Use of filters to
separate added
signals: (a) frequency
domain; (b) time
domain.

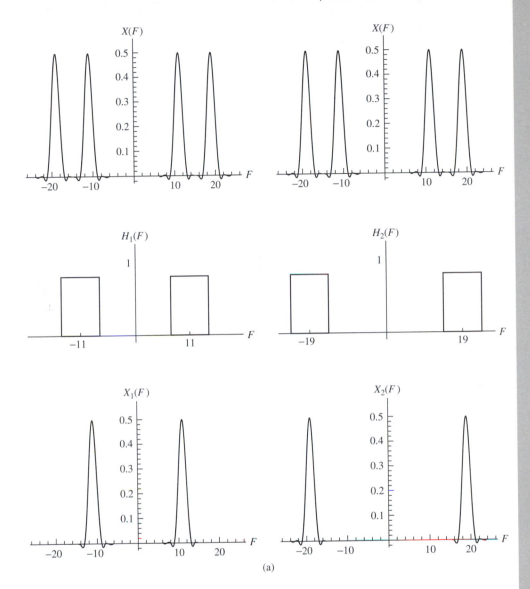

(a)

$X(F)$, you see that $X_1(F)$ and $X_2(F)$ can be separated by multiplying $X(F)$ by $H_1(F)$ and $H_2(F)$, as shown in Figure 4CT.3(a): $X(F)H_1(F) = X_1(F)$ and $X(F)H_2(F) = X_2(F)$. Complete separation is possible because the frequency bands occupied by $X_1(F)$ and $X_2(F)$ do not overlap. Each filter passes only the frequency components belonging to $x_1(t)$ or $x_2(t)$. The time domain counter part of the system in Figure 4CT.3(a) is shown in Figure 4CT.3(b).

The use of filters to separate waveforms is analogous to the use of red and blue transparencies to separate red and blue light. A red transparency permits the passage of red light, while the blue transparency permits the passage of blue light.

This example has demonstrated the analytical power of frequency domain analysis as well as its conceptual power. Think how difficult it would be, in comparison, to evaluate the convolutions $x(t) * h_1(t)$ and $x(t) * h_2(t)$!

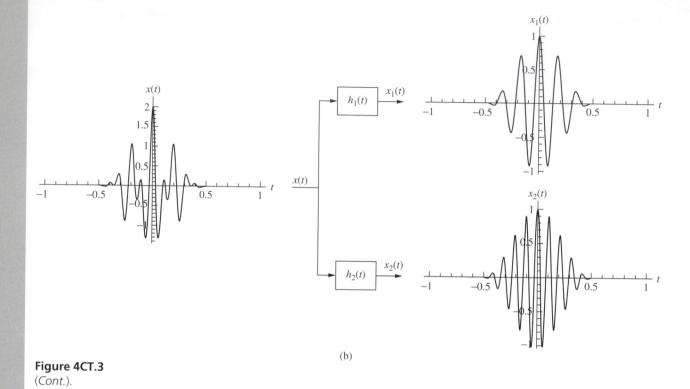

(b)

Figure 4CT.3
(*Cont.*).

4CT.3   Amplitude and Phase Spectrums

The *amplitude* and *phase spectrums* of $x(t)$ are defined as $|X(F)|$ and $\angle X(F)$, respectively. Similarly, the amplitude and phase spectrums of $y(t)$ are $|Y(F)| = |H(F)X(F)|$ and $\angle Y(F) = \angle\{H(F)X(F)\}$, respectively. By the properties of complex numbers, $|H(F)X(F)| = |H(F)||X(F)|$. Therefore, the amplitude spectrum of $y(t)$ is given by the product of the amplitude spectrum of $x(t)$ and the amplitude response characteristic of the system:

$$|Y(F)| = |H(F)||X(F)| \tag{11}$$

Similarly, by the properties of complex numbers, $\angle\{H(F)X(F)\} = \angle H(F) + \angle X(F)$. Therefore, the phase spectrum of $y(t)$ is given by the sum of the phase spectrum of $x(t)$ and the phase response characteristic of the system:

$$\angle Y(F) = \angle H(F) + \angle X(F) \tag{12}$$

4CT.4   Some Waveforms and Their Spectrums

We saw examples of Fourier transform pairs in Chapter 3CT associated with the impulse responses and frequency response characteristics of LTI systems. These FT pairs, summarized in Tables 3CT.1 and 3CT.2, apply equally to signals and their spectrums if we substitute $x(t)$ for $h(t)$ and $X(F)$ for $H(F)$. Let's now develop some important new transform pairs.

EXAMPLE 4CT.2   Rectangular Pulse

The rectangular pulse introduced in Chapter 2

$$x(t) = \sqcap\left(\frac{t}{\tau}\right) \tag{13}$$

is widely used in engineering. Its spectrum can be obtained easily. The substitution of Eq. (13) into (1) yields

$$X(F) = \mathcal{F}\left\{\sqcap\left(\frac{t}{\tau}\right)\right\} = \int_{-\infty}^{\infty} \sqcap\left(\frac{t}{\tau}\right) e^{-j2\pi Ft}\,dt = \int_{-\tau/2}^{\tau/2} e^{-j2\pi Ft}\,dt$$

$$= \left.\frac{-1}{2\pi F}e^{-j2\pi Ft}\right|_{-\tau/2}^{\tau/2} = \frac{\sin(\pi F\tau)}{\pi F} = \tau\,\mathrm{sinc}(F\tau) \tag{14}$$

The resulting Fourier transform pair is

$$\sqcap\left(\frac{t}{\tau}\right) \xleftrightarrow{\text{FT}} \tau\,\mathrm{sinc}(F\tau) \tag{15}$$

The rectangular pulse and its spectrum are plotted in Figure 4CT.4.

Notice that the width of the major lobe of the spectrum is $2/\tau$. The *bandwidth* of the rectangular pulse is defined as

$$W_{\sqcap} = \frac{k}{\tau} \tag{16}$$

where k is a constant. Typical values of k are in the range $1 \le k \le 5$.

Let's explore the insight provided by the equation $Y(F) = X(F)H(F)$ when a rectangular pulse is applied to an arbitrary low-pass filter, as illustrated in Figure 4CT.5. For simplicity, the figure assumes that the filter's frequency response characteristic $H(F)$ is real. An inspection of the figure reveals that $Y(F)$ becomes a closer approximation to $X(F)$ as the filter bandwidth W is increased. Consequently, $y(t)$ becomes a closer approximation to the input $x(t)$ as W is increased.

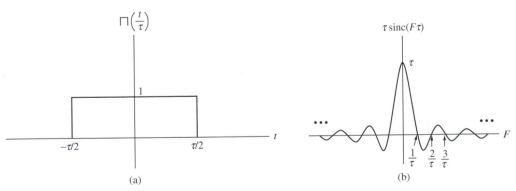

Figure 4CT.4
Rectangular pulse and its spectrum: (a) $\sqcap(t/\tau)$; (b) $\tau\,\mathrm{sinc}(F\tau)$

Figure 4CT.5
Frequency domain
view of rectangular
pulse applied to a
low-pass filter:
(a) $W = 1/\tau$;
(b) $W = 5/\tau$.

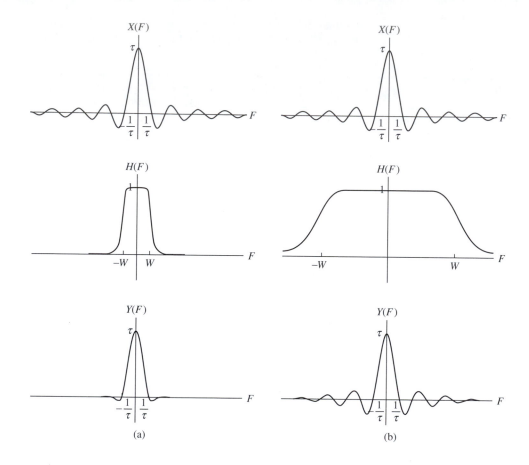

(a)   (b)

Figure 4CT.5 provides an explanation for the pulse response of the RC filter we saw in Figure 2CT.29, where the output of an RC low-pass filter became a better approximation to an applied rectangular pulse as $\tau_c = RC$ was decreased. The frequency response characteristic of this system is shown in Figure 3CT.4. We see from this figure that a decrease in τ_c increases the filter's bandwidth.

EXAMPLE 4CT.3   Unit Impulse

The unit impulse is another important signal whose spectrum can be obtained easily by direct evaluation of the Fourier transform integral. The substitution of $\delta(t)$ into Eq. (1) yields

$$X(F) = \mathcal{F}\{\delta(t)\} = \int_{-\infty}^{\infty} \delta(t)e^{-j2\pi Ft}dt = 1 \tag{17}$$

where the last step follows from the sifting property of the impulse. Therefore

$$\delta(t) \xleftrightarrow{\text{FT}} 1 \tag{18}$$

The unit impulse and its spectrum are shown in Figure 4CT.6. Notice that a unit impulse has infinite bandwidth. Its spectrum is flat.

Figure 4CT.6
Unit impulse and its
spectrum.

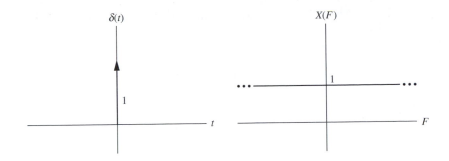

We know from Chapter 2CT that the response of any LTI system to a unit impulse is $h(t)$. Let's derive this result using the result of the preceding example. The substitution of $X(F) = 1$ into Eq. (8) yields

$$Y(F) = 1H(F) = H(F) \tag{19}$$

Inverse Fourier transformation yields

$$y(t) = \delta(t) * h(t) = h(t) \quad \checkmark \tag{20}$$

EXAMPLE 4CT.4   DC Waveform

Sometimes we need singularity functions to describe the spectrum of a waveform. An example is given by the DC waveform $x(t) = 1$. If we substitute $x(t) = 1$ into Eq. (1), we find that the integral does not converge

$$X(F) = \mathcal{F}\{1\} = \int_{-\infty}^{\infty} e^{-j2\pi Ft}\,dt = \frac{-1}{2\pi F}e^{-j2\pi Ft}\Big|_{-\infty}^{\infty} = ? \tag{21}$$

To obtain $X(F)$, remember that the notation $\int_{-\infty}^{\infty}$ means a limit:

$$X(F) = \mathcal{F}\{1\} = \lim_{T\to\infty} \int_{-T/2}^{T/2} e^{-j2\pi Ft}\,dt = \lim_{T\to\infty} \frac{-1}{2\pi F}e^{-j2\pi Ft}\Big|_{-T/2}^{T/2} \tag{22}$$

which becomes, as in Eq. (14)

$$X(F) = \lim_{T\to\infty} T \operatorname{sinc}(FT) = \delta(F) \tag{23}$$

Here, $X(0)$ does not exist as a number, but we can still represent $X(F)$ with an impulse.

We can regard the DC waveform, $x(t) = 1$, as a limiting case of a rectangular pulse $\sqcap(t/\tau)$, where $\tau\to\infty$. In this limit, the rectangular pulse becomes a constant, 1, and its spectrum becomes an impulse as in Eq. (23). The resulting DTFT pair is

$$1 \overset{\text{FT}}{\longleftrightarrow} \delta(F) \tag{24}$$

You can confirm this result by evaluating the inverse Fourier transform $\mathcal{F}^{-1}\{\delta(F)\}$, using the sifting property of the impulse. The DC waveform and its spectrum are shown in Figure 4CT.7. Notice that a DC waveform has infinite "pulse width" and zero bandwidth. The only sinusoidal component present in a DC waveform is the one having frequency 0, i.e., the DC component at $F = 0$.

Figure 4CT.7
DC waveform and its spectrum.

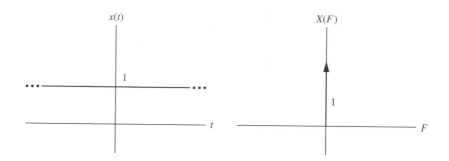

Let's use the result of the preceding example to find the response of an LTI system to the DC waveform $x(t) = 1$. The substitution of $X(F) = \delta(F)$ into Eq. (8) yields:

$$Y(F) = \delta(F)H(F) = H(0)\delta(F) \tag{25}$$

Inverse Fourier transformation yields

$$y(t) = H(0) \tag{26}$$

which is a constant. We see that if we apply the DC waveform $x(t) = 1$ to an LTI system, the response is the DC waveform $y(t) = H(0)$. Consequently, $H(0)$ is sometimes referred to as the system's *DC gain*.

EXAMPLE 4CT.5   Raised Cosine

The raised cosine pulse is used in data transmission systems and in signal and system design. It is shown in Figure 4CT.8(a), and defined by the formula

$$x(t) = \begin{cases} \dfrac{A}{2}\left\{1 + \cos\left(\dfrac{2\pi t}{\tau}\right)\right\} & \text{for } |t| \le 0.5\tau \\ 0 & \text{for } |t| > 0.5\tau \end{cases}$$

$$= \frac{A}{2}\left\{1 + \cos\left(\frac{2\pi t}{\tau}\right)\right\} \sqcap\left(\frac{t}{\tau}\right) \tag{27}$$

It takes considerable labor to find the spectrum by direct evaluation of the Fourier transform integral.

$$X(F) = \int_{-\tau/2}^{\tau/2} \frac{A}{2}\left\{1 + \cos\left(\frac{2\pi t}{\tau}\right)\right\} e^{-j2\pi Ft} dF$$

$$= \int_{-\tau/2}^{\tau/2} \frac{A}{2} e^{-j2\pi Ft} dF + \frac{A}{4}\int_{-\tau/2}^{\tau/2} e^{-j2\pi(F-\frac{1}{\tau})t} dF + \frac{A}{4}\int_{-\tau/2}^{\tau/2} e^{-j2\pi(F+\frac{1}{\tau})t} dF \tag{28}$$

The integrals work out easily, but many algebraic steps are required to reach the final result

$$X(F) = \frac{A\tau}{2} \frac{\text{sinc}(F\tau)}{(1 - (F\tau)^2)} \tag{29}$$

We describe a faster way to derive (29) in Section 4CT.7. The Fourier transform pair is

$$\frac{A}{2}\left\{1 + \cos\left(2\pi\frac{t}{\tau}\right)\right\} \sqcap \left(\frac{t}{\tau}\right) \longleftrightarrow \frac{A\tau}{2} \frac{\text{sinc}(F\tau)}{(1 - (F\tau)^2)} \tag{30}$$

The spectrum is plotted in Figure 4CT.8(b). The main lobe lies in the range $|F| \leq 2/\tau$. Although this region is twice that for a rectangular pulse of Example 4CT.2, the factor $(1 - (F\tau)^2)$ causes the amplitude spectrum to decrease much more rapidly, as $|F|$ increases beyond $2/\tau$ for the raised cosine, than the rectangular pulse. This rapid decrease in $|X(F)|$ for the raised cosine can be attributed its smoothness compared to the sharp discontinuities in the rectangular pulse.

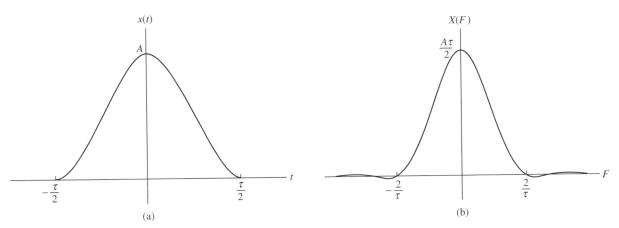

Figure 4CT.8
Raised cosine and its spectrum: (a) $(A/2)\{1 + \cos(2\pi t/\tau)\} \sqcap (t/\tau)$; (b) $(A\tau/2)[\text{sinc}(F\tau)/(1 - (F\tau)^2)]$.

EXAMPLE 4CT.6   Periodic Impulse Train

The *periodic impulse train*

$$\delta_T(t) = \sum_{k=-\infty}^{\infty} \delta(t - kT) \tag{31}$$

is used extensively in the study of sampled signals. We can find its spectrum by substituting Eq. (31) into the Fourier integral (1), taking care to evaluate the sum as a limit

$$\mathcal{F}\{\delta_T(t)\} = \int_{-\infty}^{\infty} \left\{\lim_{K\to\infty} \sum_{k=-K}^{K} \delta(t - kT)\right\} e^{-j2\pi Ft} dt$$

$$= \lim_{K\to\infty} \sum_{k=-K}^{K} \int_{-\infty}^{\infty} \delta(t - kT)e^{-j2\pi Ft} dt = \lim_{K\to\infty} \sum_{k=-K}^{K} e^{-j2\pi FkT} \tag{32}$$

The right-hand side of Eq. (32) can be written in closed form (identity (B.3) of Appendix B):

$$\sum_{k=-K}^{K} e^{-j2\pi FkT} = (2K+1)\,\text{cinc}(FT; K) \tag{33}$$

where we have defined the *circular sync* function

$$\text{cinc}(\lambda; \beta) = \frac{\sin((2\beta+1)\pi\lambda)}{(2\beta+1)\sin(\pi\lambda)} \tag{34}$$

$\text{cinc}(\lambda; K)$ is periodic with period 1 for $K = 1, 2, 3, \ldots$. A plot of $\text{cinc}(\lambda; K)$ is given in Figure 4CT.9 for $K = 3$.

Figure 4CT.9
Circular sync function
cinc $(\lambda; K)$ for $K = 3$.

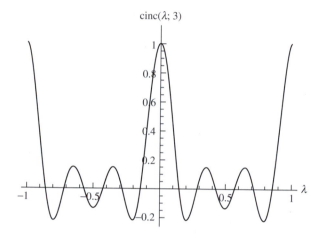

The peak values of $\text{cinc}(\lambda; K)$ equal 1 and the widths of the main lobes are inversely proportional to $2K + 1$. Unlike the sinc function, the circular sinc function is not widely known. In the limit $K \to \infty$

$$\lim_{K\to\infty} (2K+1)\,\text{cinc}(FT; K) = \frac{1}{T}\sum_{k=-\infty}^{\infty} \delta\left(F - \frac{k}{T}\right) \tag{35}$$

The resulting Fourier transform pair is

$$\sum_{k=-\infty}^{\infty} \delta(t - kT) \longleftrightarrow \frac{1}{T}\sum_{k=-\infty}^{\infty} \delta\left(F - \frac{k}{T}\right) \tag{36}$$

The periodic impulse train and its spectrum are plotted in Figure 4CT.10 where $F_s = 1/T$. It is interesting that the spectrum of a periodic impulse train is itself a periodic impulse train.

Another expression for the periodic impulse train can be obtained by substituting $(1/T)\delta_{1/T}(F) = (1/T)\sum_{k=-\infty}^{\infty} \delta(F - (k/T))$ into the inverse Fourier transform integral (2). The following result follows directly from this substitution by the use of the sifting property of the impulse (Problem 4CT.4.3):

$$\delta_T(t) = \sum_{n=-\infty}^{\infty} \delta(t - nT) = \frac{1}{T}\sum_{k=-\infty}^{\infty} e^{j2\pi \frac{k}{T}t} \tag{37}$$

Figure 4CT.10
Periodic impulse train
and its spectrum:
(a) $\sum_{k=-\infty}^{\infty} \delta(t - kT)$;
(b) $\frac{1}{T} \sum_{k=-\infty}^{\infty} \delta(F - k/T)$.
$F_s = 1/T$.

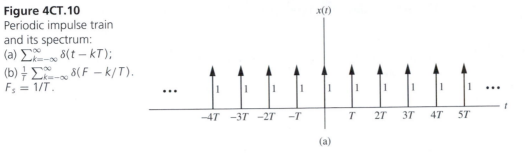

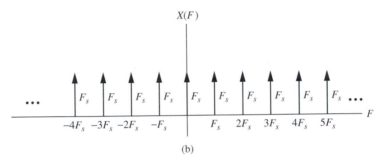

(a)

(b)

4CT.5   Properties of the Fourier Transform

Many properties of signals and systems are described best in terms of the Fourier transform. The properties can be divided into two categories: *symmetry* properties of signals and systems and *operational* properties of systems.

4CT.5.1
Symmetry
Properties

The most important symmetry properties are listed in Table 4CT.1. Properties 4CT.SP1–SP4 are familiar from Chapter 3CT, where they were applied to the frequency response characteristics of real LTI systems. The proofs of the symmetry properties are straightforward (Problem 4CT.5.21).

Examples of property 4CT.SP5 are given by the rectangular pulse, the impulse, the DC waveform, and the raised cosine pulse of Section 4CT.4. Each of these waveforms is real and even, and each has a spectrum that is real and even. The one-sided decaying exponential waveform of Example 3CT.1 is real, but neither even nor odd. It has an even

Table 4CT.1
Symmetry Properties

For Real or Complex $x(t)$	
Conjugation: $x^*(t) \longleftrightarrow X^*(-F)$	(4CT.SP1)
For Real $x(t)$	
$X(-F) = X^*(F)$	(4CT.SP2)
$\lvert X(-F) \rvert = \lvert X(F) \rvert$	(4CT.SP3)
$\angle X(-F) = -\angle X(F)$	(4CT.SP4)
$X(F)$ is real and even iff $x(t)$ is real and even.	(4CT.SP5)
$X(F)$ is imaginary and odd iff $x(t)$ is real and odd.	(4CT.SP6)

amplitude spectrum and an odd phase spectrum (Figure 3CT.4), in agreement with properties (4CT.SP3) and (4CT.SP4).

4CT.5.2 Operational Properties

Operational properties are helpful in understanding how signals interact with systems. They also provide a convenient way to derive new Fourier transform pairs.

Linearity

$$a_1 x_1(t) + a_2 x_2(t) \longleftrightarrow a_1 X_1(F) + a_2 X_2(F) \qquad \text{(4CT.OP1)}$$

The *linearity property* (4CT.OP1) states that the spectrum of the sum of two scaled signals is the sum of the signals' corresponding scaled spectrums. The proof follows from the definition of the Fourier transform.

$$\int_{-\infty}^{\infty} \{a_1 x_1(t) + a_2 x_2(t)\} e^{-j2\pi Ft} dt = a_1 \int_{-\infty}^{\infty} x_1(t) e^{-j2\pi Ft} dt + a_2 \int_{-\infty}^{\infty} x_2(t) e^{-j2\pi Ft} dt \tag{38}$$

which is the linearity property

$$\mathcal{FT}\{a_1 x_1(t) + a_2 x_2(t)\} = a_1 \mathcal{FT}\{x_1(t)\} + a_2 \mathcal{FT}\{x_2(t)\} \tag{39}$$

An application of the linearity property was given in Example 4CT.1, where the spectrum of the signal $x(t) = x_1(t) + x_2(t)$ was seen to be $X(F) = X_1(F) + X_2(F)$.

Duality

$$X(t) \longleftrightarrow x(-F) \qquad \text{(4CT.OP2)}$$

Suppose we have an FT pair $x(t) \longleftrightarrow X(F)$ and define a *waveform* $X(t)$ from spectrum $X(F)$ by substituting t for F. The *duality property* (4CT.OP2) states that the spectrum of $X(t)$ is $x(-F)$. This property can be established by starting with the definition of the Fourier transform of $X(t)$.

$$\mathcal{F}\{X(t)\} = \int_{-\infty}^{\infty} X(t) e^{-j2\pi Ft} dt = \int_{-\infty}^{\infty} X(F') e^{-j2\pi FF'} dF' = x(-F) \tag{40}$$

The second integral follows by changing the variable of integration from t to F'. The second integral is the inverse Fourier transform of $X(F')$, where $t = -F$.

EXAMPLE 4CT.7   Use of Duality to Derive Spectrums

We illustrate the duality property by showing an alternative way to derive the spectrum of a rectangular pulse $A \sqcap (t/\tau)$. We know from Table 3CT.1 that

$$2W \, \text{sinc}(2Wt) \xrightarrow{\text{FT}} \sqcap \left(\frac{F}{2W} \right) \tag{41}$$

The duality property tells us that

$$\mathcal{F}\left\{ \sqcap \left(\frac{t}{2W} \right) \right\} = 2W \, \text{sinc}(2W(-F)) = 2W \, \text{sinc}(2WF) \tag{42}$$

where the last step follows from the fact that $\text{sinc}(\lambda)$ is an even function. We obtain the Fourier transform of $A \sqcap (t/\tau)$ by setting $2W = \tau$ in the preceding equation. The result is

$$\sqcap \left(\frac{t}{\tau}\right) \overset{\text{FT}}{\longleftrightarrow} \tau \, \text{sinc}(\tau F) \tag{43}$$

which agrees with the result obtained in Example 4CT.2.

Convolution

$$x_a(t) * x_b(t) \longleftrightarrow X_a(F)X_b(F) \tag{4CT.OP3}$$

The *convolution property* states that convolution in the time domain corresponds to multiplication in the frequency domain. This property was established in Section 4CT.2. Example 4CT.1 used this property to solve the problem of separating two added signals whose spectrums were disjoint.

EXAMPLE 4CT.8   Spectrum of a Triangular Pulse

We can use the convolution property to derive the spectrum of a triangular pulse $\wedge(t/\tau)$. Equation (2CT.63) tells us that

$$\wedge\left(\frac{t}{\tau}\right) = \frac{\sqcap\left(\frac{t}{\tau}\right) * \sqcap\left(\frac{t}{\tau}\right)}{\tau} \tag{44}$$

It follows from the convolution property and Example 4CT.2 that

$$\mathcal{FT}\left\{\wedge\left(\frac{t}{\tau}\right)\right\} = \tau \, \text{sinc}^2(F\tau) \tag{45}$$

Therefore,

$$\wedge\left(\frac{t}{\tau}\right) \longleftrightarrow \tau \, \text{sinc}^2(F\tau) \tag{46}$$

The triangular pulse and its spectrum are plotted in Figure 4CT.11.

Figure 4CT.11
Triangular pulse and
its spectrum:
(a) $\wedge(t/\tau)$;
(b) $\tau \, \text{sinc}^2(F\tau)$.

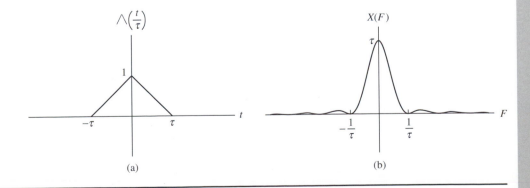

(a)   (b)

Multiplication

$$x_1(t)x_2(t) \longleftrightarrow X_1(F) * X_2(F) \qquad \text{(4CT.OP4)}$$

where

$$X_1(F) * X_2(F) = \int_{-\infty}^{\infty} X_1(\lambda)X_2(F-\lambda)\, d\lambda \qquad (47)$$

The *multiplication property* (4CT.OP4) states that multiplication in the time domain corresponds to convolution in the frequency domain. This property is the dual of the convolution property. We leave the derivation to Problem 4CT.5.22. Here, we consider the special case that $x_2(t) = x_1(t) \equiv x(t)$, for which the multiplication property becomes:

$$x^2(t) \stackrel{\text{FT}}{\longleftrightarrow} X(F) * X(F) \qquad (48)$$

Squaring is a nonlinear operation. We see from Eq. (48) that, when a signal is squared, its spectrum convolves with itself. The following example demonstrates that this self-convolution introduces frequency components into $x^2(t)$ that did not exist in $x(t)$.

EXAMPLE 4CT.9   Spectrum of a Squared Signal

Figure 4CT.12 illustrates the multiplication property for $x_2(t) = x_1(t) \equiv x(t) = \operatorname{sinc}(t)$, for which $X(F) = \sqcap(F)$. Notice that squaring a waveform introduces frequency components that were not present in the original waveform.

Figure 4CT.12
Effect of squaring a signal on its spectrum, illustrated for $x(t) = \operatorname{sinc}(t)$: (a) $x(t)$ and its spectrum; (b) $x^2(t)$ and its spectrum.

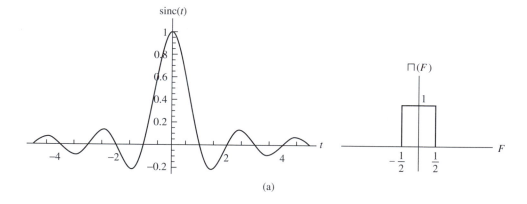

(a)

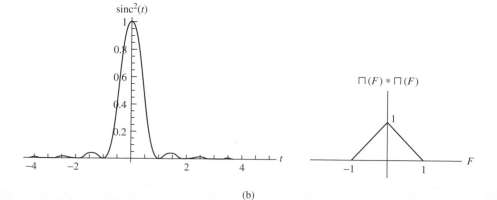

(b)

Time Scale

$$x(at) \longleftrightarrow \frac{1}{|a|} X(F/a) \qquad \text{(4CT.OP5)}$$

The *time scale property* (4CT.OP5) follows by changing the variable of integration from t to $v = at$ in the Fourier transform integral:

$$F\{x(at)\} = \int_{-\infty}^{\infty} x(at)e^{-j2\pi Ft}dt = \frac{1}{|a|}\int_{-\infty}^{\infty} x(v)e^{-j2\pi Fv/a}dv = \frac{1}{|a|}X\left(\frac{F}{a}\right) \qquad (49)$$

The time-scaled property states that if we *compress* a waveform in time ($a > 1$), then we *expand* its spectrum in frequency. Similarly, if we *expand* a waveform in time ($0 < a < 1$), then we *compress* its spectrum in frequency. If we set $a = -1$, we *reverse* the waveform in both time and frequency.

When we make a waveform shorter, we are compressing it in time. According to the time-scaled property, the inverse relationship between pulse width and bandwidth seen in Examples 4CT.2–4CT.5 is a universal rule.

EXAMPLE 4CT.10   ASK Revisited

An important practical implication of the time-scaled property is that there is a limit to the rate at which data can be transmitted through any band-limited channel. Let's reconsider the baseband *amplitude shift keying* (ASK) system of Figure 2CT.30, where rectangular pulses having duration τ were transmitted at rate $R = 1/\tau$ pulses per second. Recall from Eq. (16) that the bandwidth of a rectangular pulse is $W_\sqcap = k/\tau$ where k is a constant, typically in the range $1 \leq k \leq 5$. It follows that the bandwidth of a transmitted pulse is proportional to R.

$$W_\sqcap = kR \qquad (50)$$

If $W_\sqcap$ exceeds the bandwidth of the communications channel, then a portion of the pulse's spectrum is rejected by the channel. This rejection causes distortion, resulting in intersymbol interference in the received pulses, as illustrated in Figure 2CT.31. The time-scaled property implies that intersymbol interference occurs in every low-pass channel, no matter what shape the transmitted pulse has, if R is made too large.

Time Shift

$$x(t - \tau) \longleftrightarrow e^{-j2\pi\tau F} X(F) \qquad \text{(4CT.OP6)}$$

The time-shift property (4CT.OP6), states that a time shift by τ corresponds to a multiplication by $e^{-j2\pi\tau F}$ in the frequency domain. This property can be derived by recognizing that a time shift of τ occurs when $x(t)$ is applied to an ideal time-shift element. We know from Table 3CT.2 that the frequency response characteristic of a time-shift element is $H(F) = e^{-j2\pi\tau}$. It follows from the equation $Y(F) = H(F)X(F)$ that a time-shifted waveform $y(t) = x(t - \tau)$ has a spectrum $Y(F) = e^{-j2\pi\tau F}X(F)$.

Notice that the time shift simply adds a linear phase term to the phase spectrum of a signal: $\angle(e^{-j2\pi\tau F}X(F)) = \angle X(F) - 2\pi\tau F$. The time shift does not affect the amplitude spectrum: $|e^{-j2\pi\tau F}X(F)| = |X(F)|$. It is a *distortionless* operation. A delayed version $x(t - \tau)$ of an audio signal $x(t)$ sounds the same as the original.

Frequency Translation

$$x(t)e^{j2\pi F_o t} \longleftrightarrow X(F - F_o) \tag{4CT.OP7}$$

The frequency translation property (4CT.OP7) is the dual of property (4CT.OP6). (4CT.OP7) can be derived by evaluating the Fourier transform of $x(t)e^{j2\pi F_o t}$.

$$\mathcal{F}\{x(t)e^{j2\pi F_o t}\} = \int_{-\infty}^{\infty} x(t)e^{j2\pi F_o t}e^{-j2\pi F t} dt = \int_{-\infty}^{\infty} x(t)e^{-j2\pi(F-F_o)t} dt = X(F - F_o) \tag{51}$$

EXAMPLE 4CT.11   Rotating Phasors and Multiple Sinusoidal Component Waveforms

We can use the frequency translation property to find the spectrum of the rotating phasor (complex sinusoid) $Xe^{j2\pi F_1 t}$. Recall from Example 4CT.3 that $x(t) = 1 \longleftrightarrow X(F) = \delta(F)$. Therefore, by setting $x(t) = 1$ in (4CT.OP7), we obtain $e^{j2\pi F_1 t} \longleftrightarrow \delta(F - F_1)$. We can use this result and the linearity property of the Fourier transform to obtain

$$Xe^{j2\pi F_1 t} \longleftrightarrow X\delta(F - F_1) \tag{52}$$

Therefore, the spectrum (Fourier transform) of a complex sinusoid having complex amplitude X and frequency F_1 is an impulse located on the frequency axis at the frequency F_1. The area of the impulse is the complex amplitude X.

Real sinusoids are related to complex sinusoids by Euler's formula.

$$A\cos(2\pi F_1 t + \phi) = Xe^{j2\pi F_1 t} + X^*e^{-j2\pi F_1 t} \tag{53}$$

where $X = \frac{1}{2}A\angle\phi$. It then follows from Eq. (52) and superposition that

$$A\cos(2\pi F_1 t + \phi) \longleftrightarrow X\delta(F - F_1) + X^*\delta(F + F_1) \tag{54}$$

For the multiple sinusoidal component waveform of (CT3.51), we have

$$x(t) = A_0 + \sum_{i=1}^{I} A_i \cos(2\pi F_i t + \phi_i) = \sum_{i=-I}^{I} X_i e^{j2\pi F_i t} \tag{55}$$

where

$$X_i = \begin{cases} A_0 & i = 0 \\ \frac{1}{2}A_i e^{j\phi_i} & i = 1, 2, \ldots, I \\ X_{-i}^* & i = -1, -2, \ldots, -I \end{cases} \tag{56}$$

and

$$F_i = \begin{cases} 0 & i = 0 \\ -F_{-i} & i = -1, -2, \ldots, -I \end{cases} \tag{57}$$

and it follows from Eq. (52) and superposition that

$$A_0 + \sum_{i=1}^{I} A_i \cos(2\pi F_i t + \phi_i) = \sum_{i=-I}^{I} X_i e^{j2\pi F_i t} \longleftrightarrow \sum_{i=-I}^{I} X_i \delta(F - F_i) \tag{58}$$

Therefore the spectrum of a multiple sinusoidal component waveform is composed of impulses located at the frequencies of its complex sinusoidal components. The areas of the impulses are the complex amplitudes of the complex sinusoidal components.

Modulation

$$x(t)\cos(2\pi F_o t) \longleftrightarrow \tfrac{1}{2}X(F - F_o) + \tfrac{1}{2}X(F + F_o) \qquad \text{(4CT.OP8)}$$

The modulation property states that if you *modulate* $\cos(2\pi F_o t)$ with $x(t)$, then $X(F)$ is divided by 2 and translated by $\pm F_o$. The property can be derived from the frequency translation property and superposition. The frequency translation property tells us that

$$x(t)e^{j2\pi F_o t} \longleftrightarrow X(F - F_o) \qquad (59)$$

If we replace F_o with $-F_o$, we obtain

$$x(t)e^{-j2\pi F_o t} \longleftrightarrow X(F + F_o) \qquad (60)$$

Superposition yields

$$\tfrac{1}{2}x(t)e^{j2\pi F_o t} + \tfrac{1}{2}x(t)e^{-j2\pi F_o t} \longleftrightarrow \tfrac{1}{2}X(F - F_o) + \tfrac{1}{2}X(F + F_o) \qquad (61)$$

which simplifies to the modulation property, as stated.

EXAMPLE 4CT.12   DSB Modulation

An example of the modulation property is provided by double sideband (DSB) modulation, used in communication systems. A typical transmitted DSB waveform has the form

$$s(t) = Ax(t)\cos(2\pi F_o t) \qquad (62)$$

where A is a constant, and $x(t)$ is a baseband waveform that modulates $\cos(2\pi F_o t)$. A baseband waveform is a waveform that can be transmitted through a low-pass communications channel. Modulation enables the transmitted waveform $s(t)$ to be transmitted through a band-pass communications channel (one having center frequency $F_o \gg W$, where W is the bandwidth of $x(t)$).

In digital communications, $x(t)$ is a pulse or a sequence of pulses. In analog communications, $x(t)$ is an analog waveform, such as voice or video.

Figure 4CT.13 illustrates $s(t)$ and $S(F)$ for a single rectangular pulse $x(t) = \sqcap(t/\tau)$. This waveform is referred to as a "chip" in data communications. The spectrum of $x(t)$ is shown in Figure 4CT.4. Notice that the spectrum of $s(t)$, $S(F)$, is concentrated about $\pm F_o$, We also see from the figure that the bandwidth of $s(t)$ is twice that of $x(t)$ given by Eq. (16): $B = 2W_\sqcap$. $S(F)$ is called a *double sideband* spectrum because it contains bands of sinusoidal components located on both sides of the center frequency F_o.

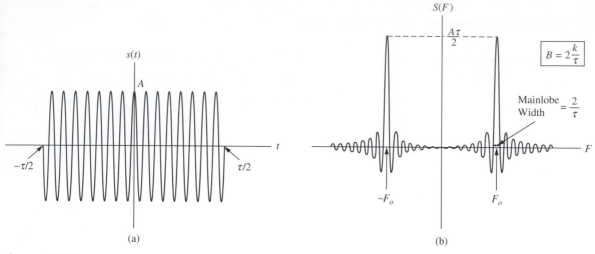

Figure 4CT.13
Chip and its spectrum.

EXAMPLE 4CT.13 Impulse Response of Ideal Band-Pass Filter

The modulation property provides a quick way to obtain the impulse response of an ideal band-pass filter. The band-pass filter frequency response characteristic is given by $H(F) = \sqcap((F - F_o)/2W) + \sqcap((F + F_o)/2W)$. Notice that $H(F)$ can be expressed as the right-hand side of (4CT.OP8) if we set $X(F) = 2\sqcap(F/2W)$. The waveform corresponding to $2\sqcap(F/2W)$ is $x(t) = 4W \operatorname{sinc}(2Wt)$. Thus, from the modulation property, $h(t) = 4W \operatorname{sinc}(2Wt)\cos(2\pi F_o t)$.

Differentiation

$$\frac{d}{dt}x(t) \longleftrightarrow j2\pi F X(F)$$

(4CT.OP9)

The differentiation property (4CT.OP9) follows in a manner similar to the time-shifted property. We can differentiate $x(t)$ by applying it to an ideal differentiator. We know from Table 3CT.2 that the frequency response characteristic of a differentiator is $H(F) = j2\pi F$. It follows from the equation $Y(F) = H(F)X(F)$ that the derivative of $x(t)$, $y(t) = (d/dt)x(t)$, has a spectrum $Y(F) = j2\pi F X(F)$.

The frequency response characteristic of a differentiator was shown in Figure 3CT.11. Notice from the figure that the amplitude characteristic $|j2\pi F|$ increases with $|F|$. Therefore, the effect of differentiating a waveform on the waveform's amplitude spectrum is to amplify high-frequency components relative to low-frequency components. It is therefore unwise to differentiate a waveform that contains wanted information in its low-frequency components when noise or interference is present in its high-frequency components.

Multiplication by t

$$tx(t) \longleftrightarrow \frac{j}{2\pi}\frac{d}{dF}X(F)$$

(4CT.OP10)

The multiplication by t property (4CT.OP10) is the dual of the differentiation property. We leave the derivation to Problem 4CT.5.22.

EXAMPLE 4CT.14   The Spectrum of $te^{-t/\tau}u(t)$

For an example of the multiplication by t property, let's find the spectrum of $te^{-t/\tau}u(t)$. It follows from Example 3CT.1 that

$$x(t) = e^{-t/\tau}u(t) \longleftrightarrow X(F) = \frac{\tau}{1 + j2\pi F\tau}$$

(63)

If we apply the multiplication by t property to this FT pair, we obtain the result

$$te^{-t/\tau}u(t) \overset{\text{FT}}{\longleftrightarrow} \frac{\tau^2}{(1 + j2\pi F\tau)^2}$$

(64)

Integration

$$\int_{-\infty}^{t} x(\lambda)\,d\lambda \longleftrightarrow \frac{1}{2}\delta(F)X(0) + \frac{1}{j2\pi F}X(F)$$

(4CT.OP11)

The integration property (4CT.OP11) can be derived in a manner similar to the time-shifted and differentiation properties. We leave this derivation to Problem 4CT.5.22.

Sampling

$$x(t)\sum_{k=-\infty}^{\infty}\delta(t - kT_s) \longleftrightarrow F_s\sum_{k=-\infty}^{\infty}X(F - kF_s)$$

where $F_s = 1/T_s$.

(4CT.OP12)

The *sampling property* (4CT.OP12) is called such because, if we multiply $x(t)$ by a periodic impulse train, we generate a sampled signal.

$$x_s(t) = x(t)\sum_{k=-\infty}^{\infty}\delta(t - kT_s) = \sum_{k=-\infty}^{\infty}x(t)\delta(t - kT_s)$$

$$= \sum_{k=-\infty}^{\infty}x(kT_s)\delta(t - kT_s)$$

(65)

where $x(kT_s)$ is the kth sample and T_s is the time between samples. The sampling rate, in samples per second, is $F_s \overset{\Delta}{=} 1/T_s$. The final step in Eq. (65), $x(t)\delta(t - kT_s) = x(kT_s)$

Figure 4CT.14
Impulse train
sampling of $x(t)$:
(a) block diagram;
(b) waveforms $x(t)$,
$\sum_{k=-\infty}^{\infty} \delta(t - kT_s)$
and $x_s(t)$.

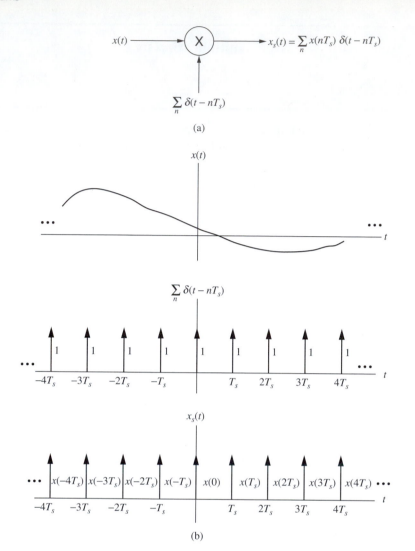

$\delta(t - kT_s)$, follows from the sampling property of the unit impulse. Figure 4CT.14 shows the block diagram and corresponding waveforms associated with (65). The sampler shown in the figure is sometimes referred to as an *ideal sampler*.

Property OP12 shows how the spectrum $X_s(F)$ of the sampled signal is related to the spectrum $X(F)$ of the unsampled signal. We see from the right hand side of OP12 that the spectrum of the sampled signal is

$$X_s(F) = F_s \sum_{k=-\infty}^{\infty} X(F - kF_s) = \cdots + F_s X(F - F_s) + F_s X(F) + F_s X(F + F_s) + \cdots$$

(66)

The preceding equation shows that $X_s(F)$ is periodic: it is composed of all translations of $F_s X(F)$ by integer multiples of F_s. Representative plots of $X(F)$ and $X_s(F)$ are shown in Figure 4CT.15(a) and (b) respectively for a real valued $X(F)$. The middle three terms in Eq. (66) are shown in Figure 4CT.15(b). We refer to these plots in Section 4CT.6 were we consider the problem of reconstructing $x(t)$ from $x_s(t)$.

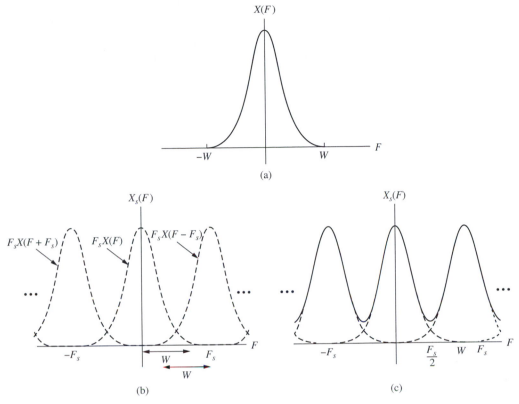

Figure 4CT.15
Spectrum of a sampled waveform: (a) spectrum $X(F)$ of unsampled waveform $x(t)$; (b) the translations $F_s X(F - kF_s)$ $k = 0, \pm 1, \pm 2, \ldots$; (c) $X_s(F)$ (solid) with tails of translations (dashed).

We can derive (4CT.OP12) with the aid of the multiplication property (4CT.OP4). Because $x_s(t)$ is a product, its Fourier transform is given by a convolution, i.e.

$$\underbrace{x(t) \sum_{k=-\infty}^{\infty} \delta(t - kT_s)}_{x_s(t)} \longleftrightarrow \underbrace{X(F) * \left\{ F_s \sum_{k=-\infty}^{\infty} \delta(F - kF_s) \right\}}_{X_s(F)} \qquad (67)$$

where we have used Fourier transform pair (4CT.36). An interchange in the order of summation and convolution yields the result

$$X(F) * \left\{ F_s \sum_{k=-\infty}^{\infty} \delta(F - kF_s) \right\} = F_s \sum_{k=-\infty}^{\infty} X(F) * \delta(F - kF_s) = F_s \sum_{k=-\infty}^{\infty} X(F - kF_s)$$
$$(68)$$

The operational properties for the Fourier transform are summarized in Table 4CT.2.

Table 4CT.2
Operational
Properties of the
Fourier Transform

	Property	Signal	Spectrum
1	Linearity	$a_1x_1(t) + a_2x_2(t)$	$a_1X_1(F) + a_2X_2(F)$
2	Duality	$X(t)$	$x(-F)$
3	Convolution	$x_1(t) * x_2(t)$	$X_1(F)X_2(F)$
4	Multiplication	$x_1(t)x_2(t)$	$X_1(F) * X_2(F)$
5	Time scale	$x(at)$	$\dfrac{1}{|a|}X(F/a)$ for $a \neq 0$
6	Time shift	$x(t - \tau)$	$e^{-j2\pi F\tau}X(F)$
7	Frequency shift	$x(t)e^{j2\pi F_o t}$	$X(F - F_o)$
8	Modulation	$x(t)\cos(2\pi F_o t)$	$\frac{1}{2}X(F - F_o) + \frac{1}{2}X(F + F_o)$
9	Differentiation	$\dfrac{d}{dt}x(t)$	$j2\pi FX(F)$
10	Multiply by t	$tx(t)$	$\dfrac{j}{2\pi}\dfrac{d}{dF}X(F)$
11	Integration	$\displaystyle\int_{-\infty}^{t} x(\lambda)\,d\lambda$	$\frac{1}{2}\delta(F)X(0) + \dfrac{1}{j2\pi F}X(F)$
12	Sampling	$x(t)\displaystyle\sum_{k=-\infty}^{\infty}\delta(t - kT_s)$	$F_s\displaystyle\sum_{k=-\infty}^{\infty}X(F - kF_s)$, where $F_s = \dfrac{1}{T_s}$

4CT.6 The Sampling Theorem

The sampling theorem describes one of the most important and surprising results in signal and systems theory. It states that under certain conditions, we can *exactly* reconstruct a waveform from its samples. This theorem provides a solid theoretical basis for virtually all modern applications in which a waveform is sampled for storage, transmission, or processing (Chapter 7).

Sampling Theorem

A waveform $x(t)$ having a spectrum limited to W Hz

$$X(F) = 0 \qquad \text{for } |F| > W \tag{69}$$

can be perfectly reconstructed from its samples $x(kT_s)$, $k = 0, \pm1, \pm2, \ldots$, if the samples are taken at a rate

$$F_s > 2W \tag{70}$$

The minimum sampling rate that must be exceeded $2W$ is called the *Nyquist rate*.

We can prove the sampling theorem by referring to the sampling property OP12. This property states that the spectrum $X_s(F)$ of $x_s(t)$ is the sum of scaled, periodic replicas of $X(F)$. In general, these periodic repetitions overlap, as was shown in Figure 4CT.15. This overlap, called aliasing, makes it impossible to extract $X(F)$ from $X_s(F)$ and consequently to reconstruct $x(t)$ from its samples.[1] The overlap does not occur if $x(t)$ is band limited

---

[1] You can obtain an intuitive understanding of aliasing by reading Section 3DT.2.1, pages 122–126 of Chapter 3DT. There, aliasing is described in terms of the strobe effect in movies.

Figure 4CT.16
Frequency domain proof of the sampling theorem: (a) filter input spectrum $X_s(F)$ with $F_s > 2W$; (b) frequency response characteristic of reconstruction filter, $H(F)$; (c) spectrum at output of reconstruction filter.

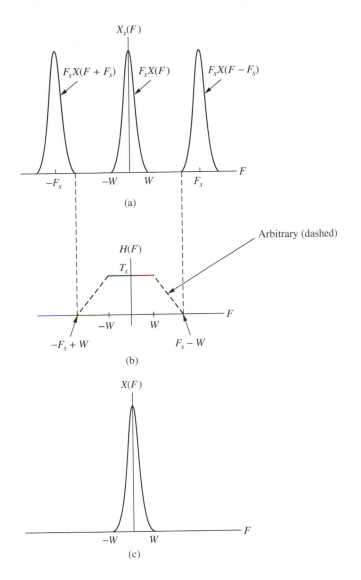

(4CT.69) and if F_s satisfies the bound, $F_s > 2W$, stated in the sampling theorem. We can obtain this bound from an inspection of Figure 4CT.15(b), where there would be no overlap if $F_s > W + W$, or from Figure 4CT.15(c), where there would be no overlap if $\frac{1}{2}F_s > W$. When $F_s > 2W$, it is possible to separate $X(F)$ from $X_s(F)$ by means of a low-pass filter, as shown in Figure 4CT.16. This means that we can reconstruct $x(t)$ from $x_s(t)$.

The frequency response characteristic of the low-pass filter is

$$H(F) = \begin{cases} T_s & |F| \leq W \\ \text{arbitrary} & W < |F| < F_s - W \\ 0 & F_s - W \leq |F| \end{cases} \tag{71}$$

The spectrum of the input to the filter is $X_s(F)$ of Eq. (66). As shown in the figure, the spectrum of the output of the low-pass filter equals $X(F)$.

$$Y(F) = X_s(F)H(F) \equiv X(F) \tag{72}$$

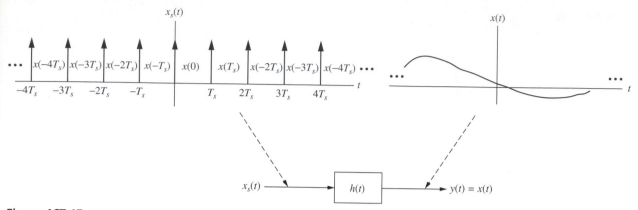

Figure 4CT.17
Reconstruction of $x(t)$ from $x_s(t)$: (a) system block diagram; (b) waveforms.

when $F_s > 2W$. The time domain equation corresponding to (72) is

$$y(t) = \left\{ \sum_{k=-\infty}^{\infty} x(kT_s)\delta(t - kT_s) \right\} * h(t) = \sum_{k=-\infty}^{\infty} x(kT_s)h(t - kT_s) = x(t) \qquad (73)$$

The time domain block diagram and waveforms corresponding to Eq. (73) are illustrated in Figure 4CT.17. The low-pass filter is sometimes called an *ideal interpolation filter* because it ideally interpolates among the samples $x(kT_s)$ to yield $x(t)$. Similarly, the second sum in Eq. (73) is sometimes called an *interpolation formula*. In this section we have considered ideal sampling and reconstruction. Some properties of practical sampling and reconstruction are considered in Problems 4CT.6.4–4CT.6.6.

4CT.6.1
The Nyquist-Shannon Interpolation Formula

A particularly simple interpolation formula is obtained if we choose

$$H(F) = \begin{cases} T_s & |F| \le 0.5F_s \\ 0 & |F| > 0.5F_s \end{cases} \qquad (74)$$

This $H(F)$ is the Nyquist-Shannon reconstruction filter. The corresponding impulse response is

$$h(t) = \mathrm{sinc}\left(\frac{t}{T_s}\right) \qquad (75)$$

The substitution of Eq. (75) into Eq. (73) yields

$$y(t) = \sum_{k=-\infty}^{\infty} x(kT_s)\,\mathrm{sinc}\left(\frac{t - kT_s}{T_s}\right) = x(t) \qquad (76)$$

This result is called the *Nyquist-Shannon interpolation formula*. Figure 4CT.18 illustrates how the sinc functions interpolate among the samples to reconstruct $x(t)$.

4CT.6.2
Antialiasing Filters

Engineers are typically faced with a trade-off in choosing the rate at which a waveform is to be sampled. On the one hand, a high value of F_s is desirable to reduce or prevent aliasing. On the other hand, a low value of F_s is desirable: In communications, a lower sampling rate leads to a lower data rate and a lower intersymbol interference, as illustrated in Figure 2CT.31. Another example of having a low value of F_s occurs in signal processing;

Figure 4CT.18
Nyquist-Shannon
reconstruction of
$x(t)$: The
interpolation function
is $h(t) = \text{sinc}(t/T_s)$.

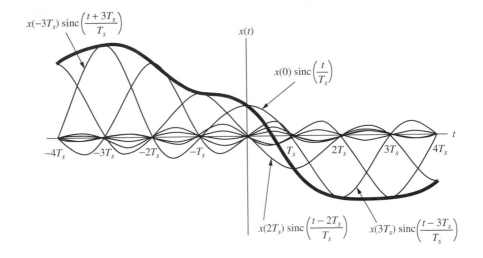

Figure 4CT.19
Sampler with
antialiasing filter.

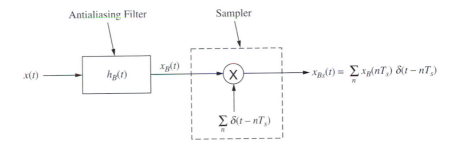

here, a lower sampling rate corresponds to fewer samples to be processed and eases the memory requirements of the processor. In view of this trade-off, an engineer may consider sampling a waveform below the Nyquist rate. When the bandwidth of a waveform W exceeds $\frac{1}{2}F_s$, an *antialiasing filter* can be used to prevent or reduce aliasing.

The block diagram of the system is shown in Figure 4CT.19. Here, the input waveform, $x(t)$, has significant frequency components for $|F| > \frac{1}{2}F_s$. To reduce or prevent aliasing, a low-pass filter, called an antialiasing filter, is placed before the sampler. The antialiasing filter rejects frequency components outside the range $|F| \leq B$, where $B < \frac{1}{2}F_s$. Because the filter output $x_B(t)$ has a bandwidth $2B < F_s$, it can be reconstructed from its samples. The antialiasing filter does not provide recovery of $x(t)$. It only provides a means to trade the distortion that would be created sampling $x(t)$ below the Nyquist rate with the distortion that results from low-pass filtering $x(t)$.

4CT.7 Deriving Spectrums by Differentiation

It is sometimes easier to derive spectrums using differentiation instead of integration [Mason, S.J. and Zimmerman, 1960]. The basic idea is to use the differentiation property to obtain an equation that can be solved for $X(F)$.

EXAMPLE 4CT.15   Triangular Pulse

Consider the triangular pulse $x(t) = \wedge(t/\tau)$. If we differentiate it twice, as shown in Figure 4CT.20, we obtain:

$$\frac{d^2x(t)}{dt^2} = \frac{1}{\tau}\{\delta(t+\tau) - 2\delta(t) + \delta(t-\tau)\} \tag{77}$$

Figure 4CT.20
Triangular pulse and
its derivatives.

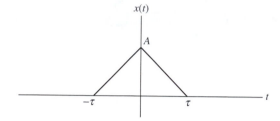

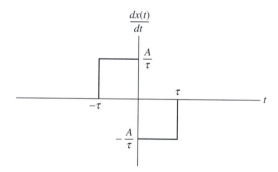

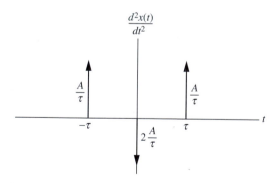

This equation has frequency domain counterpart

$$(j2\pi F)^2 X(F) = \frac{1}{\tau}\{e^{j2\pi F\tau} - 2 + e^{-j2\pi F\tau}\} = -\frac{2}{\tau}[1 - \cos(2\pi F\tau)] \tag{78}$$

from which we obtain

$$X(F) = \frac{2[1 - \cos(2\pi F\tau)]}{\tau(2\pi F)^2} = \tau \operatorname{sinc}^2(F\tau) \tag{79}$$

The method of differentiation works for all signals composed of straight-line segments. For other signals, the method occasionally avoids the extensive labor of alternative techniques.

EXAMPLE 4CT.16   Raised Cosine

Consider the raised cosine pulse of Example 4CT.5:

$$x(t) = \frac{A}{2}\left\{1 + \cos\left(2\pi\frac{t}{\tau}\right)\right\}\sqcap\left(\frac{t}{\tau}\right) \tag{80}$$

The second derivative of this waveform is

$$\frac{d^2x(t)}{dt^2} = -\frac{A}{2}\left(\frac{2\pi}{\tau}\right)^2\cos\left(2\pi\frac{t}{\tau}\right)\sqcap\left(\frac{t}{\tau}\right) \tag{81}$$

It follows that

$$x(t) + \left(\frac{\tau}{2\pi}\right)^2\frac{d^2x(t)}{dt^2} = \frac{A}{2}\sqcap\left(\frac{t}{\tau}\right) \tag{82}$$

which transforms into

$$X(F) - \left(\frac{2\pi F\tau}{2\pi}\right)^2 X(F) = \frac{A\tau}{2}\frac{\sin(\pi F\tau)}{\pi F\tau} \tag{83}$$

Therefore,

$$X(F) = \frac{A\tau}{2}\frac{\sin(\pi F\tau)}{(1 - (F\tau)^2)\pi F\tau} \tag{84}$$

which agrees with Eq. (4CT.30).

EXAMPLE 4CT.17   Gaussian Pulse

Consider a gaussian pulse $x(t) = (1/\sqrt{2\pi})e^{-\frac{1}{2}t^2}$. Its derivative is

$$\frac{d}{dt}x(t) = -tx(t) \tag{85}$$

This equation transforms to

$$j2\pi F X(F) = -\frac{j}{2\pi}\frac{d}{dF}X(F) \tag{86}$$

This equation can be rearranged to

$$\frac{d}{dF}X(F) = -(2\pi)^2 F X(F) \tag{87}$$

The similarity between Eqs. (87) and (85) suggests that $X(F)$ is itself a gaussian function. You can easily confirm that

$$X(F) = Ke^{-\frac{1}{2}(2\pi F)^2} \tag{88}$$

satisfies (87) for any K. It can be shown (Problem 4CT.8.2) that $K = 1$. If we apply the time-scaled property to Eq. (88), we obtain the more general result

$$\frac{1}{\sqrt{2\pi\sigma^2}}e^{-\frac{t^2}{2\sigma^2}} \overset{\text{FT}}{\longleftrightarrow} e^{-\frac{1}{2}(2\pi F\sigma)^2} \tag{89}$$

	Name	$x(t)$	$X(F)$
FT1	Impulse	$\delta(t)$	1
FT2	Constant	1	$\delta(F)$
FT3	Step	$u(t)$	$\frac{1}{2}\delta(F) + \frac{1}{j2\pi F}$
FT4	Complex sinusoid	$e^{j2\pi F_o t}$	$\delta(F - F_o)$
FT4	Cosine	$\cos(2\pi F_o t)$	$\frac{1}{2}\delta(F - F_o) + \frac{1}{2}\delta(F + F_o)$
FT6	Sine	$\sin(2\pi F_o t)$	$\frac{1}{2j}\delta(F - F_o) - \frac{1}{2j}\delta(F + F_o)$
FT7	Cosine with phase	$\cos(2\pi F_o t + \phi)$	$\frac{1}{2}e^{j\phi}\delta(F - F_o) + \frac{1}{2}e^{-j\phi}\delta(F + F_o)$
FT8	Multiple sinusoids	$\sum_{i=-I}^{I} X_i e^{j2\pi F_i t}$	$\sum_{i=-I}^{I} X_i \delta(F - F_i);$
FT9	Decaying exponential	$Ae^{-\frac{t}{\tau}}u(t)$	$\frac{A\tau}{1 + j2\pi F\tau}$
FT10	Exponential times ramp	$te^{-t/\tau}u(t)$	$\frac{\tau^2}{(1 + j2\pi F\tau)^2}$
FT11	Rectangular pulse	$A \sqcap \left(\frac{t}{\tau}\right)$	$A\tau\,\text{sinc}(F\tau)$
FT12	Triangular pulse	$A \wedge \left(\frac{t}{\tau}\right)$	$A\tau\,\text{sinc}^2(F\tau)$
FT13	Half-wave cosine pulse	$A \sqcap \left(\frac{t}{\tau}\right)\cos\left(\frac{\pi t}{\tau}\right)$	$\frac{2A\tau}{\pi}\frac{\cos(\pi F\tau)}{1 - (2F\tau)^2}$
FT14	Raised cosine pulse	$\frac{A}{2}\left\{1 + \cos\left(2\pi\frac{t}{T}\right)\right\}\sqcap\left(\frac{t}{T}\right)$	$\frac{AT}{2}\frac{\sin(\pi FT)}{(1-(FT)^2)\pi FT}$
FT15	Gaussian pulse	$\frac{1}{\sqrt{2\pi\sigma^2}}e^{-\frac{t^2}{2\sigma^2}}$	$e^{-\frac{1}{2}(2\pi F\sigma)^2}$
FT16	Periodic Impulse Train	$\sum_{i=-\infty}^{\infty}\delta(t - iT) = \frac{1}{T}\sum_{k=-\infty}^{\infty}e^{j2\pi\frac{k}{T}t}$	$\frac{1}{T}\sum_{k=-\infty}^{\infty}\delta(F - \frac{k}{T})$

Table 4CT.3

Fourier Transform Pairs

Table 4CT.3 summarizes the Fourier transform pairs we have derived in this chapter and lists the FT pairs of other familiar signals. Additional Fourier transform pairs have been given in Tables 3CT.1 and 3CT.2.

4CT.8   Energy Waveforms, Energy Density Spectrums, and Parseval's Theorem

Pulse signals, like signals $Ae^{-\frac{t}{\tau}}u(t)$, $A \sqcap (t/\tau)$, and $A \wedge (t/\tau)$, are members of a class of waveforms called *energy waveforms*. The *energy* in a real or complex waveform $x(t)$ is defined as

$$E_x = \int_{-\infty}^{\infty} |x(t)|^2 dt \tag{90}$$

where E_x is the integral of the instantaneous power carried by $x(t)$. The terms "energy" and "energy waveform" are motivated by electrical engineering. Physically, E_x would equal the energy dissipated in the form of heat if $x(t)$ were a real voltage applied across a 1 Ω resistor.

A waveform $x(t)$ is called an *energy waveform* if E_x is a nonzero finite number. The sinusoidal waveform $A\cos(2\pi Ft + \phi)$ is not an energy waveform because the energy it carries is infinite (the integral (4CT.90) diverges). Periodic waveforms are never energy waveforms.

The energy density spectrum of an energy waveform is defined as

---

Energy Density Spectrum

$$\Phi_x(F) = |X(F)|^2 \tag{91}$$

where $X(F) = \mathcal{FT}\{x(t)\}$.

A fundamental property of energy density spectrums is that the total energy carried by a waveform is equal to the area under the waveform's energy density spectrum. This property is called *Parseval's theorem for energy signals*.

Parseval's Theorem for Energy Signals

$$E_x = \int_{-\infty}^{\infty} |x(t)|^2 dt = \int_{-\infty}^{\infty} \Phi_x(F) dF \tag{92}$$

We can derive Parseval's theorem from the convolution property (Table 4CT.2) by setting $x_1(t) = x(t)$ and $x_2(t) = x^*(-t)$. By the conjugate symmetry and time-scaled properties, the spectrum of $x^*(-t)$ is $X^*(F)$. The convolution property then tells us that

$$\phi_x(t) = \int_{-\infty}^{\infty} x(\lambda) x^*(\lambda - t) \, d\lambda = \int_{-\infty}^{\infty} X(F) X^*(F) e^{j2\pi Ft} dF = \int_{-\infty}^{\infty} \Phi_x(F) e^{j2\pi Ft} dF \tag{93}$$

$\phi_x(t)$ is the *autocorrelation function* of $x(t)$. DT versions of autocorrelation functions are computed in cell phones and other voice encoders and decoders (codecs). Equation (93) states the important theoretical result that the autocorrelation function $\phi_x(t)$ and energy density function $\Phi_x(F)$ are a Fourier transform pair. Parseval's theorem is a special case of Eq. (93), where $t = 0$.

The energy density spectrum $\Phi_x(F)$ describes the distribution of the energy in $x(t)$ in frequency. This means that the energy $E_x(\mathcal{B})$ contained in any set of frequencies $\mathcal{B}$ is given by:

$$E_x(\mathcal{B}) = \int_{\mathcal{B}} \Phi_x(F) dF \tag{94}$$

Figure 4CT.21
Ideal filter $H_{\mathcal{B}}(F)$.

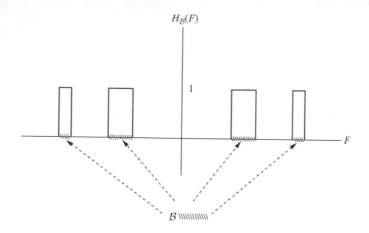

where the integration is over the frequencies in $\mathcal{B}$. We can prove this statement by applying $x(t)$ to an ideal filter $H_{\mathcal{B}}(F)$, which passes all frequency components in $\mathcal{B}$ and rejects all other frequency components:

$$H_{\mathcal{B}}(F) = \begin{cases} 1; & F \in \mathcal{B} \\ 0; & F \notin \mathcal{B} \end{cases} \tag{95}$$

where $\in$ and $\notin$ denote "is in" and "is not in," respectively. An example of $H_{\mathcal{B}}(F)$ is illustrated in Figure 4CT.21. When $x(t)$ is applied to this filter, the spectrum of the output is $Y(F) = H_{\mathcal{B}}(F)X(F)$. The energy density spectrum of the output therefore is given by

$$\Phi_y(F) = |H_{\mathcal{B}}(F)X(F)|^2 = |H_{\mathcal{B}}(F)|^2 \Phi_x(F) = \begin{cases} \Phi_x(F) & F \in \mathcal{B} \\ 0 & F \notin \mathcal{B} \end{cases} \tag{96}$$

The energy in $y(t)$ is given by Parseval's theorem.

$$E_y = \int_{-\infty}^{\infty} |y(t)|^2 dt = \int_{-\infty}^{\infty} \Phi_y(F) dF = \int_{\mathcal{B}} \Phi_x(F) dF = E_x(\mathcal{B}) \tag{97}$$

The final equality of the preceding equation is (94).

EXAMPLE 4CT.18   Rectangular Pulse Bandwidth

Consider the choice for k in the definition of the bandwidth $W_{\sqcap} = k/\tau$ for a rectangular pulse of Example 4CT.2. The ratio of the energy contained in the frequency band $-W_{\sqcap} < F < W_{\sqcap}$ to the total energy in the pulse is

$$\rho = \frac{E_x(W_{\sqcap})}{E_x} = \frac{\int_{-W_{\sqcap}}^{W_{\sqcap}} A^2 \tau^2 \operatorname{sinc}^2(F\tau) dF}{A^2 \tau} = \int_{-W_{\sqcap}\tau}^{W_{\sqcap}\tau} \operatorname{sinc}^2(\lambda) \, d\lambda \tag{98}$$

Numerical integration can be used to evaluate Eq. (98). The numerical integration reveals that the range $1 \leq k \leq 5$ mentioned in Example 2 is reasonable: 90% of the energy in the rectangular pulse is carried by the frequency components in the band $-\frac{1}{\tau} < F < \frac{1}{\tau}$, 95% by those in the band $-\frac{2}{\tau} < F < \frac{2}{\tau}$, and 98% by those in the band $-\frac{5}{\tau} < F < \frac{5}{\tau}$. Figure 4CT.22 shows a plot of ρ versus $k = W_{\sqcap}\tau$ for $0 \leq k \leq 3$.

Figure 4CT.22
Ratio of the rectangular pulse's energy in a band $-W_\sqcap < F < W_\sqcap$ to its total energy:

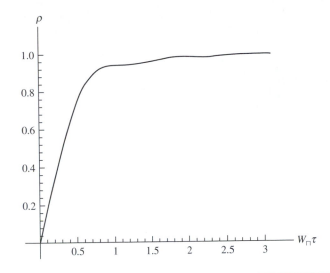

4CT.9   Dirichlet Conditions

Sinusoids are continuous functions. It is therefore remarkable that the integral of sinusoids $\mathcal{F}^{-1}\{X(F)\}$ of Eq. (2) can in fact represent a discontinuous function, such as a pulse $x(t) = \sqcap(t/\tau)$, as derived in Example 4CT.2. There are functions, however, that cannot be represented by the inverse Fourier transform. It is important therefore to know the conditions that guarantee that a waveform can be represented by an integral of sinusoids $\mathcal{F}^{-1}\{X(F)\}$. The Dirichlet conditions, which we state without proof, provide this guarantee. If all three conditions are satisfied then the inverse Fourier transform equals $x(t)$ except where $x(t)$ is discontinuous. At the values of t for which $x(t)$ is discontinuous, the inverse Fourier transform converges to the average of the values of $x(t)$ that are on either side of the discontinuity.

Dirichlet Conditions

1. $x(t)$ is absolutely integrable, i.e.

$$\int_{-\infty}^{\infty} |x(t)|dt < \infty \tag{99}$$

2. $x(t)$ does not contain an infinite number of minima or maxima within any finite time interval.
3. $x(t)$ does not contain an infinite number of discontinuities within any finite time interval.

If a waveform does not satisfy the Dirichlet conditions, then it cannot be displayed on a monitor. Examples are shown (necessarily incompletely) in Figure 4CT.23.

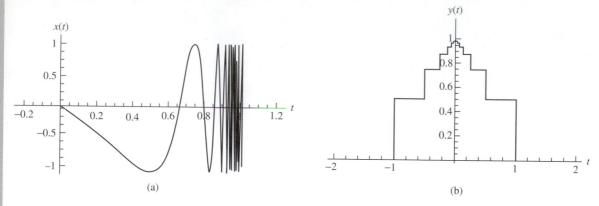

Figure 4CT.23

Two functions that cannot be represented as integrals of sinusoids: (a) $x(t) = \cos^{-1}\{(x)/[2(t-1)]\}$, $0 \le t < 1$, and $x(t) = 0$ for all other t; (b) $y(t) = \frac{1}{2} \sqcap (2t) + (\frac{1}{2})^2 \sqcap (2^2 t) + (\frac{1}{2})^3 \sqcap (2^3 t) + (\frac{1}{2})^4 \sqcap (2^4 t) + \cdots$.

 ## Convergence of the Fourier Transform and its Inverse

The Fourier transform integral (1) is a limit

$$X(F) = \lim_{T \to \infty} X_T(F) \tag{100}$$

where

$$X_T(F) = \int_{-T}^{T} x(t) e^{-j2\pi Ft} dt \tag{101}$$

$X(F)$ exists when $X_T(F)$ converges as $T \to \infty$. Two types of convergence are of interest. The first, called *uniform convergence*, is the strongest.

Definition: Uniform Convergence

$X_T(F)$ converges uniformly to $X(F)$ if, for every $\epsilon > 0$, there is a T_o such that

$$\underbrace{|X(F) - X_T(F)| < \epsilon}_{\text{This inequality holds } \textit{for all } F.} \qquad \text{for all } T > T_o \tag{102}$$

Before considering an example of uniform convergence, we prove the following theorem:

> ## Theorem
>
> $X_T(F)$ converges uniformly to $X(F)$ if $x(t)$ is absolutely integrable.

To begin the proof, we write the absolute value in Eq. (102) as

$$|X(F) - X_T(F)| =$$

$$\left| \int_{-\infty}^{\infty} x(t) e^{-j2\pi Ft} dt - \int_{-T}^{T} x(t) e^{-j2\pi Ft} dt \right| = \left| \int_{|t|>T} x(t) e^{-j2\pi Ft} dt \right| \qquad (103)$$

We can obtain an upper bound on $|X(F) - X_T(F)|$ if we interchange the orders of integration and absolute value in the last term in the preceding equation:

$$\left| \int_{|t|>T} x(t) e^{-j2\pi Ft} dt \right| \leq \int_{|t|>T} |x(t) e^{-j2\pi Ft}| \, dt$$

$$= \int_{|t|>T} |x(t)| \, dt$$

$$= \int_{-\infty}^{\infty} |x(t)| \, dt - \int_{|t|\leq T} |x(t)| \, dt \qquad (104)$$

It follows that

$$|X(F) - X_T(F)| \leq \int_{-\infty}^{\infty} |x(t)| \, dt - \int_{|t|\leq T} |x(t)| \, dt \qquad (105)$$

We can write this as inequality (102) where

$$\epsilon = \int_{-\infty}^{\infty} |x(t)| \, dt - \int_{|t|\leq T} |x(t)| \, dt \qquad (106)$$

is independent of F. The value of ϵ can be made arbitrarily small by increasing T if

$$\int_{-\infty}^{\infty} |x(t)| \, dt < \infty \qquad (107)$$

QED

EXAMPLE 4CT.19   Uniform Convergence

Consider

$$x(t) = e^{-|t|} \qquad (108)$$

It is easy to show that $x(t)$ is absolutely integrable. Therefore, by the theorem, $X_T(F)$ converges uniformly to $X(F)$. Figure 4CT.24 shows plots of $X(F)$ and $X_T(F)$ where (Problem 4CT.9.1)

$$X_T(F) = \int_{-T}^{T} e^{-|t|} e^{-j2\pi Ft} \, dt = 2 \frac{1 - e^{-T}\{\cos(2\pi FT) + 2\pi F \sin(2\pi FT)\}}{1 + (2\pi F)^2} \qquad (109)$$

and

$$X(F) = \frac{2}{1 + (2\pi F)^2} \qquad (110)$$

Figure 4CT.24
Uniform convergence
$(X_T(F)$ —,
$X(F)$ —): (a) $T = 1$;
(b) $T = 2$; (c); $T = 3$;
(d) $T = 4$.

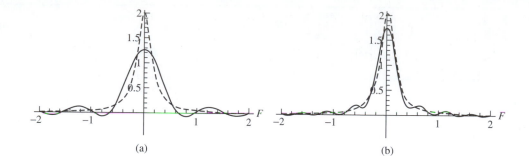

(a)   (b)

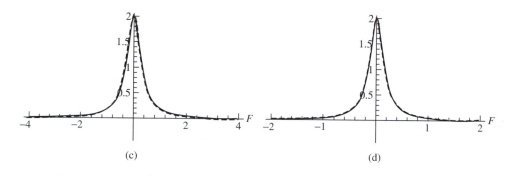

(c)   (d)

The second type of convergence is *integral square convergence*.

Definition: Integral Square Convergence

$X_T(F)$ converges to $X(F)$ in integral square if

$$\lim_{T \to \infty} \int_{-\infty}^{\infty} |X(F) - X_T(F)|^2 dF = 0 \tag{111}$$

Before considering an example of integral square convergence, we prove the following theorem.

Theorem

$X_T(F)$ converges to $X(F)$ in integral square if $x(t)$ is an energy waveform.

To begin the proof, we use Parseval's Theorem to write

$$\int_{-\infty}^{\infty} |X(F) - X_T(F)|^2 dF = \int_{-\infty}^{\infty} \left| x(t) - x(t) \sqcap \left(\frac{t}{2T} \right) \right|^2 dt$$

$$= \int_{-\infty}^{\infty} |x(t)|^2 dt - \int_{|t| \leq T} |x(t)|^2 dt \tag{112}$$

If

$$\int_{-\infty}^{\infty} |x(t)|^2 \, dt < \infty \tag{113}$$

then the right-hand side of Eq. (113) vanishes in the limit $T \to \infty$. This completes the proof.

EXAMPLE 4CT.20   Ingegral Square Convergence

Consider

$$x(t) = \text{sinc}(t) \tag{114}$$

A sinc function is not absolutely integrable but it does have finite energy. We have seen earlier that

$$X(F) = \sqcap(F) \tag{115}$$

The convergence of $X_T(F)$ to $X(F)$ is not uniform but in the integral square sense.

Figure 4CT.25 contains plots of $X_T(F)$. The function $X_T(F)$ was computed numerically for these plots. The 9 percent overshoot that appears near the discontinuities in $\sqcap(F)$ is called *Gibbs phenomenon* and is a characteristic

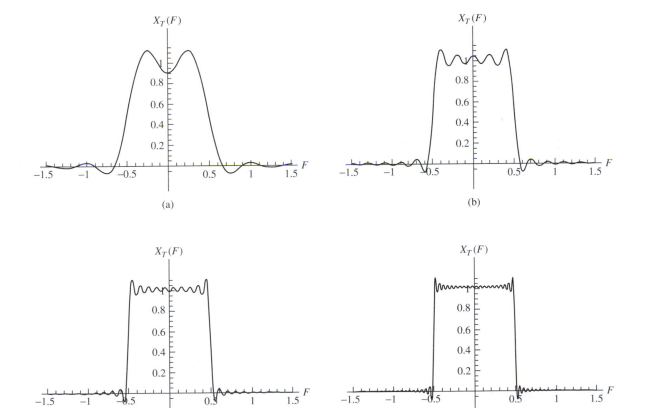

Figure 4CT.25
Integral square convergence: (a) $T = 2$, (b) $T = 5$, (c) $T = 10$, (d) $T = 20$

feature of integral squared convergence. The amplitude of the overshoot does not decrease as T increases. However, the integral square of $|X(F) - X_T(F)|$ vanishes in the limit $T \to \infty$, in agreement with the theorem.

The preceding analyses have been concerned with the convergence of the Fourier transform. Similar results apply to the inverse Fourier transform. If we define

$$x_F(t) = \int_{-F}^{F} X(F)e^{j2\pi Ft} dF \qquad (116)$$

then we can show that $x_F(t)$ converges uniformly to $x(t)$ as $F \to \infty$, if

$$\int_{-\infty}^{\infty} |X(F)| dF < \infty \qquad (117)$$

The uniform convergence of $x_F(t)$ to $x(t)$ means that for every $\epsilon > 0$, there is a F_o such that

$$\underbrace{|x(t) - x_F(t)| < \epsilon}_{\text{This inequality holds } \textit{for all } t.} \qquad \text{for all } F > F_o \qquad (118)$$

We can show $x_F(t)$ converges to $x(t)$ in integral square as $F \to \infty$, if

$$\int_{-\infty}^{\infty} |X(F)|^2 dF < \infty \qquad (119)$$

The integral square convergence of $x_F(t)$ means that.

$$\lim_{F \to \infty} \int_{-F/2}^{F/2} |x(t) - x_F(t)|^2 dt = 0 \qquad (120)$$

4CT.10.1
Fourier
Transforms
Modeled with
Impulses

If $X_T(F)$ does not converge as $T \to \infty$, we can sometimes define $X(F)$ with the aid of impulses. This statement was illustrated by the DC waveform of Example 4CT.4. The waveform $x(t) = 1$ has infinite energy and we saw that $\mathcal{F}\{1\}$ did not converge. By taking a limit, however, we were able to define $X(F) = \delta(F)$. Another example of the use of impulses to model $X(F)$ is given by multiple component sinusoids (Table 4CT.3). The DC and multiple component sinusoidal waveforms have finite average power. We will see in Chapter 5CT that we can use impulses to model $X(F)$ for every periodic signal having finite power.

It is also possible to use impulses to model the spectrums of some waveforms that have infinite average power. An example is the periodic impulse train of Table 4CT.3.

There are, however, routinely used signal models that cannot be represented by Fourier transforms. Examples are the unit ramp $r(t) = tu(t)$ and the rising exponential $x(t) = e^{at}u(t)$, where $a > 1$. The Laplace transform, (Chapter 6CT), works for these signals.

4CT.11   Summary

In this chapter we showed that a waveform $x(t)$ can be represented as an integral of complex sinusoidal components. This representation is the inverse Fourier transform of $x(t)$. The waveform's spectrum $X(F)$ describes the complex amplitudes of the infinitesimal sinusoidal components. All the properties that apply to multiple sinusoidal component waveforms developed in Chapter 3CT extend to Fourier transformable waveforms. The principle property is that the output $y(t) = x(t) * h(t)$ of an LTI system has spectrum $Y(F) = X(F)H(F)$, where $H(F)$ is the system's frequency response characteristic. Other important properties are listed in Table 4CT.2.

Summary of Definitions and Formulas

Fourier Transform (Spectrum of $x(t)$)

$$X(F) = \int_{-\infty}^{\infty} x(t)e^{-j2\pi Ft}dt$$

Inverse Fourier Transform

The inverse Fourier transform represents a waveform as an integral of complex sinusoids.

$$x(t) = \int_{-\infty}^{\infty} X(F)e^{j2\pi Ft}dF$$

Signals and LTI Systems

Time domain:

$$x(t) \xrightarrow{h(t)} y(t) = x(t) * h(t)$$

Frequency domain:

$$X(F) \xrightarrow{H(F)} Y(F) = X(F)H(F)$$

Amplitude and Phase Spectrums

$$|Y(F)| = |H(F)||X(F)|$$
$$\angle Y(F) = \angle H(F) + \angle X(F)$$

CINC Function

$$\text{cinc}(\lambda; K) = \frac{1}{2K+1}\sum_{k=-K}^{K} e^{-j2\pi FkT} = \frac{\sin((2K+1)\pi\lambda)}{(2K+1)\sin(\pi\lambda)}$$

$$\lim_{K\to\infty}(2K+1)\text{cinc}(FT; K) = F_s\sum_{k=-\infty}^{\infty}\delta(F-kF_s), \text{ where } F_s = \frac{1}{T}$$

Fourier Transform Pairs and Properties

See Tables 4CT.1–4CT.3.

Sampling Property

$$x(t)\sum_{k=-\infty}^{\infty}\delta(t-kT_s) \longleftrightarrow F_s\sum_{k=-\infty}^{\infty}X(F-kF_s), \text{ where } F_s = \frac{1}{T_s}$$

Sampling Theorem

Perfect reconstruction of $x(t)$ from its samples is possible if F_s exceeds $2W$, where W is the highest frequency in $x(t)$.

$$2W \text{ is called the } Nyquist \ rate.$$

Energy Waveforms, Energy Density Functions, and Parseval's Theorem for Energy Waveforms

$$x(t) \text{ is an energy waveform if } \int_{-\infty}^{\infty} |x(t)|^2 dt < \infty.$$

$E = \displaystyle\int_{-\infty}^{\infty} |x(t)|^2 dt$ is the energy in $x(t)$.

$\Phi_x(F) = |X(F)|^2$ is the *energy density spectrum* of $x(t)$.

$\phi_x(t) \xleftrightarrow{\text{FT}} \Phi_x(F)$

$\phi_x(t) = \displaystyle\int_{-\infty}^{\infty} x(\lambda)x^*(\lambda - t)\, d\lambda$ is the *autocorrelation function* of $x(t)$.

The energy contained in any band $\mathcal{B}$ of frequencies is

$$E_x(\mathcal{B}) = \int_{\mathcal{B}} \Phi_x(F)\, dF$$

Parseval's theorem for energy waveforms:

$$E = \int_{-\infty}^{\infty} |x(t)|^2 dt = \int_{-\infty}^{\infty} \Phi_x(F)\, dF$$

Dirichlet Conditions

The inverse Fourier transform equals $x(t)$ except where $x(t)$ is discontinuous if the following three conditions are met:

1. $x(t)$ is absolutely integrable.
2. $x(t)$ does not contain an infinite number of minima or maxima within any finite time interval.
3. $x(t)$ does not contain an infinite number of discontinuities within any finite time interval.

At the values of t for which $x(t)$ is discontinuous, the inverse Fourier transform converges to the average of the values of $x(t)$ that are on either side of the discontinuity.

Convergence

Let $X_T(F) = \int_{-T}^{T} x(t)e^{-j2\pi Ft} dt$.

Definition: $X_T(F)$ converges uniformly to $X(F)$ if, for every $\epsilon > 0$, there is a T_o such that

$$\underbrace{|X(F) - X_T(F)| < \epsilon}_{\text{This inequality holds } for \ all \ F.} \qquad \text{for all } T > T_o$$

Theorem: $X_T(F)$ converges uniformly to $X(F)$ if $x(t)$ is absolutely integrable.

Definition: $X_T(F)$ converges to $X(F)$ in integral square if

$$\lim_{T \to \infty} \int_{-\infty}^{\infty} |X(F) - X_T(F)|^2 dF = 0$$

Theorem: $X_T(F)$ converges to $X(F)$ in integral square if $x(t)$ is an energy waveform.

4CT.12   Problems

4CT.2   Spectrums and LTI Systems

4CT.2.1 Individual signals $x_1(t)$, $x_2(t)$, $\cdots$, $x_n(t)$ are added to produce the waveform

$$x(t) = x_1(t) + x_2(t) + \cdots + x_n(t) \tag{121a}$$

a) Show that the spectrum of $x(t)$ is

$$X(F) = X_1(F) + X_2(F) + \cdots + X_n(F) \tag{121b}$$

where the $X_i(F)$ are the spectrums of $x_i(t), i = 1, 2, \ldots, n$.

b) Show that if a spectrum has the form Eq. (121b), then the corresponding waveform has the form Eq. (121a).

c) Assume that the individual waveforms $x_i(t)$ overlap in time but that the corresponding spectrums $X_i(F)$ do not overlap in frequency. Invent a system having input $x(t)$ that recovers the individual waveforms $x_i(t)$. Draw the system that performs this separation. Is your system LTI?

d) Assume that the individual spectrums $X_i(F)$ overlap in frequency but that the corresponding waveforms $x_i(t)$ do not overlap in time. Invent a system that recovers the individual spectrums $X_i(F)$ by suitable processing of $x(t)$. Draw the system that performs this separation. Is your system LTI?

4CT.2.2 Use the inverse Fourier transform to find $x(t)$ for $X(F) = X\delta(F - F_o)$ where $X = \frac{1}{2} Ae^{j\phi}$.

4CT.2.3 a) Use the inverse Fourier transform to find $x(t)$ for $X(F) = X\delta(F - F_o) + X^*\delta(F + F_o)$, where $X = \frac{1}{2} Ae^{j\phi}$. Express your answer as a real sinusoidal function of t.

b) Use your result from part (a) to write down the spectrum, $X(F)$, of $x(t) = 3 \cos(2\pi 100t + 22°)$. Plot $X(F)$.

4CT.2.4 a) Use the inverse Fourier transform to find $x(t)$ for

$$X(F) = X_1\delta(F - F_1) + X_1^*\delta(F + F_1) + X_2\delta(F - F_2) + X_2^*\delta(F + F_2) \tag{122}$$

where $X_i = \frac{1}{2} A_i e^{j\phi_i}$; $i = 1, 2$. Express your answer as a real function of t.

b) Use your result from part (a) to write down the spectrum $X(F)$ of

$$x(t) = A_1 \cos(2\pi F_1 t + \phi_1) + A_2 \cos(2\pi F_2 t + \phi_2) \tag{123}$$

Plot $X(F)$.

c) Recall that in Figure 3CT.10(a) we plotted the two-sided spectrum of

$$x(t) = A_1 \cos(2\pi F_1 t + \phi_{x1}) + A_2 \cos(2\pi F_2 t + \phi_{x2}) \tag{124}$$

Describe how the line spectrum in Figure 3CT.10(a) differs from your plot of part (b). In which plot(s) are impulses used? Which of the plots is a density spectrum?

4CT.2.5 The spectrum of a waveform $x(t)$ is given by

$$X(F) = \sum_{i=-3}^{3} X_i\delta(F - F_i) \tag{125}$$

where

$$X_i = \begin{cases} A_0 & i = 0 \\ \frac{1}{2} A_i e^{j\phi_i} & i = 1, 2, 3 \\ X^*_{-i} & i = -1, -2, -3 \end{cases} \tag{126}$$

and

$$F_i = \begin{cases} 0 & i = 0 \\ -F_{-i} & i = -1, -2, -3 \end{cases} \tag{127}$$

a) Use the inverse Fourier transform to find $x(t)$. Express your answer as a sum of real sinusoidal components plus a constant (DC) term.

b) Use sketches to show that a judicious use of filters allows you to separate the individual sinusoidal components from $x(t)$. Draw the system that accomplishes this separation.

c) Sketch the input and output spectrums of every filter in your system.

4CT.2.6 Assume that an input $x(t) = \delta(t)$ is applied to the system of Figure 4CT.1. Find $X(F)$, $Y(F)$, and $y(t)$. Assume that $H(F)$ is arbitrary.

4CT.2.7 Repeat Problem 4CT.2.3 for an input $x(t) = \delta(t - \tau)$.

4CT.2.8 Assume that $x(t) = Xe^{j2\pi F_o t}$ is applied to the system of Figure 4CT.1. Find $X(F)$, $Y(F)$, and $y(t)$. Assume that $H(F)$ is arbitrary. Hint: $F\{Xe^{j2\pi F_o t}\} = X\delta(F - F_o)$.

4CT.3   Amplitude and Phase Spectrums

4CT.3.1 Do Eq. (11) and (12) hold even if $x(t)$ and $h(t)$ are not real? Justify your answer.

4CT.4   Some Waveforms and Their Spectrums

4CT.4.1 Confirm the Fourier transform pair $\delta(t) \xrightarrow{\text{FT}} 1$ by substituting $\delta(t)$ into the Fourier transform integral.

4CT.4.2 Confirm the Fourier transform pair $1 \xrightarrow{\text{FT}} \delta(F)$ by substituting $\delta(F)$ into the inverse Fourier transform integral.

4CT.4.3 Establish Eq. (37), which is repeated here.

$$\sum_{n=-\infty}^{\infty} \delta(t - nT) = \frac{1}{T} \sum_{k=-\infty}^{\infty} e^{j2\pi \frac{k}{T} t} \tag{37}$$

Hint: Read the text immediately prior to Eq. (37) in Example 4CT.6.

 4CT.4.4 a) Show that

$$\sqcap\left(\frac{t}{\tau}\right) * \sum_{n=-\infty}^{\infty} \delta(t - nT) = \sum_{n=-\infty}^{\infty} \sqcap\left(\frac{t - nT}{\tau}\right) \tag{128}$$

The right-hand side of the equation is called a *periodic rectangular pulse train*. Plot its right-hand side. Assume in your plot that $T > \tau$.

b) Refer to (37) and show that

$$\sqcap\left(\frac{t}{\tau}\right) * \frac{1}{T} \sum_{k=-\infty}^{\infty} e^{j2\pi \frac{k}{T} t} = \frac{1}{T} \sum_{k=-\infty}^{\infty} \left\{ \sqcap\left(\frac{t}{\tau}\right) * e^{j2\pi \frac{k}{T} t} \right\}$$

$$= \frac{1}{T} \sum_{k=-\infty}^{\infty} \tau \operatorname{sinc}\left(k\frac{\tau}{T}\right) e^{j2\pi \frac{k}{T} t} \tag{129}$$

Hint: Recall the fundamental result from Chapter 3CT that $e^{j2\pi Ft} * h(t) = H(F)e^{j2\pi Ft}$. Let $h(t) = \sqcap(t/\tau)$.

c) Conclude from parts (a) and (b) that a periodic rectangular pulse train can be expressed as a sum of complex sinusoids.

$$\sum_{n=-\infty}^{\infty} \sqcap\left(\frac{t-nT}{\tau}\right) = \frac{1}{T}\sum_{k=-\infty}^{\infty} \tau \operatorname{sinc}\left(k\frac{\tau}{T}\right)e^{j2\pi \frac{k}{T}t} \qquad (130)$$

The right-hand side of Eq. (130) is called the *Fourier series representation* of the periodic rectangular pulse train. The Fourier series is studied in depth in Chapter 5CT.

 4CT.4.5 Repeat Problem 4CT.4.4 with $\sqcap(t/\tau)$ replaced with an arbitrary pulse $p(t)$ to show that a periodic pulse train can be expressed as a Fourier series::

$$\sum_{n=-\infty}^{\infty} p(t-nT) = \frac{1}{T}\sum_{k=-\infty}^{\infty} P\left(k\frac{\tau}{T}\right)e^{j2\pi \frac{k}{T}t} \qquad (131)$$

where $P(F)$ is the spectrum of $p(t)$. This result is established in another way in Section 5CT.5.

4CT.5   Properties of the Fourier Transform

4CT.5.1 Show that $\int_{-\infty}^{\infty} x(t)\,dt = X(0)$.

4CT.5.2 Show that $\int_{-\infty}^{\infty} X(F)\,dF = x(0)$.

4CT.5.3 Show which of the following signals is even, which is odd, and which is neither.
a) $x_1(t) = t$
b) $x_2(t) = x_o(t) * x_o(t)$, where $x_o(t)$ is any odd function
c) $x_3(t) = \cos(2\pi t + 45°)$
d) $x_4(t) = F\{\sin(t)\}$, where $F\{\lambda\}$ is any real function of λ
e) $x_5(t) = F\{\cos(t)\}$, where $F\{\lambda\}$ is any real function of λ

4CT.5.4 Refer to Problem 4CT.5.3 and state the symmetry properties that apply to spectrums $X_1(F)$ through $X_5(F)$, corresponding to waveforms $x_1(t)$ through $x_5(t)$.

4CT.5.5 **a)** Show that $X(F)$ is real and even if and only if $x(t)$ is real and even.
b) Give an example for part (a).

4CT.5.6 **a)** Show that $X(F)$ is imaginary and odd if and only if $x(t)$ is real and odd.
b) Given an example for part (b).

4CT.5.7 The waveform $x(t) = 10\cos(2\pi 25t + 22°) + 255\cos(2\pi 125t)$ is applied to an ideal low-pass filter. Assume that the filter cutoff frequency W, equals 40 Hz.
a) Find $X(F)$ and plot it and $H(F)$ on the same plot.
b) Find and plot $Y(F)$.
c) Use the inverse Fourier transform to find $y(t)$. Write your answer as a real sinusoidal function.
d) Make a detailed comparison of your analysis and answer with that of Example 3CT.5 for $W = 40$.

4CT.5.8 Three DSB waveforms are received simultaneously. The received waveform is given by

$$y(t) = A_1 x_1(t)\cos(2\pi F_1 t) + A_2 x_2(t)\cos(2\pi F_2 t) + A_3 x_3(t)\cos(2\pi F_3 t)$$

where $x_1(t)$, $x_2(t)$, and $x_3(t)$ are three baseband "message" waveforms. The spectrum of each message signal is 0 for $|F| > W$ Hz. The frequencies F_1, F_2, and F_3 are all large compared to W, and they are separated by at least $2W$ Hz.
a) Sketch $Y(F)$. For simplicity in illustration, assume that the message spectrums are real and even. You can use any message spectrums that satisfy the condition stated.

b) Show that if you recover $x_2(t)$ from $y(t)$ if you multiply $y(t)$ by $\cos(2\pi F_2 t)$ and apply the resulting product to a low-pass filter. The receiver is called a *coherent demodulator*. It requires a *local oscillator* to generate the waveform $\cos(2\pi F_2 t)$. Because of the need for a local oscillator, coherent demodulators are relatively expensive compared to the receivers used in commercial AM (Problem 4CT.5.9).

c) Show what happens to the receiver of part (b) if the local oscillator has a phase error of $90°$. Specifically, find the output of the low-pass filter when $y(t)$ is multiplied by $\sin(2\pi F_2 t)$ instead of $\cos(2\pi F_2 t)$.

4CT.5.9 The transmitted signal from a commercial AM radio station has the form

$$y(t) = A_c(1 + mx_n(t))\cos(2\pi F_1 t) = A_c \cos(2\pi F_1 t) + A_c m x_n(t)\cos(2\pi F_1 t)$$

where $x_n(t)$ is a baseband waveform with bandwidth $W < 5$ kHz. F_1 is in the range of 540 to 1,700 kHz; $x_n(t)$ is normalized for $|x_n(t)| \le 1$; m is the *modulation index* $(0 < m \le 1)$. Because m and $x_n(t)$ have absolute values limited to unity, the factor $(1 + mx_n(t))$ in $y(t)$ is never negative. In effect, this factor modulates the amplitude of $\cos(2\pi F_1 t)$. This factor can be extracted from $y(t)$ using an inexpensive circuit called an *envelope detector*.

a) Sketch $Y(F)$. You can use any message spectrum that satisfies the condition stated. For simplicity in illustration, assume that the message spectrum is real and even.

b) Assume that $x_n(t) = \cos(2\pi 440 t)$. Find the total average power carried by $y(t)$ and the average power carried by each term $A_c \cos(2\pi F_1 t)$ and $A_c m x_n(t)\cos(2\pi F_1 t)$. Find the percentage of the total average power that is carried by the DSB term $A_c m x_n(t)\cos(2\pi F_1 t)$ and the percentage carried by the "carrier" $A_c \cos(2\pi F_1 t)$. Use your result to show that AM is inefficient in terms of the amount of power "wasted" on the carrier.

c) Make a rough sketch of $y(t)$ for $A_c = 100$, $x_n(t) = \cos(2\pi W t)$ with $W = 1$ kHz and $F_1 = 1,000$ kHz. Assume that $m = 0.7$. Use the form $y(t) = A_c(1 + mx_n(t))\cos(2\pi F_1 t)$ to show how the factor $(1 + mx_n(t))$ modulates the amplitude of the cosine. Hint: There are $1,000$ cycles of $\cos(2\pi F_1 t)$ within one cycle of $\cos(2\pi W t)$. Do not try to include all these cycles. Just give a rough indication of them on your sketch.

4CT.5.10 Apply time scaling $x(t) \to x(at)$, where $a > 0$, to the waveform $x(t) = A\cos(2\pi F_o t)$ and describe what happens to $X(F)$. Sketch the waveform and spectrum before and after the time scaling.

4CT.5.11 Time-shift the waveform $x(t) = A\cos(2\pi F_o t)$ by τ and describe what happens to $X(F)$. Sketch the waveform and spectrum before and after the time shifting.

4CT.5.12 a) Use the FT pair $e^{-\alpha t}u(t) \longleftrightarrow 1/(j2\pi F + \alpha)$ and Table 4CT.2 to show that $x(t) = e^{-\alpha|t|}$ has spectrum $X(F) = 2\alpha/((2\pi F)^2 + \alpha^2)$.
b) Sketch $x(t)$ and $X(F)$.

4CT.5.13 a) Use the FT pair $e^{-\alpha t}u(t) \longleftrightarrow 1/(j2\pi F + \alpha)$ and Table 4CT.2 to find the spectrum of $x(t) = e^{-\alpha t}u(t) - e^{\alpha t}u(-t)$.
b) Sketch $x(t)$ and $X(F)$.

4CT.5.14 Use the multiplication property and the transform pair $e^{-\alpha|t|} \longleftrightarrow 2\alpha/((2\pi F)^2 + \alpha^2)$ to evaluate the convolution $2\alpha/((2\pi F)^2 + \alpha^2) * 2\alpha/((2\pi F)^2 + \alpha^2)$. Comment on the interesting features of your result.

4CT.5.15 Use the FT pair $e^{-\alpha t}u(t) \longleftrightarrow 1/(j2\pi F + \alpha)$ and Table 4CT.2 to find the Fourier transforms of the following signals:
a) $e^{-\alpha(t-1)}u(t-1)$
b) $e^{-\alpha(t-1)}u(t)$
c) $\sum_{k=-K}^{K} e^{-\alpha(t-k)}u(t-k)$.

4CT.5.16 Derive the impulse response of an ideal high-pass filter. Hint: $H_{hp}(F) = 1 - H_{lp}(F)$.

4CT.5.17 Derive the impulse response of an ideal band-rejection filter. Hint: $H_{br}(F) = 1 - H_{bp}(F)$.

4CT.5.18 Let $g(t)$ be a real, even waveform having spectrum $G(F)$. Define a new waveform $x(t) = g(t) + G(t)$. Find $X(F)$. Comment on the interesting features of your result.

4CT.5.19 An LTI system has input $x(t) = \sqcap(t/\tau)$ and output $y(t) = \tau\{\wedge(\frac{t-0.5\tau}{\tau}) + \wedge(\frac{t+0.5\tau}{\tau})\}$.
a) Sketch $x(t)$ and $y(t)$.
b) Find $H(F)$.
c) Is this system causal?

4CT.5.20 Recall that a rectangular pulse can be expressed as $\sqcap(t/\tau) = u(t + (\tau/2)) - u[t - (\tau/2)]$. Use the time-shift property and the pair $u(t) \longleftrightarrow 1/2\delta(F) + 1/j2\pi F$ to find the spectrum of $\sqcap(t/\tau)$.

4CT.5.21 Derive the following symmetry properties of the Fourier Transform:
a) 4CT.SP1
b) 4CT.SP2
c) 4CT.SP3
d) 4CT.SP4
e) 4CT.SP5
f) 4CT.SP6

4CT.5.22 Derive the following operational properties of the Fourier transform:
a) 4CT.OP4
b) 4CT.OP8 by using the multiplication property
c) 4CT.OP10
d) 4CT.OP11

4CT.5.23 Derive the sampling property with the use of Eq. (37) and the frequency translation property.

4CT.5.24 Show that:
a)

$$\int_{-\infty}^{\infty} \text{sinc}(t)\, dt = 1 \tag{132}$$

b)

$$\int_{-\infty}^{\infty} \text{sinc}(\lambda)\, \text{sinc}(t - \lambda)\, d\lambda = \text{sinc}(t) \tag{133}$$

c)

$$\int_{-\infty}^{\infty} \text{sinc}(t - n)\, \text{sinc}(t - m)\, dt = 0 \qquad \text{for } n \neq m \tag{134}$$

where n and m are integers.
d)

$$\int_{-\infty}^{\infty} \text{sinc}^2(t)\, dt = 1 \tag{135}$$

4CT.6 The Sampling Theorem

4CT.6.1 The waveform $x(t)$ contains frequency components up to 20 kHz.
a) What is the minimum sampling rate that allows the waveform to be reconstructed from its samples?
b) Suppose that $x(t)$ is sampled at a rate $F_s = 38$ kHz. What part of $X(F)$, if any, can be recovered without error from the samples?

4CT.6.2 A particular audio source produces frequency components up to 30 kHz. An adult with good hearing can hear frequencies only between 20 and 20,000 Hz. A test is made in which a listener compares the audio produced by two systems. In system 1 the audio is sampled at the Nyquist rate, and the samples are applied to a reconstruction filter. In system 2, the source output is first applied to an antialiasing filter designed to pass frequency components only up to 20,000 Hz. The antialiasing filter output is sampled and the samples are applied to a reconstruction filter.

a) Draw the block diagram of each system.

b) Give the specifications for system 2 so that, in theory, the listener is unable to distinguish it from system 1.

4CT.6.3 A source produces frequency components up to 1 MHz. Write down the Nyquist rate corresponding to this source.

4CT.6.4 The output of an ideal sampler, $x_s(t) = \sum_n x(nT_s)\delta(t - nT_s)$, is applied to an LTI system having unit impulse response $h(t) = \sqcap((t - 0.5T_s)/T_s)$. This system, called a sample-and-hold reconstruction system, has output $x_{sh}(t)$ where $x_{sh}(t)$ is an approximation to $x(t)$.

a) Sketch $h(t)$ and show how to implement $h(t)$ using an adder, a delay, and an integrator.

b) Draw the sample and hold reconstruction system.

c) Use sketches to compare $x_{sh}(t)$ with $x(t)$.

d) Find the expression for $X_{sh}(F)$.

e) Sketch $X(F)$, $X_s(F)$, and $X_{sh}(F)$ to find conditions for which $x(t)$ can be exactly reconstructed by applying $x_{sh}(t)$ to a suitably designed low-pass filter. Specify the conditions and plot the low-pass filter frequency response characteristics. Hint: The amplitude characteristic of the low-pass filter is not constant within the passband.

4CT.6.5 A popular method of interpolating between data points is straight-line, or linear, interpolation: With this method, adjacent samples are simply connected by straight lines, as illustrated in Figure 2CT.33. In that figure, the samples are $x(-5)$, $x(-4)$, ... $x(3)$, $x(4)$.

a) Show by means of sketches that a straight-line–interpolated waveform can be expressed as

$$x_{stl}(t) = x_s(t) * \wedge\left(\frac{t}{T_s}\right)$$

where $x_s(t) = \sum_{n=-\infty}^{\infty} x(nT_s)\delta(t - nT_s)$.

b) Show that the spectrum of $x_{stl}(t)$ is given by

$$X_{stl}(F) = X_s(F)T_s \operatorname{sinc}^2\left(\frac{F}{F_s}\right)$$

where $F_s = 1/T_s$.

c) Sketch $X(F)$, $X_s(F)$, $T_s \operatorname{sinc}^2(\frac{F}{F_s})$, and $X_{stl}(F)$ to show how $X_{stl}(F)$ is related to $X(F)$. Chose a convenient real-even function $X(F)$ for your sketch.

d) Describe what happens to $x_{stl}(t)$ and $X_{stl}(F)$ as F_s is increased.

e) Show that $x(t)$ can, under certain conditions, be exactly reconstructed by applying $x_{stl}(t)$ to a low-pass filter. Specify the conditions and plot the low-pass filter frequency response characteristics. Hint: The amplitude characteristic of the low-pass filter is not constant within the passband.

4CT.6.6 Refer to Problems 4CT.6.4 and 4CT.6.5 and show you can exactly reconstruct $x(t)$ by sampling $x_{sh}(t)$ or $x_{stl}(t)$ and applying the samples to an ideal low-pass filter. State the conditions under which perfect reconstruction is possible.

4CT.6.7 In addition to its use as a reconstruction formula, Nyquist-Shannon interpolation can be used to generate a band-limited signal from an arbitrary data sequence $\{w[k]\}$ by replacing $x(kT_s)$ with $w[k]$. Show that Nyquist-Shannon interpolation always produces a band-limited waveform, regardless of the data sequence $w[k]$ used.

Figure 4CT.26
Reconstruction filter.

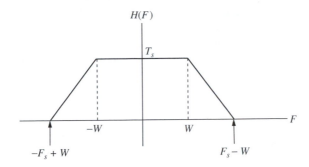

4CT.6.8 Find a closed form for $x(t)$ in the Nyquist-Shannon interpolation formula if $x(nT_s) = A\cos(2\pi F_o nT_s)$ and

a) $F_o < \frac{1}{2}F_s$.

b) $\frac{1}{2}F_s < F_o < F_s$.

Hint: The Nyquist-Shannon interpolation formula works like an ideal low-pass filter having cutoff frequency $\frac{1}{2}F_s$, gain T_s and input $x_s(t) = \sum_{n=-\infty}^{\infty} x(nT_s)\delta(t - nT_s)$. Sketch the spectrum of $x_s(t)$.

4CT.6.9 An engineer considers two ways to construct a waveform having no frequency components above W Hz. Each method starts with the same band-unlimited waveform $x(t)$. Method 1 is to filter $x(t)$ with an ideal low-pass filter having the cutoff frequency W. Method 2 is to sample $x(t)$ at a rate $F_s = 2W$ and use the samples in a Shannon-Nyquist interpolation formula.

a) Are the waveforms produced by the two systems identical? If not, how do they differ?

b) Assume that $x(t) = A\cos(2\pi F_1 t) + B\cos(2\pi F_2 t)$, where $F_1 = 0.7W$ and $F_2 = 1.2W$. Find the waveform produced by each method.

4CT.6.10 A waveform $x(t)$ having highest significant frequency W is sampled at a rate $F_s > 2W$. The sampled waveform $x_s(t)$ is then applied to a reconstruction filter having the frequency response characteristic shown in Figure 4CT.26.

a) Show, by means of frequency domain sketches, that the output of the reconstruction filter equals $x(t)$.

b) Find the impulse response of the filter.

c) Write down the interpolation formula associated with the filter.

d) State what happens to the filter and the interpolation formula if $F_s = 2W$.

4CT.7   Deriving Spectrums by Differentiation

4CT.7.1 Use the method of differentiation to find the spectrum of $x(t) = A\sqcap(t/T)$.

4CT.7.2 Consider the pulse

$$x(t) = \begin{cases} 0 & t < 0 \\ t & 0 \le t < 1 \\ 0 & 1 \le t \end{cases} \tag{136}$$

a) Sketch $x(t)$.

b) Find $X(F)$ using the method of differentiation.

4CT.7.3 You are given an even signal $x(t)$ specified by

$$x(t) = \begin{cases} 1 & 0 \le t < t_1 \\ 1 - \left(\dfrac{t - t_1}{t_2 - t_1}\right) & t_1 \le t \le t_2 \\ 0 & t_2 < t \end{cases} \tag{137}$$

This signal is called a *trapezoidal pulse*.

a) Plot $x(t)$.

b) Find $X(F)$ using the method of differentiation.

c) Show that your answer to part (b) reduces to the spectrum of a rectangular pulse for $t_2 = t_1$.

d) Show that your answer to part (b) reduces to the spectrum of a triangular pulse for $t_1 = 0$.

4CT.7.4 Consider the signal

$$x(t) = A \cos\left(2\pi \frac{t}{T}\right) \sqcap \left(\frac{2t}{T}\right) \tag{138}$$

a) Plot $x(t)$.

b) Find $X(F)$ using the method of differentiation.

c) Sketch $X(F)$. If you prefer, write a computer program to plot $X(F)$.

4CT.8   Energy Waveforms, Energy Density Spectrums, and Parseval's Theorem

4CT.8.1 Show that

$$\int_{-\infty}^{\infty} x(t) y^*(t)\,dt = \int_{-\infty}^{\infty} X(F)Y^*(F)\,dF \tag{139}$$

where $X(F)$ and $Y(F)$ are the Fourier transforms of $x(t)$ and $y(t)$, respectively. This result is known as the *generalized Parseval's theorem*. Hint: Set $x_1(t) = x(t)$ and $x_2(t) = y^*(-t)$ in the convolution property. The function

$$\phi_{xy}(t) = \int_{-\infty}^{\infty} x(\lambda) y^*(\lambda - t)\,d\lambda = \mathcal{F}^{-1}\{X(F)Y^*(F)\} \tag{140}$$

is called the *cross-correlation function* of $x(t)$ and $y(t)$. Cross-correlation functions are widely used in signal detection and pattern recognition systems.

4CT.8.2 Show that $K = 1$ in Eq. (88). Hint: Use Parseval's theorem. A change of variables of integration enables you to cancel the integrals of the gaussian functions on both sides of the Parseval's theorem equation.

4CT.10   Convergence of the Fourier Transform and Its Inverse

4CT.10.1 a) Derive $X_T(F)$ of Eq. (109).

b) Take the limit $T \to \infty$ to confirm $X(F)$ given by Eq. (110).

Miscellaneous Problems

4CT.1. The *RMS bandwidth* of a waveform $x(t)$ is defined as

$$W_{\mathrm{RMS}} = \sqrt{\frac{1}{E} \int_{-\infty}^{\infty} F^2 |X(F)|^2\,dF} \tag{141}$$

where $X(F)$ is the spectrum of $x(t)$ and E is the energy in $x(t)$.

a) Show that

$$W_{\mathrm{RMS}} = \frac{1}{2\pi} \sqrt{\frac{1}{E} \int_{-\infty}^{\infty} \left|\frac{\partial x(t)}{\partial t}\right|^2 dt} \tag{142}$$

b) Notice that this result expresses W_{RMS} directly in terms of $x(t)$. Describe the relationships between W_{RMS}, $X(F)$, and $x(t)$ in your own words.

c) Find the RMS bandwidth of a triangular pulse $x(t) = \wedge(t/\tau)$ and comment on your answer.

d) Find the RMS bandwidth of a trapezoidal pulse of Problem 4CT.7.3. What happens as $t_2 \to t_1$?

Comment on your answer.

4CT.2. Show that

$$T \sum_{n=-\infty}^{\infty} x(nT) = \sum_{n=-\infty}^{\infty} X\left(\frac{n}{T}\right) \qquad \text{where } T > 0 \qquad (143)$$

where $x(t)$ is any Fourier transformable waveform. This result is known as *Poisson's formula*.

4CT.3. Show that

$$2W_1 \operatorname{sinc}(2W_1t)^*2W_2 \operatorname{sinc}(2W_2t) = \begin{cases} 2W_1 \operatorname{sinc}(2W_1t) & \text{if } W_1 \leq W_2 \\ 2W_2 \operatorname{sinc}(2W_2t) & \text{if } W_2 < W_1 \end{cases} \qquad (144)$$

4DT Spectrums

Signals, like LTI systems, can be described in either the sequence domain or the frequency domain. The frequency domain representation of a sequence is called its *spectrum*. We obtain a sequence's spectrum by discrete-time Fourier transforming the sequence. We recover a sequence from a spectrum by inverse discrete-time Fourier transforming the spectrum.

Relation Between a Signal and Its Spectrum

Spectrum:

$$X(e^{j2\pi f}) = \sum_{m=-\infty}^{\infty} x[m]e^{-j2\pi fm} \tag{1}$$

Sequence:

$$x[n] = \int_{-0.5}^{0.5} X(e^{j2\pi f})e^{j2\pi fn}\,df \tag{2}$$

$X(e^{j2\pi f})$ is sometimes called a *density spectrum* because, if $x[n]$ has the unit volt, then $X(e^{j2\pi f})$ has the unit volts per normalized frequency. Every sequence that can be completely plotted on paper or displayed on a monitor has a spectrum.

Spectrums play a central and simplifying role in signals and systems theory. When a sequence $x[n]$ is applied to an LTI system, the spectrum of the output sequence $y[n]$ is given by the product: $Y(e^{j2\pi f}) = X(e^{j2\pi f})H(e^{j2\pi f})$. The relation between $Y(e^{j2\pi f})$ and $X(e^{j2\pi f})$ is central to signal processing. It provides a new and powerful way to find the output of an LTI system. As we shall see, we can often find $y[n]$ by an inspection of $X(e^{j2\pi f})H(e^{j2\pi f})$.

4DT.2   Spectrums and LTI Systems

The concept of a spectrum follows directly from material we have presented so far. This section presents two derivations of Eq. (1). Each derivation discloses the property $Y(e^{j2\pi f}) = X(e^{j2\pi f})H(e^{j2\pi f})$ that is central to LTI signal processing.

In the first derivation we simply substitute

$$h[n] = \int_{-0.5}^{0.5} H(e^{j2\pi f}) e^{jn2\pi f} \, df \tag{3}$$

from Chapter 3CT into

$$y[n] = \sum_{m=-\infty}^{\infty} h[n-m] x[m] \tag{4}$$

from Chapter 2DT. The substitution gives us

$$y[n] = \sum_{m=-\infty}^{\infty} \left\{ \int_{-0.5}^{0.5} H(e^{j2\pi f}) e^{j2\pi f(n-m)} \, df \right\} x[m]$$

$$= \int_{-0.5}^{0.5} H(e^{j2\pi f}) \left\{ \sum_{m=-\infty}^{\infty} x[m] e^{-j2\pi fm} \right\} e^{j2\pi fn} \, df \tag{5}$$

The term in braces is the spectrum of $x[n]$:

$$X(e^{j2\pi f}) = \sum_{m=-\infty}^{\infty} x[m] e^{-j2\pi fm} = \mathcal{DTFT}\{x[n]\} \tag{6}$$

when we substitute Eq. (6) into Eq. (5), we obtain the important result

$$y[n] = \int_{-0.5}^{0.5} H(e^{j2\pi f}) X(e^{j2\pi f}) e^{j2\pi fn} \, df \tag{7}$$

The integral on the right-hand side of Eq. (7) is an inverse discrete-time Fourier transform. This means that $H(e^{j2\pi f}) X(e^{j2\pi f})$ is the discrete-time Fourier transform of $y[n]$.

$$Y(e^{j2\pi f}) = X(e^{j2\pi f}) H(e^{j2\pi f}) \tag{8}$$

Thus, the output spectrum equals the product of the input spectrum and the system's frequency response characteristic. We can obtain $y[n]$ by inverse Fourier transforming $Y(e^{j2\pi f})$ as in Eq. (7). The sequence and frequency domain input-output relationships (4) and (8) are summarized in the following two flow graphs.

Sequence domain:

$$x[n] \xrightarrow{h[n]} y[n] = x[n] * h[n] \tag{9a}$$

Frequency domain:

$$X(e^{j2\pi f}) \xrightarrow{H(e^{j2\pi f})} Y(e^{j2\pi f}) = X(e^{j2\pi f}) H(e^{j2\pi f}) \tag{9b}$$

Figure 4DT.1 shows the relationships between sequences and spectrums given by Eqs. (6)–(8) in pictorial form. The first stage of processing Fourier transforms the waveform $x[n]$ into its spectrum $X(e^{j2\pi f})$, as in Eq. (6). The second stage performs the multiplication $X(e^{j2\pi f}) H(e^{j2\pi f})$ indicated by Eq. (8). In the final stage, $y[n]$ is obtained by inverse Fourier transforming the product $Y(e^{j2\pi f}) = X(e^{j2\pi f}) H(e^{j2\pi f})$ as in Eq. (7). The three-step processing shown is equivalent to a convolution $y[n] = x[n] * h[n]$. This block diagram cannot be realized in hardware without modification. However, it provides motivation and a starting point for the practical signal processing techniques developed in Chapter 5DT.

Figure 4DT.1
Relationships
between sequences
and spectrums:
Eqs. (6)–(8).

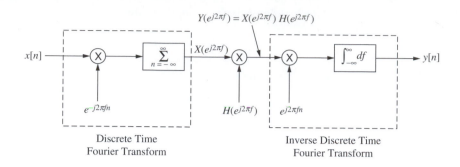

The second derivation shows that the filtering concepts connected with multiple sinu-soids apply directly to filtering any Fourier transformable sequence $x[n]$. The derivation starts with Eq. (3DT.45), which is repeated here.

$$\underbrace{\sum_{i=-I}^{I} X_i e^{j2\pi f_i n}}_{x[n]} \xrightarrow{h[n]} \underbrace{\sum_{i=-I}^{I} X_i H(e^{j2\pi f_i}) e^{j2\pi f_i n}}_{y[n]} \tag{3DT.45}$$

Recall that this I-O relation is a special case of the superposition principle. It states that the response of an LTI system to a sum of sinusoids is the sum of the responses to the individual sinusoids. Recall also that X_i tells us the amplitude and phase of the input sinusoid $X_i e^{j2\pi f_i n}$, and $X_i H(e^{j2\pi f_i})$ tells us the amplitude and phase of the corresponding output sinusoid $X_i H(e^{j2\pi f_i}) e^{j2\pi f_i n}$. If we set $f_i = i\Delta f$ and $X_i = X(e^{j2\pi i \Delta f})\Delta f$ in Eq. (3DT.45) and take the limit $\Delta f \to 0, i\Delta f \to f$, and $I\Delta f \to 0.5$, then Eq. (3DT.45) becomes

$$\underbrace{\int_{-0.5}^{0.5} X(e^{j2\pi f}) e^{j2\pi f n} df}_{x[n]} \xrightarrow{h[n]} \underbrace{\int_{-0.5}^{0.5} H(e^{j2\pi f}) X(e^{j2\pi f}) e^{j2\pi f n} df}_{y[n]} \tag{10}$$

The inverse discrete-time Fourier transforms in Eq. (10) are limiting forms of (3DT.45). We can interpret the integral on the left-hand side of Eq. (10) as a sum of infinitesimal sinusoids, each having the form $X(e^{j2\pi f}) e^{j2\pi f n} df$. The input spectrum $X(e^{j2\pi f})$ describes the amplitudes and phases of these sinusoidal components. Similarly, we can interpret the right-hand side of Eq. (10) as a sum of infinitesimal sinusoids, each having the form $H(e^{j2\pi f}) X(e^{j2\pi f}) e^{j2\pi f n} df$. The output spectrum is $Y(e^{j2\pi f}) = H(e^{j2\pi f}) X(e^{j2\pi f})$. The term $H(e^{j2\pi f})$ in Eq. (10) acts the same way, as the term $H(e^{j2\pi f_i})$ in Eq. (3DT.45) to filter the sinusoidal components in the input.

EXAMPLE 4DT.1   Distinguishing Added Sequences

Figure 4DT.2(a) a shows a sequence $x_1[n]$ and its spectrum $X_1(e^{j2\pi f})$. Looking at the spectrum, we see that $x_1[n]$ is com-posed of bands of complex sinusoidal components centered at $f = \pm 0.15$. Similarly, Figure 4DT.2(b) shows a sequence $x_2[n]$ and its spectrum $X_2(e^{j2\pi f})$. $x_2[n]$ is composed of complex sinusoidal components whose frequencies occupy a band of sinusoidal components centered at $f = \pm 0.35$. Figure 4DT.2(c) shows the sum $x[n] = x_1[n] + x_2[n]$ and

Figure 4DT.2
Two sequences and
their Spectrums:
(a) $x_1[n]$ and its
spectrum; (b) $x_2[n]$
and its spectrum;
(c) $x[n] = x_1[n] + x_2[n]$
and its spectrum. (For
clarity, the data
points of each
sequence have been
connected by straight
lines.)

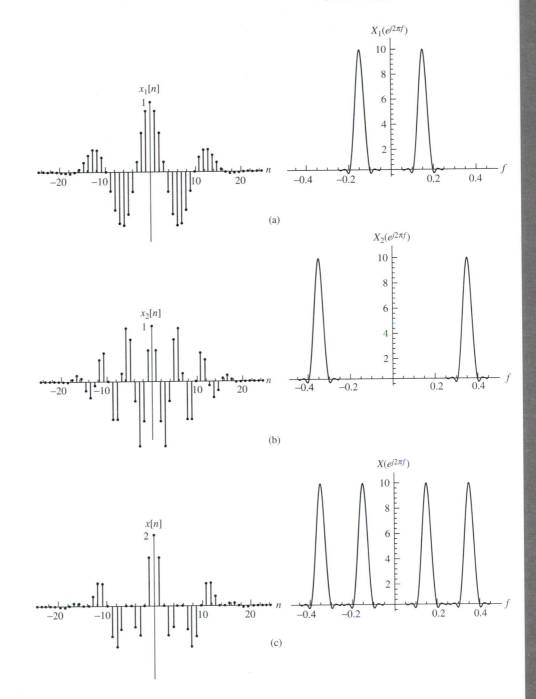

(a)

(b)

(c)

its spectrum $X(e^{j2\pi f})$. The spectrum $X(e^{j2\pi f})$ equals $X_1(e^{j2\pi f}) + X_2(e^{j2\pi f})$ because the sinusoidal components in $x[n]$ are the sum of the sinusoidal components in $x_1[n]$ and $x_2[n]$. Notice that $X_1(e^{j2\pi f})$ and $X_2(e^{j2\pi f})$ are clearly distinguishable by inspection of $X(e^{j2\pi f})$, even though $x_1[n]$ and $x_2[n]$ are not clearly distinguishable by inspection of $x[n]$. This figure shows us that it can be much easier to understand a sequence by looking at its spectrum than by looking at the sequence itself. This understanding is made possible by the DTFT's ability to distinguish sinusoidal components having different frequencies.

Suppose you wanted to process $x[n]$ in such a way as to recover the sequences $x_1[n]$ and $x_2[n]$. It would be difficult to see how to accomplish this separation if you consider only the plot of $x[n]$. However, when you look at the spectrum $X(e^{j2\pi f})$, you see that $X_1(e^{j2\pi f})$ and $X_2(e^{j2\pi f})$ can be separated by multiplying $X(e^{j2\pi f})$ by $H_1(e^{j2\pi f})$ and $H_2(e^{j2\pi f})$, shown in Figure 4DT.3(a): $X(e^{j2\pi f})H_1(e^{j2\pi f}) = X_1(e^{j2\pi f})$ and $X(e^{j2\pi f})H_2(e^{j2\pi f}) = X_2(e^{j2\pi f})$. Complete separation is possible because the frequency bands occupied by $X_1(e^{j2\pi f})$ and $X_2(e^{j2\pi f})$ do not overlap. Each filter passes only those frequency components belonging to $x_1[n]$ or $x_2[n]$. The sequence domain counterpart of Figure 4DT.3(a) is shown in Figure 4DT.3(b).

The use of filters to separate sequences is analogous to the use of red and blue transparencies to separate red and blue light. A red transparency permits the passage of red light while the blue transparency permits the passage of blue light.

This example has demonstrated the analytical power of frequency domain analysis as well as its conceptual power. Think how difficult it would be, in comparison, to evaluate the convolutions $x[n] * h_1[n]$ and $x[n] * h_2[n]$!

Figure 4DT.3
Use of filters to separate added signals: (a) frequency domain; (b) sequence domain.

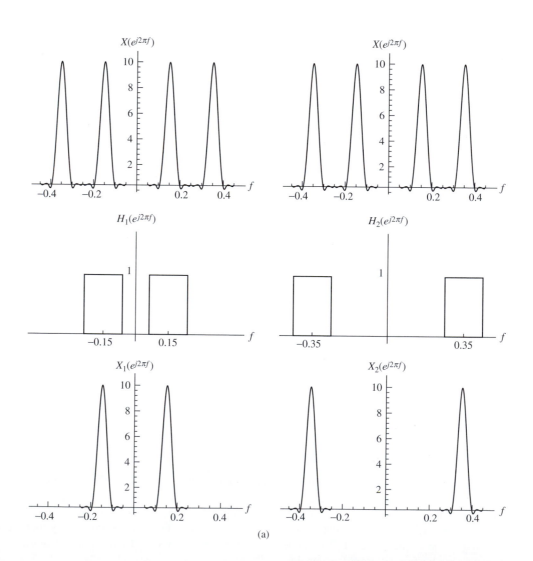

(a)

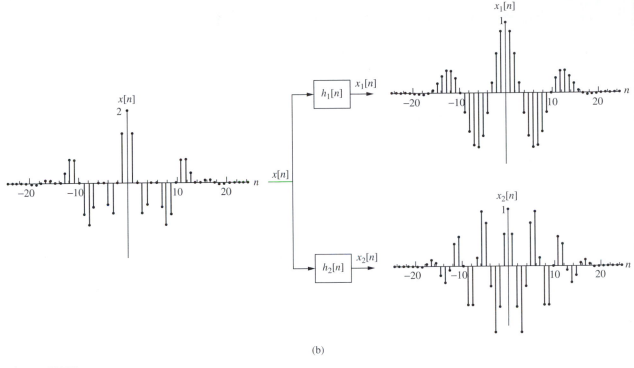

(b)

Figure 4DT.3
(*Cont.*)

4DT.3   ## Amplitude and Phase Spectrums

The *amplitude and phase spectrums* of $x[n]$ are defined as $|X(e^{j2\pi f})|$ and $\angle X(e^{j2\pi f})$, respectively. Similarly, the amplitude and phase spectrums of $y[n]$ are $|H(e^{j2\pi f})X(e^{j2\pi f})|$ and $\angle\{H(e^{j2\pi f})X(e^{j2\pi f})\}$, respectively. By the properties of complex numbers, $|H(e^{j2\pi f})X(e^{j2\pi f})| = |H(e^{j2\pi f})||X(e^{j2\pi f})|$. Therefore, the amplitude spectrum of $y[n]$ is given by the product of the amplitude spectrum of $x[n]$ and the amplitude response characteristic of the system:

$$|Y(e^{j2\pi f})| = |H(e^{j2\pi f})||X(e^{j2\pi f})| \tag{11}$$

Similarly, by the properties of complex numbers, $\angle\{H(e^{j2\pi f})X(e^{j2\pi f})\} = \angle H(e^{j2\pi f}) + \angle X(e^{j2\pi f})$. Therefore, the phase spectrum of $y[n]$ is given by the sum of the phase spectrum of $x[n]$ and the phase response characteristic of the system:

$$\angle Y(e^{j2\pi f}) = \angle H(e^{j2\pi f}) + \angle X(e^{j2\pi f}) \tag{12}$$

4DT.4   Some Sequences and Their Spectrums

We saw examples of discrete-time Fourier transform pairs in Chapter 3DT in connection with the DT impulse response and frequency response characteristic of an LTI system. These DTFT pairs, which are summarized in Tables 3DT.1 and 3DT.2, apply equally to signals and their spectrums if we substitute $x[n]$ for $h[n]$ and $X(e^{j2\pi f})$ for $H(e^{j2\pi f})$. Let's now develop some important new transform pairs.

EXAMPLE 4DT.2   Rectangular Pulse

The rectangular pulse

$$x[n] = \sqcap[n; K] \tag{13}$$

introduced in Chapter 2, is widely used in engineering. Its spectrum can be obtained easily by evaluating the discrete-time Fourier transform sum. The substitution of Eq. (13) into (1) yields

$$X(e^{j2\pi f}) = \sum_{n=-\infty}^{\infty} \sqcap[n; K]e^{-j2\pi fn} = \sum_{n=-K}^{K} e^{-j2\pi fn}$$

$$= \frac{\sin((2K+1)\pi f)}{\sin(\pi f)} = (2K+1)\operatorname{cinc}(f; K) \tag{14}$$

The resulting DTFT pair is

$$\sqcap[n; K] \stackrel{\text{DTFT}}{\longleftrightarrow} (2K+1)\operatorname{cinc}(f; K) \tag{15}$$

In obtaining Eq. (14) we used identity (B.3) of Appendix B. We define the *circular sinc function* as

$$\operatorname{cinc}(f; \beta) = \frac{\sin((2\beta+1)\pi f)}{(2\beta+1)\sin(\pi f)} \tag{16}$$

Like all discrete-time Fourier transforms, $\operatorname{cinc}(f; K)$ is a periodic function of f having period 1: $\operatorname{cinc}(f - 1; K) = \operatorname{cinc}(f; K)$ $K = 1, 2, \ldots$. Unlike the sinc function found in CT chapters, the cinc function is not well-known in science and engineering. Figure 4DT.4 shows a plot of $\operatorname{cinc}(f; K)$ for $K = 3$. Notice from the figure that $\operatorname{cinc}(0; 3) = 1$ and $\operatorname{cinc}(f; 3) = 0$ for $f = m/7$, where $m = \pm 1, \pm 2, \ldots, \pm 6$. In general, $\operatorname{cinc}(0; K) = 1$ and $\operatorname{cinc}(f; K) = 0$ for $f = m/(2K+1)$, where $m = \pm 1, \pm 2, \ldots, \pm 2K$. (These values repeat with period 1.) It can be seen from Eq. (16) that the width of the major lobe of the cinc function is $2/(2K+1)$ (see Figure 4DT.4).

The rectangular pulse and its spectrum are plotted in Figure 4DT.5 for $K = 3$ for the normalized frequency range

$$-0.5 \le f < 0.5$$

Notice that the width of the major lobe of the spectrum is $2/(2K+1) = 2/7$. The *bandwidth* of the rectangular pulse is proportional to this width

$$w_\sqcap = \frac{k}{2K+1} \tag{17}$$

where the value of the constant k depends on the application. Notice that the unit of $w_\sqcap$ is the same as the unit of f, namely cycles per sample.

Figure 4DT.4
cinc(f; 3).

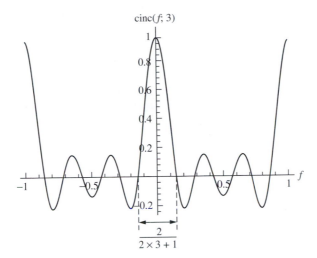

Let's explore the insight provided by the equation $Y(e^{j2\pi f}) = X(e^{j2\pi f})H(e^{j2\pi f})$ when the rectangular pulse is applied to an arbitrary low-pass filter, as illustrated in Figure 4DT.6. For simplicity, the figure assumes that the filter frequency response characteristic $H(e^{j2\pi f})$ is real. An inspection of the figure reveals that $Y(e^{j2\pi f})$ becomes a closer approximation to $X(e^{j2\pi f})$ as the filter bandwidth w is increased. Consequently $y[n]$ becomes a closer approximation to the input $x[n]$ as w is increased.

Figure 4DT.6 provides an explanation for the pulse response of the first-order system we saw in Figure 2DT.25, where the output became a better approximation to the input as the parameter a approached zero The frequency response characteristic of this system is shown in Figure 3DT.10. We see from this figure that the bandwidth of the filter increases as $|a|$ decreases. In the limit $a \rightarrow 0$, $H(e^{j2\pi f}) = 1$ and $h[n] = \delta[n]$.

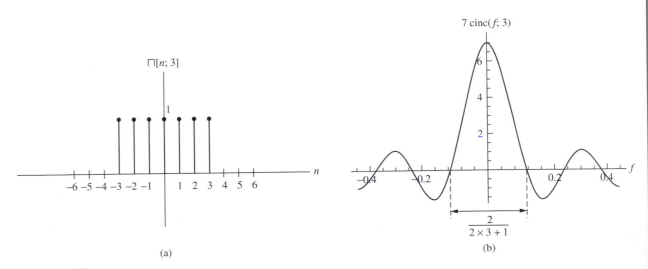

Figure 4DT.5
Rectangular pulse and Its Spectrum for $K = 3$: (a) $\sqcap[n; K]$ (the rectangular pulse contains $2K + 1$ nonzero values); (b) its spectrum $(2K+1)$ cinc($f; K$) [the width of the main lobe equals $2/(2K + 1)$].

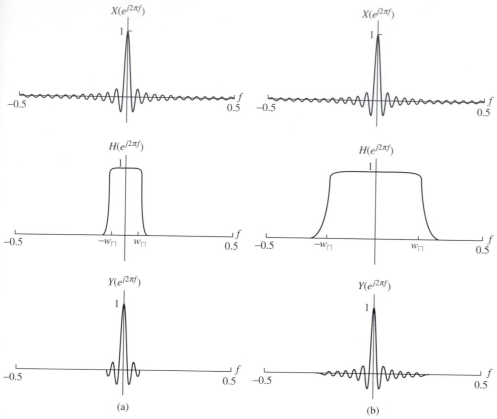

Figure 4DT.6
Frequency domain view of rectangular pulse applied to a low-pass filter.

EXAMPLE 4DT.3   Unit Impulse

The unit impulse is another important sequence whose spectrum can be obtained easily by direct evaluation of the DTFT sum. The substitution of $\delta[n]$ into Eq. (1) yields

$$X(e^{j2\pi f}) = \mathcal{DTFT}\{\delta[n]\} = \sum_{n=-\infty}^{\infty} \delta[n]e^{-j2\pi fn} = 1$$

where the last step follows from the sifting property of the impulse. Therefore

$$\delta[n] \overset{\text{DTFT}}{\longleftrightarrow} 1 \tag{18}$$

We can also derive Eq. (18) by setting $K = 0$ in Eq. (15).

The unit impulse and its spectrum are shown in Figure 4DT.7. Notice that a discrete-time unit impulse has bandwidth 1 (the maximum possible) as measured in the normalized frequency range $-0.5 \leq f < 0.5$. Its spectrum is flat.

Figure 4DT.7
Discrete-time impulse
and its spectrum:
(a) $x[n]= \delta[n]$;
(b) $X(e^{j2\pi f}) = 1$.

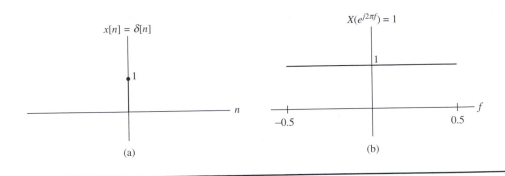

We know from Chapter 2DT that the response of any LTI system to a unit impulse is $h[n]$. Let's check this using the result of the preceding example. The substitution of $X(e^{j2\pi f}) = 1$ into Eq. (8) yields

$$Y(e^{j2\pi f}) = 1H(e^{j2\pi f}) = H(e^{j2\pi f}) \tag{19}$$

Inverse discrete-time Fourier transformation yields

$$y[n] = \delta[n] * h[n] = h[n] \quad \checkmark \tag{20}$$

EXAMPLE 4DT.4   DC Sequence

Sometimes we need singularity functions to describe the spectrum of a sequence. An example is given by the DC sequence, $x[n] = 1$. If we substitute $x[n] = 1$ into Eq. (1), we find that the sum does not converge

$$X(e^{j2\pi f}) = \mathcal{DTFT}\{1\} = \sum_{m=-\infty}^{\infty} e^{-j2\pi fm} = ? \tag{21}$$

To obtain $X(e^{j2\pi f})$, remember that the notation $\sum_{m=-\infty}^{\infty}$ means a limit.

$$X(e^{j2\pi f}) = \lim_{K\to\infty} \sum_{n=-K}^{K} e^{-j2\pi fn} = \lim_{K\to\infty} (2K+1)\,\mathrm{cinc}(f; K) \tag{22}$$

where we have used the same steps that led to Eq. (14). The major lobes in $(2K+1)\,\mathrm{cinc}(f; K)$ become narrower and higher as K increases. It can be shown that in the limit $K \to \infty$, $(2K+1)\,\mathrm{cinc}(f; K)$ is a periodic impulse train having period 1. The result is

$$x[n] = 1 \overset{\text{DTFT}}{\longleftrightarrow} X(e^{j2\pi f}) = \sum_{m=-\infty}^{\infty} \delta(f-m) \tag{23}$$

Notice that the values of $X(e^{j2\pi m})$, where m is an integer, do not exist as numbers, but we can still represent $X(e^{j2\pi f})$ using impulses.

We can regard the DC sequence $x[n] = 1$ as a limiting case of the rectangular pulse $\sqcap[n; K]$, where $K \to \infty$. We see from Figure 4DT.5 that in this limit, the pulse sequence becomes constant and its spectrum becomes an impulse on the interval $-0.5 \le f < 0.5$. The resulting DTFT pair is

$$1 \overset{\text{DTFT}}{\longleftrightarrow} \delta(f) \quad -0.5 \le f < 0.5 \tag{24}$$

You can confirm this result by evaluating the integral $\mathcal{DTFT}^{-1}\{\delta(f)\}$ using the sifting property of the impulse described in Problem 3DT.5.4. The DC sequence and its spectrum are shown in Figure 4DT.8. Notice that a DC sequence has infinite "pulse width" and zero bandwidth. The only sinusoidal component present in a DC sequence for the period $-0.5 \leq f < 0.5$ is the one having frequency 0.

Figure 4DT.8
DC Sequence and its
spectrum:
(a) $x[n] = 1$;
(b) $X(e^{j2\pi f})$,
$-0.5 \leq f < 0.5$

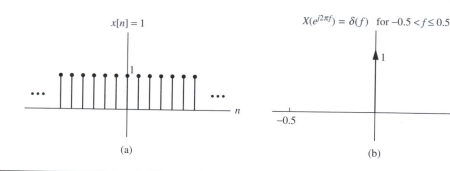

Let's use the result of the preceding example to find the response of an LTI system to the DC sequence $x[n] = 1$. The substitution of $X(e^{j2\pi f}) = \delta(f)$, $-0.5 \leq f < 0.5$, into Eq. (8) yields:

$$Y(e^{j2\pi f}) = \delta(f)H(e^{j2\pi f}) = H(e^{j2\pi 0})\delta(f) \quad -0.5 \leq f < 0.5 \tag{25}$$

Inverse Fourier transformation yields

$$y[n] = H(e^{j2\pi 0}) \tag{26}$$

which is a constant. We see that if we apply the DC sequence $x[n] = 1$ to an LTI system, the response is the DC sequence $y[n] = H(e^{j2\pi 0})$. Consequently, $H(e^{j2\pi 0})$ is sometimes referred to as the system's *DC gain*.

EXAMPLE 4DT.5   Raised Cosine

The raised cosine pulse is used in the design of digital filters. It is shown in of Figure 4DT.9(a) and defined by the formula

$$x[n] = \begin{cases} \dfrac{1}{2}\left\{1 + \cos\left(\dfrac{\pi n}{K+1}\right)\right\} & \text{for } |n| \leq K \\ 0 & \text{for } |n| > K \end{cases}$$

$$= \dfrac{1}{2}\left\{1 + \cos\left(\dfrac{\pi n}{K+1}\right)\right\} \sqcap (n; K) \tag{27}$$

We can find the spectrum by direct evaluation of the discrete-time Fourier transform sum.

$$X(e^{j2\pi f}) = \frac{1}{2}\sum_{n=-K}^{K}\left\{1 + \cos\left(\frac{\pi n}{K+1}\right)\right\}e^{-j2\pi fn} \tag{28a}$$

Euler's formula let's us express $X(e^{j2\pi f})$ in terms of real functions

$$X(e^{j2\pi f}) = 1 + \sum_{n=1}^{K}\left\{1 + \cos\left(\frac{\pi n}{K+1}\right)\right\}\cos(2\pi fn) \quad \text{for } K \geq 1 \tag{28b}$$

We can also find a closed form for Eq. (28a). If we use Euler's formula to write the cosine in (28a) as the sum of two complex exponentials and refer to Eq. (14), we obtain:

$$X(e^{j2\pi f}) = \frac{1}{2}(2K+1)\left\{\text{cinc}(f; K) + \frac{1}{2}\text{cinc}\left(f + \frac{1}{2(K+1)}; K\right) + \frac{1}{2}\text{cinc}\left(f - \frac{1}{2(K+1)}; K\right)\right\} \tag{28c}$$

Either Eq. (28b) or Eq. (28c) can be used for computer-generated plots. Figure 4DT.9(b) contains a plot of $X(e^{j2\pi f})$. The main lobe lies in the range $|f| \leq 2/(2K+1)$. Although this region is twice that for the rectangular pulse of Example 4DT.1, the amplitude spectrum decreases much more rapidly as $|f|$ increases for the raised cosine than for the rectangular pulse. Figure 4DT.9(c) shows how the three terms in Eq. (28c) work to produce small sidelobes.

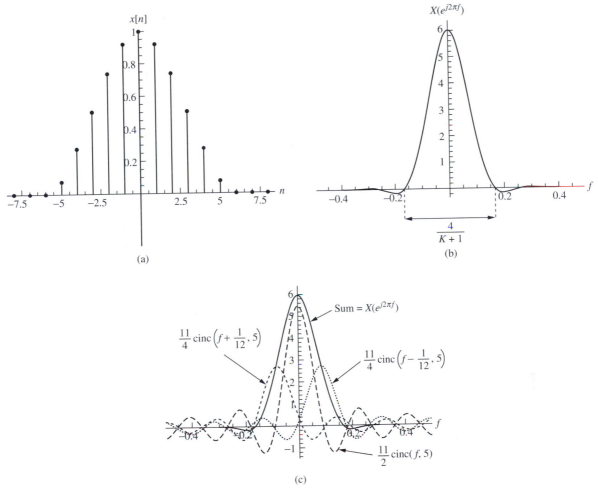

Figure 4DT.9
Raised cosine and its spectrum for $K = 5$: (a) $\frac{1}{2}\{1 + \cos(\pi n/(K+1))\} \sqcap (n, K)$ (the raised cosine has $2K+1$ nonzero values); (b) $X(e^{j2\pi f})$ (the width of the main lobe equals $4/(2K+1)$); (c) the terms in Eq. (28c) (notice how the dashed curves approximately cancel the secondary lobes of the dotted curve).

4DT.5 Properties of the Discrete-Time Fourier Transform (DTFT)

Many properties of signals and systems are described best in terms of the DTFT. The properties can be divided into two categories: *symmetry* properties of signals and systems and *operational* properties of systems.

Table 4DT.1

Symmetry Properties
of the Fourier
Transform

> **For Real or Complex** $x[n]$
>
>   Periodicity: $X(e^{j2\pi(f-1)}) = X(e^{j2\pi f})$ (4DT.SP1)
>   Conjugation: $x^*[n] \longleftrightarrow X^*(e^{-j2\pi f})$ (4DT.SP2)
>
> **For Real** $x[n]$
>
> $X(e^{-j2\pi f}) = X^*(e^{j2\pi f})$ (4DT.SP3)
> $|X(e^{-j2\pi f})| = |X(e^{j2\pi f})|$ (4DT.SP4)
> $\angle X(e^{-j2\pi f}) = -\angle X(e^{j2\pi f})$ (4DT.SP5)
> $X(e^{j2\pi f})$ is real and even iff $x[n]$ is real and even.   (4DT.SP6)
> $X(e^{j2\pi f})$ is imaginary and odd iff $x[n]$ is real and odd.   (4DT.SP7)

**4DT.5.1
Symmetry
Properties**

The most important symmetry properties for spectrums are listed in Table 4DT.1. Properties (4DT.SP1), and (4DT.SP3)–(4DT.SP5) are familiar from Chapter 3DT, where they were applied to the frequency response characteristics of LTI systems. The proofs to the symmetry properties are straightforward (Problem 4DT.5.21).

Examples of property (4DT.SP6) are given by the rectangular pulse, the impulse, the DC sequence, and the raised cosine pulse of Examples (4DT.2)–(4DT.5). Each of these sequences is real and even,[1] and each has a spectrum that is real and even. For an example of (4DT.SP4)–(4DT.SP5), consider the one-sided exponential sequence $(1 - a)a^n u[n]$, whose spectrum is plotted in Figure 3DT.10. The sequence is real, but neither even nor odd. Its amplitude spectrum is even and its phase spectrum is odd in agreement with (4DT.SP4) and (4DT.SP5).

**4DT.5.2
Operational
Properties**

The operational properties are particularly helpful in understanding how signals interact with systems. They also provide a convenient way to derive new DTFT pairs.

Linearity

$$a_1 x_1[n] + a_2 x_2[n] \longleftrightarrow a_1 X_1(e^{j2\pi f}) + a_2 X_2(e^{j2\pi f}) \qquad \text{(4DT.OP1)}$$

The *linearity property* (4DT.OP1), states that the spectrum of the sum of two scaled sequences is the sum of the sequences' corresponding scaled spectrums. The proof follows from the definition of the DTFT

$$\sum_{n=-\infty}^{\infty} \{a_1 x_1[n] + a_2 x_2[n]\}e^{-j2\pi fn} = a_1 \sum_{n=-\infty}^{\infty} x_1[n]e^{-j2\pi fn} + a_2 \sum_{n=-\infty}^{\infty} x_2[n]e^{-j2\pi fn} \qquad (29)$$

which is the linearity property

$$\mathcal{DTFT}\{a_1 x_1[n] + a_2 x_2[n]\} = a_1 \mathcal{DTFT}\{x_1[n]\} + a_2 \mathcal{DTFT}\{x_2[n]\} \qquad (30)$$

---

[1] A sequence $x[n]$ is even if $x[-n] = x[n]$. A sequence $x[n]$ is *odd* if $x[-n] = -x[n]$. Similarly, a function $g(\psi)$ is *even* if $g(-\psi) = g(\psi)$ and *odd* if $g(-\psi) = -g(\psi)$.

An application of the linearity property was given in Example 4DT.1, where the spectrum of the composite signal $x[n] = x_1[n] + x_2[n]$ was seen to be $X(e^{j2\pi f}) = X_1(e^{j2\pi f}) + X_2(e^{j2\pi f})$.

Convolution

$$x_a[n] * x_b[n] \longleftrightarrow X_a(e^{j2\pi f})X_b(e^{j2\pi f}) \tag{4DT.OP2}$$

The *convolution property* states that convolution in the sequence domain corresponds to multiplication in the frequency domain. This property was established in Section 4DT.2. Example 4DT.1 used this property to solve the problem of separating two added signals whose spectrums were disjoint.

EXAMPLE 4DT.6   Spectrum of Triangular Pulse

We can also use the convolution property to derive the spectrum of a triangular pulse $\wedge[n; M]$ defined in Eq. (2DT.76). Equation (2DT.77) tells us that

$$\wedge[n; 2K] = \frac{\sqcap[n; K] * \sqcap[n; K]}{(2K + 1)} \tag{31}$$

It follows from the convolution property and Example 4DT.2 that

$$\mathcal{DTFT}\{\wedge[n; 2K]\} = (2K + 1)\,\text{cinc}^2(f; K) \tag{32}$$

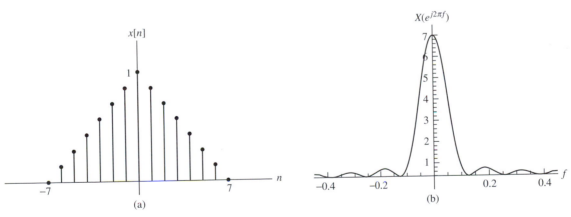

Figure 4DT.10
Triangular pulse and its spectrum: (a) $\wedge[n; M]$, (b) $(M + 1)\,\text{cinc}^2(f; \frac{1}{2}M)$ with $M = 6$.

Therefore

$$\wedge[n; M] \longleftrightarrow (M + 1)\,\text{cinc}^2\left(f; \frac{1}{2}M\right) \tag{33}$$

The triangular pulse and its spectrum are plotted in Figure 4DT.10 for $M = 6$. Although our derivation assumes that M is even, $M = 2K$, it can be shown that the result (33) applies for all $M = 0, 1, 2, \ldots$ (Problem 4DT.5.25).

Multiplication

$$x_1[n]x_2[n] \longleftrightarrow X_1(e^{j2\pi f}) \circledast X_2(e^{j2\pi f}) \tag{4DT.OP3}$$

where

$$X_1(e^{j2\pi f}) \circledast X_2(e^{j2\pi f}) = \int_{-0.5}^{0.5} X_1(e^{j2\pi \lambda})X_2(e^{j2\pi(f-\lambda)})\, d\lambda \tag{34}$$

The integral on the right-hand side of Eq. (34), denoted by $\circledast$, is called a *circular* or *periodic* convolution. The *multiplication property* (4DT.OP3) states that multiplication in the sequence domain corresponds to circular convolution in the frequency domain. This property is the dual of the convolution property. We can prove it by taking the DTFT of $x_1[n]x_2[n]$.

$$\sum_{n=-\infty}^{\infty} x_1[n]x_2[n]e^{-j2\pi fn} = \sum_{n=-\infty}^{\infty} x_2[n]e^{-j2\pi fn} \int_{-0.5}^{0.5} X_1(e^{j2\pi\lambda})\, e^{j2\pi\lambda n}d\lambda$$

$$= \int_{-0.5}^{0.5} X_1(e^{j2\pi\lambda}) \left\{ \sum_{n=-\infty}^{\infty} x_2[n]e^{-j2\pi(f-\lambda)n} \right\} d\lambda$$

$$= \int_{-0.5}^{0.5} X_1(e^{j2\pi\lambda})X_2(e^{j2\pi(f-\lambda)})\, d\lambda \tag{35}$$

An important application of the multiplication property, called *windowing*, is described in Chapter 7.

Another important application occurs when $x_2[n] = x_1[n] \equiv x[n]$. For this special case, the multiplication property becomes

$$x^2[n] \overset{\text{DTFT}}{\longleftrightarrow} \int_{-0.5}^{0.5} X(e^{j2\pi\lambda})X(e^{j2\pi(f-\lambda)})\, d\lambda \tag{36}$$

Squaring is a nonlinear operation. We see from Eq. (36) that, when a signal is squared, its spectrum convolves with itself. The following example demonstrates that this self-convolution introduces frequency components into $x^2[n]$ that did not exist in $x[n]$.

EXAMPLE 4DT.7   The Spectrum of a Squared Sequence

Problem

It can be shown that

$$2w\,\text{sinc}(2wn) \overset{\text{DTFT}}{\longleftrightarrow} \sqcap\left(\frac{f}{2w}\right) \quad -0.5 \le f < 0.5 \tag{37}$$

Find the spectrum of

$$y[n] = (2w\,\text{sinc}(2wn))^2 \tag{38}$$

Solution

We can solve this problem by substituting $x[n] = 2w \, \text{sinc}(2wn)$ and $X(e^{j2\pi f}) = \sqcap(f/2w)$, $-0.5 \le f < 0.5$, into Eq. (36) and evaluating the convolution.

We can draw the functions to be convolved on a circle, as illustrated in Figure 4DT.11. In Figure 4DT.11(a), $X(e^{j2\pi\lambda})$ is plotted above a circle having radius 1 in the complex plane. The argument $e^{j2\pi\lambda}$ is a point on this circle. As λ increases, the point $e^{j2\pi\lambda}$ travels counterclockwise around the circle, and the values of $X(e^{j2\pi\lambda})$ are repeated periodically each revolution. Figure 4DT.11(b) contains a similar plot of $X(e^{j2\pi(f-\lambda)})$. To evaluate Eq. (36), we need to find the area of the product $X(e^{j2\pi\lambda})X(e^{j2\pi(f-\lambda)})$. The product depends on the value for f. Figure 4DT.11(c) shows the product $X(e^{j2\pi\lambda})X(e^{j2\pi(f-\lambda)})$ for f in the range $0 \le f \le 2w$. We can see from the Figure 4DT.11(c) that

$$X(e^{j2\pi\lambda})X(e^{j2\pi(f-\lambda)}) = \begin{cases} 1 & f - w \le \lambda \le w \\ 0 & \text{otherwise} \end{cases} \qquad (39)$$

where $0 \le f \le 2w$.

Figure 4DT.11
Evaluation of the
circular convolution
$X(e^{j2\pi f}) \circledast X(e^{j2\pi f})$:
(a) $X(e^{j2\pi\lambda})$;
(b) $X(e^{j2\pi(f-\lambda)})$;
(c) $X(e^{j2\pi\lambda})X(e^{j2\pi(f-\lambda)})$;
(d) $X(e^{j2\pi f}) \circledast X(e^{j2\pi f})$.

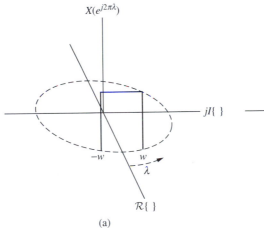

(a)

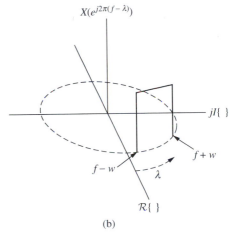

(b)

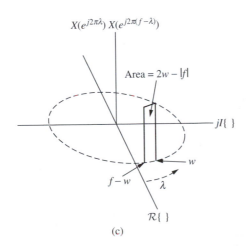

(c)

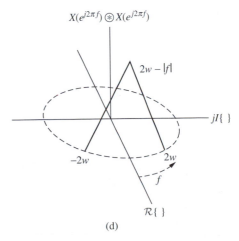

(d)

The area of the product, Eq. (39) equals

$$Y(e^{j2\pi f}) = X(e^{j2\pi f}) \circledast X(e^{j2\pi f}) = 2w - f \tag{40}$$

where $0 \le f \le 2w$.

If we increase f beyond $2w$ the functions $X(e^{j2\pi \lambda})$ and $X(e^{j2\pi(f-\lambda)})$ do not overlap and the area of their product is zero.

$$Y(e^{j2\pi f}) = X(e^{j2\pi f}) \circledast X(e^{j2\pi f}) = 0 \qquad \text{for } 2w < f \le 0.5 \tag{41}$$

Similar results hold for $-0.5 \le f \le 0$.

$$Y(e^{j2\pi f}) = X(e^{j2\pi f}) \circledast X(e^{j2\pi f}) = 2w + f \tag{42}$$

where $-2w \le f \le 0$.

In summary

$$Y(e^{j2\pi f}) = X(e^{j2\pi f}) \circledast X(e^{j2\pi f}) = \begin{cases} 2w - |f| & \text{for } 0 \le |f| \le 2w \\ 0 & \text{for } 2w < |f| \le 0.5 \end{cases} \tag{43}$$

which is the triangular function plotted in Figure 4DT.11(d). Like all DTFTs, $Y(e^{j2\pi f})$ is periodic with period 1.

Notice that the spectrum of $x^2[n]$ is nonzero in a range $0 \le |f| \le 2w$, twice as wide as the spectrum of $x[n]$. The squaring operation has introduced frequency components in the band $w < |f| \le 2w$ that were not present in the original sequence.

In the preceding example, we described circular convolution using a circular coordinate system. We can also use a rectangular coordinate system. Problem 4DT.5.19 considers the use of a rectangular coordinate system for the preceding example.

Index Reversal

$$x[-n] \longleftrightarrow X(e^{-j2\pi f}) \tag{4DT.OP4}$$

The index reversal property states that reversing the index in the sequence domain $(n \to -n)$ corresponds to reversing the normalized frequency in the frequency domain $(f \to -f)$. We can prove this property directly from the inverse discrete-time Fourier transform.

$$x[n] = \int_{-0.5}^{0.5} X(e^{j2\pi f}) e^{j2\pi fn} df \tag{44}$$

If we replace n by $-n$, Eq. (44) becomes

$$x[-n] = \int_{-0.5}^{0.5} X(e^{j2\pi f}) e^{-j2\pi fn} df = \int_{-0.5}^{0.5} X(e^{-j2\pi f}) e^{j2\pi fn} df \tag{45}$$

where the final step follows by changing the variable of integration from f to $-f$.

EXAMPLE 4DT.8   Spectrum of Two-Sided Exponential

The spectrum of the two-sided exponential

$$x[n] = a^{|n|} \quad |a| < 1 \tag{46}$$

can be obtained by writing it as

$$a^{|n|} = a^n u[n] + a^{-n} u[-n] - \delta[n] \tag{47}$$

We can now find the spectrum of $a^{|n|}$ by applying the index reversal property and superposition.

$$\mathcal{DTFT}\{a^{|n|}\} = \frac{1}{1 - ae^{-j2\pi f}} + \frac{1}{1 - ae^{j2\pi f}} - 1 \tag{48}$$

where $|a| < 1$. After some algebra we obtain the result

$$\mathcal{DTFT}\{a^{|n|}\} = \frac{1 - a^2}{1 + a^2 - 2a\cos(2\pi f)} \quad |a| < 1 \tag{49}$$

Index Shift

$$x[n - m] \longleftrightarrow e^{-j2\pi fm} X(e^{j2\pi f}) \tag{4DT.OP5}$$

The index shift property states that an index shift by m corresponds to a multiplication by $e^{-j2\pi fm}$ in the frequency domain. Proof of the index shift property for an arbitrary integer m can be obtained in a straightforward way from the definition of the DFT and its inverse (Problem 4DT.5.22). The property can be derived for $m \geq 0$ by recognizing that an index shift by $m \geq 0$ results when $x[m]$ is applied to a cascade of m unit delay elements. We know from Table 4DT.2 of Chapter 3DT that the frequency response characteristic of a unit delay is $H(e^{j2\pi f}) = e^{-j2\pi f}$. The frequency response characteristic a cascade of m such elements is $H^m(e^{j2\pi f}) = e^{-j2\pi fm}$. It follows from the equation $Y(e^{j2\pi f}) = H^m(e^{j2\pi f})X(e^{j2\pi f})$ that an index-shifted sequence, $y[n] = x[n - m]$, has spectrum $Y(e^{j2\pi f}) = e^{-j2\pi fm}X(e^{j2\pi f})$, which is what the property states.

Notice that an index shift simply adds a linear phase term to the phase spectrum of a sequence: $\angle(e^{-j2\pi fm}X(e^{j2\pi f})) = \angle X(e^{j2\pi f}) - 2\pi fm$. Index shift does not effect the amplitude spectrum: $|e^{-j2\pi fm}X(e^{j2\pi f})| = |X(e^{j2\pi f})|$. It is a *distortionless* operation. The index-shifted version of a sequence is identical to the original except for the shift.

Frequency Translation

$$x[n]e^{j2\pi f_o n} \longleftrightarrow X(e^{j2\pi(f - f_o)}) \tag{4DT.OP6}$$

The frequency-translation property states that multiplication of a sequence by $e^{j2\pi f_o n}$ causes the spectrum of the sequence to shift by f_o. This property can be derived by taking the $\mathcal{DTFT}$ of $x[n]e^{j2\pi f_o n}$.

$$\mathcal{DTFT}\{x[n]e^{j2\pi f_o n}\} = \sum_{n=-\infty}^{\infty} x[n]e^{j2\pi f_o n} e^{-j2\pi fn} = \sum_{n=-\infty}^{\infty} x[n]e^{-j2\pi(f - f_o)n}$$

$$= X(e^{j2\pi(f - f_o)}) \tag{50}$$

EXAMPLE 4DT.9   DT Rotating Phasors and Multiple Sinusoidal Component Waveforms

We can use the frequency translation property to find the spectrum of the DT rotating phasor (complex sinusoid) $Xe^{j2\pi f_1 n}$, where $-0.5 \le f_1 < 0.5$. Recall from Example 4DT.4 that $x[n] = 1 \longleftrightarrow X(e^{j2\pi f}) = \delta(f)$ for $-0.5 \le f < 0.5$. Therefore, by setting $x[n] = 1$ in (4DT.OP6), we obtain $e^{j2\pi f_1 n} \longleftrightarrow \delta(f - f_1)$, $-0.5 \le f < 0.5$. We can use this result and the linearity property of the DTFT to obtain

$$Xe^{j2\pi f_1 n} \longleftrightarrow X\delta(f - f_1) \quad -0.5 \le f < 0.5 \tag{51}$$

Therefore, the spectrum (DTFT) of a complex sinusoid having complex amplitude X and normalized frequency f_1 is an impulse located on the frequency axis at the frequency f_1, $-0.5 \le f < 0.5$. The area of the impulse is the complex amplitude X. Like all DTFTs, the spectrum on the right-hand side of Eq. (51) repeats periodically with period 1.

Real sinusoids are related to complex sinusoids by Euler's formula

$$A \cos(2\pi f_1 n + \phi) = Xe^{j2\pi f_1 n} + X^* e^{-j2\pi f_1 n} \tag{52}$$

where $X = \frac{1}{2} A \angle \phi$. It then follows from Eq. (51) and superposition that

$$A \cos(2\pi f_1 n + \phi) \longleftrightarrow X\delta(f - f_1) + X^* \delta(f + f_1) \quad -0.5 \le f < 0.5 \tag{53}$$

For the multiple sinusoidal component waveform of Eq. (3DT.39), we have:

$$x[n] = A_0 + \sum_{i=1}^{I} A_i \cos(2\pi f_i n + \phi_i) = \sum_{i=-I}^{I} X_i e^{j2\pi f_i n} \tag{54}$$

where

$$X_i = \begin{cases} A_0 & i = 0 \\ \frac{1}{2} A_i e^{j\phi_i} & i = 1, 2, \ldots, I \\ X_{-i}^* & i = -1, -2, \ldots, -I \end{cases} \tag{55}$$

and

$$f_i = \begin{cases} 0 & i = 0 \\ -f_{-i} & i = -1, -2, \ldots, -I \end{cases} \tag{56}$$

and it follows from Eq. (53) and superposition that

$$A_0 + \sum_{i=1}^{I} A_i \cos(2\pi f_i n + \phi_i) = \sum_{i=-I}^{I} X_i e^{j2\pi f_i n} \longleftrightarrow \sum_{i=-I}^{I} X_i \delta(f - f_i) \quad -0.5 \le f < 0.5 \tag{57}$$

Therefore the spectrum of a multiple sinusoidal component waveform is composed of impulses located at the frequencies of its complex sinusoidal components. The preceding equation assumes that $|f_i| < 0.5$ for $i = 1, 2, \ldots, I$. The areas of the impulses are the complex amplitudes of the complex sinusoidal components. The spectrum on the right-hand side of Eq. (57) repeats periodically with period 1.

Modulation

$$x[n] \cos(2\pi f_o n) \longleftrightarrow \frac{1}{2} X(e^{j2\pi(f - f_o)}) + \frac{1}{2} X(e^{j2\pi(f + f_o)}) \tag{4DT.OP7}$$

The modulation property states that if you *modulate* $\cos(2\pi f_o n)$ with $x[n]$, then $X(e^{j2\pi f})$ is divided by 2 and translated by $\pm f_o$. This property can be derived from frequency translation property and superposition. The frequency translation property tells us that

$$x[n]e^{j2\pi f_o n} \longleftrightarrow X(e^{j2\pi(f-f_o)}) \tag{58}$$

If we replace f_o with $-f_o$ we obtain

$$x[n]e^{-j2\pi f_o n} \longleftrightarrow X(e^{j2\pi(f+f_o)}) \tag{59}$$

Superposition yields

$$\tfrac{1}{2}x[n]e^{j2\pi f_o n} + \tfrac{1}{2}x[n]e^{-j2\pi f_o n} \longleftrightarrow \tfrac{1}{2}X(e^{j2\pi(f-f_o)}) + \tfrac{1}{2}X(e^{j2\pi(f+f_o)}) \tag{60}$$

which simplifies to the modulation property as stated.

EXAMPLE 4DT.10   "Digital Chip"

Consider a digital "digital chip"

$$y[n] = \sqcap(n, K)\cos(2\pi f_o n) \tag{61}$$

Here

$$x[n] = \sqcap(n, K) \tag{62}$$

The digital chip is illustrated in Figure 4DT.12(a), where $K = 40$ and $f_o = 1/16$.

The modulation property and Eq. (4DT.15) then give us the Fourier transform pair

$$\sqcap(n, K)\cos(2\pi f_o n) \longleftrightarrow \frac{2K+1}{2}\,\text{cinc}(f - f_o; K) + \frac{2K+1}{2}\,\text{cinc}(f + f_o; K) \tag{63}$$

The spectrum of the digital chip is plotted in Figure 4DT.12(b) for $K = 40$, and $f_o = 1/16$. The digital chip is an example of a band-pass sequence: It would be (approximately) passed by an appropriately designed band-pass filter.

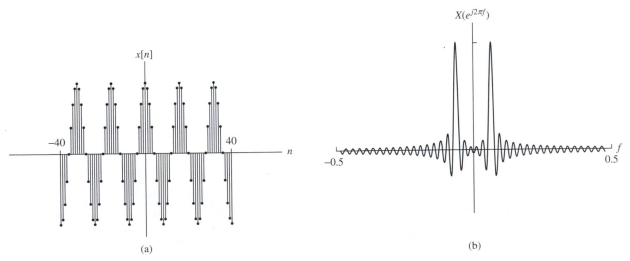

(a)   (b)

Figure 4DT.12
A "digital chip" and its spectrum: (a) $\sqcap(n, 40)\cos((2\pi/16)n)$; (b) $\frac{81}{2}\,\text{cinc}(f - 1/16; 40) + \frac{81}{2}\,\text{cinc}(f + 1/16; 40)$.

EXAMPLE 4DT.11 Impulse Response of Ideal Band-Pass Filter

The modulation property provides a quick way to obtain the impulse response of an ideal band-pass filter. The band-pass filter frequency response characteristic is given by $H(e^{j2\pi f}) = \sqcap((f - f_o)/2w) + \sqcap((f + f_o)/2w)$ where $-0.5 \leq f < 0.5$. Notice that $H(e^{j2\pi f})$ can be expressed as the right-hand side of OP7 if we set $X(e^{j2\pi f}) = 2\sqcap(f/2w)$. The sequence corresponding to $2\sqcap(f/2w)$ is $x[n] = 4w \operatorname{sinc}[2wn]$. Thus, from the modulation property, $h[n] = 4w \operatorname{sinc}[2wn] \cos(2\pi f_o n)$.

Multiplication by n

$$nx[n] \longleftrightarrow \frac{j}{2\pi} \frac{d}{df} X(e^{j2\pi f}) \qquad \text{(4DT.OP8)}$$

We can derive the multiplication by n property easily from the definition of the DTFT of $nx[n]$:

$$\mathcal{DTFT}\{nx[n]\} = \sum_{n=-\infty}^{\infty} nx[n] e^{-j2\pi fn} = \sum_{n=-\infty}^{\infty} x[n] \frac{j}{2\pi} \frac{d}{df} e^{-j2\pi fn}$$

$$= \frac{j}{2\pi} \frac{d}{df} \sum_{n=-\infty}^{\infty} x[n] e^{-j2\pi fn} = \frac{j}{2\pi} \frac{d}{df} X(e^{j2\pi f}) \qquad (64)$$

Problem 4DT.5.11 uses the multiplication by n property to find the spectrum of $na^n u[n]$, where $|a| < 1$.

Accumulate

$$\sum_{m=-\infty}^{n} x[m] \longleftrightarrow \pi\delta(f)X(e^{j2\pi 0}) + \frac{1}{1 - e^{-j2\pi f}} X(e^{j2\pi f}) \qquad \text{for } -\tfrac{1}{2} \leq f < \tfrac{1}{2} \qquad \text{(4DT.OP9)}$$

The accumulation property can be derived in a manner similar to the index shift property. We leave this derivation as a problem (Problem 4DT.5.22).

Downsampling (Decimation)

$$x_d[n; M] \longleftrightarrow X_d(e^{j2\pi f}) = \frac{1}{M} \sum_{k=0}^{M-1} X\left(e^{j\frac{2\pi}{M}(f-k)}\right) \qquad \text{(4DT.OP10)}$$

The *downsampled*[2] or *decimated* version of the sequence $x[n]$ is defined as

$$x_d[n; M] = \mathcal{DOWN}\{x[n]; M\} = x[nM] \qquad (65)$$

---

[2] Readers of Chapter 4CT might notice that decimation and upsampling are roughly (and only roughly) analogous to time scaling.

Figure 4DT.13
Lossless
downsampling:
(downsampling is
lossless if $Mw < 0.5$)
(a) $x[n]$ and its
spectrum;
(b) $x_d[n; M]$ and its
spectrum, where
$M = 3$.

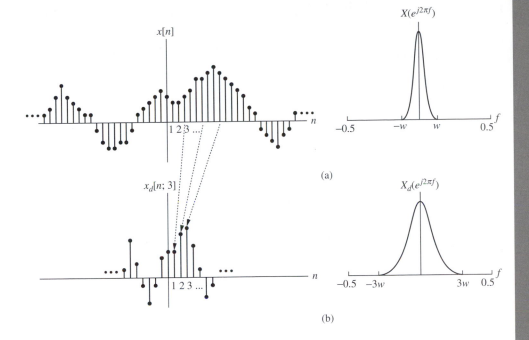

(a)

(b)

where M is a positive integer. The downsampling property, (4DT.OP10), describes the spectrum of $x_d[n; M]$ in terms of the spectrum $X(e^{j2\pi f})$ of $x[n]$.

Downsampling can be either lossless or lossy. For *lossless* downsampling, the spectrum of the downsampled sequence $x_d[n; M]$ is identical to the spectrum of $x[n]$, except for a factor $1/M$ and frequency expansion by a factor M. Lossless downsampling occurs when $Mw < 0.5$, where w is the bandwidth of $x[n]$. Figure 4DT.13 illustrates lossless downsampling, where the condition $Mw < 0.5$ is satisfied for $M = 3$.

Figure 4DT.14 illustrates *lossy* downsampling. Here $M = 8$, which is too large for the condition $Mw < 0.5$ to be satisfied. The spectrum of the downsampled sequence differs from a scaled version of the spectrum of $x[n]$.

Downsampling is used routinely in the field of *data compression* to reduce the number samples of a sequence that need to be stored or transmitted. When downsampling is lossless, the original sequence $x[n]$ can be reconstructed perfectly from its decimated version $x_d[n; M]$. The technique used for reconstructing $x[n]$ is described later in connection with upsampling and sampling.

We can derive the downsampling property using the following identity obtained from Eqs. (B.6)–(B.8) of Appendix B.

$$\Delta[m; M] = \frac{1}{M} \sum_{k=0}^{M-1} e^{j\frac{2\pi}{M}mk} = \begin{cases} 1 & m = \text{an integer multiple of } M \\ 0 & m \neq \text{an integer multiple of } M \end{cases} \tag{66}$$

Our first step is to write down the discrete-time Fourier transform of $x[nM]$

$$X_d(e^{j2\pi f}) = \mathcal{DTFT}\{x[nM]\} = \sum_{n=-\infty}^{\infty} x[nM]e^{-j2\pi fn} = \sum_{\substack{m=-\infty \\ m=\text{integer} \\ \text{multiple of } M}}^{\infty} x[m]e^{-j2\pi fm/M} \tag{67}$$

Figure 4DT.14
Lossy downsampling
(lossy downsampling
occurs if $Mw > 0.5$.):
(a) $x[n]$ and its
spectrum; (b) $x_d[n]$
and its spectrum
where $M = 8$.

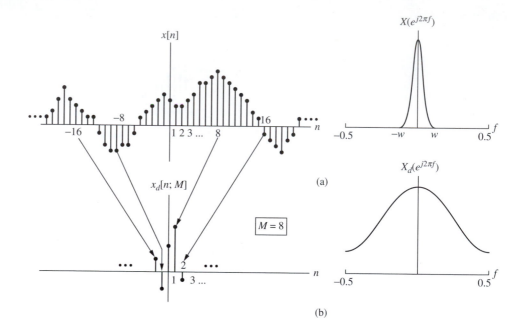

(a)

(b)

The right-hand side of this equation follows by changing the summation index from n to $m = nM$. The identity (4DT.66) lets us rewrite the right-hand side of Eq. (67) as a summation over all integer values of m.

$$X_d(e^{j2\pi f}) = \sum_{m=-\infty}^{\infty} x[m] \underbrace{\left\{ \frac{1}{M} \sum_{k=0}^{M-1} e^{j\frac{2\pi}{M}mk} \right\}}_{\substack{\text{This factor} =0 \text{ unless} \\ m \text{ is an integer} \\ \text{multiple of } M.}} e^{-j2\pi f m M} \tag{68}$$

If we interchange the orders of summation, Eq. (68) becomes

$$X_d(e^{j2\pi f}) = \frac{1}{M} \sum_{k=0}^{M-1} \left\{ \sum_{m=-\infty}^{\infty} x[m] e^{-j\frac{2\pi}{M}(f-k)m} \right\} \tag{69}$$

The quantity inside the braces is $X(e^{j\frac{2\pi}{M}(f-k)})$. Therefore

$$X_d(e^{j2\pi f}) = \frac{1}{M} \sum_{k=0}^{M-1} X(e^{j\frac{2\pi}{M}(f-k)}) \tag{70}$$

which establishes (4DT.OP10).

We can better understand the spectrum of Figure 4DT.14 if we plot Eq. (70) for f outside the range $-0.5 \le f < 0.5$, as shown in Figure 4DT.15. The figure assumes that $M = 8$ and shows the three terms $\frac{1}{8}X(e^{j2\pi f/8})$, $\frac{1}{8}X(e^{j2\pi(f-1)/8})$ and $\frac{1}{8}X(e^{j2\pi(f-7)/8})$. Each of these terms has period 8.[3] Their sum, $X_d(e^{j2\pi f})$, has period 1. We can see from the

---

[3] The term $\frac{1}{8}X(e^{j2\pi(f-7)/8})$ appears where it is shown in the figure because it has period 8: $\frac{1}{8}X(e^{j2\pi(f-7)/8}) \equiv \frac{1}{8}X(e^{j2\pi(f+1)/8})$.

Figure 4DT.15
Detailed view of the spectrum of figure 4DT.14: Spectrum foldover occurs if $Mw > 0.5$, causing the resultant spectrum to be different from a scaled version of $X(e^{j2\pi f})$.

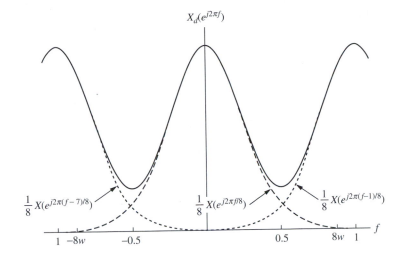

figure that the tails of the terms $\frac{1}{8} X(e^{j2\pi(f-1)/8})$ and $\frac{1}{8} X(e^{j2\pi(f-7)/8})$ *wrap around or fold over* into the region $-0.5 \leq f < 0.5$ because, in this illustration, $Mw > 0.5$ for $M = 8$. Spectrum foldover (aliasing) causes the resultant spectrum to differ from a scaled version of $X(e^{j2\pi f})$.

Upsampling

$$x_u[n; M] \longleftrightarrow X_u(e^{j2\pi f}) = X(e^{j2\pi Mf}) \tag{4DT.OP11}$$

Upsampling consists of inserting $M-1$ zero samples between the samples of a sequence $y[n]$. The upsampled sequence is given by

$$x_u[n; M] = \mathcal{UP}\{x[n]; M\} = \begin{cases} x\left[\frac{n}{M}\right] & \text{for } n = lM, \ l = \pm 1, \pm 2, \ldots \\ 0 & \text{otherwise} \end{cases} \tag{71}$$

where M is a positive integer. The upsampling property (4DT.OP11) describes the spectrum of $x_u[n; M]$ in terms of the spectrum $X(e^{j2\pi f})$ of $x[n]$. An example of a sequence and its spectrum is shown in Figure 4DT.16(a). Figure 4DT.16(b) shows the corresponding upsampled sequence $\mathcal{UP}\{x[n]; M\}$ and its spectrum $X(e^{j2\pi Mf})$ for $M = 3$. Notice from the figure that upsampling causes a compression of the spectrum. Upsampling does not cause spectrum foldover.

We can derive the upsampling property by starting with the formula for the spectrum of $\mathcal{UP}\{x[n]; M\}$.

$$\mathcal{DTFT}\{\mathcal{UP}\{x[n]; M\}\} = \sum_{n=-\infty}^{\infty} \mathcal{UP}\{x[n]; M\}e^{-j2\pi fn} = \sum_{\substack{n=-\infty \\ n=lM}}^{\infty} x\left[\frac{n}{M}\right]e^{-j2\pi fn} \tag{72}$$

Figure 4DT.16
Upsampling:
(a) original sequence
$x[n]$ and its spectrum;
(b) upsampled
Version $\mathcal{UP}\{x[n]; M\}$
and its spectrum,
where $M = 3$.

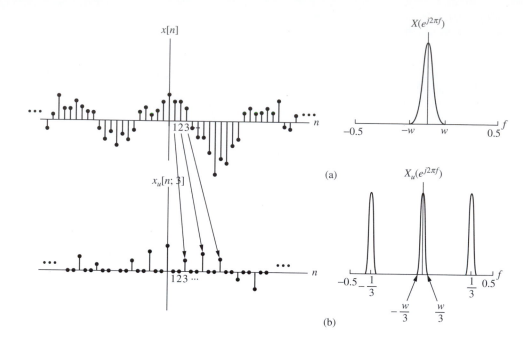

(a)

(b)

If we replace n with lM in the terms in the sum and change the index of summation to l, we obtain

$$\mathcal{DTFT}\{\mathcal{UP}\{x[n]; M\}\} = \sum_{l=-\infty}^{\infty} x[l]e^{-j2\pi Mfl} = X(e^{j2\pi Mf}) \qquad (73)$$

which is the result stated in (4DT.OP5).

Sampling

$$x_s[n] \longleftrightarrow X_s(e^{j2\pi f}) = \frac{1}{M} \sum_{k=0}^{M-1} X\left(e^{j2\pi(f-\frac{k}{M})}\right) \qquad \text{(4DT.OP12)}$$

The *sampled version* of $x[n]$ is defined as

$$x_s[n] = x[n]\Delta[n; M] = \begin{cases} x[n] & n = \text{an integer multiple of } M \\ 0 & n \neq \text{an integer multiple of } M \end{cases} \qquad (74)$$

where, from Eq. (66),

$$\Delta[n; M] = \begin{cases} 1 & n = \text{an integer multiple of } M \\ 0 & n \neq \text{an integer multiple of } M \end{cases} \qquad (75)$$

The sampling property describes the spectrum of $x_s[n]$ in terms of the spectrum of $x[n]$. The sampling property is illustrated in Figure 4DT.17. It can be seen from the figure that spectrum foldover does not occur if $w < (1/M) - w$, or equivalently, if $wM < 0.5$. This is the same condition we encountered in connection with downsampling $x[n]$.

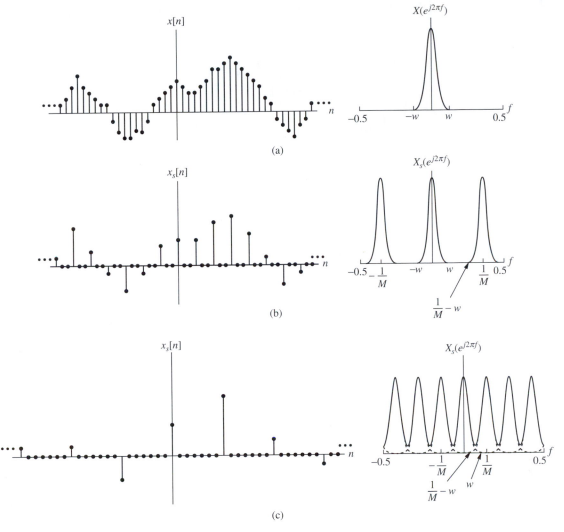

Figure 4DT.17
Sampled sequence and spectrum: (a) sequence $x[n]$ and its spectrum; (b) sampled sequence $x_s[n]$ and its spectrum where $wM < 0.5$, with $M = 3$; (c) sampled sequence $x_s[n]$ and its spectrum where $wM > 0.5$, with $M = 7$.

We can derive (4DT.OP12) by direct evaluation of the DTFT of $x[n]\Delta[n; M]$:

$$\mathcal{DTFT}\{x[n]\Delta[n; M]\} = \sum_{n=-\infty}^{\infty} x[n]\Delta[n; M]e^{-2\pi fn} = \sum_{n=-\infty}^{\infty} x[n]\left\{\frac{1}{M}\sum_{k=0}^{M-1}e^{j\frac{2\pi}{M}nk}\right\}e^{-2\pi fn}$$
(76)

where we substituted identity (66). A change in the order of summation in the right-hand side of the preceding equation yields

$$\mathcal{DTFT}\{x[n]\Delta[n; M]\} = \frac{1}{M}\sum_{k=0}^{M-1}\sum_{n=-\infty}^{\infty} x[n]e^{-j2\pi n(f-\frac{k}{M})} = \frac{1}{M}\sum_{k=0}^{M-1}X(e^{j2\pi(f-\frac{k}{M})})$$
(77)

which establishes the sampling property.

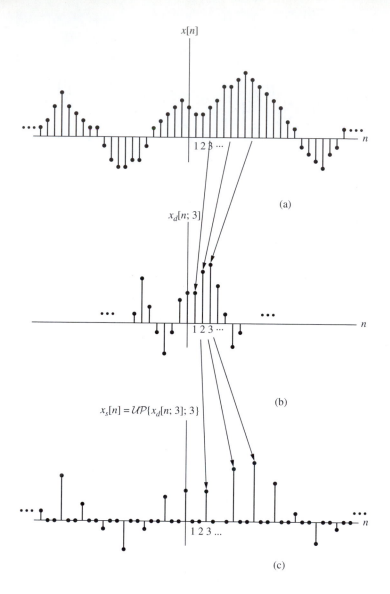

The relationship between sampling and downsampling is illustrated in Figure 4DT.18. As can be seen from the figure, one way to generate $x_s[n]$ is to upsample $x_d[n; M]$ by M.

$$x_s[n] = \mathcal{UP}\{x_d[n; M]; M\} \tag{78}$$

This relationship can be used, along with (4DT.OP10) and (4DT.OP11) to obtain $X_s(e^{j2\pi f})$, as given by (4DT.OP12).

We noted earlier that downsampling is used for data compression and that the original sequence can be recovered from $x_d[n; M]$ if $wM < 0.5$. The first step in recovering $x[n]$ from $x_d[n]$ is to generate $x_s[n]$ by upsampling, as in Eq. (78). The next step is to apply $x_s[n]$ to an interpolation filter. This step is described in the next section.

	Name	Sequence	DTFT
4DT.OP1	Linearity	$a_1 x_1[n] + a_2 x_2[n]$	$a_1 X_1(e^{j2\pi f}) + a_2 X_2(e^{j2\pi f})$
4DT.OP2	Convolution	$x_1[n] * x_2[n]$	$X_1(e^{j2\pi f}) X_2(e^{j2\pi f})$
4DT.OP3	Multiplication	$x_1[n] x_2[n]$	$X_1(e^{j2\pi f}) \circledast X_2(e^{j2\pi f})$
4DT.OP4	Index Reversal	$x[-n]$	$X(e^{-j2\pi f})$
4DT.OP5	Index Shift	$x[n - m]$	$e^{-j2\pi f m} X(e^{j2\pi f})$
4DT.OP6	Frequency Translation	$x[n] e^{j2\pi f_o n}$	$X(e^{j2\pi(f-f_o)})$
4DT.OP7	Modulation	$x[n]\cos(2\pi f_o n)$	$\frac{1}{2} X(e^{j2\pi(f-f_o)}) + \frac{1}{2} X(e^{j2\pi(f+f_o)})$
4DT.OP8	Multiply by n	$n x[n]$	$\dfrac{j}{2\pi}\dfrac{d}{df} X(e^{j2\pi f})$
4DT.OP9	Accumulate	$\displaystyle\sum_{m=-\infty}^{n} x[m]$	$X(e^{j2\pi f})\left(\frac{1}{2}\delta(f) + \dfrac{1}{1 - e^{-j2\pi f}}\right)$ $-0.5 \le f < 0.5^*$
4DT.OP10	Downsampling	$x_d[n; M] = \mathcal{DOWN}\{x[n]; M\} = x[nM]$	$X_d(e^{j2\pi f}) = \dfrac{1}{M}\displaystyle\sum_{m=0}^{M-1} X(e^{j\frac{2\pi}{M}(f-m)})$
4DT.OP11	Upsampling	$x_u[n; M] = \mathcal{UP}\{x[n]; M\} =$ $\begin{cases} x\left[\dfrac{n}{M}\right] & \text{for } n = lM,\ l = \pm 1, \pm 2, \dots \\ 0 & \text{otherwise} \end{cases}$	$X_u(e^{j2\pi f}) = X(e^{j2\pi M f})$
4DT.OP12	Sampling	$x_s[n] = x[n]\Delta[n; M]$	$X_s(e^{j2\pi f}) = \dfrac{1}{M}\displaystyle\sum_{k=0}^{M-1} X(e^{j2\pi(f-\frac{k}{M})})$

* Recall that every DTFT is a periodic function of f with period 1.

Table 4DT.2

Operational Properties of the Discrete-Time Fourier Transform

The operational properties for the Fourier transform are summarized in Table 4DT.2.

4DT.6   The Sampling Theorem

The DT sampling theorem describes an important and surprising result in signal and systems theory. It states that, under certain conditions, we can *exactly* reconstruct a sequence from its sampled version. This theorem provides a solid theoretical basis for reconstructing a sequence that has been compressed by downsampling for efficient storage or transmission.

DT Sampling Theorem

A sequence $x[n]$ having a spectrum limited to w

$$X(e^{j2\pi f}) = 0 \qquad \text{for } |f| > w, \text{ on the interval } -0.5 \le f < 0.5 \tag{79}$$

can be perfectly reconstructed from its samples $x(nM), k = 0, \pm 1, \pm 2, \dots$ if, and only if

$$\frac{1}{M} > 2w \tag{80}$$

Figure 4DT.19
Frequency domain
proof of the sampling
theorem (drawn for
$M = 3$): (a) filter
input spectrum
$X_s(e^{j2\pi f})$ with
$1/M > 2w$;
(b) frequency
response
characteristic of
reconstruction filter
$H(e^{j2\pi f})$;
(c) spectrum at
output of
reconstruction filter.

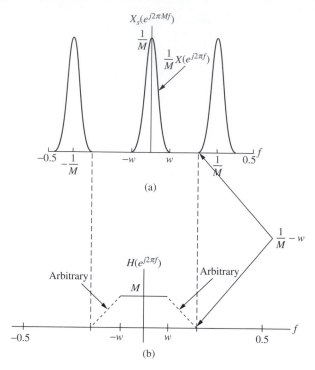

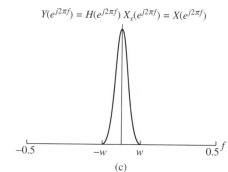

We can prove the sampling theorem by referring to the sampling property (4DT.OP12). This property states that the spectrum $X_s(e^{j2\pi f})$ of $x_s[n]$ is the sum of scaled, periodic replicas of $X(e^{j2\pi f})$. In general, these periodic repetitions overlap, as was shown in Figure 4DT.17c. This overlap, called aliasing, makes it impossible to extract $X(e^{j2\pi f})$ from $X_s(e^{j2\pi f})$ and consequently to reconstruct $x[n]$ from its samples. The overlap does not occur if M satisfies (80). When (80) is satisfied it is possible to separate $X(e^{j2\pi f})$ from $X_s(e^{j2\pi f})$ by means of a low-pass filter, as shown in Figure 4DT.19. This figure is drawn for $M = 3$. This means that we can perfectly reconstruct $x[n]$ from $x_s[n]$.

The frequency response characteristic of the low-pass filter is,

$$H(e^{j2\pi f}) = \begin{cases} M & |f| \leq w \\ \text{arbitrary} & w < |f| < 1/M - w \\ 0 & 1/M - w \leq |f| < 0.5 \end{cases} \qquad (81)$$

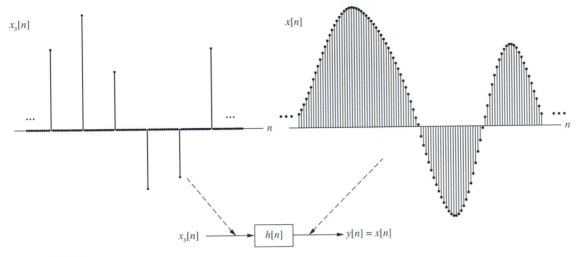

Figure 4DT.20
Reconstruction of $x[n]$ from $x_s[n]$.

for $-0.5 \leq f < 0.5$. The spectrum of the input to the filter is, by (4DT.OP12), $X_s(e^{j2\pi f})$. As shown in the figure, the spectrum of the output of the low-pass filter equals $X(e^{j2\pi f})$.

$$Y(e^{j2\pi f}) = X_s(e^{j2\pi f})H(e^{j2\pi f}) \equiv X(e^{j2\pi f}) \tag{82}$$

when $1/M > 2w$. The time domain equation corresponding to Eq. (82) is

$$y[n] = \sum_{k=-\infty}^{\infty} x_s[k]h[n-k] = x[n] \tag{83}$$

The preceding equation shows the output of the filter is $x[n]$ when the input is $x_s[k]$. The sequence domain block diagram and sequences corresponding to Eq. (83) are illustrated in Figure 4DT.20. If we substitute Eq. (74) into Eq. (83) we obtain

$$y[n] = \sum_{m=-\infty}^{\infty} x[mM]h[n-mM] = x[n] \tag{84}$$

which shows explicitly how $x[n]$ is obtained from the samples $x[mM]$. The low-pass filter is sometimes called an *interpolation filter* because it interpolates among the samples $x[mM]$ to yield $x[n]$. Similarly, the summation in Eq. (84) is sometimes called an *interpolation formula*.

4DT.6.1
The DT
Nyquist-Shannon
Interpolation
Formula

A particularly simple interpolation formula is obtained if we chose

$$H(e^{j2\pi f}) = M \sqcap (fM) = \begin{cases} M & |f| \leq \dfrac{1}{2M} \\ 0 & |f| > \dfrac{1}{2M} \end{cases} \tag{85}$$

on the interval $-0.5 \leq f < 0.5$. We call $H(e^{j2\pi f})$ the *DT Nyquist-Shannon reconstruction filter* because the frequency response characteristic given by Eq. (85) is analogous to the

Figure 4DT.21
DT Nyquist-Shannon reconstruction of $x[n]$ with $M = 15$: The interpolation sequence is $h[n] =$ sinc(n/M) (assumes only three nonzero samples in $x_s[n]$: $x[2M] = x[30] = 5$, $x[3M] = x[45] = 10$, and $x[4M] = x[60] = 8$).

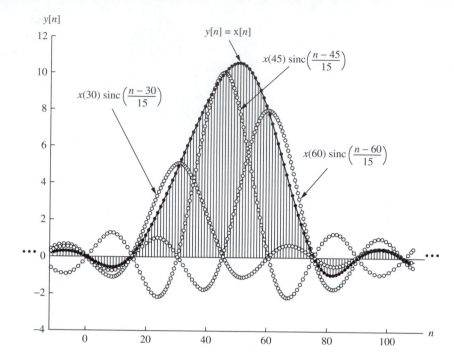

Legend:

  o 3 nonzero terms in Eq. (87)

  • Sum of 3 sinc functions $y[n]$

Nyquist-Shannon filter used to reconstruct a CT waveform from its samples Eq. (4CT.74). The impulse response corresponding to Eq. (85) is

$$h[n] = \text{sinc}\left(\frac{n}{M}\right) \tag{86}$$

If we substitute Eq. (86) into Eq. (84) we obtain

$$y[n] = \sum_{m=-\infty}^{\infty} x[mM]\, \text{sinc}\left(\frac{n - mM}{M}\right) = x[n] \tag{87}$$

which is the DT version of the Nyquist-Shannon interpolation formula of Chapter 4CT. Figure 4DT.21 illustrates how the sinc sequences interpolate among the samples to reconstruct $x[n]$.

4DT.7 Additional DTFT Pairs

Some commonly used DTFT pairs are listed in Table 4DT.3. Additional transform pairs are listed in Table 3DT.1. The spectrum of the periodic impulse train, DT.FT14, follows from the sampling property (4DT.OP12) upon setting $x[n] = 1$ (for which $X(e^{j2\pi f}) = \delta(f)$ for $-0.5 \leq f < 0.5$ by Example 4DT.4). All DTFT pairs are periodic with period 1.

	Name	$x[n]$	$X(e^{j2\pi f})$
DTFT1	Impulse	$\delta[n]$	1
DTFT2	Constant	1	$\delta(f) \quad -0.5 \leq f < 0.5$
DTFT3	Step	$u[n]$	$U(e^{j2\pi f}) = \pi\delta(f) + \dfrac{1}{1-e^{-j2\pi f}}; \quad -0.5 \leq f < 0.5$
DTFT4	Complex sinusoid	$e^{j2\pi f_o n}; -0.5 < f_o < 0.5$	$\delta(f - f_o); \quad -0.5 \leq f < 0.5$
DTFT5	Cosine	$\cos(2\pi f_o n); -0.5 < f_o < 0.5$	$\frac{1}{2}\delta(f - f_o) + \frac{1}{2}\delta(f + f_o); \quad -0.5 \leq f < 0.5$
DTFT6	Sine	$\sin(2\pi f_o n); -0.5 < f_o < 0.5$	$\dfrac{1}{2j}\delta(f - f_o) - \dfrac{1}{2j}\delta(f + f_o); \quad -0.5 \leq f < 0.5$
DTFT7	sinusoid	$\cos(2\pi f_o n + \phi); -0.5 < f_o < 0.5$	$\dfrac{1}{2}e^{j\phi}\delta(f - f_o) + \dfrac{1}{2}e^{-j\phi}\delta(f + f_o); \quad -0.5 \leq f < 0.5$
DTFT8	Exponential	$a^n u[n]$	$\dfrac{1}{1-ae^{-j2\pi f}} \quad$ for $|a| < 1$
DTFT9	Exponential times ramp	$n a^n u[n]$	$\dfrac{ae^{-j2\pi f}}{(1-ae^{-j2\pi f})^2} \quad$ for $|a| < 1$
DTFT10	Two-sided exponential	$a^{|n|}$	$\dfrac{1-a^2}{1+a^2-2a\cos(2\pi f)} \quad$ for $|a| < 1$
DTFT11	Rectangular pulse	$\sqcap[n; K]$	$(2K+1)\,\mathrm{cinc}(f; K)$
DTFT12	Triangular pulse	$\wedge[n; M]$	$(M+1)\,\mathrm{cinc}^2(f; M/2)$
DTFT13	Raised cosine pulse	$\frac{1}{2}\left\{1 + \cos\left(\pi\dfrac{n}{K+1}\right)\right\}\sqcap[n; K]$	Eq. (28)
DTFT14	Periodic impulse train	$\Delta[n; M]$	$\dfrac{1}{M}\displaystyle\sum_{k=0}^{M-1}\delta\left(f - \dfrac{k}{M}\right)$

Table 4DT.3

Discrete-Time Fourier Transform Pairs. All spectrums are periodic with period 1.

4DT.8 Energy Sequences, Energy Density Spectrums, and Parseval's Theorem

Pulse sequences like $a^n u[n]$, $\sqcap(n; K)$, and $\wedge(n; M)$ are members of a class of sequences called *energy sequences*. The *energy* in a real or complex waveform $x[n]$ is defined as

$$E_x = \sum_{n=-\infty}^{\infty} |x[n]|^2 \tag{88}$$

A *sequence* $x[n]$ is called an *energy sequence* if E_x is nonzero and finite. The sinusoidal sequence $A\cos(2\pi fn + \phi)$ is not an energy sequence because the energy it carries is infinite (the sum (88) diverges). Periodic sequences are never energy sequences.

The energy density spectrum of an energy sequence is defined as

Energy Density Spectrum

$$\Phi_x(e^{j2\pi f}) = |X(e^{j2\pi f})|^2 \tag{89}$$

where $X(e^{j2\pi f}) = \mathcal{DTFT}\{x[n]\}$.

A fundamental property of energy density spectrums is that the total energy carried by a sequence is equal to the area under the sequence's energy density spectrum. This property is called *Parseval's theorem* for energy sequences.

Parseval's Theorem for Energy Sequences

$$E_x = \sum_{n=-\infty}^{\infty} |x[n]|^2 = \int_{-0.5}^{0.5} \Phi_x(e^{j2\pi f})\, df \tag{90}$$

We can derive Parseval's theorem from the convolution property (Table 4DT.2) by setting $x_1[n] = x[n]$ and $x_2[n] = x^*[n]$. By the conjugate symmetry and index reversal properties, the spectrum of $x^*[-n]$ is $X^*(e^{j2\pi f})$. The convolution property then yields the result

$$\phi_x[n] = \sum_{k=-\infty}^{\infty} x[k]x^*[k-n] = \int_{-0.5}^{0.5} X(e^{j2\pi f})X^*(e^{j2\pi f})e^{j2\pi fn} df$$

$$= \int_{-0.5}^{0.5} \Phi_x(e^{j2\pi f})e^{j2\pi fn} df \tag{91}$$

The sequence $\phi_x[n]$ is the *autocorrelation sequence* of $x[n]$. Autocorrelation sequences are computed in cell phones and other voice encoders and decoders (codecs).[4] Equation (91) states the important theoretical result that the autocorrelation sequence $\phi_x[n]$ and energy density function $\Phi_x(e^{j2\pi f})$ are a DTFT pair. Parseval's theorem is a special case of Eq. (91), where $n = 0$.

The energy density spectrum $\Phi_x(e^{j2\pi f})$ describes the distribution of the energy in $x[n]$ in frequency. This means that the energy $E_x(\mathcal{B})$ contained in any set of frequencies $\mathcal{B}$ is given by

$$E_x(\mathcal{B}) = \int_{\mathcal{B}} \Phi_x(e^{j2\pi f})\, df \tag{92}$$

where the integration is over the frequencies in $\mathcal{B}$. We can prove this statement by applying $x[n]$ to an ideal filter $H_{\mathcal{B}}(e^{j2\pi f})$, which filter passes all frequency components in $\mathcal{B}$ and rejects all other frequency components.

$$H_{\mathcal{B}}(e^{j2\pi f}) = \begin{cases} 1 & f \in \mathcal{B} \\ 0 & f \notin \mathcal{B} \end{cases} \tag{93}$$

where $\in$ and $\notin$ denote "is in" and "is not in," respectively. An example of $H_{\mathcal{B}}(e^{j2\pi f})$ is shown in Figure 4DT.22.

When $x[n]$ is applied to this filter, the spectrum of the output is $Y(e^{j2\pi f}) = H_{\mathcal{B}}(e^{j2\pi f}) X(e^{j2\pi f})$. The energy density spectrum of the output, therefore, is given by

$$\Phi_y(e^{j2\pi f}) = |H_{\mathcal{B}}(e^{j2\pi f})X(e^{j2\pi f})|^2$$

$$= |H_{\mathcal{B}}(e^{j2\pi f})|^2\Phi_x(e^{j2\pi f}) = \begin{cases} \Phi_x(e^{j2\pi f}) & f \in \mathcal{B} \\ 0 & f \notin \mathcal{B} \end{cases} \tag{94}$$

---

[4] The theoretical framework for this and related applications, known as *linear prediction theory*, is presented in advanced signal processing texts.

Figure 4DT.22
Ideal filter $H_B(e^{j2\pi f})$.

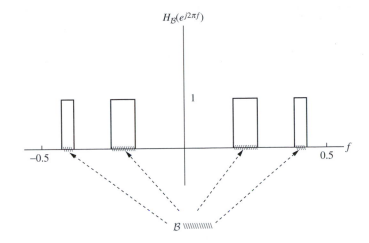

The energy in $y[n]$ is given by Parseval's theorem.

$$E_y = \sum_{n=-\infty}^{\infty} |y[n]|^2 = \int_{-\infty}^{\infty} \Phi_y(e^{j2\pi f})\, df = \int_{B} \Phi_x(e^{j2\pi f})\, df = E_x(B) \qquad (95)$$

The final equality of the preceding equation is Eq. (92).

4DT.9 Convergence of the DTFT

The DTFT is a limit.

$$X(e^{j2\pi f}) = \lim_{K \to \infty} X_K(e^{j2\pi f}) \qquad (96)$$

where

$$X_K(e^{j2\pi f}) = \sum_{n=-K}^{K} x[n]e^{-j2\pi fn} \qquad (97)$$

$X(e^{j2\pi f})$ exists when $X_K(e^{j2\pi f})$ converges as $K \to \infty$. Two types of convergence are of interest. The first, called *uniform convergence*, is the stronger.

Definition: Uniform Convergence

$X_K(e^{j2\pi f})$ converges uniformly to $X(e^{j2\pi f})$ if, for each $\epsilon > 0$, there is an integer K_o, such that

$$\underbrace{|X(e^{j2\pi f}) - X_K(e^{j2\pi f})| < \epsilon}_{\text{This inequality holds } for\ all\ f.} \qquad \text{for all } K > K_o \qquad (98)$$

Before considering an example of uniform convergence, we prove the following theorem.

Theorem

$X_K(e^{j2\pi f})$ converges uniformly to $X(e^{j2\pi f})$ if $x[n]$ is absolutely summable, i.e., if

$$\sum_{n=-\infty}^{\infty} |x[n]| < \infty \tag{99}$$

To begin the proof, we write the absolute value in Eq. (98) as

$$|X(e^{j2\pi f}) - X_K(e^{j2\pi f})| =$$

$$\left| \sum_{n=-\infty}^{\infty} x[n]e^{-j2\pi fn} - \sum_{n=-K}^{K} x[n]e^{-j2\pi fn} \right| = \left| \sum_{|n|>K} x[n]e^{-j2\pi fn} \right| \tag{100}$$

We can obtain an upper bound on $|X(e^{j2\pi f}) - X_K(e^{j2\pi f})|$ if we interchange the orders of summation and absolute value on the right-hand side of the preceding equation.

$$\left| \sum_{|n|>K} x[n]e^{-j2\pi fn} \right| \le \sum_{|n|>K} |x[n]e^{-j2\pi fn}|$$

$$= \sum_{|n|>K} |x[n]|$$

$$= \sum_{n=-\infty}^{\infty} |x[n]| - \sum_{|n|\le K} |x[n]| \tag{101}$$

It follows that

$$|X(e^{j2\pi f}) - X_K(e^{j2\pi f})| \le \sum_{n=-\infty}^{\infty} |x[n]| - \sum_{|n|\le K} |x[n]| \tag{102}$$

We can write the above as inequality (98) where

$$\epsilon = \sum_{n=-\infty}^{\infty} |x[n]| - \sum_{|n|\le K} |x[n]| \tag{103}$$

is independent of f. The value of ϵ can be made arbitrarily small by increasing K if

$$\sum_{n=-\infty}^{\infty} |x[n]| < \infty \tag{104}$$

This completes the proof.

EXAMPLE 4DT.12   Uniform Convergence

Consider the double-sided decaying exponential

$$x[n] = a^{|n|} \tag{105}$$

where a is a real number with $|a| < 1$. It can be shown that

$$\sum_{n=-\infty}^{\infty} |a^{|n|}| = \frac{1+a}{1-a} < \infty \tag{106}$$

Therefore $X_K(e^{j2\pi f})$ converges uniformly to $X(e^{j2\pi f})$. The spectrums are given by (Problem 4DT.5.10) and Example 4DT.8:

$$X_K(e^{j2\pi f}) = \frac{1 - \alpha^2 + 2a^{K+2}\cos(2\pi f K) - 2a^{K+1}\cos(2\pi(K+1)f)}{1 + \alpha^2 - 2\alpha\cos(2\pi f)} \tag{107a}$$

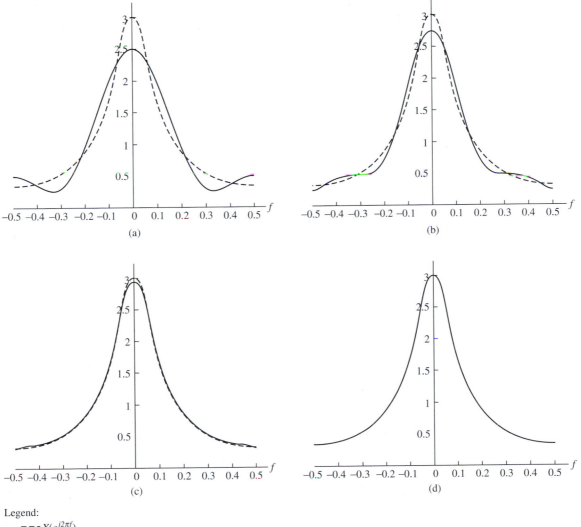

Legend:

$---X(e^{j2\pi f})$

$\underline{\qquad}\, X_K(e^{j2\pi f})$

Figure 4DT.23
Uniform convergence $|X(e^{j2\pi f}) - X_K(e^{j2\pi f})| < \epsilon$ for all f: (a) $K = 2$ ($\epsilon = 0.5$); (b) $K = 3$ ($\epsilon = 0.25$); (c) $K = 5$ ($\epsilon = 0.0625$); (d) $K = 7$ ($\epsilon = 0.0156$).

and

$$X(e^{j2\pi f}) = \frac{1 - a^2}{1 + a^2 - 2a\cos(2\pi f)} \tag{107b}$$

Figure 4DT.23 contains representative plots of $X(e^{j2\pi f})$ and $X_K(e^{j2\pi f})$ for $a = 0.5$. The values of ϵ were computed using Eq. (103). Note that, in each plot, $|X(e^{j2\pi f}) - X_K(e^{j2\pi f})| < \epsilon$ for all f.

The second type of convergence is *integral square convergence*.

Definition: Integral Square Convergence

$X_K(e^{j2\pi f})$ converges to $X(e^{j2\pi f})$ in integral square if the integral square error (ISE)

$$\text{ISE} \triangleq \lim_{K \to \infty} \int_{-0.5}^{0.5} |X(e^{j2\pi f}) - X_K(e^{j2\pi f})|^2 \, df = 0 \tag{108}$$

Before considering an example of integral square convergence, we prove the following theorem:

Theorem

$X_K(e^{j2\pi f})$ converges to $X(e^{j2\pi f})$ in integral square if $x[n]$ is an energy sequence.

To begin the proof, we use Parseval's theorem to write

$$\text{ISE} = \int_{-0.5}^{0.5} |X(e^{j2\pi f}) - X_K(e^{j2\pi f})|^2 df = \sum_{n=-\infty}^{\infty} |x[n] - x[n] \sqcap (n; K)|^2$$

$$= \sum_{n=-\infty}^{\infty} |x[n]|^2 - \sum_{|n| \leq K} |x[n]|^2 \tag{109}$$

If

$$\sum_{n=-\infty}^{\infty} |x[n]|^2 < \infty \tag{110}$$

then the right-hand side of Eq. (109) vanishes in the limit $K \to \infty$. This completes the proof.

EXAMPLE 4DT.13   Integral Square Convergence

Consider

$$x[n] = 0.5 \, \text{sinc}(0.5n) \tag{111}$$

A sinc sequence has finite energy but is not absolutely summable. We have from Table 3DT.1 that

$$X(e^{j2\pi f}) = \sqcap (2f) \quad -0.5 \leq f < 0.5 \tag{112}$$

so $X(e^{j2\pi f})$ certainly exists. Here, the convergence of the DTFT is not uniform but in the integral square sense.

Figure 4DT.24 contains plots of $X_K(e^{j2\pi f})$ for various K. The function $X_K(e^{j2\pi f})$ is computed numerically for these plots. It can be seen from these plots that $X_K(e^{j2\pi f})$ approaches $\sqcap (2f)$ as K increases. Notice the 9 percent

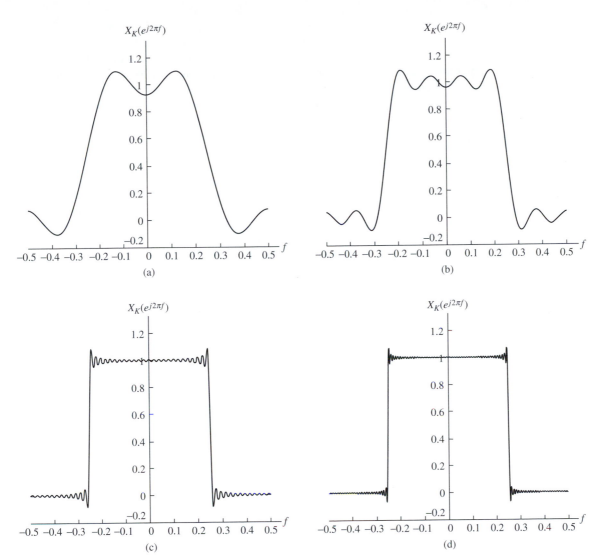

Figure 4DT.24
Integral square convergence: (a) $K = 3$, ISE = 0.025; (b) $K = 7$, ISE = 0.012; (c) $K = 50$, ISE = 0.002; (d) $K = 100$, ISE = 0.001.

overshoot and oscillations that appear near the discontinuity in $\sqcap (2f)$. This is called *Gibbs phenomenon*. The amplitude of the overshoot does not decrease as K increases. However, the energy in the error vanishes in the limit $K \rightarrow \infty$, in agreement with the theorem.

4DT.9.1
DTFTs
Modeled with
Impulses

If $X_K(e^{j2\pi f})$ does not converge as $K \rightarrow \infty$ we can sometimes define $X(e^{j2\pi f})$ with the aid of impulses. This statement is illustrated by Example 4DT.4, where the sequence $x[n] = 1$ had infinite energy. We saw that $\mathcal{DTFT}\{1\}$ did not converge. By taking a limit, however, we are able to define $X(e^{j2\pi f}) = \delta(f)$, $-0.5 \leq f < 0.5$. Another example of the use of impulses to model $X(e^{j2\pi f})$ is given by $\mathcal{DTFT}\{A\cos(2\pi f_o n + \phi)\}$. We will see in Chapter 5CT that we can use impulses to model $X(e^{j2\pi f})$ for every periodic sequence.

There are, however, routinely used sequences that cannot be represented by Fourier transforms. Examples are the unit ramp $r[n] = nu[n]$ and the rising exponential $x[n] = a^n u[n]$, where $|a| > 1$. The z-transform (Chapter 6DT) works for these signals.

4DT.10   Summary

In this chapter we showed that a sequence $x[n]$ can be represented as an integral of complex sinusoidal components. This representation is the inverse discrete-time Fourier transform of $x[n]$. The sequence's spectrum $X(e^{j2\pi f})$ describes the complex amplitudes of the infinitesimal sinusoidal components. All the properties that applied to multiple sinusoidal component sequences developed in Chapter 3DT extend to sequences that can be represented as inverse DTFTs. The principal property is that the output $y[n] = x[n] * h[n]$ of an LTI system has spectrum $Y(e^{j2\pi f}) = X(e^{j2\pi f})H(e^{j2\pi f})$, where $H(e^{j2\pi f})$ is the system's frequency response characteristic. Other important properties are listed in Table 4DT.2.

Summary of Definitions and Formulas

Discrete-Time Fourier Transform (Spectrum of x[n])

$$X(e^{j2\pi f}) = \sum_{m=-\infty}^{\infty} x[m]e^{-j2\pi fm}$$

All DTFTs repeat periodically with period 1. It is therefore a standard practice to specify them for only one period, such as $-\frac{1}{2} \le f < \frac{1}{2}$.

Inverse DTFT

The inverse DTFT represents a sequence as an integral of sinusoids.

$$x[n] = \int_{-0.5}^{0.5} X(e^{j2\pi f})e^{j2\pi fn}df$$

Signals and LTI Systems

Sequence domain:

$$x[n] \xrightarrow{h[n]} y[n] = x[n] * h[n]$$

Frequency domain:

$$X(e^{j2\pi f}) \xrightarrow{H(e^{j2\pi f})} Y(e^{j2\pi f}) = X(e^{j2\pi f})H(e^{j2\pi f})$$

Amplitude and Phase Spectrums

$$|Y(e^{j2\pi f})| = |H(e^{j2\pi f})||X(e^{j2\pi f})|$$
$$\angle Y(e^{j2\pi f}) = \angle H(e^{j2\pi f}) + \angle X(e^{j2\pi f})$$

CINC Function

$$\text{cinc}(f; K) = \frac{1}{2K+1} \sum_{n=-K}^{K} e^{-j2\pi fn} = \frac{\sin((2K+1)\pi f)}{(2K+1)\sin(\pi f)}$$

$$\lim_{K \to \infty} (2K+1)\,\text{cinc}(f; K) = \sum_{m=-\infty}^{\infty} \delta(f - m)$$

Periodic Impulse Train

$$\Delta[n; M] = \frac{1}{M} \sum_{k=0}^{M-1} e^{j\frac{2\pi}{M}nk} = \begin{cases} 1 & n = \text{an integer multiple of } M \\ 0 & n \neq \text{an integer multiple of } M \end{cases}$$

Some Useful Fourier Transform Pairs

$$\sqcap[n; K] \overset{\text{DTFT}}{\longleftrightarrow} (2K+1)\,\text{cinc}(f; K)$$

$$\wedge[n; M] \overset{\text{DTFT}}{\longleftrightarrow} (M+1)\,\text{cinc}^2(f; M/2)$$

$$e^{j2\pi f_o n} \overset{\text{DTFT}}{\longleftrightarrow} \delta(f - f_o) \quad -0.5 \leq f < 0.5, \; -0.5 < f_o < 0.5$$

Downsampling (Decimation)

The *downsampled* or *decimated* version of the sequence $x[n]$ is defined as

$$x_d[n; M] = \mathcal{DOWN}\{x[n]; M\} = x[nM]$$

where M is a positive integer. The spectrum of $x_d[n; M]$ is given by

$$X_d(e^{j2\pi f}) = \frac{1}{M} \sum_{k=0}^{M-1} X\left(e^{j\frac{2\pi}{M}(f-k)}\right)$$

Upsampling

The *upsampled* version of the sequence $y[n]$ is defined as

$$x_u[n; M] = \mathcal{UP}\{x[n]; M\} = \begin{cases} x[\frac{n}{M}] & \text{for } n = lM, \; l = \pm1, \pm2, \dots \\ 0 & \text{otherwise} \end{cases}$$

where M is a positive integer. The spectrum of $x_u[n; M]$ is given by

$$X_u(e^{j2\pi f}) = X(e^{j2\pi Mf})$$

Sampling

The sampled version of $x[n]$ is defined as

$$x_s[n] = x[n]\Delta[n; M] = \begin{cases} x[n] & n = \text{an integer multiple of } M \\ 0 & n \neq \text{an integer multiple of } M \end{cases}$$

where M is a positive integer. The spectrum of $x_s[n; M]$ is given by

$$X_s(e^{j2\pi f}) = \frac{1}{M} \sum_{k=0}^{M-1} X\left(e^{j2\pi(f-\frac{k}{M})}\right)$$

Relationship Between Sampling and Downsampling

$$x_s[n] = \mathcal{UP}\{x_d[n; M]; M\}$$

The DT Sampling Theorem

A sequence $x[n]$ having a spectrum limited to w

$$X(e^{j2\pi f}) = 0 \qquad \text{for } |f| > w, \text{ on the interval } -0.5 \le f < 0.5$$

can be perfectly reconstructed from its samples $x(nM), k = 0, \pm 1, \pm 2, \dots$ if, and only if,

$$\frac{1}{M} > 2w$$

DTFT Transform Properties and Pairs

See Tables 4DT.1–4DT.3.

Energy Sequences, Energy Density Functions, and Parseval's Theorem for Energy Sequences

$$x[n] \text{ is an energy sequence if } \sum_{n=-\infty}^{\infty} |x[n]|^2 < \infty.$$

$$E = \sum_{n=-\infty}^{\infty} |x[n]|^2 \text{ is the energy in } x[n].$$

$$\Phi_x(e^{j2\pi f}) = |X(e^{j2\pi f})|^2 \text{ is the } \textit{energy density spectrum} \text{ of } x[n].$$

$$\phi_x[n] \overset{\text{DTFT}}{\longleftrightarrow} \Phi_x(e^{j2\pi f})$$

$$\phi_x[n] = \sum_{k=-\infty}^{\infty} x[k]x^*[k-n] \text{ is the } \textit{autocorrelation sequence} \text{ of } x[n].$$

The energy contained in any band $\mathcal{B}$ of frequencies is given by

$$E_x(\mathcal{B}) = \int_{\mathcal{B}} \Phi_x(e^{j2\pi f}) \, df$$

Parseval's theorem for energy sequences

$$E = \sum_{n=-\infty}^{\infty} |x[n]|^2 = \int_{-1/2}^{1/2} \Phi_x(e^{j2\pi f}) \, df$$

Convergence

Let $X_K(e^{j2\pi f}) = \sum_{n=-K}^{K} x[n]e^{-j2\pi f n}$

Definition: $X_K(e^{j2\pi f})$ converges uniformly to $X(e^{j2\pi f})$ if, for every $\epsilon > 0$, there is an integer K_o, such that

$$\underbrace{|X(e^{j2\pi f}) - X_K(e^{j2\pi f})| < \epsilon}_{\text{This inequality holds } \textit{for all } f} \qquad \text{for all } K > K_o$$

Theorem: $X_K(e^{j2\pi f})$ converges uniformly to $X(e^{j2\pi f})$ if $x[n]$ is absolutely summable.

Definition: $X_K(e^{j2\pi f})$ converges to $X(e^{j2\pi f})$ in integral square if

$$\lim_{K \to \infty} \int_{-0.5}^{0.5} |X(e^{j2\pi f}) - X_K(e^{j2\pi f})|^2 df = 0$$

Theorem: $X_K(e^{j2\pi f})$ converges to $X(e^{j2\pi f})$ in integral square if $x[n]$ is an energy sequence.

4DT.11 Problems

4DT.2 Spectrums and LTI Systems

4DT.2.1 Individual signals $x_1[n]$, $x_2[n]$, $\cdots$, $x_n[n]$ are added to produce the sequence

$$x[n] = x_1[n] + x_2[n] + \cdots + x_n[n] \tag{113a}$$

a) Show that the spectrum of $x[n]$ is

$$X(e^{j2\pi f}) = X_1(e^{j2\pi f}) + X_2(e^{j2\pi f}) + \cdots + X_n(e^{j2\pi f}) \tag{113b}$$

where the $X_i(e^{j2\pi f})$ are the spectrums of $x_i[n], i = 1, 2, \ldots, n$.

b) Show that if a spectrum has the form (113b), then the corresponding sequence has the form (113a).

c) Assume that the individual sequences $x_i[n]$ overlap one another but that the corresponding spectrums $X_i(e^{j2\pi f})$ do not overlap one another. Invent a system having input $x[n]$ that recovers the individual sequences $x_i[n]$. Draw the system that performs this separation. Is your system LTI?

d) Assume that the individual spectrums $X_i(e^{j2\pi f})$ overlap one another but that the corresponding sequences $x_i[n]$ do not overlap one another. Invent a system that recovers the individual spectrums $X_i(e^{j2\pi f})$ by suitable processing of $x[n]$. Draw the system that performs this separation. Is your system LTI?

4DT.2.2 You are given $X(e^{j2\pi f}) = X\delta(f - f_o), -0.5 \leq f < 0.5$, where $-0.5 \leq f_o < 0.5$.

a) Plot $X(e^{j2\pi f})$ for $-2 \leq f < 2$.

b) Write an equation for $X(e^{j2\pi f})$ for the given spectrum that applies for all f.

c) Find $x[n]$.

4DT.2.3 **a)** Find $x[n]$ for $X(e^{j2\pi f}) = X\delta(f - f_o) + X^*\delta(f + f_o)$, where $-0.5 \leq f_o < 0.5$, and $X = \frac{1}{2}Ae^{j\phi}$.

Express your answer as a real sinusoidal sequence.

b) Use your result from part (a) to write down the spectrum $X(e^{j2\pi f})$ of $x[n] = 3\cos(2\pi 0.15n + 22°)$. Plot $X(e^{j2\pi f})$.

4DT.2.4 **a)** Use the inverse Fourier transform to find $x[n]$ for

$$X(e^{j2\pi f}) = X_1\delta(f - f_1) + X_1^*\delta(f + f_1) + X_2\delta(f - f_2) + X_2^*\delta(f + f_2); |f| < 0.5 \tag{114}$$

where $X_i = \frac{1}{2}A_i e^{j\phi_i}$ and $0 < f_i < 0.5, i = 1, 2$. Express your answer as a real function of n.

b) Use your result from part (a) to write down the spectrum $X(e^{j2\pi f})$ of

$$x[n] = A_1\cos(2\pi f_1 n + \phi_1) + A_2\cos(2\pi f_2 n + \phi_2) \tag{115}$$

Plot $X(f)$.

c) Recall that in Figure 3DT.13(a) we plotted the two-sided spectrum of

$$x[n] = A_1\cos(2\pi f_1 n + \phi_{x1}) + A_2\cos(2\pi f_2 n + \phi_{x2}) \tag{116}$$

Describe how the line spectrum in Figure 3DT.13(a) differs from your plot of part (b). In which plot(s) are impulses used? Which of the plots is a density spectrum?

4DT.2.5 The spectrum of a sequence $x[n]$ is given by

$$X(e^{j2\pi f}) = \sum_{i=-3}^{3} X_i\delta(f - f_i); \quad |f| < 0.5 \tag{117}$$

where

$$X_i = \begin{cases} A_0 & i = 0 \\ \frac{1}{2}A_i e^{j\phi_{xi}} & i = 1, 2, 3 \\ X_i^* & i = -1, -2, -3 \end{cases} \tag{118}$$

and

$$f_i = \begin{cases} 0 & i = 0 \\ -f_{-i} & i = -1, -2, -3 \end{cases} \tag{119}$$

with $0 < f_i < 0.5, i = 1, 2, 3$.

a) Use the inverse DTFT to find $x[n]$. Express your answer as a sum of real sinusoidal components plus a constant (DC) term.

b) Use sketches to show that a judicious use of filters allows you to separate the individual sinusoidal components from $x[n]$. Draw the system that accomplishes this separation.

c) Sketch the input and output spectrums of every filter in your system.

4DT.2.6 Assume that an input $x[n] = \delta[n]$ is applied to the system of Figure 4DT.1. Find $X(e^{j2\pi f})$, $Y(e^{j2\pi f})$ and $y[n]$. Assume that $H(e^{j2\pi f})$ is arbitrary.

4DT.2.7 Repeat Problem 4DT.2.3 for an input $x[n] = \delta[n - m]$.

4DT.2.8 Assume that $x[n] = Xe^{j2\pi f_o n}$, where $-0.5 \leq f_o < 0.5$, is applied to the system of Figure 4DT.1. Specifically, Find $X(e^{j2\pi f})$, $Y(e^{j2\pi f})$, and $y[n]$. Assume that $H(e^{j2\pi f})$ is arbitrary. Hint: $\mathcal{F}\{Xe^{j2\pi f_o n}\} = X\delta(f - f_o)$, for $-0.5 \leq f < 0.5$.

4DT.3   Amplitude and Phase Spectrums

4DT.3.1 Do Eq. (11) and (12) hold even if $x[n]$ and $h[n]$ are not real? Justify your answer.

4DT.4   Some Sequences and Their Spectrums

4DT.4.1 Confirm the Fourier transform pair $\delta[n] \overset{\text{DTFT}}{\longleftrightarrow} 1$ by substituting $\delta[n]$ into the DTFT sum.

4DT.4.2 Confirm the Fourier transform pair $1 \overset{\text{DTFT}}{\longleftrightarrow} \delta(f)$, $-0.5 \leq f < 0.5$, by substituting $\delta(f)$ into the inverse DTFT integral.

4DT.4.3 a) Write a MATLAB program to plot the absolute value $|X(e^{j2\pi f})|$ of the spectrum of the rectangular pulse of Example 4DT.2. Use the value for K given in the example.

b) Write a MATLAB program to plot the absolute value $|X(e^{j2\pi f})|$ of the spectrum of the raised cosine sequence of Example 4DT.5. Use the value for K given in the example.

c) Redraw your plots of parts (a) and (b) with $K = 25$.

d) A decibel (dB) scale provides a means to describe numbers having very large or small magnitudes using numbers in more familiar, moderate ranges. The dB value of $|X(e^{j2\pi f})|$ is defined as $20 \log_{10} |X(e^{j2\pi f})|$. Use MATLAB to plot the dB values of the amplitude spectrums of the rectangular and raised cosine sequences versus f using the value $K = 25$. Describe in your own words how the decibel plots compare to the linear scale plots of part (c).

4DT.5   Properties of the DTFT

4DT.5.1 Show that $\sum_{-\infty}^{\infty} x[n] = X(e^{j2\pi 0})$.

4DT.5.2 Show that $\int_{-0.5}^{0.5} X(e^{j2\pi f}) \, df = x[0]$.

4DT.5.3 Show which of the following sequences is even, which is odd, and which is neither.
    a) $x_1[n] = n$
    b) $x_2[n] = x_o[n] * x_o[n]$, where $x_o[n]$ is any odd function
    c) $x_3[n] = \cos(2\pi 0.1n + 45°)$
    d) $x_4[n] = F\{\sin[n]\}$, where $F\{\xi\}$ is any function of ξ
    e) $x_5[n] = F\{\cos[n]\}$, where $F\{\xi\}$ is any function of ξ

4DT.5.4 Refer to Problem 4DT.5.3 and state the symmetry properties that apply to spectrums $X_1(e^{j2\pi f})$ through $X_5(e^{j2\pi f})$, corresponding to sequences $x_1[n]$ through $x_5[n]$.

4DT.5.5 a) Show that $X(e^{j2\pi f})$ is real and even if and only if $x[n]$ is real and even.
    b) Give an example for part (a).

4DT.5.6 a) Show that $X(e^{j2\pi f})$ is imaginary and odd if and only if $x[n]$ is real and odd.
    b) Give an example for part (b).

4DT.5.7 The sequence $x[n] = 10\cos(2\pi 0.1n + 22°) + 255\cos(2\pi 0.25n)$ is applied to an ideal low-pass filter. Assume that the filter cutoff frequency w equals 0.2.
    a) Find $X(e^{j2\pi f})$ and plot it and $H(e^{j2\pi f})$ on the same plot.
    b) Find and plot $Y(e^{j2\pi f})$.
    c) Use the inverse Fourier transform to find $y[n]$. Write your answer as a real sinusoidal function.
    d) Make a detailed comparison of your analysis and answer with that of Example 3DT.3 for $w = 0.2$.

4DT.5.8 The sequence $x[n] = 10\cos(2\pi 0.05n)$ is applied to a system that shifts it by five samples. Find the spectrum before and after the delay.

4DT.5.9 Use the DTFT pair $a^n u[n] \longleftrightarrow 1/(1 - \alpha e^{-j2\pi f})$ and Table 4DT.2 to find the spectrum of $x[n] = a^n u[n] - a^{-n}u[-n]$.

4DT.5.10 a) Find the DTFT of $\sqcap[n; K]a^{|n|}$.
    b) Use a computer to plot $\sqcap[n; K]a^{|n|}$ and $X(e^{j2\pi f})$ for $a = 0.95$ $K = 7$.

4DT.5.11 Use the multiply-by-n property to find the spectrum of $na^n u[n]$.

4DT.5.12 Use the multiplication property and the transform pair $a^{|n|} \longleftrightarrow (1 - a^2)/(1 + a^2 - 2a\cos(2\pi f))$ to evaluate the circular convolution $(1 - a^2)/(1 + a^2 - 2a\cos(2\pi f)) \circledast (1 - a^2)/(1 + a^2 - 2a\cos(2\pi f))$. Comment on the interesting features of your result.

4DT.5.13 Use the DTFT pair $a^n u[n] \longleftrightarrow 1/(1 - \alpha e^{-j2\pi f})$ and Table 4DT.2 to find the Fourier transforms of the following signals.
    a) $a^{n-1}u[n - 1]$
    b) $a^{n-1}u[n]$
    c) $\sum_{k=-K}^{K} a^{n-k}u[n - k]$.

4DT.5.14 Derive the impulse response of an ideal high-pass filter. Hint: $H_{hp}(e^{j2\pi f}) = 1 - H_{lp}(e^{j2\pi f})$.

4DT.5.15 Derive the impulse response of an ideal band-stop filter. Hint: $H_{bf}(e^{j2\pi f}) = 1 - H_{bp}(e^{j2\pi f})$.

4DT.5.16 An LTI system has input $x[n] = \sqcap[n; K]$ and output $y[n] = (2K + 1)(\wedge[n - K; 2K] + \wedge[n + K; 2K])$.
    a) Sketch $x[n]$ and $y[n]$ for $K = 3$.
    b) Find $H(e^{j2\pi f})$.
    c) Is this system causal?

4DT.5.17 Recall that a rectangular pulse can be expressed as $\sqcap[n; K] = u[n+K] - u[n-K-1]$. Use the time shift property and the pair $u[n] \longleftrightarrow \pi\delta(f) + 1/(1 - e^{-j2\pi f})$ for $-\frac{1}{2} \le f < \frac{1}{2}$ to find the spectrum of $\sqcap[n; K]$.

4DT.5.18 In Example 4DT.7 we stated that $2w \, \text{sinc}(2wn) \overset{\text{DTFT}}{\longleftrightarrow} \sqcap(f/2w), -0.5 \leq f < 0.5$. Derive this DTFT pair by substituting $\sqcap(f/2w)$ into the inverse DTFT integral.

4DT.5.19 Circular coordinates were used in Example 4DT.7 to evaluate a circular convolution. Circular convolutions can also be evaluated using rectangular coordinates by bearing in mind that the functions to be convolved are periodic. Consider $X(e^{j2\pi f}) \circledast X(e^{j2\pi f})$, where $X(e^{j2\pi f}) = \sqcap(f/2w), -0.5 \leq f < 0.5$, as in Example 4DT.7.
  a) Plot $X(e^{j2\pi\lambda})$ on a rectangular coordinate system having horizontal coordinate λ, where $-1.5 \leq \lambda \leq 1.5$.
  b) Make a similar plot of $X(e^{j2\pi(f-\lambda)})$ for $0 \leq f \leq 2w$.
  c) Plot the product $X(e^{j2\pi\lambda})X(e^{j2\pi(f-\lambda)})$ for $-1.5 < \lambda < 1.5$.
  d) Show that the area of the product in the range $-0.5 \leq \lambda \leq 0.5$ is

$$\int_{-0.5}^{0.5} X(e^{j2\pi\lambda})X(e^{j2\pi(f-\lambda)}) \, d\lambda = 2w - f$$

where $0 \leq f \leq 2w$, which is the same result as in Example 4DT.7.
  e) Continue in a similar way to evaluate $X(e^{j2\pi f}) \circledast X(e^{j2\pi f})$ for all f.

4DT.5.20 Consider $y[n] = x^2[n]$ where $x[n] = A \cos(2\pi f_o n)$ and $0 < f_o < 0.25$.
  a) Find $Y(e^{j2\pi f})$ by evaluating the circular convolution $X(e^{j2\pi f}) \circledast X(e^{j2\pi f})$.
  b) Check your result using the identity $\cos^2\theta = \frac{1}{2} + \frac{1}{2}\cos(2\theta)$.
  c) Compare the frequency components in $x^2[n]$ with those in $x[n]$.

4DT.5.21 Prove the following symmetry properties of the following DTFTs:
  a) 4DT.SP1
  b) 4DT.SP2
  c) 4DT.SP3
  d) 4DT.SP4
  e) 4DT.SP5
  f) 4DT.SP6
  g) 4DT.SP7

4DT.5.22 **a)** Prove the index shift property two ways: (1) using the definition of the DTFT and (2) using the inverse DTFT.
  b) Derive (4DT.OP9).

4DT.5.23 Use the equation $x_s[n] = \mathcal{UP}\{x_d[n; M]; M\}$ along with (4DT.OP10) and (4DT.OP11) to derive $X_s(e^{j2\pi f})$. Does your result agree with that in (4DT.OP12)?

4DT.5.24 Show that:
  a)

$$\int_{-0.5}^{0.5} \text{cinc}(f; K) \, df = \frac{1}{2K+1} \tag{120}$$

  b)

$$\int_{-0.5}^{0.5} \text{cinc}(\lambda; K) \, \text{cinc}(f - \lambda; K) \, d\lambda = \frac{1}{2K+1} \text{cinc}(f; K) \tag{121}$$

  c)

$$\int_{-0.5}^{0.5} \text{cinc}\left(\lambda - \frac{n}{2K+1}; K\right) \text{cinc}\left(\lambda - \frac{m}{2K+1}; K\right) d\lambda = 0 \tag{122}$$

if $n - m$ is not an integer multiple of $2K + 1$.
  d)

$$\int_{-0.5}^{0.5} \text{cinc}^2(f; K) \, df = \frac{1}{2K+1} \tag{123}$$

4DT.5.25 In Example 4DT.6, we used the convolution property to establish the DTFT pair

$$DT\mathcal{FT}\{\wedge[n; M]\} = (M + 1)\,\text{cinc}^2\left(f; \frac{1}{2}M\right)$$

for even values of M : $M = 2K$, $K = 0, 1, \ldots$ Show, by direct evaluation of $DT\mathcal{FT}$ $[\wedge[n; M]]$, that the result this result actually holds for both even and odd $M = 0, 1, 2, \ldots$. Hint: The derivation is lengthy. Use the following identity, which can be obtained from identity 5 of Appendix B:

$$\sum_{n=0}^{K} n\xi^n = (1 - \xi^{K+1})\frac{\xi}{(1 - \xi)^2} - \xi^{K+1}(K + 1)\frac{1}{1 - \xi}$$

4DT.6   The Sampling Theorem

4DT.6.1 The sampled sequence $x_s[n] = x[n]\Delta[n; M]$ is applied to an LTI system having the unit impulse response $h[n] = u[n] - u[n - M]$. This system, called a DT sample-and-hold reconstruction system, has output $x_{sh}[n]$ where $x_{sh}[n]$ is an approximation to $x[n]$.

a) Sketch $h[n]$ for $M = 7$, and show how to implement $h[n]$ using a shift register.

b) Draw the sample-and-hold reconstruction system.

c) Use sketches to compare $x_{sh}[n]$ with $x[n]$.

d) Find the expression for $X_{sh}(e^{j2\pi f})$. You may assume that M is an odd integer. Hint: When M is odd, $h[n]$ can be expressed in terms of a rectangular pulse $\sqcap[\cdot; \cdot]$.

e) Sketch $X(e^{j2\pi f})$, $X_s(e^{j2\pi f})$, and $X_{sh}(e^{j2\pi f})$ to find conditions for which $x[n]$ can be exactly reconstructed by applying $x_{sh}[n]$ to a low-pass filter. Specify the conditions and plot the low-pass filter amplitude and phase characteristics. Hint: The amplitude characteristic of the low-pass filter is not constant within the passband of the filter.

4DT.6.2 A popular method of interpolating between data points of $x_s[n] = x[n]\Delta[n; M]$ is straight-line, or linear, interpolation: With this method, the interpolated data values lie on straight lines running between the adjacent samples of $x_s[n]$. For example, if $M = 3$, then the interpolation between $x[0]$ and $x[3]$ is given by

$$x_{stl}[n] = x[0] + \frac{x[3] - x[0]}{3}n \quad n = 1, 2$$

Similarly, the interpolation between $x[3]$ and $x[6]$ is given by

$$x_{stl}[n] = x[3] + \frac{x[6] - x[3]}{3}(n - 3) \quad n = 4, 5$$

and so on.

a) Show by means of sketches that a straight-line interpolated sequence can be expressed as

$$x_{stl}[n] = x_s[n] * \wedge[n; M - 1]$$

b) Show that the spectrum of $x_{stl}[n]$ is given by

$$X_{stl}(e^{j2\pi f}) = X_s(e^{j2\pi f})M\,\text{cinc}^2\left(f; \frac{1}{2}(M - 1)\right)$$

c) Sketch $X(e^{j2\pi f})$, $X_s(e^{j2\pi f})$, $M\,\text{cinc}^2(f; \frac{1}{2}(M - 1))$, and $X_{stl}(e^{j2\pi f})$ to show how $X_{stl}(e^{j2\pi f})$ is related to $X(e^{j2\pi f})$. Chose a convenient real-even function $X(e^{j2\pi f})$ for your sketch.

d) Describe what happens to $x_{stl}[n]$ and $X_{stl}(e^{j2\pi f})$ as M is increased.

e) Show that $x[n]$ can, under certain conditions, be exactly reconstructed by applying $x_{stl}[n]$ to a low-pass filter. Specify the conditions and plot the low-pass filter amplitude and phase characteristics. Hint: The amplitude characteristic of the low-pass filter is not constant that passband of the filter.

Figure 4DT.25
Reconstruction filter.

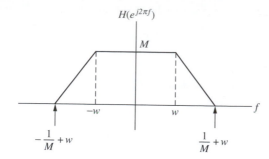

4DT.6.3 Refer to Problems 4DT.6.2 and 4DT.6.3 and you can exactly reconstruct $x[n]$ by sampling $x_{sh}[n]$ or $x_{stl}[n]$ and applying the samples to an ideal low-pass filter. State the conditions under which perfect reconstruction is possible.

4DT.6.4 In addition to its use as a reconstruction formula, DT Nyquist-Shannon interpolation can be used to generate a band-limited sequence from an arbitrary data sequence $\{w[m]\}$ by replacing $x[mM]$ in Eq. (87) by $w[m]$ for $m = 0, \pm1, \pm2, \cdots$. Show that DT Nyquist-Shannon interpolation always produces a band-limited sequence, regardless of the data sequence $w[m]$ used.

4DT.6.5 Find a closed form for $x[n]$ in the DT Nyquist-Shannon interpolation formula if $x(mM) = A\cos(2\pi f_o mM)$ and
a) $f_o < 1/(2M)$.
b) $1/(2M) < f < 1/M$.
Hint: The DT Nyquist-Shannon interpolation formula works like an ideal low-pass filter having cutoff frequency $\frac{1}{2M}$, gain M and input $x_S[n] = x[n]\Delta[n; M]$. Sketch the spectrum of $x_S[n]$.

4DT.6.6 A sequence $x[n]$ has spectrum $X(e^{j2\pi f})$, where $X(e^{j2\pi f}) = 0$ for $|f| > w$ on the interval $-0.5 \le f < 0.5$. The sequence $x[n]$ is sampled with $1/M > 2w$. The sampled sequence $x_s[n]$ is then applied to a reconstruction filter having the frequency response characteristic shown in Figure 4DT.25
a) Show, by means of frequency domain sketches, that the output of the reconstruction filter equals $x[n]$.
b) Find the impulse response of the filter. Hint: Direct evaluation of $\mathcal{DTFT}^{-1}\{H(e^{j2\pi f})\}$ is straightforward but lengthy. However, you can find $h[n]$ easily by finding $\mathcal{DTFT}^{-1}\{\frac{d^2 H(e^{j2\pi f})}{df^2}\}$ and using the multiply by n property.
c) Write down the interpolation formula associated with the filter.
d) State what happens to the filter and the interpolation formula if $1/M = 2w$.

4DT.7   Additional DTFT Pairs

4DT.7.1 Use Tables 4DT.2 and 4DT.3 to find the spectrums of
a) $\sin(2\pi f_o n)$.
b) $A\cos(2\pi f_o n + \phi)$.

4DT.7.2 A sequence $x[n] = 10\cos(2\pi 25nT_s + 22°) + 255\cos(2\pi 125nT_s)$ is applied to an ideal low-pass filter. Assume that $T_s = 0.001$ and the filter cutoff frequency $f_c = 0.05$.
a) Find $X(e^{j2\pi f})$ and plot it and $H(e^{j2\pi f})$ on the same plot.
b) Find and plot $Y(e^{j2\pi f})$.
c) Use the inverse DTFT to find $y[n]$. Write your answer as a real sequence.

4DT.7.3 Consider shifting the sequence $x[n] = A\cos(2\pi f_o n)$ by m samples. Describe what happens to $X(e^{j2\pi f})$. Sketch the sequence and the spectrum before and after the time shifting.

4DT.7.4 Derive the spectrum of the two-sided decaying exponential $a^{|n|}$ where $|a| < 1$.

4DT.8   Energy Sequences, Energy Density Spectrums, and Parseval's Theorem

4DT.8.1 Show that

$$\sum_{n=-\infty}^{\infty} x[n]y^*[n] = \int_{-0.5}^{0.5} X(e^{j2\pi f})Y^*(e^{j2\pi f})\, df \tag{124}$$

where $X(e^{j2\pi f})$ and $Y(e^{j2\pi f})$ are the Fourier transforms of $x[n]$ and $y[n]$, respectively. This result is known as the *generalized Parseval's theorem*. Hint: Set $x_1[n] = x[n]$ and $x_2[n] = y^*[-n]$ in the convolution property. The sequence

$$\phi_{vw}[n] = \sum_{m=-\infty}^{\infty} x[n]y^*[-n+m] = \mathcal{DTFT}^{-1}\{X(e^{j2\pi f})Y^*(e^{j2\pi f})\} \tag{125}$$

encountered in the derivation is called the *cross-correlation sequence* of $x[n]$ and $y[n]$.

5CT Fourier Series

5CT.1 Introduction

The Fourier series is the premier analytical aid for analyzing a periodic waveform and determining the response it produces from an LTI system. A waveform is *periodic* if it satisfies

$$x(t) = x(t + T) \tag{1}$$

for all t. *The period T* is the smallest positive quantity for which Eq. (1) applies. Every periodic waveform that can be drawn on paper or displayed on an oscilloscope can be represented by a Fourier series.

The Fourier series can be viewed as a special case of the multiple sinusoidal component representations we saw previously. For example, the multiple sinusoid representation of Eq. (3CT.51)

$$x(t) = A_0 + \sum_{k=1}^{I} A_k \cos(2\pi F_k t + \phi_k) \tag{3CT.51}$$

becomes, for periodic waveforms

$$x(t) = A_0 + \sum_{k=1}^{\infty} A_k \cos(2\pi k F_1 t + \phi_k) \tag{2}$$

Thus, the Fourier series represents a periodic waveform as an infinite sum of sinusoids whose frequencies are given by

$$F_k = k F_1 \quad k = 0, 1, 2, \ldots \tag{3a}$$

where

$$F_1 = \frac{1}{T} \tag{3b}$$

Frequencies related by Eq. (3a) are called *harmonic* frequencies: F_1 is the *fundamental* or *first* harmonic, $F_2 = 2F_1$ is the *second* harmonic, and so on. The term "harmonic" refers to the pleasant sound heard when two or more musical notes having frequencies related by ratios of small integers are played simultaneously.

Figure 5CT.1 illustrates a typical application of the Fourier series. Here, a Fourier series is used to describe the distortion introduced when an audio amplifier is driven into saturation by a sinusoidal input voltage. The input sinusoid is amplified by gain g until the amplifier output voltage saturates at $\pm A_{\max}$. When the input exceeds $A_{\max}/g$, the output is "clipped" to $\pm A_{\max}$ volts as shown. The clipped output waveform is not sinusoidal, but it can be expressed as a sum of sinusoids, i.e., a Fourier series Eq. (2). Each sinusoidal component in

Figure 5CT.1
Saturating audio amplifier with sinusoidal input and nonsinusoidal output.

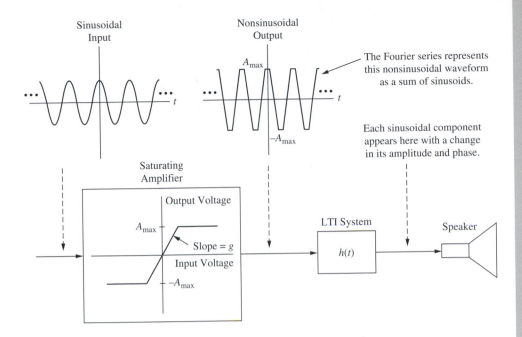

The Fourier series represents this nonsinusoidal waveform as a sum of sinusoids.

Each sinusoidal component appears here with a change in its amplitude and phase.

the clipped waveform propagates through the LTI system elements that follow the amplifier and is present in the speaker audio output with only a change in amplitude and phase. The sinusoidal components introduced by the clipping degrade the quality of the sound. This degradation is called *harmonic distortion*.

This chapter shows you how to analyze the system of Figure 5CT.1 as well as other applications involving periodic waveforms.

5CT.2   Forms of the Fourier Series

The Fourier series has three forms: complex, amplitude-phase, and sine-cosine. The complex form applies to both real and complex $x(t)$.

Complex Form of the Fourier Series

$$x(t) = \sum_{k=-\infty}^{\infty} X_k e^{j2\pi k F_1 t} \tag{4}$$

where

$$F_1 = 1/T \tag{5}$$

is called the *fundamental frequency*. The constant X_k, is called the kth complex Fourier series coefficient. It is given by

$$X_k = \frac{1}{T} \int_T x(t) e^{-j2\pi k F_1 t} dt \tag{6}$$

The notation $\int_T$ denotes integration over any period, $\int_{t_o}^{t_o+T}$, where t_o is arbitrary.

The amplitude-phase, and sine-cosine forms of the Fourier series are special cases of the complex form that apply to a real valued $x(t)$.

Amplitude-Phase Form of the Fourier Series

$$x(t) = A_0 + \sum_{k=1}^{\infty} A_k \cos(2\pi k F_1 t + \phi_k) \tag{7}$$

where $A_0 = X_0$ and

$$A_k = 2|X_k| \quad k = 1, 2, \ldots \tag{8}$$

$$\phi_k = \angle X_k \quad k = 1, 2, \ldots \tag{9}$$

Sine-Cosine Form of the Fourier Series

$$x(t) = a_0 + \sum_{k=1}^{\infty} a_k \cos(2\pi k F_1 t) + \sum_{k=1}^{\infty} b_k \sin(2\pi k F_1 t) \tag{10}$$

where $a_0 = X_0$ and

$$a_k = 2\mathcal{R}\{X_k\} = \frac{2}{T} \int_T x(t) \cos(2\pi k F_1 t)\, dt \tag{11}$$

$$b_k = -2\mathcal{I}\{X_k\} = \frac{2}{T} \int_T x(t) \sin(2\pi k F_1 t)\, dt \tag{12}$$

The three forms are equivalent for real valued $x(t)$. If you know the complex Fourier series coefficients of a real periodic waveform, you can refer to Eqs. (8)–(9) and Eqs. (11)–(12) to write down the amplitude-phase and sine-cosine forms.

Graphs of Fourier series coefficients are called *line spectrums* and are similar to the line spectrums of Chapter 3CT. The coefficients X_k can be plotted versus F, kF_1, or k.

EXAMPLE 5CT.1 Periodic Rectangular Pulse Train

Consider the periodic rectangular pulse train of Figure 5CT.2

We can find the complex Fourier series coefficients by evaluating Eq. (6) with a region of integration from $-0.5T$ to $+0.5T$. When we substitute the periodic pulse train into Eq. (6), the integral from $-0.5T$ to $+0.5T$ breaks up into the three regions, indicated in the following table:

Time Interval	$x(t)$
$-0.5T < t \leq -0.5\tau_p$	0
$-0.5\tau_p < t \leq 0.5\tau_p$	A
$0.5\tau_p < t \leq 0.5T$	0

Figure 5CT.2
Periodic rectangular
pulse train and its line
spectrum:
(a) waveform; (b) line
spectrum.

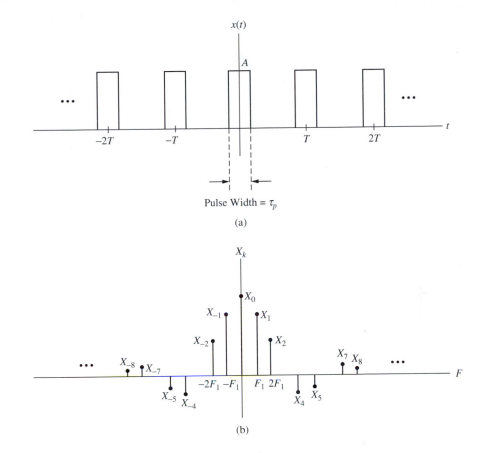

Thus

$$X_k = \frac{1}{T} \int_{-T/2}^{T/2} x(t) e^{-j2\pi k F_1 t}\, dt$$

$$= \frac{1}{T} \int_{-0.5T}^{-0.5\tau_p} 0 e^{-j2\pi k F_1 t}\, dt + \frac{1}{T} \int_{-0.5\tau_p}^{0.5\tau_p} A e^{-j2\pi k F_1 t}\, dt + \frac{1}{T} \int_{0.5\tau_p}^{0.5T} 0 e^{-j2\pi k F_1 t}\, dt \qquad (13)$$

The first and last integrals on the right-hand side of this equation equal 0. Therefore

$$X_k = \frac{1}{T} \int_{-0.5\tau_p}^{0.5\tau_p} A e^{-j2\pi k F_1 t}\, dt \qquad (14)$$

and this integral becomes

$$X_k = \frac{A}{T} \frac{-1}{j2\pi k F_1} e^{-j2\pi k F_1 t} \Big|_{-\tau_p/2}^{\tau_p/2} = \frac{A}{T} \frac{-1}{j2\pi k F_1} (e^{-j2\pi k F_1 \tau_p/2} - e^{j2\pi k F_1 \tau_p/2}) \qquad (15)$$

for $k \neq 0$. We can simplify this expression with the aid of Euler's formula.

$$\sin\theta = \frac{1}{2j}(e^{j\theta} - e^{-j\theta}) \qquad (16)$$

The result is

$$X_k = \frac{A\tau_p}{T} \frac{\sin(\pi k F_1 \tau_p)}{\pi k F_1 \tau_p} \quad k \neq 0 \tag{17a}$$

We assumed that $k \neq 0$ because, if $k = 0$, then we cannot divide by k in (15). We can evaluate the integral (6) with $k = 0$ to find X_0.

$$X_0 = \frac{1}{T} \int_{-\tau_p/2}^{\tau_p/2} A\, dt = \frac{A\tau_p}{T} \tag{17b}$$

Equations (17a) and (17b) can be written as

$$X_k = \frac{A\tau_p}{T} \, \text{sinc}(k F_1 \tau_p) \qquad \text{for all } k. \tag{18}$$

The pulse train and its (two-sided) line spectrum are plotted in Figure 5CT.2(b) for $T = 3\tau_p$.

We can use Eqs. (8), (9), and (11), Eq. (12) to write down the following three forms of the series for the periodic pulse train.

Complex Form

$$x(t) = \sum_{k=-\infty}^{\infty} \frac{A\tau_p}{T} \, \text{sinc}(k F_1 \tau_p) e^{j2\pi k F_1 t} \tag{19}$$

Amplitude-Phase Form

$$x(t) = \frac{A\tau_p}{T} + \sum_{k=1}^{\infty} \frac{2A\tau_p}{T} |\, \text{sinc}(k F_1 \tau_p)| \cos(2\pi k F_1 t + \angle X_k) \tag{20}$$

where

$$\angle X_k = \begin{cases} 0 & \text{if } \text{sinc}(k F_1 \tau_p) \geq 0 \\ -\pi & \text{if } \text{sinc}(k F_1 \tau_p) < 0 \end{cases} \tag{21}$$

Sine-Cosine Form

$$x(t) = \frac{A\tau_p}{T} + \sum_{k=1}^{\infty} \frac{2A\tau_p}{T} \, \text{sinc}(k F_1 \tau_p) \cos(2\pi k F_1 t) \tag{22}$$

EXAMPLE 5CT.2 Periodic Impulse Train

Consider next a *periodic impulse train*.

$$x(t) = \sum_{i=-\infty}^{\infty} \delta(t - iT) \tag{23}$$

The Fourier coefficients of an impulse train can be obtained by evaluation of (6). Notice that only the impulse $\delta(t)$ (i.e., the $i = 0$ term) in Eq. (23) enters the region of integration. Therefore Eq. (23) becomes

$$X_k = \frac{1}{T} \int_{-T/2}^{T/2} \delta(t) e^{-j2\pi k F_1 t}\, dt = \frac{1}{T} \qquad \text{for all } k \tag{24}$$

The periodic impulse train and its line spectrum are plotted in Figure 5CT.3

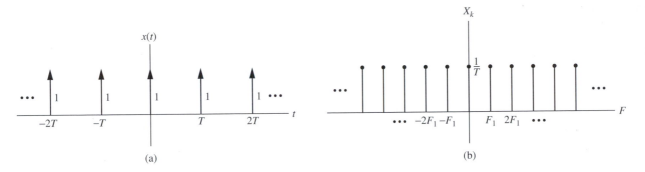

Figure 5CT.3
Periodic impulse train and its line spectrum: (a) waveform; (b) line spectrum.

The alternative Fourier series representations for a periodic impulse train are given by the complex form

$$\sum_{i=-\infty}^{\infty} \delta(t - iT) = \frac{1}{T} \sum_{n=-\infty}^{\infty} e^{j2\pi n F_1 t} \tag{25a}$$

and the amplitude phase (or sine-cosine) form

$$\sum_{i=-\infty}^{\infty} \delta(t - iT) = \frac{1}{T} + \frac{2}{T} \sum_{n=1}^{\infty} \cos(2\pi n F_1 t) \tag{25b}$$

The preceding examples illustrate how to derive the Fourier series coefficients. The next example considers the practical application shown in Figure 5CT.1.

EXAMPLE 5CT.3   Fourier Analysis of a Clipped Sinusoid

In this example we use a Fourier series to analyze the harmonic distortion caused by overdriving the amplifier in Figure 5CT.1. We can think of the input sinusoid as having a fixed amplitude A_{in} and the amplifier as having an adjustable gain control that sets the value of the gain g. If we make g too large, clipping occurs. Clipping occurs when $A_{in}g > A_{max}$, where A_{in} is the input amplitude and A_{max} is the output clip level. A close-up view of a clipped output waveform is shown in Figure 5CT.4 where $t_1 = 0.5T - t_0$.

We will use the sine-cosine form of the series Eq. (10) with a region of integration from $-T/2$ to $+T/2$ in Eqs. (11) and (12). The waveform $x(t)$ and the integrals break up into the five regions shown in the table:

Time Interval	$x(t)$
$-0.5T < t \le -t_1$	$-A_{max}$
$-t_1 < t \le -t_0$	$A \cos(2\pi F_1 t)$
$-t_0 < t \le t_0$	A_{max}
$t_0 < t \le t_1$	$A \cos(2\pi F_1 t)$
$t_1 < t \le 0.5T$	$-A_{max}$

Figure 5CT.4
Clipped cosine.

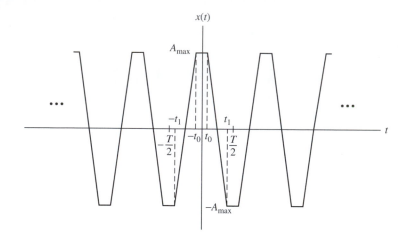

The remaining steps for evaluating the Fourier coefficients are straightforward and like those in Example 1. However, the analysis is lengthy. The results are

$$b_k = 0 \quad k = 1, 2, 3, \dots \tag{26a}$$

$$a_k = 0 \quad k = 0, 2, 4, \dots \tag{26b}$$

$$a_1 = A_{\max} \{\lambda(2 \operatorname{sinc}(\lambda/2) - g_n \operatorname{sinc}(\lambda)) + g_n(1 - \lambda)\} \tag{26c}$$

and

$$a_k = A_{\max}\lambda \{2 \operatorname{sinc}(k\lambda/2) - g_n [\operatorname{sinc}((k+1)\lambda/2) + \operatorname{sinc}((k-1)\lambda/2)]\} \quad k = 3, 5, 7, \dots \tag{26d}$$

In the preceding equation, g_n is the amplifier's normalized gain

$$\frac{g}{g_0} = g_n \tag{27}$$

where g_0 is the maximum gain for which clipping does not occur. It is defined by the equation $A_{\text{in}}g_0 = A_{\max}$. The constant λ is the fraction of time the waveform is saturated. This fraction is related to t_0 and g_n by

$$\lambda = \frac{4t_0}{T} = \frac{2}{\pi} \cos^{-1}(1/g_n) \tag{28}$$

Figure 5CT.5 shows the values of the Fourier coefficients (a_i; $i = 0, 1, \dots$) for various values of g_n. The figure assumes a $1 = $ V clip level: $A_{\max} = 1$. Figure 5CT.5(a) shows the output spectrum when the normalized gain is unity. Here, because there is no clipping, the only component in the output is the fundamental. Figure 5CT.5(b) shows the spectrum when the amplifier is overdriven by 10 percent ($g_n = 1.1$). Notice that the amplitude of the fundamental is also increased by approximately 10 percent. Higher odd-numbered harmonics (particularly the third harmonic) have been introduced by the clipping but their amplitudes are relatively small. Figure 5CT.5(c) shows the output spectrum for $g_n = 2$. Here we see a relatively large increase in the amplitudes of the third and fifth harmonics compared to the fundamental. Figure 5CT.5(d) shows the output spectrum for $g_n = 1,000$. The enormous increase in g_n from 2 to 1,000 has resulted in a relatively small increase in the amplitudes of the first few harmonics. However, it has substantially increased the amplitudes of many higher-frequency harmonics.

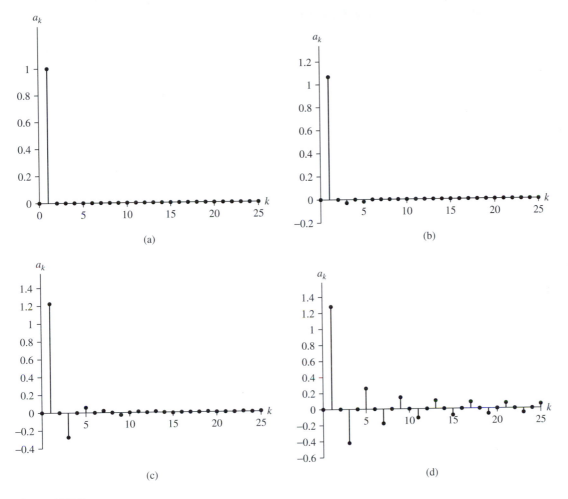

Figure 5CT.5
Fourier coefficients of clipped cosine: (a) $g_n = 1$; (b) $g_n = 1.1$; (c) $g_n = 2$; (d) $g_n = 1,000$.

5CT.3 LTI System Response to Periodic Input

We can use a Fourier series representation of a periodic waveform to determine the response of an LTI system to a periodic input. To derive this technique, notice that the Fourier series is a special case of the sum

$$x(t) = \sum_{i=-I}^{I} X_i e^{j2\pi F_i t} \tag{29}$$

that appeared in Eq. (3CT.57). It follows that all the properties of multiple sinusoidal waveforms that were developed in Chapter 3CT carry over to the Fourier series. In particular, the following I-O relationship of Chapter 3CT

$$\underbrace{\sum_{i=-I}^{I} X_i e^{j2\pi F_i t}}_{x(t)} \xrightarrow{h(t)} \underbrace{\sum_{i=-I}^{I} X_i H(F_i) e^{j2\pi F_i t}}_{y(t)} \qquad \text{(3CT.57)}$$

becomes, for periodic waveforms

$$\underbrace{\sum_{i=-\infty}^{\infty} X_i e^{j2\pi i F_1 t}}_{x(t)} \xrightarrow{h(t)} \underbrace{\sum_{i=-\infty}^{\infty} X_i H(i F_1) e^{j2\pi i F_1 t}}_{y(t)} \qquad \text{(30)}$$

The right-hand side of Eq. (30) is the Fourier series representation for $y(t)$. For a real input and real LTI system, Eq. (30) can be written as

$$\underbrace{A_0 + \sum_{i=1}^{\infty} A_i \cos(2\pi i F_1 t + \phi_i)}_{x(t)} \xrightarrow{h(t)} \underbrace{A_0 H(0) + \sum_{i=1}^{\infty} A_i |H(i F_1)| \cos(2\pi i F_1 t + \phi_i + \angle H(i F_1))}_{y(t)}$$

$$\text{(31)}$$

The following simple example illustrates this formula when the LTI system is an ideal filter.

EXAMPLE 5CT.4   Response of Ideal Filters to Periodic Input

Assume that the periodic pulse train of Example 5CT.1 is applied to an ideal low-pass filter whose bandwidth is less than F_1. Inspection shows that the filter passes only the DC component. The output is

$$y(t) = \frac{A\tau_p}{T} \qquad \text{(32)}$$

Similarly, if the pulse train is applied to an ideal band-pass filter that passes only the first harmonic, then the output is

$$y(t) = \frac{2A\tau_p}{T} \operatorname{sinc}(\pi F_1 \tau_p) \cos(2\pi F_1 t) \qquad \text{(33)}$$

5CT.4 Derivation of the Fourier Series by Approximating a Waveform

**5CT.4.1
The Minimum
Mean Square
Error (MMSE)
Criterion**

We can derive the Fourier series by starting with the basic problem of approximating a periodic waveform $x(t)$. We consider an approximation having the form

$$\hat{x}_K(t) = \sum_{k=-K}^{K} X_k e^{j2\pi k F_1 t} \qquad \text{(34)}$$

where the X_k are complex constants. How should we chose the X_k to make $\hat{x}_K(t)$ a good approximation to $x(t)$? The answer to this question depends on what we mean by "a good approximation."

In a Fourier series, the X_k are chosen to minimize the mean-square approximation error

$$\mathcal{E}_K = \frac{1}{T} \int_T |x(t) - \hat{x}_K(t)|^2 \, dt \tag{35}$$

where $\mathcal{E}_K$ is the average power carried by the error waveform

$$e_K(t) = x(t) - \hat{x}_K(t) \tag{36}$$

We show in Appendix 5CT.A that the coefficients that minimize $\mathcal{E}_K$ are given by Eq. (6). The Fourier coefficients therefore are best in the minimum mean square error (MMSE) sense.

5CT.4.2
Partial Fourier Series and Mean Square Convergence

The *partial Fourier series* is defined as $\hat{x}(t)$ Eq. (34) when the coefficients are given by Eq. (6). Appendix 5CT.A shows that the mean square error produced by the partial Fourier series equals the difference between the average powers in $x(t)$ and $\hat{x}_K(t)$.

$$\mathcal{E}_{K\,\text{min}} = P - \sum_{l=-K}^{K} |X_l|^2 = \frac{1}{T} \int_T |x(t)|^2 \, dt - \frac{1}{T} \int_T |\hat{x}_K(t)|^2 \, dt \tag{37}$$

If

$$\lim_{K \to \infty} \mathcal{E}_{K\,\text{min}} = 0 \tag{38}$$

then $\hat{x}(t)$ is said to converge to $x(t)$ in *mean square*. All periodic waveforms that can be drawn on paper or displayed on an oscilloscope have partial Fourier series that converge in mean square. This means that the average power in the error waveform $e_K(t)$ of Eq. (36) vanishes in the limit $K \to \infty$ and there is no physical way to distinguish between any waveform you can display and its Fourier series.

5CT.5 Derivation of the Fourier Series From the Fourier Transform

The Fourier series can also be derived from the Fourier transform of a periodic waveform. It is useful to understand this derivation because it provides a way to understand and obtain the Fourier coefficients of a periodic waveform from the Fourier transform of a related pulse waveform $p(t)$. The derivation starts with the observation that every periodic waveform $x(t)$ can be generated by periodically repeating a pulse waveform $p(t)$.

$$x(t) = \sum_{m=-\infty}^{\infty} p(t - mT) \tag{39}$$

where $p(t)$ is the part of $x(t)$ that is in the interval $-T/2 \le t < T/2$.

$$p(t) \overset{\Delta}{=} \begin{cases} x(t) & -T/2 \le t < T/2 \\ 0 & \text{otherwise} \end{cases} \tag{40}$$

Equations (39) and (40) are illustrated in Figure 5CT.6

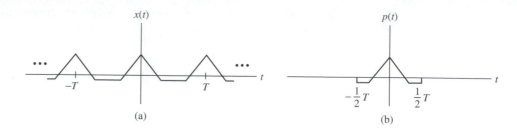

(a)   (b)

We can write Eq. (39) as a convolution:

$$x(t) = p(t) * \sum_{n=-\infty}^{\infty} \delta(t - nT) \tag{41}$$

and apply the convolution property of Fourier transforms to obtain the following expression for the Fourier transform of $X(F)$.

$$X(F) = \frac{1}{T} P(F) \sum_{k=-\infty}^{\infty} \delta(F - kF_1) = \frac{1}{T} \sum_{k=-\infty}^{\infty} P(kF_1)\delta(F - kF_1) \tag{42}$$

In the preceding equation, $P(F)$ is the Fourier transform of $p(t)$.

$$P(F) = \int_{-\infty}^{\infty} p(t)e^{-j2\pi Ft}\, dt = \int_{-T/2}^{T/2} x(t)e^{j2\pi Ft}\, dt \tag{43}$$

Equation (42) shows us that the Fourier transform of a periodic waveform is composed of impulses that occur at integer multiples of $F_1 = 1/T$. The areas of the impulses are scaled samples, $(1/T)P(kF_1)$, of the spectrum of the generating pulse $p(t)$. This result is summarized in Figure 5CT.7. It is easy to show that these areas, $(1/T)P(kF_1)$, of the impulses are the coefficients in the Fourier series representation of $x(t)$.

$$\frac{1}{T} P(kF_1) = \frac{1}{T} \int_{-\infty}^{\infty} p(t)e^{-j2\pi Ft}\, dt \bigg|_{F=kF_1} = \frac{1}{T} \int_{-T/2}^{T/2} x(t)e^{j2\pi kF_1 t}\, dt = X_k \tag{44}$$

If we substitute Eqs. (44) into (42) we obtain

$$X(F) = \sum_{k=-\infty}^{\infty} X_k \delta(F - kF_1) \tag{45}$$

This important result shows that the Fourier transform of a periodic waveform is composed of impulses whose areas are the Fourier series coefficients. Recall that the line spectrum of

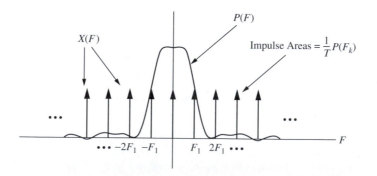

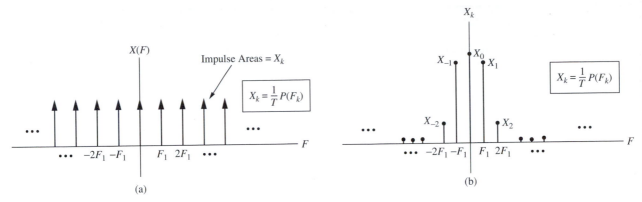

Figure 5CT.8
Fourier transform and line spectrum of a periodic waveform: (a) Fourier transform; (b) line spectrum.

a periodic waveform is a plot of the coefficients X_k versus F, or kF_1, or F. The relationship between $X(F)$ and the line spectrum is illustrated in Figure 5CT.8. We can obtain the Fourier series representation of $x(t)$ by taking the inverse Fourier transform of Eq. (45):

$$x(t) = \mathcal{F}^{-1}\left\{ \sum_{k=-\infty}^{\infty} X_k \delta(F - kF_1) \right\} = \sum_{k=-\infty}^{\infty} X_k \mathcal{F}^{-1}\{\delta(F - kF_1)\}$$

$$= \sum_{k=-\infty}^{\infty} X_k e^{j2\pi kF_1 t} \tag{46}$$

where we used the sifting property of the impulse to evaluate $\mathcal{F}^{-1}\{\delta(F - kF_1)\}$. This completes the derivation of the Fourier series from the Fourier transform.

An important benefit of the derivation is that Eq. (44) provides a way to obtain the Fourier series coefficients of a periodic waveform directly from the Fourier transform of the generating pulse $p(t)$. The following examples illustrate.

EXAMPLE 5CT.5   Periodic Rectangular Pulse Train

The rectangular pulse train of Figure 5CT.2 is composed of periodic repetitions of the rectangular pulse.

$$p(t) = A \sqcap \left(\frac{t}{\tau_p} \right) \tag{47}$$

The Fourier transform of $p(t)$ is

$$P(F) = \tau \, \text{sinc}(F\tau_p) \tag{48}$$

Therefore, the Fourier coefficients are given by Eq. (44) as

$$X_k = \frac{1}{T} P(kF_1) = \frac{\tau}{T} \, \text{sinc}(kF_1\tau_p) \tag{49}$$

in agreement with the results of Example 5CT.1.

EXAMPLE 5CT.6   Periodic Impulse Train

Consider the periodic impulse train

$$x(t) = \sum_{i=-\infty}^{\infty} \delta(t - iT) \tag{50}$$

The generating pulse is

$$p(t) = \delta(t) \tag{51}$$

for which

$$P(F) = 1 \tag{52}$$

It follows from Eq. (44) that

$$X_k = \frac{1}{T} P(kF_1) = \frac{1}{T} \tag{53}$$

Therefore

$$\sum_{i=-\infty}^{\infty} \delta(t - iT) = \frac{1}{T} \sum_{k=-\infty}^{\infty} e^{j2\pi kF_1 t} \tag{54}$$

in agreement with the results of Example 5CT.2.

If we substitute Eq. (53) into Eq. (44), we obtain the Fourier *transform* pair (4CT.FT15) of Table 4CT.3.

$$\sum_{i=-\infty}^{\infty} \delta(t - iT) \overset{\text{FT}}{\longleftrightarrow} \frac{1}{T} \sum_{i=-\infty}^{\infty} \delta(F - iF_1) \tag{55}$$

EXAMPLE 5CT.7   Rectified Cosine

The *half-wave and full-wave rectified cosines* of Figures 5CT.9(a) and (b) are found in electronic DC power supplies.

We can obtain the Fourier series coefficients of the full-wave rectified cosine directly from Eq. (44) and FT12 in Table 3 of 4CT. We set $\tau = T$ to obtain

$$X_k = \frac{2A\tau}{\pi T} \frac{\cos(\pi F\tau)}{1 - (2F\tau)^2} \Bigg|_{\substack{F = kF_1 \\ \tau = T}} = \frac{2A}{\pi} \frac{\cos(\pi k)}{1 - (2k)^2} = \frac{2A}{\pi} \frac{(-1)^k}{1 - (2k)^2} \tag{56}$$

Figure 5CT.9
Rectified cosine:
(a) Half-wave; (b) full wave.

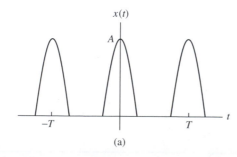

(a)

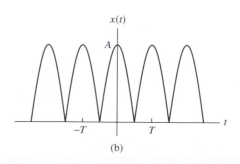

(b)

where $k = 0, \pm1, \pm2, \ldots.$ You can obtain the following Fourier series coefficients for the half-wave cosine directly from this result (Problem 5CT.7.3):,

$$X_l = \frac{A}{\pi} \frac{(-1)^{l/2}}{1 - l^2} \text{ for } l = \pm2, \pm4, \ldots \tag{57}$$

with $X_0 = A/\pi$, $X_1 = X_{-1} = A/4$, and $X_l = 0$ for $l = \pm3, \pm5, \ldots.$

We showed in this section that the Fourier series can be derived from the Fourier transform. Appendix 5CT.B strengthens this connection between the Fourier series and the Fourier transform by showing that the Fourier transform can be derived from the Fourier series.

5CT.6 Properties of the Fourier Series

As with the Fourier transform, the properties of the Fourier series can be divided into two categories: *symmetry* properties and *operational* properties.

5CT.6.1 Symmetry Properties

The most important symmetry properties of the Fourier series are listed in Table 5CT.1. To prove (5CT.SP1), we conjugate both sides of Eq. (4) to obtain

$$x^*(t) = \left(\sum_{k=-\infty}^{\infty} X_k e^{j2\pi k F_1 t} \right)^* = \sum_{k=-\infty}^{\infty} X_k^* e^{-j2\pi k F_1 t} = \sum_{m=-\infty}^{\infty} X_{-m}^* e^{j2\pi m F_1 t} \tag{58}$$

where, in the last step, we changed the index in the sum from k to $-m$. It follows from Eq. (58) that the kth Fourier coefficient of $x^*(t)$ is X_{-k}^*, as (5CT.SP1) states. Properties (5CT.SP3) through (5CT.SP7) follow easily from (5CT.SP1). Properties (5CT.SP8) and (5CT.SP9) follow from Eqs. (11)–(12) and (5CT.SP6) and (5CT.SP7).

The half-wave symmetry property (5CT.SP2) states conditions under which a waveform has only odd harmonics, i.e., $X_k = 0$ when k is an even integer. By definition, a half-wave symmetric waveform is a waveform for which

$$x\left(t - \frac{T}{2}\right) = -x(t) \tag{59}$$

Table 5CT.1
Symmetry Properties
of the Fourier Series

For Real or Complex $x(t)$	
$x^*(t) \longleftrightarrow X_{-k}^*$	(5CT.SP1)
$X_k = 0$ for even k if an only if $x(t)$ is half-wave symmetric: $x\left(t - \frac{T}{2}\right) = -x(t).$	(5CT.SP2)
For Real $x(t)$	
$X_{-k} = X_k^*$	(5CT.SP3)
$\|X_{-k}\| = \|X_k\|$	(5CT.SP4)
$\angle X_{-k} = -\angle X_k$	(5CT.SP5)
X_k is real and even iff $x(t)$ is real and even.	(5CT.SP6)
X_k is imaginary and odd iff $x(t)$ is real and odd.	(5CT.SP7)
If $x(t)$ is even, $x(-t) = x(t)$, then it contains only DC and cosine components.	(5CT.SP8)
If $x(t)$ is odd, $x(-t) = -x(t)$, then it contains only sine components.	(5CT.SP9)

Figure 5CT.10
Square wave has
half-wave symmetry:
$(t - (T/2)) = -x(t)$.

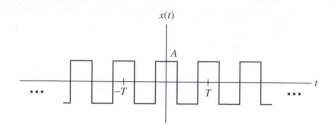

Half-wave symmetry is illustrated by the *square wave* in Figure 5CT.10. We can derive (5CT.SP2) by substituting Eq. (4) into Eq. (59) to obtain

$$\sum_{k=-\infty}^{\infty} X_k e^{j2\pi k F_1 \left(t - \frac{T}{2}\right)} = -\sum_{k=-\infty}^{\infty} X_k e^{j2\pi k F_1 t} \tag{60}$$

If we factor the exponential and simplify the left-hand side, we obtain

$$\sum_{k=-\infty}^{\infty} e^{j\pi k} X_k e^{j2\pi k F_1 t} = -\sum_{k=-\infty}^{\infty} X_k e^{j2\pi k F_1 t} \tag{61}$$

This equation is true for all t if and only if

$$e^{-j\pi k} X_k = -X_k \tag{62}$$

Equation (62) is satisfied if and only if X_k is 0 when k is even. This completes the derivation of (5CT.SP2).

**5CT.6.2
Operational
Properties**

The most important operational properties of the Fourier series are listed in Table 5CT.2. The linearity property (5CT.OP1) follows directly from the definition of the Fourier series. We leave the straightforward proof to Problem 5CT.6.4.

It will be advantageous to prove the time shift property (5CT.OP4) first. The property is obtained if we replace t by $t - \lambda$ in the Fourier series representation for $x(t)$.

$$x(t - \tau) = \sum_{k=-\infty}^{\infty} X_k e^{j2\pi k F_1 (t-\tau)} = \sum_{k=-\infty}^{\infty} X_k e^{-j2\pi k F_1 \tau} e^{j2\pi k F_1 t} \tag{63}$$

We see that the kth complex Fourier series coefficient of $x(t - \tau)$ is $X_k e^{-j2\pi k F_1 \tau}$, as stated by (5CT.OP4).

Property (5CT.OP2) introduces the new concept of *circular* or *periodic* convolution in the time domain. Let $x_1(t)$ and $x_2(t)$ denote two periodic waveforms having period T.

Table 5CT.2
Operational
Properties of the
Fourier Series

	Property	Signal	Coefficients
5CT.OP1	Linearity	$a_1 x_1(t) + a_2 x_2(t)$	$a_1 X_{1,k} + a_2 X_{2,k}$
5CT.OP2	Circular convolution	$x_1(t) \circledast x_2(t)$	$X_{1,k} X_{2,k}$
5CT.OP3	Multiplication	$x_1(t) x_2(t)$	$X_{1,k} * X_{2,k}$
5CT.OP4	Time shift	$x(t - \tau)$	$e^{-j2\pi k F_1 \tau} X_k$
5CT.OP5	Differentiation	$\dfrac{d}{dt} x(t)$	$j2\pi k F_1 X_k$
5CT.OP6	Reversal	$x(-t)$	X_{-k}

Figure 5CT.11
Circular convolution:
(a) illustrated with
Cartesian
coordinates; (b) on
a circle.

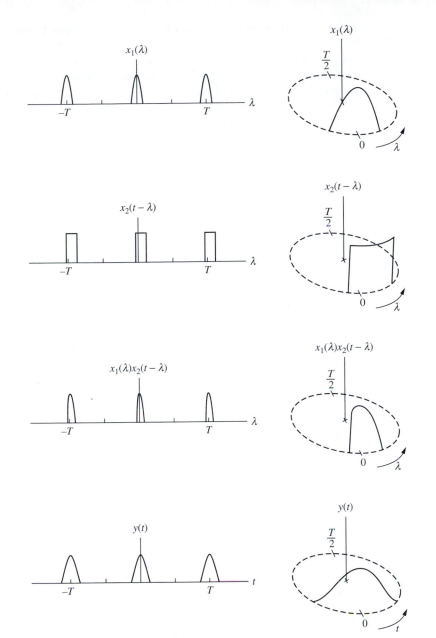

The circular convolution of $x_1(t)$ and $x_2(t)$ is defined as

$$z(t) = x_1(t) \circledast x_2(t) = \frac{1}{T} \int_T x_1(\lambda) x_2(t - \lambda)\, d\lambda \qquad (64)$$

Figure 5CT.11 illustrates the meaning of circular convolution. The figure shows how the circular convolution integral can be interpreted two ways. In Figure 5CT.11(a), it is interpreted in the straightforward way using Cartesian coordinates. Here we often choose the region of integration $-T/2$ to $T/2$. The name *circular convolution* originates with the interpretation of Figure 5CT.11(b), where $x_1(\lambda)$, $x_2(t - \lambda)$, $x_1(\lambda)x_2(t - \lambda)$, and $z(t)$ are plotted on circles having circumference T. The periodicity of the waveforms are inherent in the circular plots because the functions plotted are necessarily the same at a coordinates t and $t + T$. The integration (64) is taken once around the circle for each value of t. The function $x(t - \lambda)$ rotates as t is increased. One can interpret these plots in a manner similar to Figure 16 of Chapter 2CT, but instead of linear tapes that slide, we use circular tapes that rotate.

To prove the circular convolution property, we find the kth complex Fourier coefficient Z_k, of $z(t)$. The substitution of Eq. (64) into Eq. (6) yields

$$Z_k = \frac{1}{T} \int_T \left(\frac{1}{T} \int_T x_1(\lambda) x_2(t - \lambda) \, d\lambda \right) e^{-j2\pi k F_1 t} \, dt \tag{65}$$

$$= \frac{1}{T} \int_T \left(\frac{1}{T} \int_T x_2(t - \lambda) \, e^{-j2\pi k F_1 t} \, dt \right) x_1(\lambda) \, d\lambda \tag{66}$$

where we changed orders of integration in the last step. The integral inside the large parenthesis is, by the time shift property (5CT.OP4), $X_{2,k} e^{-j2\pi k F_1 \lambda}$. Therefore

$$Z_k = \frac{1}{T} \int_T x_1(\lambda) e^{-j2\pi k F_1 \lambda} d\lambda \, X_{2,k} = X_{1,k} X_{2,k} \tag{67}$$

which is what property (5CT.OP2) states.

Similarly, we can prove the multiplication property from Eq. (4):

$$x_1(t)x_2(t) = \sum_{n=-\infty}^{\infty} X_{1,n} e^{j2\pi n F_1 t} \sum_{m=-\infty}^{\infty} X_{2,m} e^{j2\pi m F_1 t}$$

$$= \sum_{n=-\infty}^{\infty} \sum_{m=-\infty}^{\infty} X_{1,n} X_{2,m} e^{j2\pi(n+m) F_1 t} \tag{68}$$

If we set $k = n + m$, we can write the preceding equation as

$$x_1(t)x_2(t) = \sum_{k=-\infty}^{\infty} \left(\sum_{n=-\infty}^{\infty} X_{1,n} X_{2,k-n} \right) e^{j2\pi k F_1 t} \tag{69}$$

which shows that the kth complex Fourier series coefficient of $x_1(t)x_2(t)$ is given by the convolution of $X_{1,n}$ and $X_{2,n}$, as stated by (5CT.OP3).

The proofs of the differentiation and reversal properties, (5CT.OP5) and (5CT.OP6), are left as problems.

5CT.7   The use of Properties to Derive the Fourier Series

There are several techniques are available for finding Fourier series coefficients. We illustrated the direct evaluation of Eq. (6) in Examples 5CT.1 and 5CT.2. An alternative method is to obtain the Fourier series coefficients from a table of Fourier transforms. This technique was illustrated in Examples 5CT.4–5CT.6. A third method is to use previously established Fourier series and the properties of the Fourier series. This method is often the fastest and has the advantage of showing how the Fourier series of different waveforms are related. The method is illustrated in the following examples.

EXAMPLE 5CT.8   Square Wave

The square wave of Figure 5CT.10 is found in many electronic systems. We can easily find its Fourier series using the Fourier series of the rectangular pulse train of Figure 5CT.2. Let's denote the square wave of Figure 5CT.10 as $x_{sw}(t)$ and the pulse train of Figure 5CT.2 as $x_{pt}(t)$. We see by inspection of the two figures that

$$x_{sw}(t) = 2x_{pt}(t) - A \tag{70}$$

when $\tau_p = T/2$. We can substitute the Fourier series of the pulse train (19) into (70) with $\tau_p = T/2$ to obtain the complex Fourier series of the square wave.

$$x_{sw}(t) = 2\sum_{k=-\infty}^{\infty} \frac{A\tau_p}{T}\,\mathrm{sinc}(\pi k F_1 \tau_p)e^{j2\pi k F_1 t}\bigg|_{\tau_p=\frac{T}{2}} - A \tag{71}$$

The preceding equation reduces to

$$x_{sw}(t) = \sum_{k=\pm1,\pm3,\cdots}^{\infty} \frac{2A(-1)^{\frac{k-1}{2}}}{\pi k}e^{j2\pi k F_1 t} \tag{72}$$

The corresponding real form is

$$x_{sw}(t) = \sum_{k=1,3,\cdots}^{\infty} \frac{4A(-1)^{\frac{k-1}{2}}}{\pi k}\cos(2\pi k F_1 t) \tag{73}$$

EXAMPLE 5CT.9   Triangle Wave

The triangular waveform of Figure 5CT.12 is used in communication electronics. Its Fourier series can be obtained easily using the Fourier series of the square wave of Figure 5CT.10. Let's denote the triangular wave as $x_{tw}(t)$ and denote the square wave as $x_{sw}(t)$. We see by inspection of the figures that the derivative of the triangular wave is given by

$$\frac{d}{dt}x_{tw}(t) = \frac{4}{T}x_{sw}(t) \tag{74}$$

Let's denote the kth Fourier series coefficient of $x_{tw}(t)$ as X_k. It follows from the differentiation property that the kth Fourier series coefficient of $(d/dt)x_{tw}(t)$ is $j2\pi k F_1 X_k$. If we equate $j2\pi k F_1 X_k$ to the kth Fourier series coefficient of $(4/T)x_{sw}(t)$ (See (72)), we obtain

$$j2\pi k F_1 X_k = \begin{cases} \dfrac{4}{T}\dfrac{2A(-1)^{\frac{k-1}{2}}}{\pi k} & k = \pm1, \pm3, \ldots \\ 0 & k = 0, \pm2, \pm4, \ldots \end{cases} \tag{75}$$

Figure 5CT.12
Triangular wave.

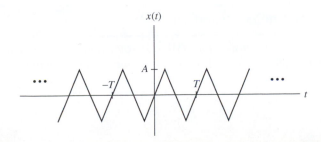

This rearranges to

$$X_k = \begin{cases} \dfrac{4A(-1)^{\frac{k-1}{2}}}{j(\pi k)^2} & k = \pm 1, \pm 3, \ldots \\ 0 & k = 0, \pm 2, \pm 4, \ldots \end{cases} \tag{76}$$

Therefore, the complex Fourier series of the triangular wave is

$$x_{\mathrm{tw}}(t) = \sum_{k=\pm 1, \pm 3, \cdots}^{\infty} \frac{4A(-1)^{\frac{k-1}{2}}}{j(\pi k)^2} e^{j2\pi k F_1 t} \tag{77}$$

The corresponding real form is

$$x_{\mathrm{tw}}(t) = \sum_{k=1,3,\cdots}^{\infty} \frac{8A(-1)^{\frac{k-1}{2}}}{(\pi k)^2} \sin(2\pi k F_1 t) \tag{78}$$

5CT.8   Parseval's Theorem for Periodic Waveforms

Parseval's theorem states that the total average power in a periodic waveform equals the sum of the average powers in the waveform's components.

Parseval's Theorem for Periodic Waveforms

Real or complex $x(t)$:

$$P = \frac{1}{T} \int_T |x(t)|^2 \, dt = \sum_{k=-\infty}^{\infty} |X_k|^2 \tag{79a}$$

Real $x(t)$:

$$P = \frac{1}{T} \int_T x^2(t) \, dt = A_0^2 + \frac{1}{2} \sum_{k=1}^{\infty} A_k^2 \tag{79b}$$

$$= a_0^2 + \frac{1}{2} \sum_{k=1}^{\infty} a_k^2 + \frac{1}{2} \sum_{k=1}^{\infty} b_k^2 \tag{79c}$$

Equation (79a) is a direct result of Parseval's theorem for multiple sinusoids of Chapter 3CT. Equations (79b) and (79c) follow from the relationships between the real and complex coefficients (8)–(9), (11)–(12).

EXAMPLE 5CT.10   Total Harmonic Distortion

A figure of merit, called the *total harmonic distortion* (THD), is sometimes used to characterize the degradation introduced by a nonlinear amplifier or other system. The input is assumed to be a pure sinusoid. The total harmonic distortion, measured at the output, is defined as

$$\rho = \frac{1}{P_1} \sum_{n=2}^{\infty} P_n \tag{80}$$

Figure 5CT.13
Total harmonic distortion ρ_{dB}, introduced by saturating amplifier of Figure 5CT.1 versus normalized gain g_n: ρ_{dB} is less than -28 dB if the amplifier is overdriven by under 10% ($g_n < 1.1$); ρ_{dB} increases rapidly to -12.6 dB as g_n increases from 1.1 to 2.

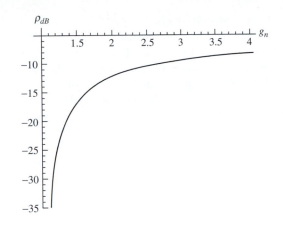

where P_1 is the average power of the output fundamental and $\sum_{n=2}^{\infty} P_n$ is the sum of the averages powers in all higher harmonics. Manufacturers typically specify total THD as a percentage.

$$\rho_\% = \rho 100\% \tag{81}$$

or in decibels (dB)

$$\rho_{dB} = 10\log_{10}(\rho)\ \text{dB} \tag{82}$$

Parseval's theorem enables us to evaluate the THD without having to sum the infinite series in Eq. (80). According to Parseval's theorem, the total average power P of a periodic waveform equals the sum of the average powers in the harmonics. Assuming that there is no DC component at the amplifier output, we have

$$P = P_1 + \sum_{n=2}^{\infty} P_n \tag{83}$$

The substitution of Eq. (83) into Eq. (80) yields

$$\rho = \frac{P - P_1}{P_1} \tag{84}$$

In this example we apply Eq. (84) to the saturating amplifier in Figure 5CT.1, where

$$P_1 = \tfrac{1}{2}\,a_1^2 \tag{85}$$

with a_1 given by Eq. (26c). Direct evaluation of the time average

$$P = \frac{1}{T}\int_T x^2(t)\,dt \tag{86}$$

for the clipped cosine (Figure 5CT.1 and Figure 5CT.4) is routine and we omit the details. The result is

$$P = \tfrac{1}{2}g_n^2 A_{\max}^2\,(1 - \lambda - \lambda\,\text{sinc}(\lambda)) + A_{\max}^2\,\lambda \tag{87}$$

A plot of ρ_{dB} is shown in Figure 5CT.13. We can see from this figure that the total harmonic distortion is below -28 dB if the amplifier is overdriven by less than 10 percent ($g_n = 1.1$) but increases rapidly to -12.6 dB as g_n increases from 1.1 to 2. ρ_{dB} tends toward a maximum of -6.3 dB as $g_n \to \infty$.

We have obtained Parseval's theorem for periodic waveforms (79) by recognizing that it is a special case of Parseval's theorem for multiple sinusoids. We can also derive it by setting $x_1(t) = x(t)$ and $x_2(t) = x^*(-t)$ in the circular convolution property (5CT.OP2). This derivation leads to the important concepts of autocorrelation function and power spectral density of a periodic waveform. The derivation begins with the observation that the kth complex Fourier coefficient of $x^*(-t)$ is, by (5CT.SP1), and (5CT.OP6), X_k^*. The circular convolution property, with $x_1(t) = x(t)$ and $x_2(t) = x^*(-t)$, then yields

$$R_x(t) = \frac{1}{T}\int_T x(\lambda)x^*(\lambda - t)\,d\lambda = \sum_{k=-\infty}^{\infty} |X_k|^2 e^{j2\pi kF_1 t} \tag{88}$$

$R_x(t)$ is the autocorrelation function of $x(t)$. The power spectrum density of $x(t)$ is defined as the Fourier transform of $R_x(t)$.

$$S_x(F) = \mathcal{F}\{R_x(t)\} = \sum_{k=-\infty}^{\infty} |X_k|^2 \delta(F - kF_1) \tag{89}$$

Consequently, $R_x(t)$ is the inverse Fourier transform of $S_x(F)$.

$$R_x(t) = \int_{-\infty}^{\infty} S_x(F)e^{j2\pi Ft}\,dF \tag{90}$$

Parseval's theorem is obtained if we set $t = 0$ in (90) and substitute (88) and (89)

$$R_x(0) = \int_{-\infty}^{\infty} S_x(F)\,dF = \frac{1}{T}\int_T |x(t)|^2\,dt = \sum_{k=-\infty}^{\infty} |X_k|^2 \tag{91}$$

The power spectral density $S_x(F)$ describes the distribution of the average power of $x(t)$ in frequency. The average power $P_x(\mathcal{B})$ contained in any set of frequencies $\mathcal{B}$ is given by

$$P_x(\mathcal{B}) = \int_{\mathcal{B}} S_x(F)\,dF \tag{92}$$

where the integration is over the frequencies in $\mathcal{B}$.

5CT.9 Dirichlet Conditions

Every periodic waveform having engineering interest can be represented as a Fourier series. However, not every periodic waveform that can be defined mathematically has a Fourier series. For example, if the functions shown in Figure 4CT.23 were periodically repeated every T seconds, the resulting periodic functions would not have Fourier series representations. Because it is important to know if the Fourier series representation of a periodic function exists, Dirichelt derived conditions guaranteeing that a periodic function can be represented by a Fourier series. This guarantee applies when all these conditions are met.

Dirichlet Conditions

1. $x(t)$ is absolutely integrable over a period, i.e.

$$\int_{-T/2}^{T/2} |x(t)|dt < \infty \tag{93}$$

2. $x(t)$ does not contain an infinite number of minima or maxima within a period.
3. $x(t)$ does not contain an infinite number of discontinuities within a period.

If all three conditions are satisfied, then the Fourier series equals $x(t)$ except where $x(t)$ is discontinuous. At the values of t for which $x(t)$ is discontinuous, the Fourier series converges to the average of the values of $x(t)$ that are on either side of the discontinuity. If a waveform does not satisfy the Dirichlet conditions, then it cannot be displayed on any physical device.

5CT.10   Convergence of the Fourier Series

When we use a computer to plot a Fourier series, it can add only a finite number of terms of the series. Therefore, the plot is always an approximation to the actual Fourier series. A plot of the partial Fourier series $\hat{x}_K(t)$ can confuse you if you do not know the convergence properties presented in this section. As with the Fourier transform, there are two types of convergence of interest: uniform convergence and mean square convergence.

5CT.10.1 Uniform Convergence

$\hat{x}_K(t)$ converges uniformly to $x(t)$ if for each $\epsilon > 0$, there is an integer K_o such that

$$|x(t) - \hat{x}_K(t)| < \epsilon \qquad \text{for all } K > K_o \tag{94}$$

for each value of t. It can be shown in a manner analogous to the derivation in Section 4CT.10 that $\hat{x}_K(t)$ converges uniformly to $x(t)$ if

$$\sum_{k=-\infty}^{\infty} |X_k| < \infty \tag{95}$$

EXAMPLE 5CT.11   Uniform Convergence

Figure 5CT.14(a)–(d) contains computer-generated plots of the partial Fourier series $\hat{x}_K(t)$ of the triangular wave of Figure 5CT.12 and the associated approximation error $e_K(t)$ for $A = 1$, and $K = 3$ and 25. The minimum mean square error $\mathcal{E}_{K\,\min}$ is plotted for a range of K in Figure 5CT.11(e). The minimum mean square error $\mathcal{E}_{K\,\min}$ is given by the formula

$$\mathcal{E}_{K\,\min} = \frac{A^2}{3} - \frac{A^2}{2} \sum_{k=1,3,\cdots}^{K} \left(\frac{8}{(\pi k)^2} \right)^2 \tag{96}$$

Figure 5CT.14
Partial Fourier series
of a triangular wave:
(a) $\hat{x}_3(t)$; (b) $\hat{x}_{25}(t)$;
(c) $\hat{e}_3(t)$; (d) $\hat{e}_{25}(t)$;
(e) $\log_{10}(\mathcal{E}_{K\min})$

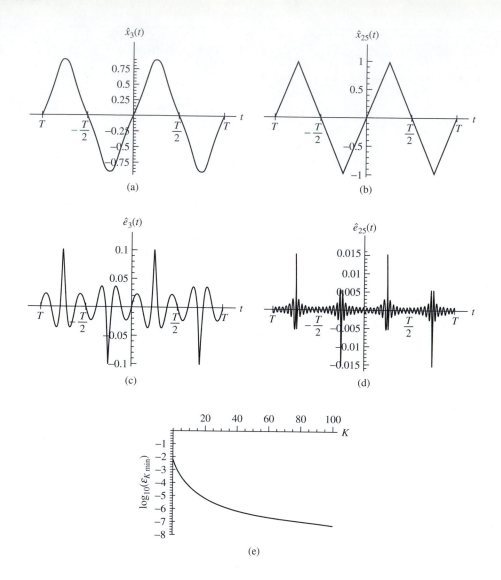

The Fourier coefficients of a triangular wave satisfy Eq. (95). Therefore the partial Fourier series converges uniformly.

**5CT.10.2
Mean Square
Convergence**

$\hat{x}_K(t)$ converges to $x(t)$ in mean square if

$$\lim_{K\to\infty} \mathcal{E}_{K\min} = \lim_{K\to\infty} \frac{1}{T} \int_T |x(t) - \hat{x}_K(t)|^2 \, dt = 0 \qquad (97)$$

It can be shown, in a manner analogous to the derivation in Section 4CT.9, that $\hat{x}_K(t)$ converges to $x(t)$ in the mean square sense if

$$\sum_{k=-\infty}^{\infty} |X_k|^2 < \infty \qquad (98)$$

EXAMPLE 5CT.12   Mean Square Convergence

Figure 5CT.15(a)–(d) contains computer-generated plots of the partial Fourier series $\hat{x}_K(t)$ of the square wave of Figure 5CT.10 and the associated approximation error $e_K(t)$ for $A = 1$ and $K = 7$ and 25. The minimum mean square error $\mathcal{E}_{K\,\min}$ is plotted for a range of K in Figure 5CT.11(e). The minimum mean square error $\mathcal{E}_{K\,\min}$ is given by the form the formula

$$\mathcal{E}_{K\,\min} = A^2 - \frac{A^2}{2} \sum_{k=1,3,\cdots}^{K} \left(\frac{4}{\pi k} \right)^2 \tag{99}$$

The coefficients for the square wave satisfy (98) but not (95). Therefore, the partial Fourier series converges in mean square, but not uniformly. The overshoot near the discontinuities of $x(t)$ is about 9 percent the value of the value of the discontinuity and *does not decrease as K increases*. The oscillations in $\hat{x}_K(t)$ become faster and the significant ripple amplitudes get closer to the edge of the discontinuity as K increases. In the limit $n \to \infty$, $\mathcal{E}_{K\,\min}$ equals 0. In this limit the overshoot has become so thin that it contains no energy. The oscillatory behavior and overshoot of the partial Fourier series near the discontinuities of $x(t)$ is called *Gibbs phenomenon*. Gibbs phenomenon appears in the partial Fourier series of every periodic waveform having discontinuities.

Figure 5CT.15
Partial Fourier series
of a square wave:
(a) $\hat{x}_7(t)$; (b) $\hat{x}_{25}(t)$;
(c) $\hat{e}_7(t)$; (d) $\hat{e}_{25}(t)$;
(e) $\log_{10}(\mathcal{E}_{K\,\min})$.

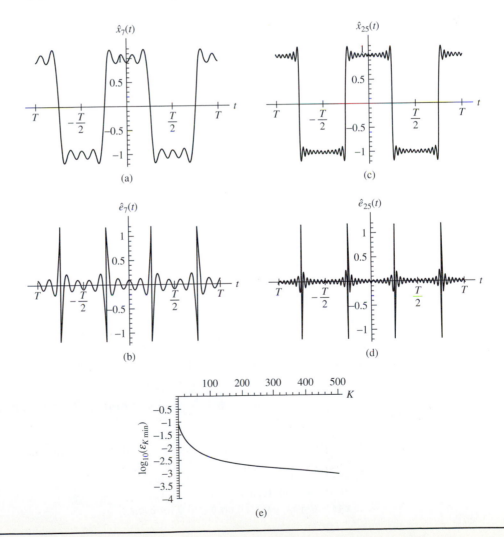

5CT.11   Summary

This chapter introduced the Fourier series as a special case of the multiple sinusoidal component representation of waveforms introduced in Chapter 3CT. The Fourier series and the (inverse) Fourier transform are strongly related: They both represent a waveform as a sum of sinusoidal components. The inverse Fourier transform is used primarily to represent pulse signals. The Fourier series is used to represent periodic signals. The multiple sinusoid representations of waveforms works well for understanding and describing the response of LTI systems because of the fundamental property $e^{j2\pi Ft} \rightarrow H(F)e^{j2\pi Ft}$ derived in Chapter 3CT.

Summary of Definitions and Formulas

Complex Fourier Series

$$x(t) = \sum_{k=-\infty}^{\infty} X_k e^{j2\pi k F_1 t}$$

where $x(t) = x(t+T)$ and $F_1 = 1/T$. The constant X_k is the kth *complex series coefficient*, given by

$$X_k = \frac{1}{T} \int_T x(t) e^{-j2\pi k F_1 t}\, dt$$

F_1 is the *fundamental frequency*; $F_1, 2F, 3F, \ldots$ are *harmonic* frequencies.

Amplitude-Phase Form of Fourier Series

If $x(t)$ is real, then the complex form can be expressed as

$$x(t) = A_0 + \sum_{k=1}^{\infty} A_k \cos(2\pi k F_1 t + \phi_k)$$

where $A_0 = X_0$, $A_k = 2|X_k|$, and $\phi_k = \angle X_k$; $k = 1, 2, \ldots$.

Sine-Cosine Form of Fourier Series

If $x(t)$ is real, then the complex form can be expressed as

$$x(t) = a_0 + \sum_{k=1}^{\infty} a_k \cos(2\pi k F_1 t) + \sum_{k=1}^{\infty} b_k \sin(2\pi k F_1 t)$$

where $a_0 = X_0$, $a_k = 2\mathcal{R}\{X_k\}$, and $b_k = -2\mathcal{I}\{X_k\}$; $k = 1, 2, \ldots$.

If $x(-t) = x(t)$, then $b_k = 0, k = 1, 2, \ldots$.
If $x(-t) = x(t)$, then $a_k = 0, k = 0, 1, \ldots$.
If $x\left(t - \frac{T}{2}\right) = -x(t)$, then $X_k = 0$ for even k.

LTI System Response to Periodic Input

$$\sum_{i=-\infty}^{\infty} X_i e^{j2\pi i F_1 t} \xrightarrow{h(t)} \sum_{i=-\infty}^{\infty} X_i H(i F_1) e^{j2\pi i F_1 t}$$

$$A_0 + \sum_{i=1}^{\infty} A_i \cos(2\pi i F_1 t + \phi_i) \xrightarrow{h(t)} A_0 H(0)$$

$$+ \sum_{i=1}^{\infty} A_i |H(i F_1)| \cos(2\pi i F_1 t + \phi_i + \angle H(i F_1))$$

Fourier Series Derived from First Principles

$$\hat{x}_K(t) = \sum_{k=-K}^{K} X_k e^{j2\pi k F_1 t}$$

$$e_K(t) = x(t) - \hat{x}_K(t)$$

The Fourier coefficients minimize the mean square approximation error.

$$\mathcal{E}_K = <|e_K(t)|^2>$$

The minimum mean square value of $\mathcal{E}_K$ is

$$\mathcal{E}_{K\,min} = P - \sum_{l=-K}^{K} |X_l|^2$$

where the X_l are the Fourier coefficients.

Fourier Series Derived from the Fourier Transform

$$x(t) = \sum_{m=-\infty}^{\infty} p(t - mT)$$

$$p(t) \triangleq \begin{cases} x(t) & -T/2 \le t < T/2 \\ 0 & \text{otherwise} \end{cases}$$

$$X(F) = \sum_{k=-\infty}^{\infty} X_k \delta(F - kF_1)$$

$$X_k = \frac{1}{T} \mathcal{F}\{p(t)\}\Big|_{F=kF_1} = \frac{1}{T} \int_{-T/2}^{T/2} x(t) e^{-j2\pi k F_1 t} \, dt$$

$$x(t) = \mathcal{F}^{-1}\{X(F)\} = \sum_{i=-\infty}^{\infty} X_i e^{j2\pi i F_1 t}$$

See Tables 1 and 2.

Parseval's Theorem

$$P = \frac{1}{T} \int_T |x(t)|^2 \, dt = \sum_{k=-\infty}^{\infty} |X_k|^2$$

Dirichlet Conditions

1. $x(t)$ is absolutely integrable over a period.
2. $x(t)$ does not contain an infinite number of minima or maxima within a period.
3. $x(t)$ does not contain an infinite number of discontinuities within a period.

If all three conditions are satisfied, then the Fourier series will equal $x(t)$ except where $x(t)$ is discontinuous. At the values of t for which $x(t)$ is discontinuous, the Fourier series converges to the average of the values of $x(t)$ that are on either side of the discontinuity.

Uniform Convergence

Definition

$\hat{x}_K(t)$ converges uniformly to $x(t)$ if for each $\epsilon > 0$, there is an integer K_o such that $|x(t) - \hat{x}_K(t)| < \epsilon$ for all $K > K_o$ for each value of t.

Theorem
$\hat{x}_K(t)$ converges uniformly to $x(t)$ if

$$\sum_{k=-\infty}^{\infty} |X_k| < \infty$$

Mean Square Convergence

Definition

$\hat{x}_K(t)$ converges uniformly to $x(t)$ if

$$\lim_{K \to \infty} \mathcal{E}_{K\,\text{min}} = 0$$

Theorem
$\hat{x}_K(t)$ converges to $x(t)$ in the mean square sense if

$$\sum_{k=-\infty}^{\infty} |X_k|^2 < \infty \qquad (100)$$

5CT.12 Problems

5CT.1   Introduction

5CT.1.1 Which of the following waveforms are periodic and which are not? Justify your answers either analytically or by plots.
(a) $x(t) = \cos(2\pi t) + \cos(\sqrt{2}\pi t)$
(b) $x(t) = \cos(2\pi \cos(2\pi t))$
(c) $x(t) = \cos(2\pi t^2)$

5CT.1.2 Find the fundamental and third harmonic frequencies, F_1 and F_3, respectively of $x(t) = 15 + 6.2 \cos(300\pi t) + 8 \sin(700\pi t)$

5CT.1.3 Find necessary and sufficient conditions for $x(t) = A \cos(2\pi F_a t + \phi_a) + B \cos(2\pi F_b t + \phi_b)$ to be periodic.

5CT.1.4 Let $g(t)$ be a periodic function having period T. Show that the value of the integral

$$\int_{t_o}^{t_o+T} g(t)\, dt \qquad (101)$$

does not depend on t_o.

Figure 5CT.16
Graphic equalizer
faceplate and the
keyboard of a grand
piano.

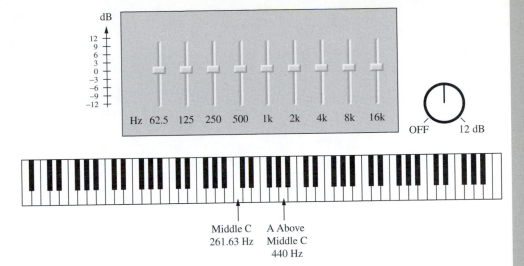

Middle C   A Above
261.63 Hz   Middle C
                    440 Hz

5CT.1.5 Figure 5CT.16 depicts the faceplate of a graphic equalizer used in an audio system and the
keyboard of a grand piano.

The fundamental frequencies produced by the strings on the piano are given by the formula $F_1(n) = F_1(0)2^{n/12}$, where the keys are numbered consecutively $n = \ldots, -2, -1, 0,$
$1, 2, \ldots$. This tuning is called *equal temperment*. The fundamental frequency of the A above
middle C is $F_1(0) = 440$ Hz. When the nth piano key is played, the corresponding piano
string produces the fundamental $F_1(n)$ and its higher harmonics $kF_1(n)$, where $k = 2, 3, \ldots$.

a) Identify the piano keys having fundamental frequencies closest to the frequencies
shown on the faceplate of the graphic equalizer.

b) Show that the second harmonic frequency $2F_1(n)$ produced by the nth key is exactly in
tune with the fundamental $F_1(n + 12)$ of the $n + 12$th key for every n.

c) Show that the third harmonic frequency $3F_1(n)$ produced by the nth key is very nearly
in tune with the fundamental $F_1(n + 19)$ of the $n + 19$th key. Find the numerical values
of these two frequencies for $n = 0$ and find the period of the beat frequency they
produce. (Refer to Problem 3CT.4.4 for the definition of beats.) Show that the ratio
$F_1(n + 19) \div 2F_1(n)$ is nearly 3/2, which is the ratio of a perfect fifth in music.

5CT.2   Forms of the Fourier Series

5CT.2.1 Find the sine-cosine form of the Fourier series for $x(t) = A\cos^2(2\pi F_o t + \phi)$ without
using the formulas for the Fourier coefficients.

5CT.2.2 Find the complex Fourier series for $x(t) = x_1(t) + jx_2(t)$, where $x_1(t) = 3\cos(200\pi t) +$
$3\cos(400\pi t + 45°)$ and $x_2(t) = 3\sin(200\pi t) + 3\sin(400\pi t + 45°)$

5CT.2.3 Find the complex Fourier series for $x(t) = 3\cos(200\pi t) + 3\sin(200\pi t) + 3\cos(400\pi t +$
$45°)$.

5CT.2.4 Find the amplitude-phase form of the Fourier series for the waveform of Problem 5CT.2.3.

5CT.2.5 Equations (8)–(9) of the text show how to obtain the amplitude phase form of the Fourier
series from the complex Fourier series coefficients. Find the equations that do the reverse;
that is, find the equations that give you the complex Fourier series coefficients from the
amplitudes and the phases.

5CT.2.6 Equations (11)–(12) of the text show how to obtain the sine-cosine form of the Fourier series from the complex Fourier series coefficients. Find the equations that do the reverse; that is, find the equations that give you the complex Fourier series coefficients from the sine-cosine form of the Fourier series.

5CT.3 LTI System Response to Periodic Input

5CT.3.1 The pulse train of Example 5CT.1 is applied to an ideal low-pass filter having the cutoff frequency $W = 3.5F_1$. Find the output.

5CT.3.2 A periodic waveform having the period 10 ms is applied to an ideal differentiator. The Fourier series coefficients of the input are X_k; $k = 0, \pm1, \pm2, \ldots$. Write down the Fourier series coefficients of the output.

5CT.3.3 The periodic impulse train of Example 5CT.2 is applied to a first-order low-pass filter having the impulse response $h(t) = (1/\tau)e^{-\frac{t}{\tau}}u(t)$.
a) Use the equation $y(t) = x(t) * h(t)$ to write down the expression for the output. Leave your answer in the form of a sum. (Do not use the Fourier series.)
b) Sketch your answer to part (a). You can assume that $\tau \ll T$ for your sketch. Hint: If the expression for $y(t)$ confuses you, just think about what happens when the impulse train is applied to the system.
c) Find the Fourier series for $y(t)$.
d) Find a closed form for $y(t)$ for $0 < t < T$ that is valid for any $\tau > 0$.

5CT.3.4 A periodic impulse train is applied to an LTI system having the impulse response $g(t)$.
a) Show that the output is periodic.
b) Find the complex Fourier series coefficients of the output waveform. Express your answer in terms of the Fourier transform of $g(t)$, $G(F)$.

5CT.3.5 a) Sketch the waveform $x_1(t) = \sum_{k=-\infty}^{\infty} \delta(t - \frac{T}{2} - kT)$.
b) Find all three forms of the Fourier series of $x_1(t)$.
c) Sketch the waveform $x_2(t) = \sum_{k=-\infty}^{\infty}(-1)^k\delta(t - k(T_o/2))$.
d) What is the period of $x_2(t)$?
e) Find all three forms of the Fourier series of $x_2(t)$.

5CT.3.6 The waveform $x_2(t)$ of Problem 5CT.3.5 is applied to an LTI system having the impulse response $g(t)$.
a) Show that the output is periodic.
b) Find the complex Fourier series coefficients of the output waveform. Express your answer in terms of the Fourier transform of $g(t)$, $G(F)$.

5CT.3.7 Assume that the clipped cosine of Example 5CT.3, with $A_{max} = 1$ V and $T = 16$ ms, is applied to the graphic equalizer shown in Figure 5CT.16. The decibel power gain of the equalizer, determined by readings on the sliders and the knob, is defined as

$$G_{dB}(F) = 10\log_{10}\left[|H(j2\pi F)|^2 G_o^2\right] \qquad (102)$$

where $H(j2\pi F)$ is the frequency response characteristic of the equalizer and G_o is a voltage gain determined by the knob shown to the right of the sliders. Assume that the knob is set to a +3-dB power gain and the sliders are set to the power gains shown in the following table:

F (kHz)	0.0625	0.125	0.250	0.500	1	2	4	8	16
Slider (dB)	+6	+6	0	0	0	0	+3	+3	−6

a) Plot the values of $G_{dB}(F)$ in the table versus $\log_2(F/1\text{ kHz})$ for 62.5 Hz $\leq F \leq 16$ kHz.
b) Assume that the amplifier that produced the clipped sinusoid was severely overdriven. Use your plot of $G_{dB}(F)$ and the spectrum shown in Figure 5CT.5d to estimate the amplitudes of harmonics at the output of the equalizer for frequencies in the range 62.5 Hz to 1.25 kHz.

5CT.5   Derivation of the Fourier Series from the Fourier Transform

5CT.5.1 Find the complex FS coefficients of a triangular wave by using the Fourier transform of a triangular pulse.

5CT.5.2 Suppose that a pulse $p(t)$ is repeated $M = 2K + 1$ times to produce the waveform

$$x(t; K) = \sum_{k=-K}^{K} p(t - kT) \tag{103}$$

a) Find the Fourier transform $X(F; K)$ of $x(t; K)$. Express your answer in terms of the Fourier transform $P(F)$ of $p(t)$ and the cinc function.

b) Assume that $P(F) = \wedge (F/W)$ and sketch $X(F; K)$ for $1/(2K + 1)T \ll W$.

c) Describe what happens in the limit $K \to \infty$.

5CT.6   Properties of the Fourier Series

5CT.6.1 Which of the waveforms shown in Figure 5CT.17 have:

a) Even symmetry?

b) Odd symmetry?

c) Half-wave symmetry?

Figure 5CT.17
Figure for Problems
5CT.6.1–5CT.6.3

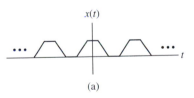

(a)

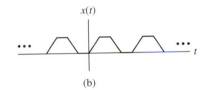

(b)

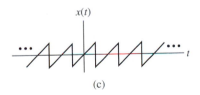

(c)

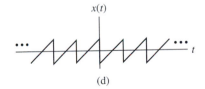

(d)

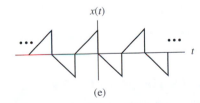

(e)

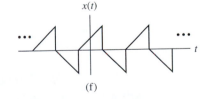

(f)

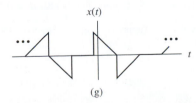

(g)

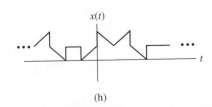

(h)

5CT.6.2 Consider the sine-cosine form for the Fourier series of the waveforms shown in Figure 5CT.17. Without evaluating any Fourier coefficients, state which waveforms contain:
  a) A DC component.
  b) No sine components.
  c) No cosine components.

5CT.6.3 Without evaluating any Fourier coefficients, state which waveforms of Figure 5CT.17 contain:
  a) Both sine and cosine components.
  b) Only odd harmonics.
  c) DC plus odd harmonics.

5CT.6.4 Use the formula $X_k = (1/T) \int_T x(t)e^{-j2\pi kF_1 t}\, dt$ to prove 5CT.OP1.

5CT.6.5 a) Prove 5CT.SP3.
  b) Prove 5CT.SP4.
  c) Prove 5CT.SP5.
  d) Prove 5CT.SP6.
  e) Prove 5CT.SP7.
  f) Prove 5CT.SP8.
  g) Prove 5CT.SP9.

5CT.6.6 a) Prove 5CT.OP1.
  b) Prove 5CT.OP5.
  c) Prove 5CT.OP6.

5CT.7   The Use of Properties to Derive the Fourier Series

5CT.7.1 Confirm the results of Example 5CT.8 by direct evaluation of Eq. (6) for the square wave of Figure 5CT.10.

5CT.7.2 Find the sine-cosine form of the Fourier series of any four waveforms of your choice among the set (a)–(g) in Figure 5CT.17. Assume that each waveform has period T and peak amplitude A. Hint: You can use the Fourier series properties to simplify your work. For example, if you differentiate waveform (d) you obtain $y(t) = (d/dt)x(t) = 2A/T - 2A\delta(t)$ for $-\frac{1}{2}T \leq t < \frac{1}{2}T$. The Fourier coefficients for $y(t)$ are easily obtained by integration: $Y_k = (1/T)\int_{-T/2}^{T/2} y(t)e^{-j2\pi kF_1 t}\, dt$. ($Y_0 = 0$ and $Y_k = 2A/T$ for $k \neq 0$.) You can then use the differentiation property $Y_k = j2\pi kF_1 X_k$ to solve for X_k.

5CT.7.3 Sketch some plots that show that a half-wave rectified cosine can be obtained by adding a cosine waveform to a full-wave rectified cosine. Use this result and the Fourier series of a full-wave rectified cosine to obtain the Fourier series of a half-wave rectified cosine.

5CT.8   Parseval's Theorem for Periodic Waveforms

5CT.8.1 Find the average output power when a square wave is applied to an ideal band-pass filter having the center frequency $F_o = 7F_1$ and bandwidth $B = 0.2F_1$.

5CT.8.2 An *RC* circuit can be used to (approximately) extract the DC component in a full-wave rectified cosine from the AC components. Let P_0 denote the average power in the DC component of the circuit's output, and P_{AC} denote the average power in the remaining (AC) components of the circuit's output.

a) Find an expression for the ratio P_{AC}/P_0. Your expression should include the filter time constant τ and the period of the full-wave rectified cosine $T = 1/F_1$ as parameters.

b) Write a computer program to plot the power ratio P_{AC}/P_0 in decibels versus τ/T:
$$(P_{AC}/P_0)_{dB} = 10 \log_{10} (P_{AC}/P_0).$$

5CT.8.3 Repeat Problem 5CT.8.2 for a half-wave rectified cosine.

5CT.8.4 Show analytically that the THD of a clipped cosine of Example 5CT.9 approaches -6.31 dB as $g_n \to \infty$. Hint: The clipped cosine becomes a square wave in the limit $g_n \to \infty$.

5CT.10   Convergence of the Fourier Series

5CT.10.1 Program a computer to plot the partial Fourier series of a periodic impulse train $x(t) = \delta_T (t) = \sum_{n=-\infty}^{\infty} \delta(t - nT)$.
a) Make plots for various values of K.
b) Comment on your results from part (a). Does the partial Fourier series converge? If so, how?

5CT.10.2 Program a computer to duplicate the plots in Example 5CT.11.

5CT.10.3 Write a computer program to plot the partial Fourier series of a half-wave rectified cosine. Make plots for various K and comment on your results.

5CT.10.4 Gibbs phenomenon can be reduced by a technique called windowing. Instead of using the Fourier series coefficients X_k in the series, we use the "windowed" coefficients

$$X'_k = W_k X_k \quad k = 0, \pm1, \pm2, \dots, \pm K \tag{104}$$

where W_k is a window sequence. A popular window sequence is the raised cosine (called the Hann or Hanning) window, defined by

$$W_k = \begin{cases} \frac{1}{2}\left\{1 + \cos\left(\dfrac{\pi k}{K+1}\right)\right\} & \text{for } |k| \le K \\ 0 & \text{for } |k| > K \end{cases} \tag{105}$$

a) Program a computer to calculate and plot the partial Fourier series of Example 5CT.12. Plot the partial Fourier series for a few values of K to demonstrate Gibbs phenomenon.
b) Modify your program so that it uses the windowed coefficients X'_k instead of the Fourier series coefficients. Make new plots using the same values of K as in part (a).
c) Describe your results in words. Are the windowed coefficients X'_k best in the MMSE sense? If not, what is the advantage of windowing?

5CT.10.5 The pulse train of Example 5CT.1 with $\tau_p = 0.5T$ is applied to an LTI system having the I-O equation

$$\frac{d^2 y(t)}{dt^2} + 2\alpha \frac{dy(t)}{dt} + \Omega_o^2 y(t) = x(t) \tag{106}$$

where $\alpha = 1$ and $\Omega_o = 2\pi 7$.
a) Find the amplitude characteristic $|H(F)|$.
b) Use hand calculations or a computer to plot $|H(F)|$.
c) Use a computer to plot the partial Fourier series for $y(t)$ for various K.

Appendix 5CT.A The Fourier Coefficients Yield Minimum Mean Square Error

This appendix derives the coefficients X_k that minimize the mean square error

$$\mathcal{E}_K = \frac{1}{T} \int_T \left| x(t) - \sum_{k=-K}^{K} X_k e^{j2\pi k F_1 t} \right|^2 dt$$

$$= \frac{1}{T} \int_T \left(x(t) - \sum_{k=-K}^{K} X_k e^{j2\pi k F_1 t} \right) \left(x(t) - \sum_{l=-K}^{K} X_l e^{j2\pi l F_1 t} \right)^* dt \qquad \text{(A1)}$$

If we multiply the two factors in the parentheses and use the property that

$$\int_T e^{j2\pi(k-l)F_1 t} dt = \begin{cases} T & k = l \\ 0 & k \neq l \end{cases} \qquad \text{(A2)}$$

we obtain

$$\mathcal{E}_K = P - 2\mathcal{R} \left\{ \sum_{l=-K}^{K} X_l^* \frac{1}{T} \int_T x(t) e^{-j2\pi l F_1 t} dt \right\} + \sum_{l=-K}^{K} |X_l|^2 \qquad \text{(A3)}$$

where P is the average power in $x(t)$ and X_l is a complex constant. It has magnitude $|X_l|$ and angle $\phi_l = \angle X_l$.

$$X_l = |X_l| e^{j\phi_l} \qquad \text{(A4)}$$

The integral is also a complex constant. It has some magnitude B_l and angle θ_l:

$$\frac{1}{T} \int_T x(t) e^{-j2\pi l F_1 t} dt = B_l e^{j\theta_l} \qquad \text{(A5)}$$

The substitution of Eqs. (A4) and (A5) into Eq. (A3) yields

$$\mathcal{E}_K = P - 2 \sum_{l=-K}^{K} B_l |X_l| \cos(\theta_l - \phi_l) + \sum_{l=-K}^{K} |X_l|^2 \qquad \text{(A6)}$$

Because P, B_k, and $|X_k|$ are all nonnegative, we minimize $\mathcal{E}_K$ by selecting

$$\angle X_l \equiv \phi_l = \theta_l \qquad \text{(A7)}$$

for $-K \leq l \leq K$. The substitution of (A7) into (A6) yields

$$\mathcal{E}_K = P - 2 \sum_{l=-K}^{K} |X_l| B_l + \sum_{l=-K}^{K} |X_l|^2 \qquad \text{(A8)}$$

The $|X_k|$ that minimize $\mathcal{E}_K$ satisfy

$$\frac{d\mathcal{E}_K}{d|X_k|} = -2B_k + 2|X_k| = 0 \quad -K \leq k \leq K \qquad \text{(A9)}$$

from which we find

$$|X_k| = B_k \quad -K \leq k \leq K \qquad \text{(A10)}$$

We conclude from Eqs. (A7), (A10), and (A5) that

$$X_k = B_k e^{j\theta_k} = \frac{1}{T} \int_T x(t) e^{-j2\pi k F_1 t} dt \qquad \text{(A11)}$$

which is the formula for the kth complex Fourier series coefficient.

Appendix 5CT.B The Fourier Transform Can Be Derived from the Fourier Series

The Fourier transform can be derived from the Fourier series. The Fourier series representation for a periodic waveform $x(t)$ is

$$x(t) = \sum_{k=-\infty}^{\infty} X_k e^{j2\pi k F_1 t} \tag{B1}$$

where

$$X_k = \frac{1}{T} \int_{-T/2}^{T/2} x(t) e^{-j2\pi k F_1 t} \, dt \tag{B2}$$

A typical periodic waveform is shown in Figure 5CT.18(a). We see from Figures 5CT.2(b) and 5CT.1(c) that the periodic signal becomes a single pulse in the limit $T \to \infty$.

The integral in Eq. (B2) remains finite in the limit $T \to \infty$. However, the values of the Fourier coefficients X_k in Eq. (B2) all equal 0 in this limit because of the factor $1/T$. Therefore, the Fourier series representation of $x(t)$ is not useful for representing a single pulse. However, if we define

$$\Delta F = F_1 = \frac{1}{T} \tag{B3}$$

and

$$X(k\Delta F) = T X_k = \int_{-T/2}^{T/2} x(t) e^{-j2\pi k F_1 t} \, dt \tag{B4}$$

then we can write Eq. (B1) as

$$x(t) = \sum_{k=-\infty}^{\infty} X(k\Delta F) e^{j2\pi k F_1 t} \Delta F \tag{B5}$$

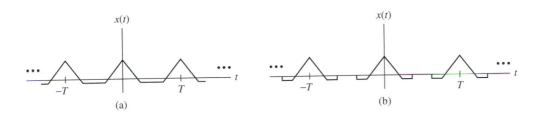

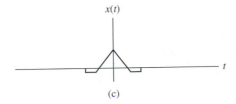

Figure 5CT.18
Periodic waveform becomes a pulse in the limit $T \to \infty$: (a) periodic waveform; (b) periodic waveform with increased T; (c) periodic waveform in the Limit $T \to \infty$.

The quantity ΔF is the spacing between the harmonic frequencies of the Fourier series. As $T \to \infty$, this spacing tends toward 0, and $kF_1 = k\Delta F$ tends toward the variable F. In the limit $T \to \infty$, Eq. (B4) becomes the Fourier transform of $x(t)$.

$$X(F) = \int_{-\infty}^{\infty} x(t)e^{-j2\pi Ft}\, dt \tag{B6}$$

and the sum (B5) becomes an integral.

$$x(t) = \int_{-\infty}^{\infty} X(F)e^{j2\pi Ft}\, dF \tag{B7}$$

The DFS and the DFT

5DT.1 Introduction

This chapter introduces the discrete Fourier series (DFS) and the discrete Fourier transform (DFT). Two transforms are included in one chapter because the equations that define them are nearly identical.

The DFS is the premier analytical and computational aid for analyzing a periodic sequence and determining the response it produces when it is applied to an LTI system. It is placed first in this chapter because it is more closely related to previous developments in this book.

The DFT is the computational counterpart of the DTFT. It is the most important *computational* aid in the entire field of spectrum analysis and signal processing. It can be computed efficiently and precisely using a *fast Fourier transform* (FFT) computer algorithm. The DFT is defined only for sequences having length N. Its primary applications include high-speed convolution (Section 5DT.9) and spectrum estimation (Section 5DT.10). If you are primarily interested in the DFT, you can skip directly to Section 5DT.4.

5DT.2 The DFS

The discrete-Fourier series (DFS) applies to periodic sequences. Recall that a sequence is *periodic* if

$$x[n + N] = x[n] \tag{1}$$

where the *period* N is defined as the smallest positive integer for which Eq. (1) applies. A typical periodic sequence is illustrated in Figure 5DT.1. In Figure 5DT.1 $N = 14$. The DFS represents a periodic sequence in a manner analogous to the multiple sinusoidal component sequence of Section 3DT.4. For a periodic sequence there are exactly N frequencies $f_0, f_1, f_2, \ldots, f_{N-1}$ and these are related by

$$f_k = kf_1 \quad k = 0, 1, 2, \ldots, N - 1 \tag{2a}$$

where

$$f_1 = \frac{1}{N} \tag{2b}$$

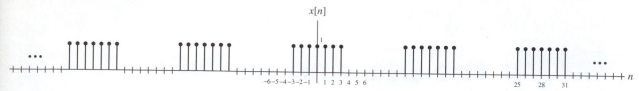

Figure 5DT.1
A typical periodic sequence.

Every periodic sequence can be expressed as a DFS.

Discrete Fourier Series

$$x[n] = \frac{1}{N} \sum_{k=0}^{N-1} X[k] e^{j2\pi \frac{k}{N} n} \quad \forall n \tag{3}$$

Inverse Discrete Fourier Series

$$X[k] = \sum_{n=0}^{N-1} x[n] e^{-j2\pi \frac{k}{N} n} \quad \forall k \tag{4}$$

In the preceding equations, $X[k]$, is the kth *discrete Fourier series coefficient*. The symbol $\forall$ means "for all." We emphasize that the DFS applies for *all* n because this property distinguishes the DFS from the DFT.

We can confirm that the formula for $X[k]$, given by Eq. (3), is correct if we substitute Eq. (4) into the right-hand side of Eq. (3).

$$\frac{1}{N} \sum_{k=0}^{N-1} \left\{ \sum_{n'=0}^{N-1} x[n'] e^{-j2\pi \frac{k}{N} n'} \right\} e^{j2\pi \frac{k}{N} n} = \frac{1}{N} \sum_{n'=0}^{N-1} x[n'] \sum_{k=0}^{N-1} e^{-j2\pi \frac{k}{N} (n'-n)} \tag{5}$$

We can simplify the summation over k by referring to identity (B.6) in Appendix B, which tells us that

$$\frac{1}{N} \sum_{k=0}^{N-1} e^{j\frac{2\pi}{M}(n'-n)} = \Delta[n'-n; N] = \begin{cases} 1 & n' = n + lN \\ 0 & n' \neq n + lN \end{cases} \tag{6}$$

where l is any integer. The substitution of Eq. (6) into the right-hand side of Eq. (5) yields

$$\sum_{n'=0}^{N-1} x[n'] \Delta[n'-n; N] \equiv x[n] \tag{7}$$

Therefore, the formula for $X[k]$ given by Eq. (4) is correct. We will show in Section 5DT.3 that the sequence $X[k]$, like $x[n]$, is periodic with period N

$$X[k+N] = X[k] \tag{8}$$

A major advantage of writing $x[n]$ as a DFS is to determine the response of an LTI system to a periodic input: As in Section 3DT.4, we obtain the I-O relationship simply by superimposing the LTI system responses to each sinusoidal component in the DFS representation of $x[n]$. The result is

---

I-O Relation for LTI System with Periodic Input

$$x[n] = \frac{1}{N}\sum_{k=0}^{N-1} X[k]e^{j2\pi\frac{k}{N}n} \xrightarrow{h[n]} y[n] = \frac{1}{N}\sum_{k=0}^{N-1} Y[k]e^{j2\pi\frac{k}{N}n} \tag{9a}$$

where

$$Y[k] = X[k]H\left(e^{j2\pi\frac{k}{N}}\right) \quad k = 0, 1, \ldots, N-1 \tag{9b}$$

DFS pairs are written as

$$x[n] \xleftrightarrow{\text{DFS}} X[k] \tag{10a}$$

or

$$x[n] = \mathcal{DFS}\{X[k]\} \tag{10b}$$

$$X[k] = \mathcal{IDFS}\{x[n]\} \tag{10c}$$

where $\mathcal{DFS}\{X[k]\}$ denotes the discrete Fourier series having coefficients $X[k]$, and $\mathcal{IDFS}\{x[n]\}$ denotes the coefficients belonging to $x[n]$.

EXAMPLE 5DT.1   Band-Pass Filtering of a Periodic Pulse Train

Let's use a DFS to find the output of an ideal band-pass filter to the periodic pulse train of Figure 5DT.1. To obtain the DFS, we note from the figure that

$$x[n] = \begin{cases} 1 & n = 0, 1, 2, 3 \\ 0 & n = 4, 5, 6, 7, 8, 9, 10 \\ 1 & n = 11, 12, 13 \end{cases} \tag{11a}$$

for $0 \le n \le 13$, and the period is $N = 14$. The DFS coefficients are given by Eq. (4)

$$X[k] = \sum_{n=0}^{13} x[n]e^{-j2\pi\frac{k}{14}n} = \sum_{n=0}^{3} e^{-j2\pi\frac{k}{14}n} + \sum_{n=11}^{13} e^{-j2\pi\frac{k}{14}n} \tag{11b}$$

We will simplify the expression for $X[k]$ if we change n to $m = n - 14$ in the last sum. This substitution yields

$$\sum_{n=11}^{13} e^{-j2\pi\frac{k}{14}n} = \sum_{m=-3}^{-1} e^{-j2\pi\frac{k}{14}(m+14)} = \sum_{m=-3}^{-1} e^{-j2\pi\frac{k}{14}m} \tag{12}$$

where we used the identity $e^{-j2\pi\frac{k}{14}14} = 1$. When we substitute Eq. (12) into (11b), we obtain

$$X[k] = \sum_{n=-3}^{3} e^{-j2\pi\frac{k}{14}n} = \frac{\sin\left(7k\frac{\pi}{14}\right)}{\sin\left(\pi\frac{k}{14}\right)} = 7\,\text{cinc}\left(\frac{k}{14};3\right) \quad \forall k \tag{13}$$

where we have used identity (B.3) in Appendix B. A plot of the DFS coefficients is shown in Figure 5DT.2. The coefficient, $X[k]$ refers to frequency $f = k/N$ because $X[k]$ is the complex amplitude of the complex sinusoid

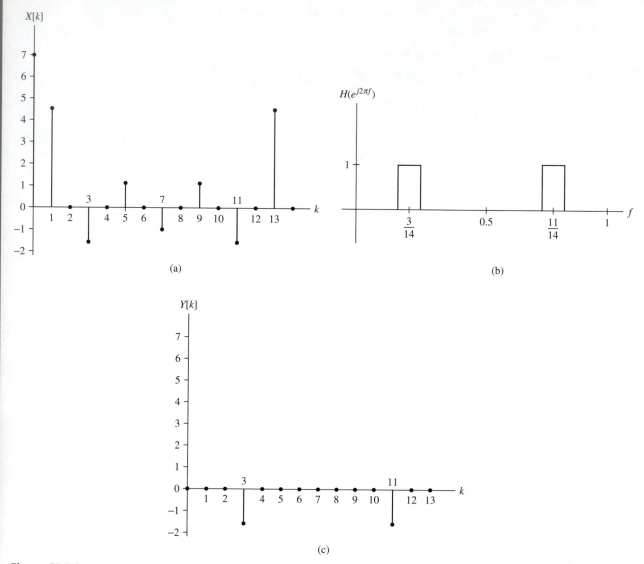

Figure 5DT.2
Frequency domain view of filtering a periodic sequence: (a) $X[k]$; (b) $H(e^{j2\pi f})$; (c) $Y[k]$.

$e^{j2\pi \frac{k}{N}n}$. Figure 5DT.2 also shows the frequency response characteristic of a band-pass filter. We can see from the plots and Eq. (9b) that the filter passes only the complex sinusoids having frequencies $f = 3/14$ and $f = 11/14$. The output is, by Eq. (9a)

$$\frac{1}{14}X[3]e^{j2\pi \frac{3}{14}n} + \frac{1}{14}X[11]e^{j2\pi \frac{11}{14}n} \equiv -\frac{1}{14}(1.6039)e^{j2\pi \frac{3}{14}n} - \frac{1}{14}(1.6039)e^{-j2\pi \frac{3}{14}n}$$

$$= -\frac{1}{14}(3.2078)\cos\left(2\pi \frac{3}{14}n\right) = -0.2291\cos\left(2\pi \frac{3}{14}n\right) \tag{14}$$

where we used the identity $e^{j2\pi \frac{11}{14}n} = e^{j2\pi \frac{11}{14}n}e^{-j2\pi \frac{14}{14}n}$ to replace $e^{j2\pi \frac{11}{14}n}$ by its alias $e^{-j2\pi \frac{3}{14}n}$.

We calculated the DFS coefficients by hand in the preceding example to show the analytical steps. Once you know how the hand calculations work, you would generally prefer to use a computer. The software for computing the DFS coefficients (as well as the DFT of Section 5DT.4) is called the *fast Fourier transform* (FFT). MATLAB Problem 5DT.2.2 introduces you to the FFT.

We have observed that the DFS is analogous to the multiple sinusoidal component representation of sequences introduced in Chapter 3DT. We show in Appendix 5DT.A that the DFS can obtained from the DTFT.

5DT.3   **Properties of the DFS**

The key properties of the DFS are listed in Tables 5DT.1 and 5DT.2. All the properties of the DFS can be proven directly from Eq. (2) and (3).

5DT.3.1 Symmetry Properties

Symmetry property (5DT.SP1) states that both $x[n]$ and $X[k]$ are periodic sequences with period N. The periodicity of the coefficients can be established from Eq. (4) as follows

$$X[k+N] = \sum_{n=0}^{N-1} x[n]e^{-j\frac{2\pi}{N}(k+N)n} = \sum_{n=0}^{N-1} x[n]e^{-j\frac{2\pi}{N}kn}e^{-j2\pi n} = X[k] \qquad (15)$$

where the last step follows from the fact that $e^{-j2\pi n} = 1$ for every integer n.

Symmetry property (5DT.SP2) states that the coefficients have conjugate symmetry for every real periodic sequence $x[n]$. Conjugate symmetry is defined by $X[-k] = X^*[k]$. Since the coefficients here are periodic, $X[-k] = X[N-k]$. Thus, as stated in property (5DT.SP2)

$$X[N-k] = X^*[k] \quad 0 \le k < N-1 \qquad (16)$$

Table 5DT.1
Discrete Fourier Series
Symmetry Properties

	$x[n]$	$X[k]$
5DT.S1	$x[n]$ is periodic with period N	$X[k]$ is periodic with period N
5DT.S2	Real	$X[N-k] = X^*[k]$
5DT.S3	Real even: $x[N-n] = x[n]$	Real even: $X[N-k] = X[k]$
5DT.S4	Real odd: $x[N-n] = -x[n]$	Imaginary odd: $X[N-k] = -X[k]$

Table 5DT.2
Discrete Fourier Series
Operational
Properties

The sequences $x[n]$ and $X[k]$ are periodic with period N. Convolution property (5DT.OP2a) applies for nonperiodic $h[n]$: $h[n] \overset{\text{DTFT}}{\longleftrightarrow} H(e^{j2\pi f})$. Periodic convolution property (5DT.OP2b) applies for periodic $h[n]$: $h[n] \overset{\text{DFS}}{\longleftrightarrow} H[k]$.

	Name	Sequence	Coefficients
5DT.OP1	Linearity	$a_1x_1[n] + a_2x_2[n]$	$a_1X_1[k] + a_2X_2[k]$
5DT.OP2a	Convolution	$x[n] * h[n]$	$X[k]H\left(e^{jk\frac{2\pi}{N}}\right)$
5DT.OP2b	Periodic convolution	$x[n] \overset{N}{*} h[n]$	$X[k]H[k]$
5DT.OP3	Sequence shift	$x[n-m]$	$e^{-j\frac{2\pi m}{N}k}X[k]$
5DT.OP4	Coefficient shift	$x[n]e^{j\frac{2\pi m}{N}n}$	$X[k-m]$
5DT.OP5	Modulation	$x[n]\cos\left(2\pi\frac{m}{N}n\right)$	$\frac{1}{2}X[k-m] + \frac{1}{2}X[k+m]$

We can prove property (5DT.SP2) from Eq. (3) as follows

$$X[N-k] = \sum_{n=0}^{N-1} x[n]e^{j2\pi \frac{(N-k)}{N}n} = \sum_{n=0}^{N-1} x[n]e^{-j2\pi \frac{k}{N}n}$$

$$= \left(\sum_{n=0}^{N-1} x[n]e^{j2\pi \frac{k}{N}n} \right)^* = X^*[k] \tag{17}$$

Symmetry properties (5DT.SP3) and (5DT.SP4) apply for real-valued even and real-valued odd periodic sequences, respectively. A sequence $x[n]$ is even if $x[-n] = x[n]$ for all n. Since $x[N-n] = x[-n]$ for a periodic sequence, we can define a periodic sequence to be even if $x[N-n] = x[n]$. Similarly, a sequence $x[n]$ is odd if $x[-n] = -x[n]$ for all n, and we define a periodic sequence to be odd if $x[N-n] = -x[n]$. Examples of even and odd periodic sequences are $\cos(2\pi k/Nn)$ and $\sin(2\pi k/Nn)$, respectively, where k is any integer. The proofs of (5DT.SP3) and (5DT.SP4) are straightforward (Problems 5DT.3.3 and 5DT.3.4).

5DT.3.2
Operational
Properties

Table 5DT.2 lists the most important operational properties of the DFS.

Operational property (5DT.OP1), linearity, follows by summing the individual DFS representations of $a_1 x_1[n]$ and $a_2 x_2[n]$.

$$a_1 x_1[n] + a_2 x_2[n] = a_1 \frac{1}{N} \sum_{k=0}^{N-1} X_1[k]e^{j2\pi \frac{k}{N}n} + a_2 \frac{1}{N} \sum_{k=0}^{N-1} X_2[k]e^{j2\pi \frac{k}{N}n}$$

$$= \frac{1}{N} \sum_{k=0}^{N-1} \{a_1 X_1[k] + a_2 X_2[n]\}e^{j2\pi \frac{k}{N}n} \tag{18}$$

We see from the right-hand side of Eq. (18) that $a_1 x_1[n] + a_2 x_2[n]$ can be represented as a DFS having coefficients $a_1 X_1[k] + a_2 X_2[k]$, which is what the property states.

We derived operational property (5DT.OP2a), in Section 5DT.2. In operational property (5DT.OP2b), the operator $\overset{N}{*}$ denotes *periodic convolution*, which is defined as

$$x[n] \overset{N}{*} h[n] = \sum_{m=0}^{N-1} x[m]h[n-m] \tag{19}$$

where both $x[n]$ and $h[n]$ have period N. Operational property (5DT.OP2b) states that $x[n] \overset{N}{*} h[n]$ is a periodic sequence having DFS coefficients $X[k]H[k]$ where the $X[k]$ and $H[k]$ are the DFS coefficients of $x[n]$ and $h[n]$, respectively. This property is easily proven by direct calculation of the inverse DFS of $X[k]H[k]$.

$$\frac{1}{N} \sum_{k=0}^{N-1} X[k]H[k]e^{j2\pi \frac{k}{N}n} = \frac{1}{N} \sum_{n=0}^{N-1} \left\{ \sum_{m=0}^{N-1} x[m]e^{-j2\pi \frac{k}{N}m} \right\} H[k]e^{j2\pi \frac{k}{N}n}$$

$$= \sum_{m=0}^{N-1} x[m] \frac{1}{N} \sum_{n=0}^{N-1} H[k]e^{j2\pi \frac{k}{N}(n-m)} = \sum_{m=0}^{N-1} x[m]h[n-m] \tag{20}$$

Operational property (5DT.OP3) can be proven by recognizing that sequence shift is the result of a convolution $x[n] * h[n]$ where $h[n] = \delta[n-m]$. We know from Section 3DT.6, that $H(e^{j2\pi f}) = \mathcal{DTFT}\{\delta[n-m]\} = e^{-j2\pi mf}$. The application of (5DT.OP2a) then yields property (5DT.OP3).

We can prove (5DT.OP4) by direct evaluation of the inverse DFS of $x[n]e^{j2\pi \frac{m}{N}n}$, as follows.

$$\mathcal{IDFS}\{x[n]e^{j2\pi \frac{m}{N}n}\} = \sum_{n=0}^{N-1} x[n]e^{-j2\pi \frac{k}{N}n}e^{j\frac{2\pi m}{N}n} = \sum_{n=0}^{N-1} x[n]e^{-j2\pi \frac{(k-m)}{N}n} = X[k-m] \quad (21)$$

Operational property (5DT.OP5) follows from (5DT.OP1) and (5DT.OP4) (Problem 5DT.3.5).

**5DT.3.3
Parseval's
Theorem for
Periodic
Sequences**

A final property is *Parseval's theorem:*

Parseval's Theorem for Periodic Sequences

$$\sum_{n=0}^{N-1} |x[n]|^2 = \frac{1}{N} \sum_{k=0}^{N-1} |X[k]|^2 \qquad (22)$$

Parseval's theorem states that the energy in one period of $x[n]$ can be computed from the DFS coefficients. This property can be derived easily from the definition of the DFS (Problem 5DT.3.6).

5DT.4   The DFT

The DFT is the most useful computational tool in all of signals and systems theory. It transforms an N-point data sequence x[0], x[1], ..., x[$N-1$] into an N-point coefficient sequence X[0], X[1], ..., X[$N-1$]. The DFT and its inverse are defined as follows.

Discrete Fourier Transform (DFT)

$$X[k] = \sum_{n=0}^{N-1} x[n]W_N^{kn} \quad k = 0, 1, \ldots, N-1 \qquad (23)$$

Inverse Discrete Fourier Transform (Inverse DFT)

$$x[n] = \frac{1}{N} \sum_{k=0}^{N-1} X[k]W_N^{-kn} \quad n = 0, 1, \ldots, N-1 \qquad (24)$$

where W_N is the complex constant

$$W_N = e^{-j\frac{2\pi}{N}} \qquad (25)$$

We use Roman letters x and X in denoting the DFT. The Roman letters emphasize that x[n] and X[k] are defined *only* for $0 \le n, k \le N - 1$. The italicised letters, $x[n]$ and $X[k]$, that we have used previously, denote sequences that are defined for *all* n and k. We see from Eq. (23) that the DFT is simply a system of N linear algebraic equations. These equations transform N variables x[0], x[1], ..., x[N − 1] into N variables X[0], X[1], ..., X[N − 1]. The inverse DFT (5DT.24) is a similar system of equations. These equations transform X[0], X[1], ..., X[N − 1] back into x[0], x[1], ..., x[N − 1].

When a computer is used to evaluate a DFT or its inverse, the values of $W_N^{\pm kn}$, $n = 0, 1, \ldots, N - 1$ can be evaluated "off-line," i.e., before reading the x[n] or X[k] into in the sums (5DT.23) or (5DT.24). The term *fast Fourier transform* (FFT) denotes a family of computer subroutines that compute the direct and inverse DFT efficiently. Maximum computational efficiency is achieved when N is a power of 2.

The symbol W_N Eq. (25) is standard notation for $e^{-j\frac{2\pi}{N}}$. It has the following properties.

Properties of W_N

$$W_N^* = W_N^{-1} \tag{26}$$

$$W_N^{mN} = 1 \qquad \text{where } m \text{ is any integer} \tag{27}$$

and (Identity (5DT.B6) in Appendix B)

$$\sum_{n=0}^{N-1} W_N^{kn} = N\Delta_N[k] \tag{28a}$$

where:

$$\Delta[k; N] \triangleq \begin{cases} 1 & \text{when } k \text{ is an integer multiple of } N \\ 0 & \text{otherwise} \end{cases} \tag{28b}$$

We can prove that Eq. (24) is the inverse of Eq. (23) by substituting Eq. (24) into the right-hand side of Eq. (23). We change the index of summation k in Eq. (24) to k' to distinguish it from the coefficient k in Eq. (23). An interchange in the orders of summation yields

$$\sum_{n=0}^{N-1} \left\{ \frac{1}{N} \sum_{k'=0}^{N-1} X[k'] W_N^{-k'n} \right\} W_N^{kn} = \sum_{k'=0}^{N-1} X[k'] \left\{ \frac{1}{N} \sum_{n=0}^{N-1} W_N^{(k-k')n} \right\} \tag{29}$$

The term in braces can be simplified with the help of Eq. (28).

$$\sum_{k'=0}^{N-1} X[k'] \left\{ \frac{1}{N} \sum_{n=0}^{N-1} W_N^{(k-k')n} \right\} = \sum_{k'=0}^{N-1} X[k']\Delta_N[k' - k] = X[k] \qquad k = 0, 1, \ldots, N - 1 \tag{30}$$

which shows that Eq. (24) is the inverse of Eq. (23).

Useful notations for the DFT transform pair given by Eqs. (23) and (24) are

$$X[k] = \mathcal{DFT}\{x[n]\} \tag{31a}$$

$$x[n] = \mathcal{IDFT}\{X[k]\} \tag{31b}$$

and

$$x[n] \overset{N\text{-point}}{\underset{\text{DFT}}{\longleftrightarrow}} X[k] \tag{32}$$

As we proceed, it is important to remember that x[n] and X[k] are undefined if n and k are outside the range $0 \le n, k \le N - 1$. In future discussions, we often follow common practice and assume $0 \le n, k \le N - 1$ implicitly.

The following analytical examples help you develop an understanding of the DFT. Additional analytical examples are given in Section 5DT.3. These help you relate the DFT coefficients to the DTFT of Chapter 4. You need to learn this theory to correctly apply FFT software.

EXAMPLE 5DT.2 Unit Impulse $x[n] = \delta[n]; n = 0, 1, \ldots, N - 1$

Consider the sequence

$$x[n] = \{x[0], x[1], \ldots, x[N-1]\} = \underbrace{\{1, 0, \ldots, 0\}}_{N\text{-point sequence}} \equiv \delta[n] \quad n = 0, 1, \ldots, N - 1 \tag{33}$$

The substitution of Eq. (33) into Eq. (23) yields

$$X[k] = \sum_{n=0}^{N-1} \delta[n] W_N^{kn} = 1 \quad k = 0, 1, \ldots, N - 1 \tag{34}$$

Therefore,

$$\delta[n] \overset{N\text{-point}}{\underset{\text{DFT}}{\longleftrightarrow}} 1 \tag{35}$$

EXAMPLE 5DT.3 Constant $x[n] = 1; n = 0, 1, \ldots, N - 1$

Consider

$$x[n] = 1 \quad n = 0, 1, \ldots, N - 1 \tag{36}$$

The substitution of Eq. (36) into (23) yields

$$X[k] = \sum_{n=0}^{N-1} W_N^{kn} = N \Delta_N[k] = \begin{cases} N & \text{when } k \text{ is an integer multiple of } N \\ 0 & \text{otherwise} \end{cases} \tag{37}$$

where we used identity (B.6) of Appendix B. By definition of the DFT, k is in the range $0 \le k \le N - 1$. By combining this condition with Eq. (37) we obtain the result

$$X[k] = \begin{cases} N & k = 0 \\ 0 & k = 1, 2, \ldots, N - 1 \end{cases} \tag{38}$$

Therefore,

$$1 \overset{N\text{-point}}{\underset{\text{DFT}}{\longleftrightarrow}} N\delta[k] \tag{39}$$

This result is the dual of that of Example 5DT.1.

EXAMPLE 5DT.4   Complex Sinusoid $x[n] = e^{j2\pi \frac{k_o}{N}n}; n = 0, 1, \ldots, N-1$

Consider the sequence

$$x[n] = e^{j2\pi \frac{k_o}{N}n} \equiv W_N^{-k_o n} \quad n = 0, 1, \ldots, N-1 \tag{40}$$

where k_o is an integer in the range $0 \le k_o \le N-1$. The substitution of Eq. (40) into Eq. (23) yields

$$X[k] = \sum_{n=0}^{N-1} W_N^{-k_o n} W_N^{kn} = \sum_{n=0}^{N-1} W_N^{(k-k_o)n} \tag{41}$$

Again, identity (B.6) of Appendix B provides a simplification:

$$\sum_{n=0}^{N-1} W_N^{(k-k_o)n} = N\Delta[k - k_o; N] = \begin{cases} N & \text{when } k - k_o \text{ is an integer multiple of } N \\ 0 & \text{otherwise} \end{cases} \tag{42}$$

Recall that by definition of the DFT, k is in the range $0 \le k \le N-1$. By combining this condition with Eq. (42) we have the final result

$$e^{j2\pi \frac{k_o}{N}n} \overset{N\text{-point}}{\underset{\text{DFT}}{\longleftrightarrow}} N\delta[k - k_o] \quad \text{for } 0 \le k_o \le N-1 \tag{43}$$

EXAMPLE 5DT.5   Real Sinusoid $x[n] = \cos(2\pi(k_o/N)n); n = 0, 1, \ldots, N-1$

Consider the sequence

$$x[n] = \cos\left(2\pi \frac{k_o}{N}n\right) \quad n = 0, 1, \ldots, N-1 \tag{44}$$

where k_o is an integer in the range $0 \le k_o \le N-1$. We can use Euler's identity to write $x[n]$ as

$$x[n] = \tfrac{1}{2}e^{j2\pi \frac{k_o}{N}n} + \tfrac{1}{2}e^{-j2\pi \frac{k_o}{N}n} \tag{45}$$

When we substitute Eq. (45) into Eq. (23), we obtain two sums that can be evaluated using steps similar to those in Example 5DT.4. The result is

$$\sum_{n=0}^{N-1} \cos\left(2\pi \frac{k_o}{N}n\right) W_N^{kn} = \tfrac{1}{2}N\Delta_N[k - k_o] + \tfrac{1}{2}N\Delta_N[k + k_o] \tag{46}$$

By combining Eq. (42) with Eq. (46), we have the result

$$\cos\left(2\pi \frac{k_o}{N}n\right) \overset{N\text{-point}}{\underset{\text{DFT}}{\longleftrightarrow}} \tfrac{1}{2}N\delta[k - k_o] + \tfrac{1}{2}N\delta[N - k - k_o] \quad \text{for } 0 \le k_o \le N-1 \tag{47}$$

5DT.5   Relationship of the DFT to the DTFT

An important application of the DFT is to estimate the spectrum $X(e^{j2\pi f})$ of a sequence $x[n]$. In this section, we assume that the sequence $x[n]$ equals 0 outside the range $0 \le n \le N - 1$, i.e.

$$x[n] = \ldots 0, 0, 0, \underbrace{x[0], x[1], \ldots, x[N-1]}_{x[0], x[1], \ldots, x[N-1]}, 0, 0, 0, \ldots \qquad (48)$$

The horizontal brace identifies the elements applied to the DFT. The arrow indicates the $n = 0$ element of $x[n]$.

Consider the DTFT of $x[n]$ of Eq. (48)

$$X(e^{j2\pi f}) = \sum_{n=-\infty}^{\infty} x[n] e^{-j2\pi f n} = \sum_{n=0}^{N-1} x[n] e^{-j2\pi f n} \qquad (49)$$

We see by comparing Eq. (49) with Eq. (23) that the DFT coefficients of $\{x[0], x[1], \ldots, x[N-1]\}$ are given by

$$X[k] = \sum_{n=0}^{N-1} x[n] e^{-j2\pi f n} \bigg|_{f = \frac{k}{N}} = X\left(e^{j2\pi \frac{k}{N}}\right) \quad k = 0, 1, \ldots, N-1 \qquad (50)$$

Equation (50) states that the DFT coefficients equal the values of the DTFT at the frequencies $f = k/N; k = 0, 1, \ldots, N - 1$. This result means that we can use the DFT to compute N equally spaced samples of the DTFT of any sequence having the form of (48). The following examples illustrate this relationship.

EXAMPLE 5DT.6 $x[n] = \sqcap[n - K; K]$

Consider the sequence

$$x[n] = \ldots 0, 0, 0, \underbrace{1, 1, 1, \ldots, 1, 1}_{x[0], x[1], \ldots, x[N-1]}, 0, 0, 0, \ldots \qquad (51)$$

where N is odd, $N = 2K + 1$ for some positive integer K. We can write Eq. (51) as a shifted rectangular pulse

$$x[n] = \sqcap[n - K; K] \qquad (52)$$

and use Tables 4DT.2 and 4DT.3 to obtain its spectrum.

$$X(e^{j2\pi f}) = (2K + 1)\,\text{cinc}(f; K) e^{-j2\pi f K} \qquad (53)$$

Figure 5DT.3 shows plots of the absolute values of $X(e^{j2\pi f})$ and its samples at $f = k/N, k = 0, 1, \ldots, N - 1$ where $N = 7$. The sample values equal the DFT coefficients we found in Example 5DT.2

$$X[k] = N\delta[k] \quad k = 0, 1, \ldots, N - 1 \qquad (54)$$

Figure 5DT.3
$|X(e^{j2\pi f})|$ and $|X[k]|$:
relationship for $x[n]$
and $x[n]$ given by
Eq. (50) with $N = 7$.

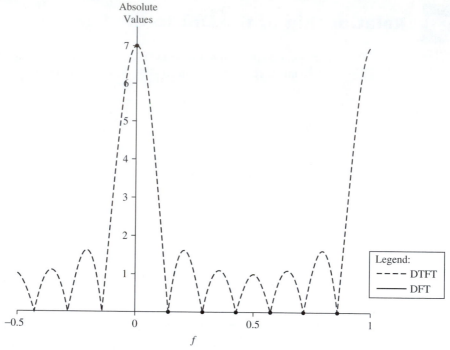

When x[n] and x[n] are related by Eq. (48), the DFT coefficients, X[0],
X[1],..., X[N − 1], equal samples of the DTFT, $X(e^{j2\pi f})$ at $f = k/N$, $k = 0, 1,..., N − 1$.

EXAMPLE 5DT.7   Chip Sequence

Consider the "chip" sequence

$$x[n] = \ldots 0, 0, 0, \cos(2\pi f_o 0), \cos(2\pi f_o 1), \ldots, \cos(2\pi f_o(N-1)), 0, 0, 0, \ldots \tag{55}$$

$\uparrow$

$x[0], x[1], \ldots, x[N-1]$

where N is odd: $N = 2K + 1$ for some positive integer K. We can write Eq. (55) as

$$x[n] = \sqcap[n - K; K] \cos(2\pi f_o n) \tag{56}$$

We can easily derive the spectrum of $x[n]$ from Eq. (53) and the modulation property of the DTFT (Table 4DT.2). The result is

$$X(e^{j2\pi f}) = \tfrac{1}{2}(2K + 1)\,\mathrm{cinc}(f - f_o; K)e^{-j2\pi(f-f_o)K}$$

$$+ \tfrac{1}{2}(2K + 1)\,\mathrm{cinc}(f + f_o; K)e^{-j2\pi(f+f_o)K} \tag{57}$$

Recall that the DFT of $x[n]$ was found in Example 5DT.5.

$$X[k] = \tfrac{1}{2}N\delta[k - k_o] + \tfrac{1}{2}N\delta[N - k - k_o] \qquad k = 0, 1, \ldots, N - 1 \tag{58}$$

Figure 5DT.4
DTFT and DFT
amplitudes for
Example 5DT.7:
relationship for x[n]
and x[n] given by
Eq. (50) with $k_o = 5$
and $K = 11$
($N = 23$).

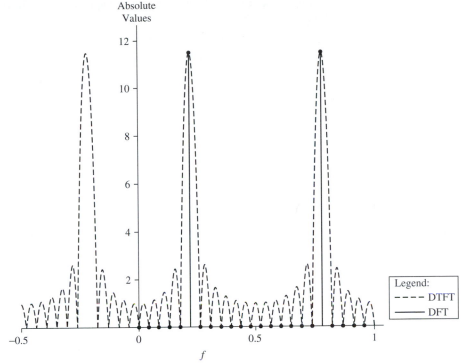

When x[n] and x[n] are related by Eq. (48), the DFT coefficients X[k] equal samples of the DTFT $X(e^{j2\pi f})$ at $f = k/N$, with $k = 0, 1,..., N-1$.

The DFT equals samples of the spectrum $X(e^{j2\pi f})$ at $f = k/N, k = 0, 1, \ldots, N-1$. This relationship is illustrated by Figure 5DT.4(a), which shows amplitude spectrum $|X(e^{j2\pi f})|$ and its samples at $f = k/N, k = 0, 1, \ldots, N-1$, where $f_o = k_o/N$ with $k_o = 5$ and $K = 11$ ($N = 23$).

5DT.6   The Picket Fence Effect

We have seen in the examples in the previous section that the DTFT contains spectrum values that are not seen in the DFT. When plotted on the same plot as the DTFT, the DFT coefficients show only values of $X(e^{j2\pi f})$ at sample points $f = k/N, k = 0, 1, \ldots, N-1$. It is as if they are glimpses of $X(e^{j2\pi f})$ through a picket fence, where the gaps between the pickets are spaced $\Delta f = 1/N$ apart. This is the *picket fence effect*.

If we want to use a DFT to see more of $X(e^{j2\pi f})$, we need to decrease the sample spacing Δf. A simple way to accomplish this decrease is to append zeros to the sequence x[0], x[1], ..., x[N − 1]. This method is called *zero padding*. The result is a new sequence x′[n] having length $N' = N + N_0$ where N_0 is the number of appended (padded) zeros:

$$\{x'[n]\} = x[0], x[1], \ldots, x[N-1], \underbrace{0, 0, \ldots, 0}_{N_0} \quad (59)$$

$$\underbrace{}_{N'}$$

Looking back at Eq. (48), we see that, by appending zeros to x[n], we are in effect just taking more samples of the sequence.

$$x[n] = \ldots 0, 0, 0, \underbrace{x[0], x[1], \ldots, x[N-1], 0, 0, \ldots, 0, 0, 0, 0, \ldots}_{\{x'[n]\}} \tag{60}$$

The DTFT of $x[n]$ is the same as that obtained in Eq. (49):

$$X'(e^{j2\pi f}) = \sum_{n=0}^{N'} x'[n]e^{-j2\pi fn} \equiv \sum_{n=0}^{N} x[n]e^{-j2\pi fn} = X(e^{j2\pi f}) \tag{61}$$

but, because we have appended the zeros, we have decreased the frequency sample spacing in the corresponding DFT from $\Delta f = 1/N$ to

$$\Delta f' = \frac{1}{N'} = \frac{1}{N + N_0} \tag{62}$$

This result follows because the sequence $\{x'[n]\}$ is an $N' = N + N_0$–point sequence. Its N'-point DFT has N' coefficients. These coefficients equal samples of $X(e^{j2\pi f})$ spaced $\Delta f' = 1/N'$ apart.

$$X'[k] = \sum_{n=0}^{N'-1} x'[n]e^{-j2\pi fn}\bigg|_{f=\frac{k}{N'}} = X(e^{j2\pi \frac{k}{N'}}) \quad k = 0, 1, \ldots, N'-1 \tag{63}$$

We can decrease the picket fence degradation as much as we wish by appending more and more zeros to the data. To take full advantage of the efficiency of the FFT algorithm, we chose N_0 so that N' is a power of 2.

EXAMPLE 5DT.8   Chip with Appended Zeros

If we append zeros to the chip of Example 5DT.7, we obtain

$$\{x'[n]\} = \underbrace{\cos(2\pi f_o 0), \cos(2\pi f_o 1), \ldots, \cos(2\pi f_o(N-1)), 0, 0, \ldots, 0}_{x[0], x[1], \ldots, x[N-1], x[N], x[N+1], \ldots, x[N'-1]} \tag{64}$$

Figure 5DT.5 shows the absolute values of the DFT coefficients of $\{x'[n]\}$, where $N = 23$ and $f_o = 5/23$. The DFT coefficients have been plotted versus f such that X[k] is plotted at $f = k/N'$, $k = 0, 1, \ldots, N'-1$. For Figure 5DT.5(a), no zeros were appended ($N' = 23$); for Figure 5DT.5(b), 41 zeros were appended ($N' = 64$); and for Figure 5DT.5(c), 105 zeros were appended ($N' = 128$). The data applied to the DFT is given by Eq. (64). As more zeros are appended, the plot reveals more of the DTFT of the sequence because the picket fence spacing is reduced.

Figure 5DT.5
Picket fence
degradation—
decreased when
zeros are appended
to a sequence (plot
range is $0 \leq f \leq 1$):
(a) $N' = 23$ (no zeros
appended);
(b) $N' = 64$;
(c) $N' = 128$.

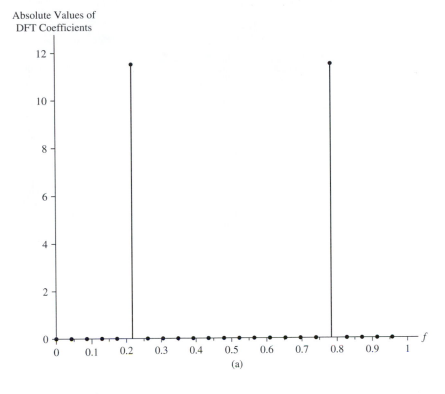

(a)

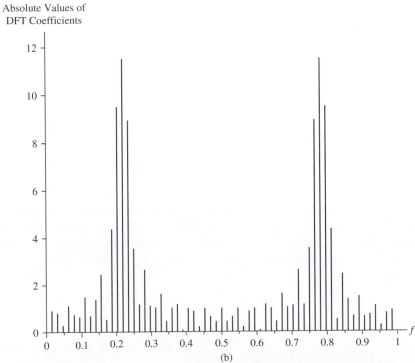

(b)

Figure 5DT.5
(*Cont.*).

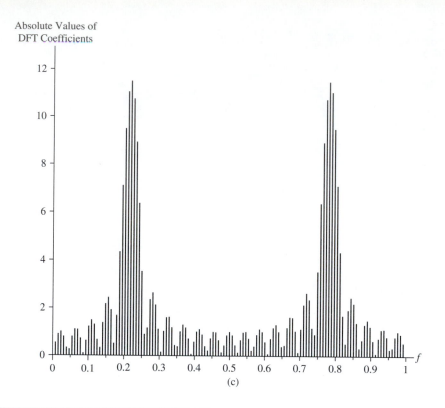

Absolute Values of
DFT Coefficients

(c)

Because the DFT is an invertible transformation, the DFT coefficients X[k] contain complete information about the N-point sequence x[0], x[1], ..., x[$N-1$]. This implies that a formula exists for reconstructing the complete spectrum $X(e^{j2\pi f})$ of $x[n]$ = ..., 0, 0, 0, x[0], x[1], ..., x[$N-1$], 0, 0, 0, ... from the DFT coefficients. This formula and its derivation are in Appendix 5DT.B.

5DT.7   Properties of the DFT

To apply the DFT correctly, we need to know its properties.

5DT.7.1
Symmetry
Properties

The most important symmetry properties of the DFT are listed in Table 5DT.3.
Symmetry property (5DT.SP1) states that the kth DFT coefficient of x*[n] is X*[$N-k$], where X[k] is the kth DFT coefficient of $x[n]$. This property applies to complex, real, and purely imaginary x[n]. We can prove (5DT.SP1) as follows.

$$\mathcal{DFT}\{x^*[n]\} = \sum_{n=0}^{N-1} x^*[n]W_N^{kn} = \left(\sum_{n=0}^{N-1} x[n]W_N^{-kn}\right)^*$$

$$= \left(\sum_{n=0}^{N-1} x[n]W_N^{(N-k)n}\right)^* = X^*[N-k] \quad k = 0, 1, \ldots, N-1 \qquad (65)$$

Table 5DT.3
Symmetry Properties
of the DFT x[n] and
X[k] are defined only
for $0 \leq n, k \leq N - 1$.

For Real or Complex x[n]

$$x^*[n] \longleftrightarrow X^*[N - k] \qquad\qquad\qquad (5DT.SP1)$$

For Real x[n]

$$X[N - k] = X^*[k] \quad 0 \leq k < N - 1 \qquad\qquad (5DT.SP2)$$
$$|X[N - k]| = |X[k]| \qquad\qquad\qquad\qquad (5DT.SP3)$$
$$\angle X[N - k] = -\angle X[k] \qquad\qquad\qquad\qquad (5DT.SP4)$$
$$\text{real and even } x[n] \overset{\text{DFT}}{\longleftrightarrow} \text{real and even } X[k] \qquad (5DT.SP5)$$
$$\text{real and odd } x[n] \overset{\text{DFT}}{\longleftrightarrow} \text{imaginary and odd } X[k] \qquad (5DT.SP6)$$

For an example of (5DT.SP1), consider the sequence $x[n] = e^{j2\pi \frac{k_o}{N} n}$, where k_o is an integer in the range $0 \leq k_o \leq N - 1$. We know from Example 5DT.4 that $\mathcal{DFT}\{e^{j2\pi \frac{k_o}{N} n}\} = N\delta[k - k_o]$. It follows from (5DT.SP1) that $\mathcal{DFT}\{e^{-j2\pi \frac{k_o}{N} n}\} = N\delta[N - k - k_o]$. This checks a result found in Example 5DT.5 in connection with the DFT of $\cos(2\pi(k_o/N))$.

Symmetry property (5DT.SP2) states that the DFT coefficients X[k] of a real sequence having length N are *conjugate symmetric*. By definition, the sequence X[0], X[1], ..., X[N − 1] is conjugate symmetric if $X[N - k] = X^*[k]$. We can prove (5DT.SP2) by specializing (5DT.SP1) to real valued sequences. For real x[n], the sequence x[n] and its conjugate $x^*[n]$ are equal. Property (5DT.SP1) then tells us that

$$\mathcal{DFT}\{x[n]\} = X^*[N - k] \qquad\qquad\qquad (66)$$

On the other hand, we also have, by definition of the DFT

$$\mathcal{DFT}\{x[n]\} = X[k] \qquad\qquad\qquad (67)$$

Property (5DT.SP2) follows directly from these two equations.

Symmetry properties (5DT.SP3) and (5DT.SP4) follow directly from (5DT.SP2).

Symmetry properties (5DT.SP5) and (5DT.SP6) apply for real even and real odd sequences, respectively. A length N sequence x[0], x[1], ..., x[N − 1], is defined as *even* if $x[N - n] = x[n], 0 \leq n \leq N - 1$. Similarly, the sequence x[0], x[1], ..., x[N − 1] is defined as *odd* if $x[N - n] = -x[n]$. Examples of even and odd length N sequences are $x[n] = \cos(2\pi(k/N)n)$ and $x[n] = \sin(2\pi((k/N)n) \ 0 \leq n \leq N - 1$, respectively. The proofs of properties (5DT.SP5) and (5DT.SP6) are straightforward and relegated to Problem 5DT.7.1.

5DT.7.2
Operational
Properties

The operational properties of the DFT involve the new mathematical concepts of *circular indexing*, *periodic extension*, *circular shift*, and *circular convolution in the sequence domain*. You need to understand these concepts before you can understand the operational properties of the DFT.

Circular Indexing

The circular index $r = (n)_N$ is defined as the remainder when n is divided by N. For example,

$$(19)_8 = 3 \qquad\qquad\qquad (68)$$

because the remainder is 3 when 19 is divided by 8. Mathematicians refer to $(n)_N$ as "n mod N" or "n modulo N."

Figure 5DT.6
Circular indexing
shown as a circular
path.

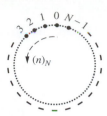

We can visualize $(n)_N$ using a circular path having N stepping-stones, as illustrated in Figure 5DT.6. The stones are numbered 0 to $N-1$. Imagine that you walk around the path starting from stone 0. As you walk, you count your steps: $n = 0, 1, 2, 3, \ldots, N-1, N, N+1, \ldots$. The circular index $(n)_N$ is the number on the stone you reach on the nth step. The number of steps you have taken to get there is the sum of an integer multiple of N plus the circular index (remainder), $(n)_N$

$$n = qN + (n)_N \tag{69}$$

where q is an integer. For example, 19 steps on an 8-step circular path brings you to stone 3, as in Eq. (68):

$$19 = 2 \times 8 + 3 \tag{70}$$

An important property of the circular index $(n)_N$ is

$$(n + kN)_N = (n)_N \quad -\infty < n < \infty \tag{71}$$

Equation (71) follows from the fact that, if you add any multiple of N to n, you do not change the remainder when you divide by N. We can use Eq. (71) to find $(n)_N$ when n is negative, for example

$$(-9)_8 = (-9 + 16)_8 = (7)_8 = 7 \tag{72}$$

Nine steps backward on an 8-step circular path take you to the same stone as seven steps forward.

Periodic Extension

If we use a circular index $(n)_N$ as the index in a length N sequence, we obtain the *periodic extension* of that sequence. The periodic extension $x[(n)_N]$ consists of periodic repetitions of $x[n]$, where the period is N. Figure 5DT.7(a) shows a length-$N = 8$ sequence $x[n]$. Figure 5DT.7(b) shows its periodic extension $x[(n)_8]$.

Figure 5DT.7(c) shows an equivalent way to view $x[(n)_N]$: Here $x[(n)_N]$ is plotted above the circular path of Figure 5DT.6.

Circular Shift

We can shift $x[(n)_N]$ on the circular path by replacing n with $n - m$ where m is the amount of shift. *Circular shift* is illustrated in Figure 5DT.8 Figure 5DT.8(a) shows an 8-point sequence plotted on a circle. Figure 5DT.8(b) shows the sequence circularly shifted by $m = 3$.

Circular Convolution

Consider two sequences, $x[n]$ and $h[n]$, both defined for $n = 0, 1, \ldots, N-1$. The *circular convolution* of $x[n]$ and $h[n]$ is defined as

Figure 5DT.7
Periodic extension of
a sequence:
(a) original sequence
x[n]; (b) periodic
extension x[(n)$_8$];
(c) x[(n)$_8$] plotted on
a circular path.

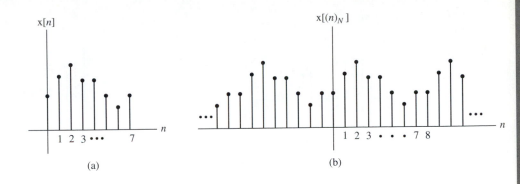

(a)   (b)

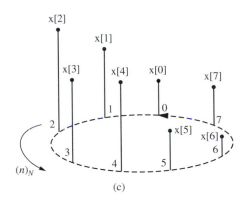

(c)

Figure 5DT.8
Circular shift of an
8-point sequence:
(a) x[(n)$_8$];
(b) x[(n − 3)$_8$].

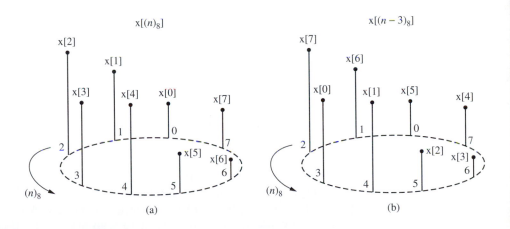

(a)   (b)

$$y[n] = x[n] \circledast h[n] = \sum_{m=0}^{N-1} x[(m)_N] h[(n-m)_N] \quad n = 0, 1, \ldots, N-1$$

$$= \sum_{m=0}^{N-1} x[m] h[(n-m)_N] \quad n = 0, 1, \ldots, N-1 \tag{73}$$

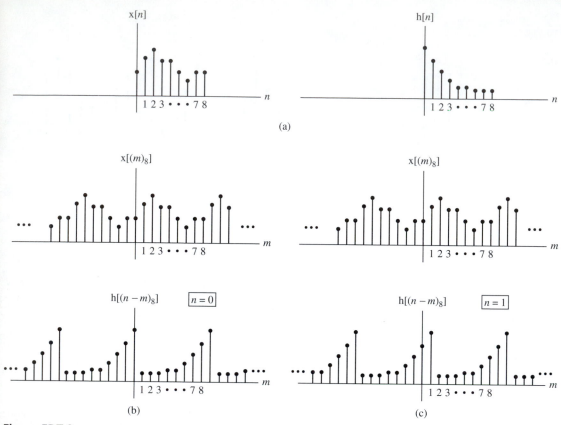

Figure 5DT.9

8-point circular convolution depicted with rectangular coordinates: (a) 8-point sequences x[n] and h[n]; (b) x[(m)₈] and h[(n − m)₈] for n = 0; (c) x[(m)₈] and h[(1 − m)₈] for n = 1.

Figures 5DT.9 and 5DT.10 show two ways to visualize circular convolution. Figure 5DT.9(a) shows two 8-point sequences, $x[n]$ and $h[n]$, plotted versus n on rectangular coordinates. Figures 5DT.9(b) and (c) show $x[(m)_8]$ and $h[(n - m)_8]$ plotted versus m on rectangular coordinates for $n = 0$ and $n = 1$, respectively. The values of $y[n]$ are obtained by multiplying values of $x[(m)_8]$ and $h[(n - m)_8]$ that are aligned vertically and summing the products for $m = 0$ to $N - 1$, as defined in Eq. (73).

In Figure 5DT.10, $x[m]$ and $h[m]$ have been recorded onto circular tapes, each having eight cells. The sequence $x[m]$ is written in a counterclockwise direction, and $h[m]$ is written in a clockwise direction.[1] The tape with the impulse response values is rotated counterclockwise as n is increased. Figures 5DT.10(a) and 5DT.10(b) illustrate the rotational positions for $n = 0$ and $n = 1$, respectively. The circular convolution $y[n] = x[n] \circledast h[n]$ is obtained by multiplying the eight data and impulse response values that are aligned radially and summing the eight products.

[1] The rotational positions of the tapes can be shown using a optional circular coordinate index $(m)_8$, as included in the figure. This coordinate index is optional because all that matters is the *relative* rotational positions of the tapes.

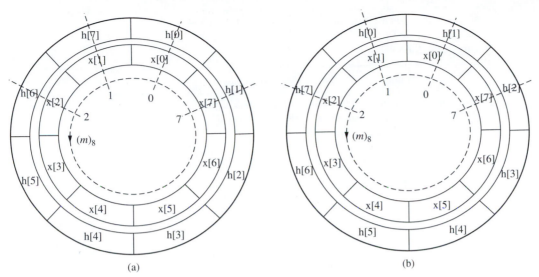

Figure 5DT.10
8-point circular convolution depicted with rotating tapes: (a) $x[(m)_8]$ and $h[(n-m)_8]$ for $n = 0$; (b) $x[(m)_8]$ and $h[(1-m)_8]$ for $n = 1$.

For example, for $n = 0$, we see from either Figure 5DT.10(b) or 5DT.10(a) that

$$y[0] = x[0]h[0] + x[1]h[7] + x[2]h[6] + x[3]h[5] + x[4]h[4] + x[5]h[3]$$
$$+ x[6]h[2] + x[7]h[1] \tag{74}$$

For $n = 1$, we see from either Figure 5DT.10(b) or 5DT.10(a) that

$$y[1] = x[0]h[1] + x[1]h[0] + x[2]h[7] + x[3]h[6] + x[4]h[5]$$
$$+ x[5]h[4] + x[6]h[3] + x[7]h[2] \tag{75}$$

The values of $y[n]$ are computed in a similar manner for each n. The following matrix equation shows the results in a compact form:

$$
\begin{bmatrix} y[0] \\ y[1] \\ y[2] \\ y[3] \\ y[4] \\ y[5] \\ y[6] \\ y[7] \end{bmatrix} =
\begin{bmatrix}
h[0] & h[7] & h[6] & h[5] & h[4] & h[3] & h[2] & h[1] \\
h[1] & h[0] & h[7] & h[6] & h[5] & h[4] & h[3] & h[2] \\
h[2] & h[1] & h[0] & h[7] & h[6] & h[5] & h[4] & h[3] \\
h[3] & h[2] & h[1] & h[0] & h[7] & h[6] & h[5] & h[4] \\
h[4] & h[3] & h[2] & h[1] & h[0] & h[7] & h[6] & h[5] \\
h[5] & h[4] & h[3] & h[2] & h[1] & h[0] & h[7] & h[6] \\
h[6] & h[5] & h[4] & h[3] & h[2] & h[1] & h[0] & h[7] \\
h[7] & h[6] & h[5] & h[4] & h[3] & h[2] & h[1] & h[0]
\end{bmatrix}
\begin{bmatrix} x[0] \\ x[1] \\ x[2] \\ x[3] \\ x[4] \\ x[5] \\ x[6] \\ x[7] \end{bmatrix} \tag{76}
$$

The entries above the principal diagonal in the preceding 8×8 matrix are present because of the circular coordinate path. These above-diagonal entries are called *wraparound* terms. The matrix itself is called a *circulant* matrix.

Table 5DT.4

DFT Operational
Properties

	$x[n]$ and $X[k]$ are DFT pairs defined for $n, k = 0, 1, \ldots, N - 1$.		
	$h[n]$ and $H[k]$ are DFT pairs defined for $n, k = 0, 1, \ldots, N - 1$.		
	Throughout this table, $n, k = 0, 1, \ldots, N - 1$.		

	Name	**Sequence**	**DFT**
5DTOP1	Linearity	$a_1 x_1[n] + a_2 x_2[n]$	$a_1 X_1[k] + a_2 X_2[k]$
5DTOP2	Circular sequence shift	$x[(n - m)_N]$	$W_N^{mk} X[k]$
5DTOP3	Circular coefficient shift	$x[n] W_N^{-mk}$	$X[(k - m)_N]$
5DTOP4	Modulation	$x[n] \cos\left(2\pi \dfrac{m}{N} n\right)$	$\frac{1}{2} X[(k - m)_N] + \frac{1}{2} X[(k - m)_N]$
5DTOP5	Circular convolution	$x[n] \circledast h[n]$	$X[k] H[k]$
5DTOP6	Multiplication	$x[n] w[n]$	$X[k] \circledast H[k]$

Now that we have described the concepts of circular indexing, periodic extensions, circular shift, and circular convolution, we can present the operational properties of the DFT. These properties are listed in Table 5DT.4.

Operational property (5DT.OP1), linearity, follows easily by definition of the DFT. We leave the proof as a problem (Problem 5DT.7.2a).

Operational property (5DT.OP2) states that a circular shift of $x[n]$ by m causes the DFT coefficients $X[k]$ to be multiplied by W^{mk}. We can prove (5DT.OP2) by substituting $(n - m)_N$ for n in the inverse DFT (5DT.24). This substitution yields

$$x[(n - m)_N] = \frac{1}{N} \sum_{k=0}^{N-1} X[k] W_N^{-(n-m)_N k} \quad n = 0, 1, \ldots, N - 1 \tag{77}$$

From Eq. (69) we know that

$$(n - m)_N = n - m - qN \tag{78}$$

for some integer q. Therefore

$$W_N^{-(n-m)_N k} = W_N^{-(n-m)k} W_N^{qNk} = W_N^{-(n-m)k} \tag{79}$$

where the last step follows from (27). The substitution of Eq. (79) into Eq. (77) yields

$$x[(n - m)_N] = \frac{1}{N} \sum_{k=0}^{N-1} X[k] W_N^{-(n-m)k} = \frac{1}{N} \sum_{k=0}^{N-1} \left(X[k] W_N^{-nk}\right) W_N^{mk} \quad n = 0, 1, \ldots, N - 1 \tag{80}$$

Therefore

$$x[(n - m)_N] \overset{\text{DFT}}{\longleftrightarrow} X[k] W_N^{mk} \tag{81}$$

which proves property (5DT.OP2).

Property (5DT.OP3), circular coefficient shift, is the dual of property (5DT.OP2). We leave the proof for a problem (Problem 5DT.2b).

Property (5DT.OP4), modulation, follows easily from property (5DT.OP3).

Property (5DT.OP5) states that the inverse DFT of the sequence, $X[k] H[k]$, $k = 1, 2, \ldots, N - 1$, equals the circular convolution of $x[n]$ and $h[n]$. To prove property

(5DT.OP5), we substitute Eq. (23) into the formula for the inverse DFT of X[k]H[k] to obtain

$$\frac{1}{N}\sum_{k=0}^{N-1}X[k]H[k]W_N^{-nk} = \frac{1}{N}\sum_{k=0}^{N-1}\left\{\sum_{m=0}^{N-1}x[m]W_N^{mk}\right\}H[k]W_N^{-nk}$$

$$= \sum_{m=0}^{N-1}x[m]\underbrace{\left\{\frac{1}{N}\sum_{k=0}^{N-1}H[k]W_N^{-(n-m)k}\right\}}_{h[(n-m)_N]} \qquad (82)$$

Consider the term in the last pair of braces in Eq. (82). Notice that m and n have values ranging from 0 to $N-1$. As a consequence, the exponential factor $(n-m)$ ranges from $-(N-1)$ to $+(N-1)$, which is outside the range 0 to N allowed for the index of $h[\cdot]$. However, the identity (5DT.79) allows us to write down the result.

$$\frac{1}{N}\sum_{k=0}^{N-1}H[k]W_N^{-(n-m)k} = \frac{1}{N}\sum_{k=0}^{N-1}H[k]W_N^{-(n-m)_N k} = h[(n-m)_N] \qquad (83)$$

as indicated by the horizontal lower brace in Eq. (82). We obtain the final result by substituting Eq. (83) back into Eq. (82).

$$\frac{1}{N}\sum_{k=0}^{N-1}X[k]H[k]e^{jk\frac{2\pi}{N}n} = \sum_{m=0}^{N-1}x[m]h[(n-m)_N] = x[n]\circledast h[n] \qquad (84)$$

Property (5DT.OP6), multiplication, is the dual of property (5DT.OP5). We leave the proof as a problem to the reader. An important application of property (5DT.OP6) is called *windowing*. Windowing is used in spectrum estimation (Section 5DT.9) and digital filter design (Chapter 7).

In addition to the symmetry and operational properties, there is a Parseval's theorem for the DFT. It states that the energy in x[n] can be found from the coefficients X[k].

$$\sum_{n=0}^{N-1}|x[n]|^2 = \frac{1}{N}\sum_{k=0}^{N-1}|X[k]|^2 \qquad (85)$$

The proof the Parseval's theorem is identical to that for the DFS.

5DT.8 Application of the DFT In Filtering

A convolution can often be evaluated much more efficiently using the DFT than by direct evaluation of the convolution sum. The increase in efficiency occurs when an FFT algorithm is used to compute the DFT. The steps involved in filtering finite-length sequences and the savings in computational time are described in the following section. Sections 5DT.8.2 and 5DT.8.3 show how to use a DFT to filter sequences that are arbitrarily long.

5DT.8.1
Computing a Linear Convolution Using the DFT

In this section we show how to use the DFT to obtain the *ordinary*, or *linear*, convolution $y[n] = x[n] * h[n]$ between two finite-length sequences $x[n]$ and $h[n]$. We assume throughout this section that $x[n]$ and $h[n]$ have lengths N_x and N_h, respectively, i.e., they equal 0 for n outside the ranges $0 \le n \le N_x - 1$ and $0 \le n \le N_h - 1$.

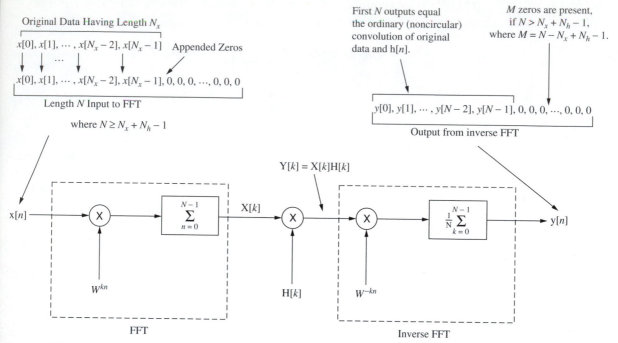

Figure 5DT.11
Computation of linear convolution by FFT.

We can perform the linear convolution using N-point FFTs, as illustrated in Figure 5DT.11.

The basic structure of the system shown is motivated by Figure 1 of Chapter 4DT, but it includes essential modifications. In Figure 5DT.11, the data fed into the FFT processor is

$$\{x[n]\} = \underbrace{x[0], x[1], \ldots, x[N_x - 1], 0, 0, \ldots, 0}_{N} \tag{86}$$

where

$$N \geq N_o \overset{\Delta}{=} N_x + N_h - 1 \tag{87}$$

In selecting the value N it is advantageous to chose the smallest power of 2 that satisfies the condition $N \geq N_o$. This choice enables us to exploit the full computational efficiency of the FFT to perform the DFT.

The N-point FFT of $x[n]$ produces coefficients $X[k], k = 0, 1, 2, \ldots, N - 1$. The filter is defined by filter coefficients $H[k], k = 0, 1, 2, \ldots, N - 1$, obtained by taking the N-point FFT of $h[n]$, where

$$\{h[n]\} = \underbrace{h[0], h[1], \ldots, h[N_h - 1], 0, 0, \ldots, 0}_{N} \tag{88}$$

Calculation of the H[k] is omitted from the figure because these coefficients can be computed "off-line" before the arrival of the data. The data coefficients X[k] and filter coefficients H[k]

Figure 5DT.12
Linear convolution by
FFT described with
matrix notation.

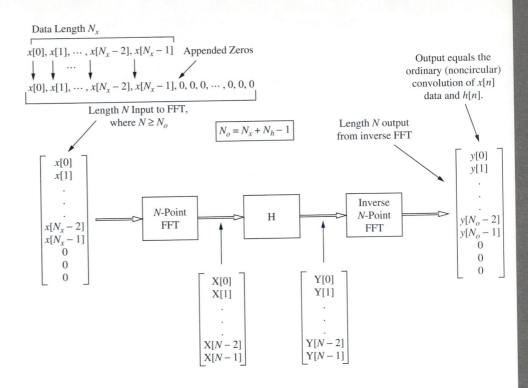

are multiplied to produce N output data coefficients, $Y[k] = X[k]H[k], k = 0, 1, \ldots, N-1$. In the final stage, an N-point inverse FFT is applied to $Y[k]$ to obtain the N-point output data sequence $y[n]$. As we see in Example 5DT.9, the appended zeros make the circular convolution described by property 5DT.OP5 equal to the linear convolution $y[n] = x[n] * h[n]$ for $n = 0, 1, 2, \ldots, N-1$.

$$y[n] = x[n] \circledast h[n] = x[n] * h[n] \quad n = 0, 1, \ldots, N-1 \tag{89}$$

Figure 5DT.12 shows another way to depict the operations of Figure 5DT.11. Here the operations are shown in matrix notation. Specifically, the FFT input is denoted by a column vector having N elements.

$$\mathbf{x} = \begin{bmatrix} x[0] \\ x[1] \\ x[2] \\ \cdot \\ \cdot \\ \cdot \\ \cdot \\ \cdot \\ \cdot \\ x[N-1] \end{bmatrix} = \begin{bmatrix} x[0] \\ x[1] \\ x[2] \\ \cdot \\ \cdot \\ x[N_x-2] \\ x[N_x-1] \\ 0 \\ 0 \\ \cdot \\ \cdot \\ 0 \end{bmatrix} \tag{90}$$

The N-point FFT of $\mathbf{x}$ yields the FFT coefficients of the data

$$\mathbf{X} = \begin{bmatrix} X[0] \\ X[1] \\ X[2] \\ \cdot \\ \cdot \\ \cdot \\ \cdot \\ \cdot \\ \cdot \\ \cdot \\ X[N-1] \end{bmatrix} \tag{91}$$

The filter is defined by an $N \times N$ diagonal matrix containing the filter coefficients.

$$\mathbf{H} = \begin{bmatrix} H[0] & 0 & 0 & \cdot & \cdot & \cdot & & 0 \\ 0 & H[1] & 0 & \cdot & \cdot & \cdot & & 0 \\ 0 & 0 & H[2] & & & & & \cdot \\ \cdot & \cdot & & \cdot & & & & \cdot \\ \cdot & \cdot & & & \cdot & & & \cdot \\ \cdot & \cdot & & & & \cdot & & 0 \\ 0 & 0 & \cdot & & \cdot & \cdot & 0 & H[N-1] \end{bmatrix} \tag{92}$$

The coefficients of the output data are given by the matrix product.

$$\begin{bmatrix} Y[0] \\ Y[1] \\ Y[2] \\ \cdot \\ \cdot \\ \cdot \\ \cdot \\ \cdot \\ \cdot \\ \cdot \\ Y[N-1] \end{bmatrix} = \begin{bmatrix} H[0] & 0 & 0 & \cdot & \cdot & \cdot & & 0 \\ 0 & H[1] & 0 & \cdot & \cdot & \cdot & & 0 \\ 0 & 0 & H[2] & & & & & \cdot \\ \cdot & \cdot & & \cdot & & & & \cdot \\ \cdot & \cdot & & & \cdot & & & \cdot \\ \cdot & \cdot & & & & \cdot & & 0 \\ 0 & 0 & \cdot & & \cdot & \cdot & 0 & H[N-1] \end{bmatrix} \begin{bmatrix} X[0] \\ X[1] \\ X[2] \\ \cdot \\ \cdot \\ \cdot \\ \cdot \\ \cdot \\ \cdot \\ \cdot \\ X[N-1] \end{bmatrix} \tag{93}$$

which can be written compactly as

$$\mathbf{Y} = \mathbf{HX} \tag{94}$$

The inverse FFT of $\mathbf{Y}$ yields the output data

$$
\mathbf{y} = \begin{bmatrix} y[0] \\ y[1] \\ y[2] \\ . \\ . \\ . \\ . \\ . \\ . \\ . \\ . \\ y[N-1] \end{bmatrix} = \begin{bmatrix} y[0] \\ y[1] \\ y[2] \\ . \\ . \\ y[N_o - 2] \\ y[N_o - 1] \\ 0 \\ . \\ . \\ . \\ 0 \end{bmatrix} \tag{95}
$$

EXAMPLE 5DT.9   Computing a Linear Convolution with a DFT

Consider data $x[n]$ with length $N_x = 5$ and a filter impulse response $h[n]$ with length $N_y = 4$. The value of $x[n]$ is nonzero only for $0 \le n \le 4$, and $h[n]$ is nonzero only for $0 \le n \le 3$. The value of N in the N-point FFTs must satisfy

$$
N \ge N_x + N_h - 1 = 5 + 4 - 1 = 8 \tag{96}
$$

We chose $N = 8$ because it is the smallest power of 2 satisfying the inequality.[2] To obtain the data coefficients $X[k]$ we apply

$$
\mathrm{x}[n] = \begin{cases} x[n] & n = 0, 1, 2, 3, 4 \\ 0 & n = 5, 6, 7 \end{cases} \tag{97}
$$

to an 8-point FFT. To obtain the filter coefficients $H[k]$, we apply

$$
\mathrm{h}[n] = \begin{cases} h[n] & n = 0, 1, 2, 3 \\ 0 & n = 4, 5, 6, 7 \end{cases} \tag{98}
$$

to an 8-point FFT. We know by (5DT.OP5) that the 8-point inverse FFT of $Y[k] = X[k]H[k]$ is the circular convolution $y[n] = x[n] \circledast h[n]$. Figure 5DT.13 uses rotating tapes to demonstrate the circular convolution. Here the $x[n]$ and $h[n]$ have been written onto circular tapes, each having a total of eight cells. Blank cells represent the appended zero values in $x[n]$, Eq. (97), and $h[n]$, Eq. (98). We can see from parts (a), (b), (c), and (d), respectively of Figure 5DT.13 that

$$
y[0] = h[0]x[0] \tag{99a}
$$

$$
y[1] = h[1]x[0] + h[0]x[1] \tag{99b}
$$

$$
y[6] = h[3]x[3] + h[2]x[4] \tag{99c}
$$

$$
y[7] = h[3]x[4] \tag{99d}
$$

Other rotations of the outer tape produce the remaining values of $y[n]$.

---

[2] The values of N_x and N_h in the example are so small that use of the FFT is unnecessary. The small values are helpful for demonstrating how a circular convolution can give the same results as a linear convolution.

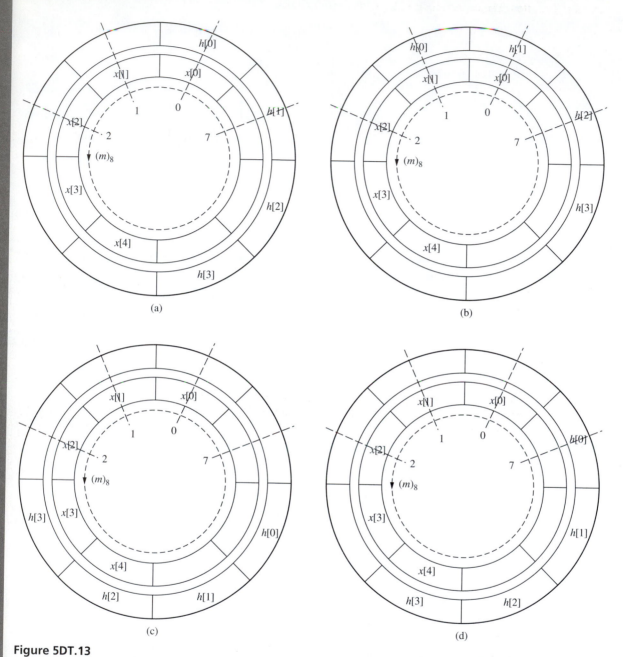

Figure 5DT.13
Circular convolution mimicing Linear Convolution: Tape positions (a), (b), (c), and (d), respectively, generate Eqs. (99a), (99b), (99c), and (99d). Wraparound terms do not occur in y[n] because zeros have been appended to x[n] and h[n].

The results can be written compactly in matrix form

$$
\begin{bmatrix} y[0] \\ y[1] \\ y[2] \\ y[3] \\ y[4] \\ y[5] \\ y[6] \\ y[7] \end{bmatrix} =
\begin{bmatrix}
h[0] & 0 & 0 & 0 & 0 & h[3] & h[2] & h[1] \\
h[1] & h[0] & 0 & 0 & 0 & 0 & h[3] & h[2] \\
h[2] & h[1] & h[0] & 0 & 0 & 0 & 0 & h[3] \\
h[3] & h[2] & h[1] & h[0] & 0 & 0 & 0 & 0 \\
0 & h[3] & h[2] & h[1] & h[0] & 0 & 0 & 0 \\
0 & 0 & h[3] & h[2] & h[1] & h[0] & 0 & 0 \\
0 & 0 & 0 & h[3] & h[2] & h[1] & h[0] & 0 \\
0 & 0 & 0 & 0 & h[3] & h[2] & h[1] & h[0]
\end{bmatrix}
\begin{bmatrix} x[0] \\ x[1] \\ x[2] \\ x[3] \\ x[4] \\ 0 \\ 0 \\ 0 \end{bmatrix}
$$

$$
=
\begin{bmatrix}
x[0]h[0] \\
x[0]h[1] + x[1]h[0] \\
x[0]h[2] + x[1]h[1] + x[2]h[0] \\
x[0]h[3] + x[1]h[2] + x[2]h[1] + x[3]h[0] \\
x[1]h[3] + x[2]h[2] + x[3]h[1] + x[4]h[0] \\
x[2]h[3] + x[3]h[2] + x[4]h[1] \\
x[3]h[3] + x[4]h[2] \\
x[4]h[3]
\end{bmatrix}
\tag{100}
$$

We can also obtain Eq. (100) as a special case of Eq. (76), where $x[n] = 0$ for $n = 5, 6, 7$, and $h[n] = 0$ for $n = 4, 5, 6, 7$. Notice that the h matrix, seen in the first equality of Eq. (100), does include wraparound terms, but, because the zeros had been appended to both the impulse response and the data, no wraparound terms appear $y[n]$. The reader is invited to show that $y[n]$, given by the last equality of Eq. (100), is identical to the linear convolution of $x[n]$ and $h[n]$ for $n = 0, 1, \ldots, 7$.

We have shown how to use the DFT to linearly convolve finite-length sequences. In this method, the DFT is applied to the entire input data sequence, and the inverse DFT yields the entire output data sequence. This *block-processing* method requires excessive computer memory when the length of the input data N_x is large. It also requires us to wait for the arrival of all input data before obtaining any output $y[n]$. In the next subsections, we describe two techniques that eliminate the excessive memory and long wait time. In these subsequent sections we assume that the nonzero values of the input data start at $n = 0$ and continue indefinitely after $n = 0$, i.e.

$$
\{x[n]\} = \ldots, 0, 0, \ldots, 0, \underset{\uparrow}{x[0]}, x[1], x[2], x[3], x[4], x[5], \ldots
\tag{101}
$$

We stated at the beginning of Section 5DT.8 that a convolution sum can often be evaluated much more efficiently using the DFT than by direct evaluation of the convolution sum. A review of the preceding paragraphs shows that the DFT method performs a convolution using 3 DFTs and N multiplications. A DFT can be computed in $N \log N$ operations (complex multiplications and additions) using an FFT when N is a power of 2. Thus, it takes $3N \log N + N$ operations to evaluate a convolution using the DFT. In contrast, direct evaluation of a convolution sum requires N^2 operations. The computational times required to evaluate a convolution using the DFT and direct methods are therefore approximately $k_1(3N \log N + N)$ and $k_2 N^2$, respectively, where the proportionality constants k_1 and k_2 depend on the software and the computer. If we compare these processing times, we see that the DFT is faster when N exceeds some minimum value N_c and that the speedup factor increases rapidly as N increases beyond N_c. The value of N_c depends on the software and the computer, but is typically a moderate value such as 28 [Stockham, T.G.Jr., 1966; Helms, H.D., 1967].

5DT.8.2 Overlap-Add Technique

In the *overlap-add* technique the data $\{x[n]\}$ is partitioned into subblocks having length N_x. The subblocks have the form

$$\{x_i[n]\} = \underbrace{\{x[iN_x], x[iN_x + 1], \ldots, x[iN_x + N_x - 1]\}}_{\text{length } N_x} \qquad (102)$$

where $i = 0, 1, \ldots$. Thus, the data in the ith subblock is given by

$$x_i[n] = x[n] \quad iN_x \leq n \leq iN_x + N_x - 1 \qquad (103)$$

for $i = 0, 1, \ldots$. This method of partitioning allows us to write (101) as

$$x[n] = \begin{cases} \sum_{i=0}^{\infty} x_i[n] & n \geq 0 \\ 0 & n < 0 \end{cases} \qquad (104)$$

and we can express the linear convolution between $x[n]$ and $h[n]$ as

$$y[n] = x[n] * h[n] = \sum_{i=0}^{\infty} x_i[n] * h[n] = \sum_{i=0}^{\infty} y_i[n] \qquad (105a)$$

where

$$y_i[n] = x_i[n] * h[n] \qquad (105b)$$

Equation (105) shows that $x[n] * h[n]$ is obtained by computing the linear convolutions between $h[n]$ and the input data subblocks $x_i[n]$ and summing the results. When we linearly convolve the length N_x data $x_i[n]$ with the length N_h impulse response $h[n]$, we obtain a length $N_o = N_x + N_h - 1$ output subblock $y_i[n]$. The ith output subblock $\{y_i[n]\}$ has the form

$$\{y_i[n]\} = \underbrace{\{y_i[iN_x], y_i[iN_x + 1], \ldots, y_i[iN_x + N_x - 1], y_i[iN_x + N_x], \ldots, y_i[iN_x + N_o - 1]\}}_{\text{length } N_o}$$

$$(106)$$

Because the length N_o of each $\{y_i[n]\}$ exceeds the length N_x of each $\{x_i[n]\}$, the sum Eq. (105a) consists of *overlapping* output data $y_i[n]$. This is the reason for the name "overlap-add" technique.

Some of the details of the overlap-add technique are best learned from an example.

EXAMPLE 5DT.10   Overlap-Add Technique

Assume that the data of Eq. (101) is partitioned into subblocks having length $N_x = 5$, and that $h[n]$ has length $N_h = 4$, i.e., $h[n]$ has 4 nonzero elements $h[0], h[1], h[2], h[3]$. The length of each output block is therefore $N_o = 5 + 4 - 1 = 8$. The computations in (105) can be visualized as follows

$$\underbrace{x[0] \;\; x[1] \;\; x[2] \;\; x[3] \;\; x[4]}_{\text{Input Block } i=0} \; \underbrace{x[5] \;\; x[6] \;\; x[7] \;\; x[8] \;\; x[9]}_{\text{Input Block } i=1} \; \underbrace{x[10] \;\; x[11] \;\; x[12] \;\; x[13] \;\; x[14] \;\; x[15]}_{\text{Input Block } i=2} \cdots$$

$$y_0[0] \; y_0[1] \; y_0[2] \; y_0[3] \; y_0[4] \; y_0[5] \; y_0[6] \; y_0[7]$$
$$y_1[5] \; y_1[6] \; y_1[7] \; y_1[8] \; y_1[9] \; y_1[10] \; y_1[11] \; y_1[12]$$
$$y_2[10] \; y_2[11] \; y_2[12] \; y_2[13] \; y_2[14] \; y_2[15] \cdots$$
$$y_3[15] \cdots$$

$$y[0] \;\; y[1] \;\; y[2] \;\; y[3] \;\; y[4] \;\; y[5] \;\; y[6] \;\; y[7] \;\; y[8] \;\; y[9] \;\; y[10] \;\; y[11] \;\; y[12] \;\; y[13] \;\; y[14] \;\; y[15] \cdots$$

The top row shows the input sequence partitioned into subblocks having length $N_x = 5$. The first three complete output data subblocks are shown in rows 2–4. Notice that each of these output subblocks is positioned to start at the beginning of each input subblock, in agreement with Eq. (106). Notice also and that the length of each output subblock is $N_o = N_x + N_h - 1 = 8$, in agreement with Eq. (106). The resulting overlap in the output subblocks is evident. The last row shows the output $y[n] = x[n] * h[n]$. Each $y[n]$ is obtained by summing the output subblock data in the rows above it. For example, $y[0] = y_0[0]$, $y[1] = y_0[1]$, $y[5] = y_0[5] + y_1[5]$, and so on.

The individual convolutions in the overlap-add method Eq. (105) can be performed with or without the use of the DFT.

**5DT.8.3
Overlap-Save
Technique**

The *overlap-save* technique is designed specifically for use with the DFT. In the overlap-save technique, the input data Eq. (101) is segmented into *overlapping* blocks $\{x_i[n]\}$, each having length N. No zeros are appended to the input data. The input data subblock length N is greater than the length N_h of $h[n]$ and is typically chosen to be a power of 2. The impulse response sequence used to obtain the filter coefficients includes appended zeros.

$$\{h[n]\} = \underbrace{h[0], h[1], \ldots, h[N_h - 1], 0, 0, \ldots, 0}_{\text{length } N} \tag{107}$$

N-point FFTs are to perform circular convolutions of the data segments and $\{h[n]\}$. The parts of the output data that correspond to linear convolutions are retained and joined to equal the linear convolution of $x[n]$ and $h[n]$. The details of the overlap-save technique are most easily understood by looking at a specific example. A general formulation of the overlap-save method appears after the example.

EXAMPLE 5DT.11   Overlap-Save Technique

This example shows how the overlap-save method applies to a filter impulse response having $N_h = 5$ nonzero elements. The first step is to chose an FFT block length $N \gg 2(N_h - 1)$. Let's chose $N = 16$ so that Eq. (107) specializes to

$$\{h[n]\} = \underbrace{h[0], h[1], \ldots, h[4], 0, 0, \ldots, 0}_{N = 16} \tag{108}$$

The input data sequence of Eq. (101) is segmented into overlapping subblocks, each having length $N = 16$. To initialize the first FFT, $N_h - 1 = 4$ zeros are inserted before $x[0]$.

$$\{x_0[n]\} = \underbrace{\overbrace{0, 0, 0, 0,}^{\substack{N_h - 1 = 4 \\ \text{initializing} \\ \text{zeros}}} x[0], x[1], \ldots, x[8], x[9], x[10], x[11]}_{\text{FFT input block length } N = 16} \tag{109}$$

The next two N-point input blocks to the FFT have the form

$$\{x_1[n]\} = \underbrace{\overbrace{x[8], x[9], x[10], x[11],}^{\substack{\text{Last } N_h - 1 = \\ \text{4 terms from} \\ \text{previous block}}} x[12], x[13], \ldots, x[20], x[21], x[22], x[23]}_{\text{FFT input block length } N = 16} \tag{110}$$

$$\{x_2[n]\} = \underbrace{\overbrace{x[20], x[21], x[22], x[23],}^{\substack{\text{Last } N_h - 1 = \\ \text{4 terms from} \\ \text{previous block}}} x[24], x[25], \ldots, x[34], x[33], x[34], x[35]}_{\text{FFT input block length } N = 16} \tag{111}$$

Note that the initial $N_h - 1$ elements of each input block equal the last $N_h - 1$ from the previous block. These are the overlap elements. High-speed convolution by the FFT produces circular convolutions between the data subblocks and the impulse response block Eq. (108) to yield output data subblocks. Figure 5DT.14 illustrates the circular convolution of $\{x_0[n]\}$ and $\{h[n]\}$. Blank cells represent zero values.

The alignment of the tapes for obtaining the first output element is shown in part (a) of the figure. Notice that this alignment contains wraparound errors, which continue as the outer tape rotates counterclockwise for the second, third, and $N_h - 1$ (fourth) output. Part (b) shows the alignment for the fourth output. The alignment that produces the first error-free output occurs for the N_h^{th} output, as shown in part (c). Additional rotations produce error-free outputs that correspond to linear convolutions. Part (d) shows the tape alignment for the Nth output. In summary, the circular convolution of $\{x_0[n]\}$ and $\{h[n]\}$ yields

$$\{y_0[n]\} = \underbrace{\overbrace{\times, \times, \times, \times,}^{\substack{\text{Discard initial} \\ N_h - 1 = 4 \\ \text{computed values}}} y[0], y[1], \ldots, y[8], y[9], y[10], y[11]}_{\text{FFT output block length } N = 16} \tag{112}$$

Figure 5DT.15 shows the circular convolution of $\{x_1[n]\}$ and $\{h[n]\}$. An examination of this figure shows that $\{y_1[n]\}$ has the form

$$\{y_1[n]\} = \underbrace{\overbrace{\times, \times, \times, \times,}^{\substack{\text{Discard initial} \\ N_h - 1 = 4 \\ \text{computed values}}} y[12], y[13], \ldots, y[20], y[21], y[22], y[23]}_{\text{FFT output block length } N = 16} \tag{113}$$

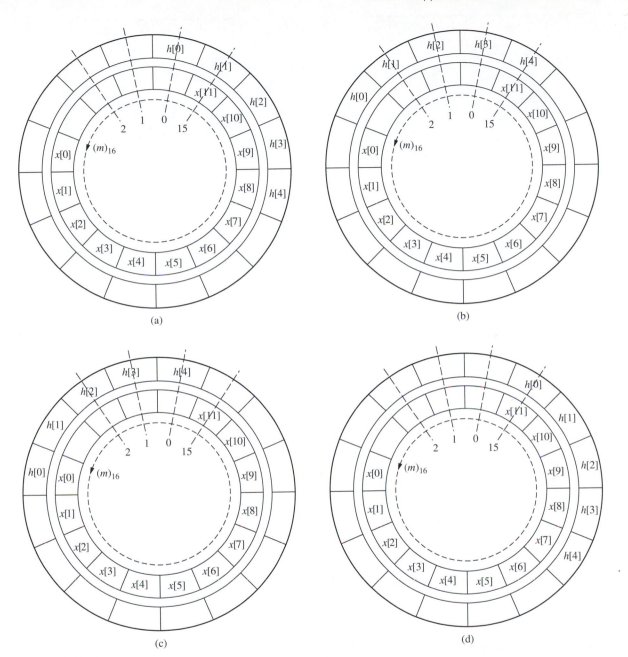

Figure 5DT.14
Circular convolution of $\{x_0[n]\}$ and $\{h[n]\}$: Tape positions (a), (b), (c), and (d), respectively, generate elements 1, 4, 5, and 16 in $\{y_0[n]\}$.

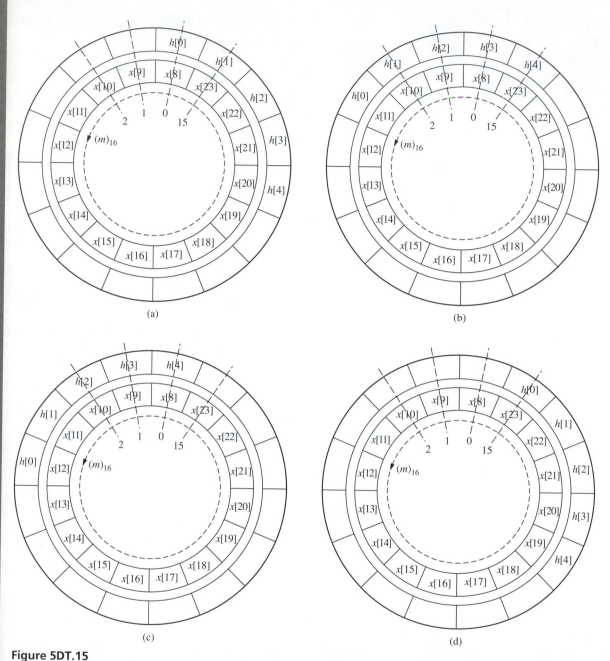

Figure 5DT.15

Circular convolution of $\{x_1[n]\}$ and $\{h[n]\}$: Tape positions (a), (b), (c), and (d), respectively, generate elements 1, 4, 5, and 16 in $\{y_1[n]\}$.

Again, it can be seen from the figure that the first $N_h - 1$ outputs are contaminated by wraparound terms. These are discarded. The remaining $N - N_h + 1$ outputs correspond to linear convolutions. We can obtain the first 24 output terms $y[0], y[1], \ldots, y[22], y[23]$ by joining the error-free elements from $\{y_0[n]\}$ and $\{y_1[n]\}$.

The next $N - N_h + 1$ outputs corresponding to linear convolution are obtained similarly, from the circular convolution of

$$\{y_2[n]\} = \overbrace{\times, \times, \times, \times,}^{\substack{\text{Discard initial} \\ N_h - 1 = 4 \\ \text{computed values}}} \underbrace{y[24], y[25], \ldots, y[32], y[33], y[34], y[35]}_{\text{FFT output block length } N = 16} \tag{114}$$

and $\{h[n]\}$, and the process continues.

If you understand this example, you are ready to proceed to the general formulation that follows.

In general, the impulse response sequence applied to the FFT has the form

$$\{h[n]\} = \underbrace{h[0], h[1], \ldots, h[N_h - 1], 0, 0, \ldots, 0}_{\text{Length } N} \tag{115}$$

The first data vector applied to the FFT has the form

$$\{x_0[n]; 0 \le n \le N - 1\} =$$

$$\underbrace{\left\{ \overbrace{0, \ldots, 0, 0,}^{N_h - 1} x[0], x[1], \ldots, x[N - 2N_h + 2], \ldots, 0, x[N - N_h] \right\}}_{\text{Length } N} \tag{116}$$

The next data vector applied to the FFT has the form

$$\{x_1[n]; 0 \le n \le N - 1\} =$$

$$\underbrace{\left\{ \overbrace{x[N - 2N_h + 2], \ldots, 0, x[N - N_h]}^{N_h - 1}, x[N - N_h + 1], \ldots, x[2N - 2N_h + 1] \right\}}_{\text{Length } N} \tag{117}$$

for $i = 1, 2, \ldots$. Notice that the first $N_h - 1$ terms in $\{x_1[n]\}$ are the last $N_h - 1$ terms from $\{x_0[n]\}$. Subsequent data input vectors to the FFT are constructed similarly. The first $N_h - 1$ terms in $\{x_i[n]\}$ are the last $N_h - 1$ terms from $\{x_{i-1}[n]\}$. For example, the subblock $\{x_2[n]\}$ has the form

$$\{x_2[n]; 0 \le n \le N - 1\} =$$

$$\underbrace{\left\{ \overbrace{x[2N - 3N_h + 3], \ldots, x[2N - 2N_h + 1]}^{N_h - 1}, x[2N - 2N_h + 2], \ldots, x[3N - 3N_h + 2] \right\}}_{\text{Length } N}$$

$$\tag{118}$$

and, in general,

$$\{x_i[n]; 0 \le n \le N - 1\} =$$

$$\left\{ \overbrace{x[iN - (i + 1)(N_h - 1)], \ldots, x[iN - iN_h + 1]}^{N_h - 1}, x[iN - iN_h + i], \ldots, x[(i + 1)(N - N_h) + i] \right\}$$

$$\underbrace{\qquad\qquad\qquad\qquad\qquad\qquad\qquad\qquad\qquad\qquad\qquad\qquad\qquad\qquad\qquad\qquad}_{\text{Length } N}$$

(119)

for $i = 1, 2, \ldots$ We circularly convolve each $\{x_i[n]\}$ with $\{h[n]\}$ and retain only the part of each circular convolution that equals a linear convolution. The first $N_h - 1$ output values from the circular convolution include wraparound terms and are therefore discarded. The remaining $N - N_h + 1$ output values agree with a linear convolution.

5DT.9   Spectrum Estimation

A primary application of the DFT is to estimate the spectrum $X(e^{j2\pi f})$ of a sequence $x[n]$. In this section we show how to use windowing, zero padding, and circular shift to prepare data for the FFT. We describe the important concepts of *truncation error*, *windowing*, *leakage*, and *frequency resolution*, and we review the picket fence effect.[3]

Perhaps the best way to begin is with an example.

EXAMPLE 5DT.12   Spectrum Estimation of the Sequence $a^{|n|}$

Consider the sequence

$$x[n] = a^{|n|} \quad |a| < 1 \tag{120}$$

shown in Figure 5DT.16 for $a = 0.7$.

Of course, we can obtain the true spectrum (DTFT) of $a^{|n|}$ analytically. The true spectrum is

$$X(e^{j2\pi f}) = \frac{1 - a^2}{1 + a^2 - 2a \cos(2\pi f)} \tag{121}$$

(See Example 4DT.8.) For the purposes of illustration, we assume that we do not know $X(e^{j2\pi f})$ and want to estimate it using an FFT.

Step 1

Recall that the FFT works with data in the range $0 \le n \le N - 1$, where N is finite. Therefore, our first step is to chose a finite range for n that contains most or all of the significant values of $x[n]$. In this example, we are working with an

---

[3] Higher-level texts describe an additional concept called *statistical convergence*. This concept is needed to understand spectrum estimates of random sequences.

Figure 5DT.16
Sequence $a^{|n|}$, where
$a = 0.7$.

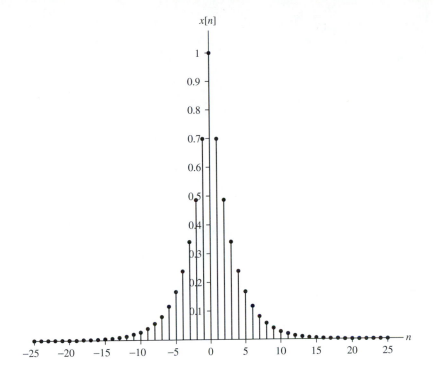

even sequence whose values decrease as $|n|$ increases. To include the most significant values of $x[n]$ in the FFT, we chose n in range $-K \leq n \leq K$, where K is some constant. Thus, the selected data is given by the

$$N = 2K + 1 \tag{122}$$

nonzero values of the truncated sequence

$$x_w[n] = x[n] \, \sqcap (n; K) = \begin{cases} x[n] & |n| \leq K \\ 0 & |n| > K \end{cases} \tag{123}$$

shown in Figure 5DT.17.

Our intuition tells us that by choosing K large enough, the spectrum of $x_w[n]$ should be a good approximation to that of $x[n]$.[4]

Step 2

The second step is to shift $x_w[n]$ K points to the right, $x_w[n] \to x_w[n-K]$, so that its nonzero values lie in the FFT index range $0 \leq n \leq N - 1$. This step is illustrated in Figure 5DT.18.

---

[4]We can appeal to more than intuition: The sequence $x[n]$ of (120) is absolutely summable. We therefore know from Section 4DT.9 that the spectrum $X_K(e^{j2\pi f})$ of the truncated sequence $x_w[n]$ converges uniformly to $X(e^{j2\pi f})$.

Figure 5DT.17
Sequence $a^{|n|}$
$\sqcap(n; K)$, where
$a = 0.7$ and $K = 8$.

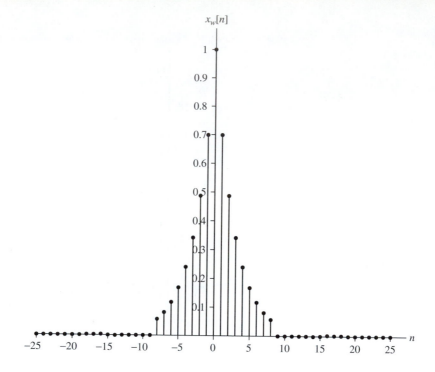

Figure 5DT.18
Shifted sequence
$x_w[n - K]$ with
$K = 8$.

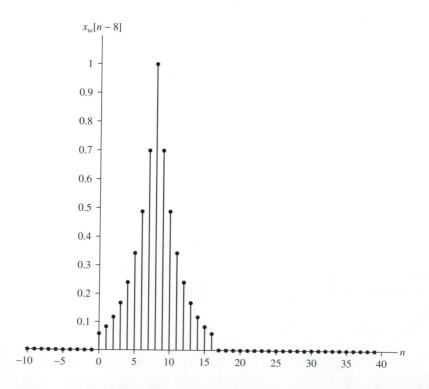

Figure 5DT.19
Length $N' = 64$
sequence x'[n].

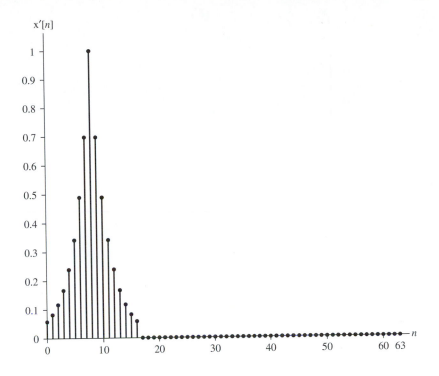

Step 3

The third step is zero padding. It is usually a good idea to append zeros to $x_w[n - K]$ to reduce the picket fence degradation, as described in Section 5DT.6. The length $N' = N + N_0$ zero-padded sequence has the form:

$$x'[n] = \left\{ \begin{array}{c} \underbrace{x_w[-K], x_w[-K+1], \ldots, x_w[K+1]}, \underbrace{0, 0, \ldots, 0}_{\substack{N_0 \\ \text{appended} \\ \text{zeros}}} \\ \begin{array}{cc} \uparrow & \uparrow \\ n = 0 & n = N - 1 \end{array} \end{array} \right\} \qquad (124)$$

$$\underbrace{\phantom{x'[n] = \left\{ x_w[-K], x_w[-K+1], \ldots, x_w[K+1], 0, 0, \ldots, 0 \right\}}}_{\text{Length } N' = N + N_0}$$

x'[n] is illustrated in Figure 5DT.19, where $N' = 64$.

In addition to the recommended zero padding of Step 3, a fourth step is also recommended: We know from the shift property of the DTFT that, when we shifted the data $x_w[n] \rightarrow x_w[n - K]$ in Step 2, we changed the phase structure of the spectrum. Specifically, we multiplied the DTFT $X_w(e^{j2\pi f})$ of $x_w[n]$ by $e^{-2\pi f K}$. The absolute value of $e^{-2\pi f K}$ is unity; so the data shift does not change the absolute value of the spectrum $X_w(e^{j2\pi f})$. On the other hand, the phase of $e^{-2\pi f K}$ is $\angle e^{-2\pi f K} = -2\pi f K$, and this linear phase adds to the phase of $X_w(e^{j2\pi f})$ as a result of the data shift. If we do nothing, the phase factor $e^{-2\pi f K}$ shows up in samples of $X_w(e^{j2\pi f})e^{-2\pi f K}$ that are computed by the DFT. According to the circular shift property of the DFT, we can correct for this phase factor by circularly shifting x'[n] leftward by K, as shown in Step 4.

Step 4

The fourth step is to circularly left-shift x'[n] by K, $x'[n] \longrightarrow x'[(n + K)_{N'}]$, to correct for the right shift performed in Step 2.

Figure 5DT.20
The Sequence x″[n].

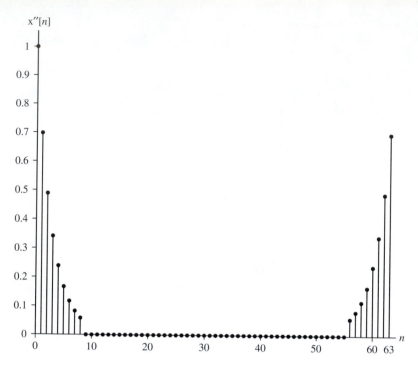

Figure 5DT.21
FFT Coefficients X″[k] and DTFT of $a^{|n|}$.

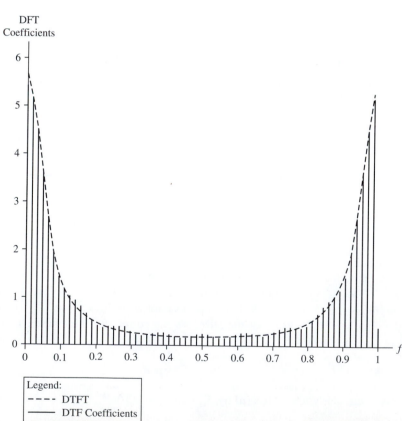

Legend:
- - - - DTFT
——— DTF Coefficients

Step 4 produces the FFT input sequence

$$x''[n] = \left\{ \underbrace{x_w[0], x_w[1], \ldots, x_w[K], \underbrace{0, 0, \ldots, 0}_{\substack{N_0 \\ \text{padded} \\ \text{zeros}}}, x_w[-K], x_w[-K+1], \ldots, x_w[-1]}_{N' = N + N_0} \right\} \tag{125}$$

where the arrows mark $n=0$ below $x_w[0]$ and $n=K$ below $x_w[K]$.

$x''[n]$ is illustrated in Figure 5DT.20.

The FFT output for the sequence $x''[n]$ of Figure 5DT.20 is shown in Figure 5DT.21. The dotted line is the DTFT (the true spectrum) of $a^{|n|}$ derived in Example 4DT.8. We see that the FFT is a good approximation to the DTFT. The discrepancy is caused by the truncation introduced in Step 1 where $K = 8$. Further experiments reveal that our intuition described at the end of Step 2 was correct: The approximation does indeed improve as K is increased.

The following example introduces spectrum *resolution* and *leakage*.

EXAMPLE 5DT.13   Spectrum Resolution and Leakage

Consider the sequence

$$x[n] = \cos(2\pi f_a n) + \cos(2\pi f_b n) \tag{126}$$

where $f_a = 0.1$ and $f_b = 0.2$. We know from previous work that, for $-0.5 \le f < 0.5$

$$X(e^{j2\pi f}) = \tfrac{1}{2}\delta(f - f_a) + \tfrac{1}{2}\delta(f + f_a) + \tfrac{1}{2}\delta(f - f_b) + \tfrac{1}{2}\delta(f + f_b) \tag{127}$$

and $X(e^{j2\pi f})$ repeats periodically with period 1. Figure 5DT.22 shows $X(e^{j2\pi f})$ plotted in the range $0 \le f < 1$. Assume that we did not know this analytical result and decide to estimate $X(e^{j2\pi f})$ using Steps 1 to 4 of Example 5DT.12. These steps include truncating $x[n]$, right-shifting the truncated sequence by K, preparing the FFT input by appending zeros, and performing a left circular shift by K. The FFT result is shown in Figure 5DT.23 for $K = 25$ (data length $N = 51$) and $N' = 256$. The FFT plot is clearly a degraded version of the true spectrum. It indicates *bands* of large sinusoidal components, not *single* sinusoidal components, in the vicinity of $f = 0.1, 0.2, 0.8$ (alias of -0.2) and 0.9 (alias of -0.1). This degradation is called *degradation in frequency resolution*. The FFT plot also

Figure 5DT.22
DTFT of $\cos(2\pi f_a n) +$ $\cos(2\pi f_b n)$ plotted for $0 \le f < 1$.

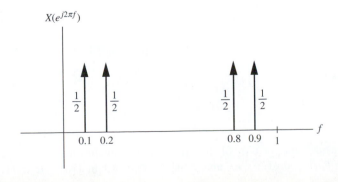

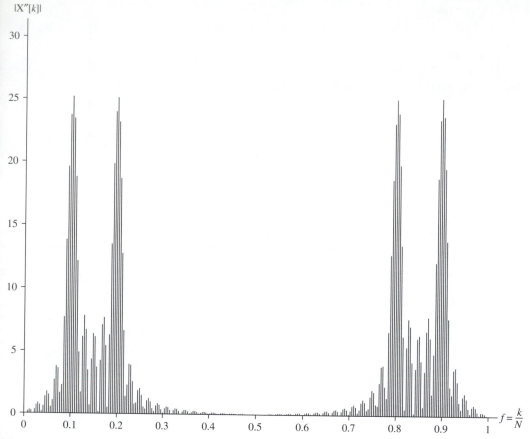

Figure 5DT.23
Absolute values of FFT coefficients.

indicates the existence of smaller sinusoidal components at values of f substantially different from $f = 0.1, 0.2, 0.8,$ and 0.9. This degradation is called *spectrum leakage*.

The frequency resolution and spectrum leakage degradation that occurred in the preceding example are consequences of limiting the range of data processed to $-K \leq n \leq K$. In Examples 5DT.12 and 5DT.13, the data range was limited by truncation, a process that is equivalent to multiplying $x[n]$ by $\sqcap[n; K]$. More generally, we can multiply $x[n]$ by any sequence $w[n]$ that equals 0 for $|n| > K$.

$$x_w[n] = w[n]x[n] \tag{128}$$

The DTFT of $x_w[n]$ is

$$X_w(e^{j2\pi f}) = \int_{-0.5}^{0.5} W(e^{j2\pi(f-\lambda)})X(e^{j2\pi\lambda})\,d\lambda \tag{129}$$

Consider the results of the preceding convolution for the sequences $x[n] = \cos(2\pi f_a n) + \cos(2\pi f_b n)$ and $w[n] = \sqcap[n; K]$ in Example 5DT.13. If we substitute of

Figure 5DT.24

Terms in Eq. (130): $f_a = 0.1$; $f_b = 0.2$ and $K = 25$.

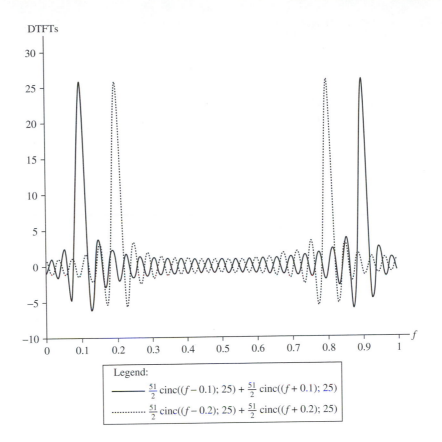

DTFTs

Legend:
—— $\frac{51}{2}$ cinc$((f - 0.1); 25) + \frac{51}{2}$ cinc$((f + 0.1); 25)$
········· $\frac{51}{2}$ cinc$((f - 0.2); 25) + \frac{51}{2}$ cinc$((f + 0.2); 25)$

$W(e^{j2\pi f}) = (2K + 1)$ cinc$(f; K)$ and $X(e^{j2\pi\lambda})$ of Eq. (127) into Eq. (129) we find

$$X_w(e^{j2\pi f}) = \tfrac{1}{2}(2K + 1) \text{ cinc}((f - f_a); K) + \tfrac{1}{2}(2K + 1) \text{ cinc}((f + f_a); K)$$
$$+ \tfrac{1}{2}(2K + 1) \text{ cinc}((f - f_b); K) + \tfrac{1}{2}(2K + 1) \text{ cinc}((f + f_b); K) \quad (130)$$

Figure 5DT.24 shows plots of the first and second pairs of terms from the right-hand side of Eq. (130) with $f_a = 0.1$ and $f_b = 0.2$ and $K = 25$, as in Example 5DT.13. This figure shows how the major lobes and oscillating tails of the cinc functions interfere constructively and destructively to produce the rather complicated-looking function $\left|X_w(e^{j2\pi f})\right|$ shown in Figure 5DT.25. We know from Section 5DT.6 that the absolute values of the DFT coefficients that were plotted in Figure 5DT.23 consist of samples of $\left|X_w(e^{j2\pi f})\right|$. Figure 5DT.24 shows that the loss in resolution and leakage degradation is caused by the major lobes and tails of the cinc functions in Eq. (130). Thus, the loss in resolution and the leakage degradation are ultimately caused by data truncation.

Choices for $w[n]$ other than $\sqcap(n; K)$ can produce better results: For example, if we use a raised cosine sequence (called a Hann or Hanning window)

$$w[n] = \frac{1}{2}\left\{1 + \cos\left(\frac{\pi n}{K + 1}\right)\right\} \sqcap (n; K) \equiv w_{\text{H}}[n]$$

in Eq. (128), we obtain the amplitude spectrum $|X_w(e^{j2\pi f})|$ and DFT coefficient plots of Figure 5DT.26(a) and (b).

Figure 5DT.25
Amplitude spectrum
$|X_w(e^{j2\pi f})|$.

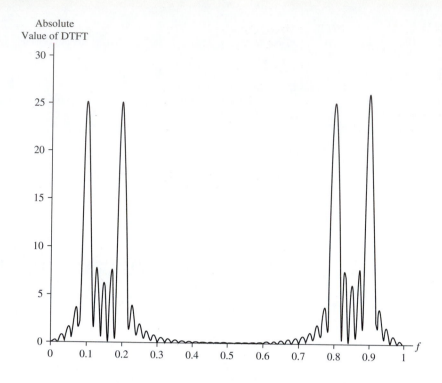

Figure 5DT.26
Absolute values of
DTFT and DFT
coefficients:
(a) $|X_w(e^{j2\pi f})|$;
(b) $|X[k]|$.

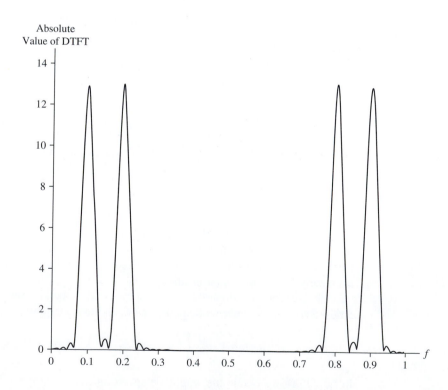

Absolute Values of
FFT Coefficients

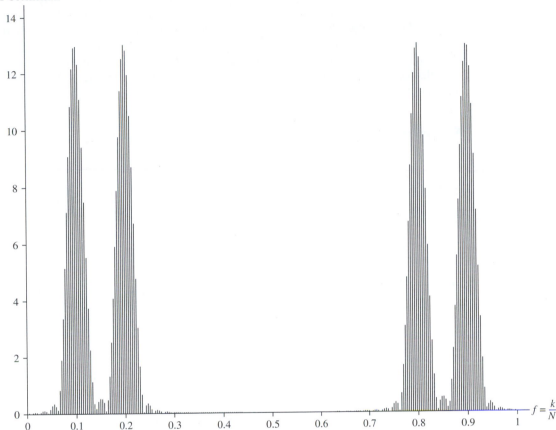

Figure 5DT.26
(*Cont.*).

Notice that the spectrum leakage in these plots is substantially less than that for the rectangular window. The frequency resolution (measured by the width of the main lobes), however, is better using the rectangular window than using the Hann window. These properties can be attributed to the reduced side lobes and wider main lobe in the spectrum of the raised cosine window compared to the rectangular window.

Descriptions of a large variety of alternative windows are contained in the MATLAB help files.

5DT.10   Summary

This chapter focused on the DFS and the DFT. These two transforms were considered together because the equations that define them are identical except for the range of their arguments. Both the DFT and the DFS, with their inverses, can be computed efficiently using an FFT.

The DFS is the premier analytical tool for analyzing periodic sequences and the responses they produce when applied to LTI systems: In the DFS, $x[n]$ and $X[k]$ are both periodic with period N, and k and n are defined for $-\infty < n, k < \infty$. The DFS is special case of the multiple sinusoid representation of sequences introduced in Chapter 3DT. All the concepts involved with the multiple sinusoid representation extend directly to the DFS.

The DFT is arguably the premier computational tool in the field of spectrum analysis and signal processing. In the DFT, $x[n]$ and $X[k]$ are defined only for $0 \le n, k \le N - 1$. Because of the limited range of the arguments of the DFT, new mathematical concepts (circular indexing, periodic extension, circular shift, and circular convolution in the sequence domain) were needed to describe the properties of the DFT.

Summary of Definitions and Formulas

DFS

DFS

$$x[n] = \frac{1}{N} \sum_{k=0}^{N-1} X[k] e^{j2\pi \frac{k}{N} n} \quad \forall n$$

where $x[n + N] = x[n]$.

Inverse DFS

$$X[k] = \sum_{n=0}^{N-1} x[n] e^{-j2\pi \frac{k}{N} n} \quad \forall k$$

where $X[k + N] = X[k]$.

Response of Stable LTI System To Periodic Input

$$x[n] = \frac{1}{N} \sum_{k=0}^{N-1} X[k] e^{j2\pi \frac{k}{N} n} \xrightarrow{h[n]} y[n] = \frac{1}{N} \sum_{k=0}^{N-1} X[k] H\left(e^{j2\pi \frac{k}{N}}\right) e^{j2\pi \frac{k}{N} n}$$

Relationship to DTFT

Let $x[n]$ be a periodic sequence with period N.
Define the pulse

$$p[n] = \begin{cases} x[n] & n = 0, 1, \ldots, N - 1 \\ 0 & \text{otherwise} \end{cases}$$

The DFS coefficients, $X[k]$, are equally spaced sample values of $P(e^{j2\pi f}) = \mathcal{DTFT}\{p[n]\}$.

$$X[k] = P\left(e^{j2\pi \frac{k}{N}}\right)$$

Periodic Convolution

Periodic convolution is defined as

$$x[n] \overset{N}{*} h[n] = \sum_{m=0}^{N-1} x[m] h[n - m]$$

where $x[n]$ and $h[n]$ have period N.

The periodic sequence $x[n] \overset{N}{*} h[n]$ has DFS coefficients $X[k]H[k]$.
See Tables 1 and 2.

Parseval's Theorem for Periodic Sequences

$$P = \sum_{n=0}^{N-1} |x[n]|^2 = \frac{1}{N} \sum_{k=0}^{N-1} |X[k]|^2$$

DFT

DFT

$$X[k] = \sum_{n=0}^{N-1} x[n] W_N^{kn} \quad k = 0, 1, \ldots, N-1$$

where W_N is the complex constant.

$$W_N = e^{-j\frac{2\pi}{N}}$$

Inverse DFT

$$x[n] = \frac{1}{N} \sum_{k=0}^{N-1} X[k] W_N^{-kn} \quad n = 0, 1, \ldots, N-1$$

Properties of W_N

$$W_N^* = W_N^{-1}$$

$$W_N^{mN} = 1$$

where m is any integer.

$$\sum_{n=0}^{N-1} W_N^{kn} = N \Delta_N[k]$$

where

$$\Delta[n; N] \triangleq \begin{cases} 1 & \text{when } k \text{ is an integer multiple of } N \\ 0 & \text{otherwise} \end{cases}$$

Relationship of DFT to DTFT

$$x[n] = \ldots 0, 0, 0, \underset{\uparrow}{x[0]}, x[1], \ldots, \underbrace{x[N-1], 0, 0, 0, \ldots}_{x[0], x[1], \ldots, x[N-1]}$$

$$X[k] = \left. \sum_{n=0}^{N-1} x[n] e^{-j2\pi fn} \right|_{f=\frac{k}{N}} = X\left(e^{j2\pi \frac{k}{N}}\right) \quad k = 0, 1, \ldots, N-1$$

The above equation shows that the DFT coefficients equal samples of the DTFT where the sample spacing is $\Delta f = \frac{1}{N}$.

Picket Fence Effect

The picket fence effect refers to the spacing Δf between the DTFT samples. We can decrease Δf by appending zeros to a sequence.

$$\{x'[n]\} = \underbrace{x[0], x[1], \ldots, x[N-1], \underbrace{0, 0, \ldots, 0}_{N_0}}_{N'}$$

$$X'[k] = \sum_{n=0}^{N'-1} x'[n]e^{-j2\pi fn}\bigg|_{f=\frac{k}{N'}} = X(e^{j2\pi \frac{k}{N'}}) \quad k = 0, 1, \ldots, N'-1$$

Here the spacing between samples is $\Delta f' = \frac{1}{N+N_0}$

Circular Convolution

$$y[n] = x[n] \circledast h[n] = \sum_{m=0}^{N-1} x[(m)_N]h[(n-m)_N] \quad n = 0, 1, \ldots, N-1$$

where the circular index $(n)_N$ is the remainder when n is divided by N.

Circular Convolution Property

If you multiply DFT coefficients and inverse DFT the product, you get the *circular* convolution of the corresponding sequences.

$$x[n] \circledast h[n] \overset{\text{DFT}}{\longleftrightarrow} X[k]H[k]$$

Parseval's Theorem

$$\sum_{n=0}^{N-1} |x[n]|^2 = \frac{1}{N} \sum_{k=0}^{N-1} |X[k]|^2$$

A Circular Convolution Mimicing a Linear Convolution

This property makes it possible to exploit the efficiency and the FFT to compute linear convolutions.

Append zeros as follows:

$$\{x[n]\} = \underbrace{x[0], x[1], \ldots, x[N_x - 1], 0, 0, \ldots, 0}_{\text{Length } N}$$

$$\{h[n]\} = \underbrace{h[0], h[1], \ldots, h[N_h - 1], 0, 0, \ldots, 0}_{\text{Length } N}$$

where

$$N \geq N_o \overset{\Delta}{=} N_x + N_h - 1$$

In selecting the value N, it is advantageous to chose the smallest power of 2 that satisfies the condition $N \geq N_o$. This choice enables us to exploit the full efficiency of the FFT to compute $X[k]$ and $H[k]$; $k = 0, 1, 2, \ldots, N-1$.

$\mathcal{IDFT}\{X[k]H[k]\} = x[n] \circledast h[n]$ is identical to the linear convolution $x[n] * h[n]$ for $0 \leq n \leq N-1$.

Overlap-Add and Overlap-Save Methods

These methods are used to convolve arbitrarily long data sequences with impulse responses that have finite length. See the text for details.

Spectrum Estimation: Picket Fence Effect

See preceding "Pickict Fence Effect."

Spectrum Estimation: Leakage

The term "leakage" refers to the tails in a spectrum estimate that are caused by data truncation. Leakage can be reduced somewhat by choosing a window other than a rectangular window.

Spectrum Estimation: Frequency Resolution

The term "frequency resolution" describes the ability to discern sinusoids that are closely spaced in frequency. Data truncation degrades frequency resolution. Alternative windows allow you to trade reduction of leakage with reduction of frequency resolution.

5DT.11 Problems

5DT.2 The DFS

5DT.2.1 Assume that the band-pass filter $H(e^{j2\pi f})$ shown in Figure 5DT.2 of Example 5DT.1 is replaced with an ideal low-pass filter having a cutoff frequency $w = 4/14$.
a) Sketch the frequency response characteristic of the ideal low-pass filter.
b) Write down the output $y[n]$. Express your answer in terms of real sequences.

5DT.2.2 A sequence $x[n] = A\cos(2\pi f_o n)$ where $0 \le f_o < \frac{1}{2}$ is applied to a system having frequency response characteristic

$$H(e^{j2\pi f}) = j2\pi F_s f \quad -\frac{1}{2} < f < \frac{1}{2} \tag{131}$$

a) Find the output $y[n]$. Express your answer as a real sequence.
b) Assume that $f_o = 3/14$. Find the DFS of $x[n]$ and $y[n]$.

The DFS coefficients can be efficiently computed using the MATLAB function `fft(x)`. Similarly, the DFS can be computed using the MATLAB function `ifft(x)`. Read the MATLAB help files for `fft(x)` and its inverse `ifft(X)`. Notice from the MATLAB help files that the sequence (data vector) x and the coefficients (coefficient vector) X are indexed from 1 to N. This is because MATLAB does not accept 0 for the argument of a vector.

5DT.2.3 Use the MATLAB `fft` and `ifft` functions to obtain the results of Example 5DT.1.

5DT.2.4 Use the MATLAB `fft` and `ifft` functions to compute output $y[n]$ in Problem 5DT.2.1 Plot one period of $y[n]$.

5DT.3 Properties of the DFS

5DT.3.1 Plot an example of a real even periodic sequence other than $\cos(2\pi(k/N)n)$.

5DT.3.2 Plot an example of a real odd periodic sequence other than $\sin[2\pi(k/N)n]$.

5DT.3.3 Demonstate (5DT.SP3) for $N = 3$ and $N = 4$.

5DT.3.4 Demonstate (5DT.SP4) for $N = 3$ and $N = 4$.

5DT.3.5 Derive operational property (5DT.OP5).

5DT.3.6 Derive Parseval's theorem for periodic sequences directly from Eqs. (3) and (4).

5DT.4   The DFT

5DT.4.1 State the meaning of the symbols $X[k]$, $\mathsf{X}[k]$, $x[n]$, $\mathsf{x}[n]$, and $X(e^{j2\pi f})$ in your own words.

5DT.4.2 Show that $W_N^* = W_N^{-1}$.

5DT.4.3 Show that $W_N^{mN} = 1$, where m is any integer.

5DT.4.4 Confirm that $\sum_{n=0}^{N-1} W_N^{kn} = N\Delta_N[k]$ works for $N = 4$ by writing down the four terms in the summation $\sum_{n=0}^{N-1} W_N^{kn}$. Simplify the terms and add them. Start with $k = 0$ and then continue for $k = 1, 2,$ and 3.

The DFT can be efficiently computed using the MATLAB function `fft(x)`. Similarly, the inverse DFT can be computed using the MATLAB function `ifft(x)`. Read the MATLAB description of `fft(x)` and its inverse `ifft(X)`. Notice from the MATLAB help files that the sequence (data vector) x and the coefficients (coefficient vector) X are be indexed from 1 to N MATLAB because MATLAB does not accept 0 for the argument of a vector. Use the `fft` and `ifft` functions to solve the following problems.

5DT.4.5 Use MATLAB to confirm that $\sum_{n=0}^{N-1} W_N^{kn} = N\Delta_N[k]$ for $N = 256$.

5DT.4.6 Use MATLAB to confirm the results of Example 5DT.2 for $N = 256$.

5DT.4.7 Use MATLAB to confirm the results of Example 5DT.3 for $N = 256$.

5DT.4.8 Use MATLAB to confirm the results of Example 5DT.4 for $k_o = 30$ and $N = 256$.

5DT.4.9 Use MATLAB to confirm the results of Example 5DT.5 for $k_o = 30$ and $N = 256$.

5DT.5   Relationship of the DFT to the DTFT

5DT.5.1 Use MATLAB to confirm the plot of Figure 5DT.3.

5DT.5.2 Use MATLAB to confirm the plot of Figure 5DT.4.

5DT.6   The Picket Fence Effect

5DT.6.1 Use the MATLAB function `fft(X)` to confirm the DFT plot of Figures 5DT.5(a)–(c) of Example 5DT.8.

5DT.6.2 Write a MATLAB program to plot the DFT and DTFT for Example 5DT.7 with $f_o = 5.5/23$ instead of $f_o = 5/23$, as in the example. Comment on your results.

5DT.7   Properties of the DFT

5DT.7.1 **a)** Prove (5DT.SP3).
b) Prove (5DT.SP4).
c) Demonstrate (5DT.SP5) for $N = 3$ and $N = 4$.
d) Demonstrate (5DT.SP6) for $N = 3$ and $N = 4$.

5DT.7.2 **a)** Prove (5DT.OP1) of the DFT.
b) Prove (5DT.OP3) of the DFT.

5DT.7.3 Let $x[n] = \{1, 1, 1\}$ and $h[n] = \{1, -1, 0\}$.
a) Plot $x[n]$ and $h[n]$.
b) Evaluate and plot the 3-point circular convolution of $x[n]$ and $h[n]$.

5DT.7.4 Let x[n] = {1, 1, 1, 0} and h[n] = {1, −1, 0, 0}.
  a) Plot x[n] and h[n].
  b) Evaluate and plot the 4-point circular convolution of x[n] and h[n]

5DT.7.5 Let x[n] = {1, 1, 1, 0, 0} and h[n] = {1, −1, 0, 0, 0}.
  a) Plot x[n] and h[n].
  b) Evaluate and plot the 5-point circular convolution of x[n] and h[n].

5DT.7.6 Consider the sequences

$$x[n] = \begin{cases} 1 & n = 0, 1, 2 \\ 0 & n \neq 0, 1, 2 \end{cases} \quad \text{and} \quad h[n] = \begin{cases} 1 & n = 0 \\ -1 & n = 1 \\ 0 & n \neq 0, 1 \end{cases} \tag{132}$$

  a) Plot $x[n]$ and $h[n]$.
  b) Evaluate and plot the ordinary convolution $x[n] * h[n]$
  c) Compare your answers to parts (a) and (b) with your answers to Problems 5DT.7.3–7.5. When was the ordinary (linear) convolution $x[n] * h[n]$ equal to the circular convolution $x[n] \circledast h[n]$?
  What was the reason for the equality?

5DT.8   Application of the DFT in Filtering

5DT.8.1 Let

$$x[n] = \begin{cases} 1 & n = 0, 1, 2, 3, 4 \\ 0 & n \neq 0, 1, 2, 3, 4 \end{cases} \quad \text{and} \quad h[n] = \begin{cases} 1 & n = 0, 1, 2, 3 \\ 0 & n \neq 0, 1, 2, 3 \end{cases} \tag{133}$$

  in Example 5DT.9.
  a) Fill in the matrixes in Example 5DT.9 for this special case.
  b) Plot y[n] and confirm to your satisfaction that the circular convolution $x[n] \circledast h[n]$ equals the linear convolution $x[n] * h[n]$ for $0 \leq n \leq 7$.

5DT.8.2 Use the MATLAB function `fft` to obtain the circular convolution $x[n] \circledast h[n]$ in the preceding problem.

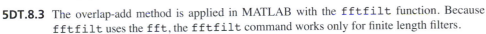

5DT.8.3 The overlap-add method is applied in MATLAB with the `fftfilt` function. Because `fftfilt` uses the `fft`, the `fftfilt` command works only for finite length filters.
  a) Read the MATLAB description of `fftfilt`.
  b) Use `fftfilt` to filter the sequence $x[n] = 1 + \cos(2\pi 0.1n) + \cos(2\pi 0.25n)$, where $0 \leq n < 255$ with the finite-length filter that has impulse response $h[n] = \cos(2\pi 0.1n)$, where $0 \leq n < 9$. Plot the filter input and output y[n].
  c) Comment on your plots. Specifically, state whether or not the filter is performing as expected. Use analytical and/or computational tools to justify your answer.

5DT.8.4 Repeat parts (b) and (c) of the preceding problem for a sequence $x[n]$ generated by the MATLAB function `randn`. You obtain the sequence using the following MATLAB statement: `x=50*randn(1,128)+100`. The sequence you obtain will vary by roughly ±50 about an average value of about 100 but will otherwise have no recognizable structure.
  a) Plot the filter input $x[n]$ and output $y[n]$.
  b) Comment on your plots.
    See Chapter 8 to develop theoretical insight into random sequences.

5DT.9   Spectrum Estimation

5DT.9.1 Write a MATLAB program to estimate the spectrum of $x[n] = a^{|n|}$ as in Example 5DT.12. Use the `fft` function. Then run your program to:
  a) Confirm Figure 5DT.21.
  b) Experiment with different choices for K and comment on the resulting truncation error.

5DT.9.2 Write a MATLAB program to duplicate Figure 5DT.23.

5DT.9.3 Write a MATLAB program to duplicate Figure 5DT.26b.

Miscellaneous Problems

5DT.1. As shown in Appendix 5DT2, the DFS can be derived from the DTFT. Show how to do the reverse; that is, show how to derive the DTFT from the DFS. Hint: Start with the DFS representation of an arbitrary periodic sequence $x[n]$ and let the period N increase to infinity to obtain a pulse sequence.

Appendix 5DT.A DFS Derived from DTFT

This appendix derives the DFS starting from the DTFT. The derivation provides insight into the relationship between the DFS and the DTFT.

We start with the DTFT of a periodic sequence, such as that shown in Figure 5DT.27.

Notice from the figure that the periodic sequence can be represented as a sum of repeated pulses.

$$x[n] = \sum_{m=-\infty}^{\infty} p[n - mN] \tag{A1}$$

where

$$p[n] = \begin{cases} x[n] & n = 0, 1, \ldots, N-1 \\ 0 & \text{otherwise} \end{cases} \tag{A2}$$

These equations state the simple fact that, if we repeat the part of $x[n]$ that is in the interval $n = 0, 1, \ldots, N-1$, we obtain a periodic sequence. Equation (A1) can be written as a convolution.

$$x[n] = p[n] * \Delta[n; N] \tag{A3}$$

where:

$$\Delta[n; N] \triangleq \begin{cases} 1 & \text{when } k \text{ is an integer multiple of } N \\ 0 & \text{otherwise} \end{cases} \tag{4DT.45}$$

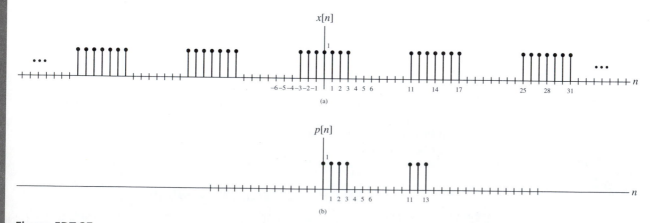

Figure 5DT.27
Periodic sequence $x[n]$ and its generating pulse $p[n]$: (a) $x[n]$; (b) $p[n]$.

We can write down the DTFT of $x[n]$ by using the convolution property (Table 4DT.2) and the DTFT for $\Delta[n; N]$ (Table 4DT.3). The result is

$$X(e^{j2\pi f}) = P(e^{j2\pi f})\frac{1}{N}\sum_{k=-\infty}^{\infty}\delta\left(f-\frac{k}{N}\right) = \frac{1}{N}\sum_{k=-\infty}^{\infty}P\left(e^{j2\pi \frac{k}{N}}\right)\delta\left(f-\frac{k}{N}\right) \quad (A4)$$

where $P(e^{j2\pi f})$ is the spectrum of $p[n]$.

$$P(e^{j2\pi f}) = \sum_{n=0}^{N-1}x[n]e^{-2\pi fn} \quad (A5)$$

Equation (A4) shows that the DTFT of a periodic sequence is composed of impulses that occur at integer multiples of $1/N$. The areas of the impulses equal $(1/N)P(e^{j2\pi \frac{k}{N}})$, where $P(e^{j2\pi f})$ is the DTFT of $p[n]$. It follows from Eq. (A5) that the DFS coefficients $X[k]$ are equally spaced sample values of the spectrum $P(e^{j2\pi f})$ of $p[n]$.

$$X[k] \triangleq P(e^{jk\frac{2\pi}{N}}) = \sum_{n=0}^{N-1}x[n]e^{-\frac{2\pi}{N}kn} \quad (A6)$$

We obtain the DFS by taking the inverse DTFT of $X(e^{j2\pi f})$ of Eq. (4). Recall from Eq. (3DT.47a) that the region of integration for an inverse DTFTs can be any interval having length 1. It is convenient here to chose the interval $0 \le f < 1$.

$$x[n] = \int_0^1 X(e^{j2\pi f})e^{j2\pi fn}df = \int_0^1 \frac{1}{N}\sum_{k=-\infty}^{\infty}X[k]\delta\left(f-\frac{k}{N}\right)e^{j2\pi fn}df$$

$$= \frac{1}{N}\sum_{k=-\infty}^{\infty}X[k]\int_0^1\delta\left(f-\frac{k}{N}\right)e^{j2\pi fn}df = \frac{1}{N}\sum_{k=0}^{N-1}X[k]e^{j2\pi \frac{k}{N}n} \quad \forall n \quad (A7)$$

which is the DFS.

In summary, we have shown that the DFS can be derived from the DTFT of a periodic sequence. In the course of this derivation we showed that the DTFT of a periodic sequence is composed of impulses that occur $f = k/N, k+0, \pm 1, \pm 2, \ldots$. The areas of the impulses equal $(1/N)P(e^{j2\pi \frac{k}{N}})$, where $P(e^{j2\pi f})$ is the DTFT of the first period of $x[n]$ given by Eq. (A2). We also saw that the DFS coefficients are sample values of samples of $P(e^{j2\pi f})$: $X[k] = P(e^{j2\pi \frac{k}{N}})$.

Appendix 5DT.B Reconstruction of DTFT from DFT

This appendix derives a formula for reconstructing the spectrum $X(e^{j2\pi f})$ of

$$x[n] = \ldots 0,0,0, x[0], x[1], \ldots, x[N-1], 0,0,0, \ldots \quad (B1)$$

from the DFT coefficients

$$X[k] = \sum_{n=0}^{N-1}x[n]e^{-j2\pi \frac{k}{N}n} \quad k = 0,1,\ldots,N-1 \quad (B2)$$

We start with the formula for the DTFT of $x[n]$.

$$X(e^{j2\pi f}) = \sum_{n=-\infty}^{\infty}x[n]e^{-j2\pi fn} \equiv \sum_{n=0}^{N-1}x[n]e^{-j2\pi fn} \quad (B3)$$

and replace x[n] with the formula for the inverse DFT.

$$x[n] = \frac{1}{N} \sum_{k=0}^{N-1} X[k] e^{j2\pi \frac{k}{N} n} \tag{B4}$$

The substitution of Eq. (B4) into Eq. (B3) yields

$$X(e^{j2\pi f}) = \sum_{n=-0}^{N-1} \left\{ \frac{1}{N} \sum_{k=0}^{N-1} X[k] e^{j2\pi \frac{k}{N} n} \right\} e^{-j2\pi f n} = \sum_{k=0}^{N-1} X[k] \left\{ \frac{1}{N} \sum_{n=0}^{N-1} e^{-j2\pi (f - \frac{k}{N}) n} \right\} \tag{B5}$$

$$= \sum_{k=0}^{N-1} X[k] \frac{\sin \left(N\pi \left(f - \frac{k}{N} \right) \right)}{N \sin \left(\pi \left(f - \frac{k}{N} \right) \right)} e^{j(N-1)\pi \left(f - \frac{k}{N} \right)} \tag{B6}$$

where the last step follows from identity (B.1) of Appendix B. The right-hand side of Eq. (B6) is an interpolation formula that shows how $X(e^{j2\pi f})$ can be reconstructed from the DFT coefficients.

6CT The Laplace Transform

6CT.1 Introduction

The Laplace transform is the premier analytical aid in CT signals and systems theory for determining the response of an LTI system to a suddenly applied input. It is an analytical extension of the Fourier transform based on the idea of *complex frequency*. The concept of a signal's spectrum (Fourier transform) $X(F)$, so important in previous chapters, is included in the framework of the Laplace transform. The Laplace transform, however, also introduces entirely new concepts. These new concepts (*poles*, *zeros*, and *regions of convergence*) enable us to solve for the response of both stable and unstable LTI systems to suddenly applied inputs using just algebra and table lookup. The new concepts provide us with a deeper understanding of (1) a signal's spectrum $X(F)$, (2) a system's frequency response characteristic $H(F)$, and (3) how the sinusoidal steady-state response and transient response are related. Thus, the Laplace transform not only simplifies determining the transient and steady-state responses of LTI systems, but also provides intuition and insight into CT signals and systems concepts.

6CT.2 Definition of the Laplace Transform

The Laplace Transform[1] of a waveform $x(t)$ is defined as

Laplace Transform

$$\mathbf{X}(s) = \mathcal{L}\{x(t)\} = \int_0^\infty x(t)e^{-st}\,dt \qquad (1)$$

where s is a *complex frequency* variable having real part σ and imaginary part $\Omega = 2\pi F$.

$$s = \sigma + j\Omega \qquad (2)$$

---

[1] There are two versions of the Laplace transform: the *unilateral* Laplace transform defined in Eq. (1) and the *bilateral* Laplace transform introduced in Problems 6CT.2.3 and 6CT.2.4. We follow popular terminology when we refer to the unilateral Laplace transform as "the" Laplace transform.

The integral defining $\mathbf{X}(s)$ ignores the time interval $t < 0$. Therefore, waveforms that are different for $t < 0$ but equal for $t \geq 0$ have the same Laplace transform. If we like, we can assume for convenience that $x(t) = 0$ for $t < 0$. This or any other assumption about $x(t)$ for $t < 0$ has no effect on $\mathbf{X}(s)$.

The class of waveforms that are Laplace transformable is considerably larger than the class of Fourier-transformable waveforms. We can establish this fact by upper-bounding the absolute value of $\mathbf{X}(s)$:

$$|\mathbf{X}(s)| = \left| \int_0^\infty x(t)e^{-st}dt \right| \leq \int_0^\infty |x(t)e^{-st}|\, dt = \int_0^\infty |x(t)|e^{-\sigma t}dt \tag{3}$$

If finite numbers M and σ_a exist such that

$$|x(t)| < Me^{\sigma_a t} \tag{4}$$

then

$$\int_0^\infty |x(t)|e^{-\sigma t}dt < M \int_0^\infty e^{-(\sigma-\sigma_a)t}dt \tag{5a}$$

The sum on the right-hand side converges for all $\sigma > \sigma_a$. Therefore, if $|x(t)|$ satisfies Eq. (4), then

$$\int_0^\infty |x(t)|e^{-\sigma t}dt < \infty \qquad \text{for } \sigma > \sigma_a \tag{5b}$$

and $\mathbf{X}(s)$ exists for values of s in the right-half of the s plane defined by $R\{s\} > \sigma_a$. σ_a is called the *abscissa of convergence*. Waveforms satisfying Eq. (4) are said to have *exponential order*. If we compare Eq. (5b) with Eq. (4CT.107), we see that the factor $e^{-\sigma t}$ can allow the Laplace transform of a waveform to exist when the Fourier transform does not.

EXAMPLE 6CT.1   DC Waveform

The substitution

$$x(t) = 1 \tag{6}$$

into Eq. (1) yields

$$\mathbf{X}(s) = \int_0^\infty e^{-st}dt = -\frac{1}{s}e^{-st}\Big|_0^\infty = -\frac{1}{s}e^{-st}\Big|_\infty + \frac{1}{s}e^{-st}\Big|_0 \tag{7}$$

To evaluate the term $e^{-st}\big|_\infty$ we write it as a limit.

$$e^{-st}\big|_\infty = \lim_{t\to\infty} e^{-(\sigma+j\Omega)t} = \lim_{t\to\infty} e^{-\sigma t}e^{-j\Omega t} = \begin{cases} 0 & \text{for } \sigma > 0 \\ \text{undefined} & \text{for } \sigma = 0 \\ \infty & \text{for } \sigma < 0 \end{cases} \tag{8}$$

The last term of Eq. (7) equals $1/s$

$$\frac{1}{s}e^{-st}\Big|_0 = \frac{1}{s}e^{-s0} = \frac{1}{s} \tag{9}$$

and we have the result

$$\mathbf{X}(s) = \frac{1}{s} \quad \text{if} \quad R\{s\} > 0 \tag{10}$$

Figure 6CT.1
Pole-zero plot for a
DC waveform: The
region of
convergence is the
unshaded half plane
to the right of the
pole.

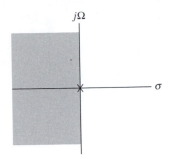

We can display $\mathbf{X}(s)$ using the *pole-zero diagram* shown in Figure 6CT.1. Here the pole-zero diagram includes one pole at $s = 0$, indicated by the cross, and no zeros. In general, a pole-zero diagram consists of both crosses and circles. The crosses indicate the values of s for which the denominator of $\mathbf{X}(s)$ equals 0 (these are the pole values). The circles indicate the values of s for which the numerator of $\mathbf{X}(s)$ equals 0 (these are the zero values).

We see from Eq. (10) that the $\mathbf{X}(s)$ is defined only for $\mathcal{R}\{s\} > 0$, i.e., for points lying in the right half of the s plane. This region, called the *region of convergence* (ROC) of $\mathbf{X}(s)$, is shown unshaded in the figure. Notice that the region of convergence is the half plane to the right of the pole.

Before leaving this example, it is a good idea for the reader to show that the Laplace transform of a unit step $u(t)$ is also given by Eq. (10)

We showed in the preceding example that the Laplace transform of the constant $x(t) = 1$ exists provided that s is a value in a region of convergence. Recall that, in contrast, the Fourier transform of the constant $x(t) = 1$ does not exist: The limit we took in Eq. (4CT.22) became unbounded as $T \to \infty$. We were able to represent $X(F)$ as an impulse $\delta(F)$, even though $\delta(0)$ does not exist as a number. A constant $x(t) = 1$ is the first of many examples you will find in this chapter for which the Laplace transform exists and the Fourier transform does not.

The complex exponential waveform in the following example provides a building block for many other waveforms.

EXAMPLE 6CT.2   Complex Exponential

Consider the complex exponential

$$x(t) = e^{s_0 t} \tag{11}$$

where s_0 is a complex constant

$$s_0 = \sigma_0 + j\Omega_0$$

The substitution of Eq. (11) into Eq. (1) yields

$$\mathbf{X}(s) = \int_0^\infty e^{s_0 t} e^{-st} dt = -\frac{1}{s} e^{-(s-s_0)t} \Big|_0^\infty \tag{12}$$

$$= -\frac{1}{s - s_0} e^{-(s-s_0)t} \Big|_\infty + \frac{1}{s - s_0} \tag{13}$$

where

$$e^{-(s-s_0)t}\Big|_\infty = \lim_{t\to\infty} e^{-(\sigma-\sigma_0)t}e^{-j(\Omega-\Omega_0)t} = \begin{cases} 0 & \text{for } \sigma > \sigma_0 \\ \text{undefined} & \text{for } \sigma = \sigma_0 \\ \infty & \text{for } \sigma < \sigma_0 \end{cases} \quad (14)$$

Therefore,

$$\mathbf{X}(s) = \frac{1}{s-s_0} \quad \text{if} \quad \mathcal{R}\{s\} > \sigma_0 \quad (15)$$

The pole-zero plot is shown in Figure 6CT.2. The region of convergence is the half plane lying to the right of the pole.

The real exponential $x(t) = e^{\sigma_0 t}$ and the complex sinusoid $x(t) = e^{j\Omega_0 t}$ are special cases of Eq. (15), obtained by setting $\Omega_0 = 0$ and $\sigma_0 = 0$, respectively. The waveform $x(t) = 1$ that was considered in the first example is a special case obtained by setting $\Omega_0 = \sigma_0 = 0$.

Figure 6CT.2
Pole-zero plot for a complex exponential: The region of convergence is the unshaded half plane to the right of the pole.

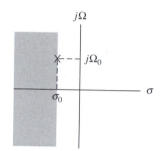

EXAMPLE 6CT.3   The Unit Impulse

The substitution

$$x(t) = \delta(t) \quad (16)$$

into Eq. (192) yields[2]

$$\mathbf{X}(s) = \int_0^\infty \delta(t)e^{-st}dt = 1 \quad (17)$$

Here the region of convergence is the entire s plane.

6CT.3   Relationship Between the Laplace and Fourier Transforms of Causal Waveforms

A waveform is called *causal* if it equals 0 for $t < 0$. We now show that the Laplace and Fourier transforms of causal waveforms are related in a fundamental way. If we set $s = \sigma + j2\pi F$ and assume that $x(t) = 0$ for $t < 0$, then Eq. (1) becomes

$$\mathbf{X}(s)\Big|_{s=\sigma+j2\pi F} = \int_0^\infty x(t)e^{-(\sigma+j2\pi F)t}dt = \int_{-\infty}^\infty (x(t)e^{-\sigma t})e^{-j2\pi Ft}dt \quad (18)$$

[2]We interpret the lower limit 0 in Eq. (1) as a limit from below: $0 = \lim_{\epsilon\to 0} -\epsilon$, where $\epsilon > 0$. This interpretation is consistent with our definition of the impulse as an even function of t.

Equation (18) shows that the Laplace transform of $x(t)$ equals the Fourier transform of the exponentially weighted version $x(t)e^{-\sigma t}$ of $x(t)$.

$$\mathbf{X}(s)\big|_{s=\sigma+j2\pi F} = \mathcal{F}\{x(t)e^{-\sigma t}\} \qquad \text{for causal } x(t) \qquad (19)$$

We noted earlier that the factor $e^{-\sigma t}$ often makes it possible for the Laplace transform to exist when the Fourier transform does not. If we set $\sigma = 0$, then Eq. (19) becomes

$$\mathbf{X}(s)\big|_{s=j2\pi F} = X(F) \qquad \text{for causal } x(t) \qquad (20)$$

where $X(F)$ is the Fourier transform of $x(t)$. Therefore, the Laplace transform of $x(t)$ evaluated on the $j\Omega$ axis of the s-plane is equal to the Fourier transform of $x(t)$. It follows from Eq. (20) that the Fourier transform (spectrum) of a causal waveform exists if and only if the Laplace transform converges for $\sigma = 0$. The notations $\mathbf{X}(j2\pi F)$ and $X(F)$ are both commonly used[3] for the Fourier transform of $x(t)$.

EXAMPLE 6CT.4   The Real Exponential

Assume

$$x(t) = e^{\sigma_0 t}u(t) \qquad (21)$$

where σ_0 is real. It follows from Eq. (15) with $s_0 = \sigma_0$ that

$$\mathbf{X}(s) = \frac{1}{s - \sigma_0} \qquad (22)$$

Figure 6CT.3
Pole-zero plot for
$e^{\sigma_0 t}u(t)$: The region
of convergence is the
unshaded half plane
to the right of the
pole.

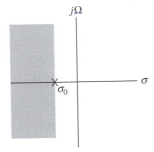

where $\mathcal{R}\{s\} > \sigma_0$. As we see in Figure 6CT.3, the region of convergence $\mathcal{R}\{s\} > \sigma_0$ includes the $j\Omega$ axis if $\sigma_0 < 0$. Therefore, if $\sigma_0 < 0$, the Fourier transform of $e^{\sigma_0 t}u(t)$ exists and we can set $s = j2\pi F$ in Eq. (22) to obtain the spectrum of $e^{\sigma_0 t}u(t)$.

$$\mathbf{X}(j2\pi F) = X(F) = \frac{1}{j2\pi F - \sigma_0} \qquad (23)$$

If $\sigma_0 \geq 0$, then the region of convergence $\mathcal{R}\{s\} > \sigma_0$ does not include the $j\Omega$ axis and the Fourier transform of $e^{\sigma_0 t}u(t)$ does not exist.

---

[3]You will encounter various definitions of the Fourier transform in the literature. Some books define the Fourier transform as $X(j\Omega) = \int_{-\infty}^{\infty} x(t)e^{-j\Omega t}dt$, where $x(t) = (1/2\pi)\int_{-\infty}^{\infty} X(j\Omega)e^{j\Omega t}d\Omega$. This definition has the advantage of notationally strengthening the connection between the Fourier and Laplace transforms. However, it is asymmetric and has the disadvantage of introducing fussy factors of 2π in the transforms of a constant and a rectangular pulse: $1 \leftrightarrow 2\pi\delta(\Omega)$ and $\sqcap(\frac{t}{\tau}) \leftrightarrow \tau \, \text{sinc}(\tau\Omega/2\pi)$. Other notations for Fourier transform are also used. For example, the MATLAB help file defines the Fourier transform of a function $f(x)$ as $F(\omega) = \int_{-\infty}^{\infty} f(x)e^{-j\omega x}dx$.

EXAMPLE 6CT.5   The Unit Step

We can regard the unit step $x(t) = u(t)$ as a special case of the causal exponential $e^{\sigma_0 t} u(t)$ of the preceding example. It follows by setting $\sigma_0 = 0$ in Eq. (22) that

$$\mathcal{L}\{u(t)\} = \mathbf{X}(s) = \frac{1}{s} \qquad \text{for } \mathcal{R}\{s\} > 0 \tag{24}$$

This result was obtained in Example 6CT.1. The region of convergence does not include the $j\Omega$ axis. Therefore, the unit step function is not Fourier transformable. This conclusion agrees with Appendix 3CT where we showed that the Fourier integral does not converge at the point $F = 0$. We showed in that appendix that, although $U(F)$ does not exist we can define it as

$$U(F) = \frac{1}{2}\delta(F) + \frac{1}{j2\pi F} \tag{25}$$

<hr>

6CT.4   Properties of the Laplace Transform

We can obtain many additional Laplace transforms by applying the properties of the Laplace transform to the preceding results. One of the most important properties is linearity.

Linearity Property

$$\mathcal{L}\{ax_1(t) + bx_2(t)\} = a\mathcal{L}\{x_1(t)\} + b\mathcal{L}\{x_2(t)\} \tag{26}$$

The linearity property follows directly from Eq. (1). As shown in the following example, the region of convergence of $\mathcal{L}\{ax_1(t) + bx_2(t)\}$ is the right half of the s-plane defined by the rightmost pole of $a\mathcal{L}\{x_1(t)\} + b\mathcal{L}\{x_2(t)\}$.

EXAMPLE 6CT.6   Sum of Two Exponentials

Consider

$$w(t) = 6e^{-2t} + 5e^{3t} \tag{27}$$

Because the Laplace transform operator is a linear operation

$$\mathbf{W}(s) = \mathcal{L}\{6e^{-2t} + 5e^{3t}\} = 6\mathcal{L}\{e^{-2t}\} + 5\mathcal{L}\{e^{3t}\} \tag{28}$$

The Laplace transforms of the terms on the right-hand side of this equation follow by setting $\sigma_0 = -2$ and in $\sigma_0 = 3$ in Eq. (22). The results are

$$\mathcal{L}\{e^{-2t}\} = \frac{1}{s+2} \tag{29}$$

where $\mathcal{R}\{s\} > -2$, and

$$\mathcal{L}\{e^{3t}\} = \frac{1}{s-3} \tag{30}$$

Figure 6CT.4
Pole-zero plot for
$6e^{-2t} + 5e^{3t}$: The
region of
convergence is the
unshaded half plane
to the right of the
rightmost pole.

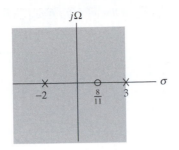

where $\mathcal{R}\{s\} > 3$. $\mathbf{W}(s)$ exists if and only if *both* $\mathcal{L}\{e^{-2t}\}$ and $\mathcal{L}\{e^{3t}\}$ exist. Therefore, the region of convergence for $\mathbf{W}(s)$ is $\mathcal{R}\{s\} > 3$ and we have

$$\mathbf{W}(s) = \frac{6}{s+2} + \frac{5}{s-3} = \frac{11\left(s - \frac{8}{11}\right)}{(s+2)(s-3)} \tag{31}$$

where $\mathcal{R}\{s\} > 3$. The pole-zero plot of $\mathbf{W}(s)$ is shown in Figure 6CT.4. Notice that the region of convergence lies in the right half plane that starts to the right of the pole that is furthest to the right. A region of convergence can never include a pole because the s-transform does not converge at a pole.

We saw in the above example that the region of convergence of the sum of two scaled exponential waveforms is the right half s plane beyond the rightmost pole. The Laplace transform of this sum was a *rational algebraic fraction*, $\mathbf{W}(s)$, given by the right-handside of (31). We can generalize this result as follows: Whenever a Laplace transform is a rational algebraic fraction, the region of convergence is the right half s plane that lies beyond rightmost pole. The general form of a rational algebraic fraction is given by (100) of Section 6CT.6.3.

In the following example, we apply the linearity property and the result of Example 6CT.2 to find the Laplace transform of a sinusoid.

EXAMPLE 6CT.7   The Sinusoid

Consider

$$x(t) = A\cos(\Omega_0 t + \phi) \tag{32}$$

We can write this as

$$x(t) = Xe^{j\Omega_0 t} + X^* e^{-j\Omega_0 t} \tag{33}$$

where

$$X = \frac{1}{2}Ae^{j\phi} \tag{34}$$

By the linearity property and the result of Example 6CT.2, we obtain

$$\mathbf{X}(s) = X\mathcal{L}\{e^{j\Omega_0 t}\} + X^*\mathcal{L}\{e^{-j\Omega_0 t}\}$$

$$= \frac{X}{s - j\Omega_0} + \frac{X^*}{s + j\Omega_0} \qquad \mathcal{R}\{s\} > 0 \tag{35}$$

A second useful important property is the *s-shift property*. It states that if we multiply $x(t)$ by $e^{s_0 t}$, then we shift $\mathbf{X}(s)$ by s_0.

---

s-Shift Property

$$\mathcal{L}\{x(t)e^{s_0 t}\} = \mathbf{X}(s - s_0) \tag{36}$$

We can easily prove the s-shift property by noticing that because

$$\mathcal{L}\{x(t)\} = \int_0^\infty x(t)e^{-st}\,dt = \mathbf{X}(s) \tag{37}$$

then

$$\mathcal{L}\{x(t)e^{s_0 t}\} = \int_0^\infty x(t)e^{-(s-s_0)t}\,dt = \mathbf{X}(s - s_0) \tag{38}$$

If $\mathbf{X}(s)$ has ROC $\mathcal{R}\{s\} = \sigma > \sigma_x$, then $\mathbf{X}(s - s_0)$ has ROC $\mathcal{R}\{s - s_0\} = \sigma - \sigma_0 > \sigma_x$ or, equivalently, $\sigma_x > \sigma + \sigma_0$. In the next example, we apply the s-shift property to find the Laplace transform of a damped sinusoid

EXAMPLE 6CT.8   A Damped Sinusoid

Consider the damped sinusoid

$$x(t) = Ae^{\sigma_0 t}\cos(\Omega_0 t + \phi) \tag{39}$$

If we apply the s-shift property with $s_0 = \sigma_0$ to the result of Example 6CT.7, we obtain

$$\mathbf{X}(s) = \frac{X}{s - \sigma_0 - j\Omega_0} + \frac{X^*}{s - \sigma_0 + j\Omega_0} \quad \mathcal{R}\{s\} > \sigma_0 \tag{40}$$

A third useful property is the "multiply by t property."

Multiply by t Property

$$tx(t) \overset{LT}{\longleftrightarrow} -\frac{d}{ds}X(s) \tag{41}$$

The proof follows directly from Eq. (1), for if

$$\mathbf{X}(s) = \mathcal{L}\{x(t)\} = \int_0^\infty x(t)e^{-st}\,dt \tag{42}$$

then

$$-\frac{d}{ds}\mathbf{X}(s) = -\int_0^\infty x(t)\frac{d}{ds}e^{-st}\,dt = \int_0^\infty tx(t)e^{-st}\,dt = \mathcal{L}\{tx(t)\} \tag{43}$$

EXAMPLE 6CT.9   A Ramp

Consider

$$x(t) = r(t) = tu(t) \tag{44}$$

We showed earlier that

$$u(t) \xleftrightarrow{\text{LT}} \frac{1}{s} \quad \mathcal{R}\{s\} > 0 \tag{45}$$

The "multiply by t" property yields

$$r(t) \xleftrightarrow{\text{LT}} -\frac{d}{ds}\frac{1}{s} = \frac{1}{s^2} \quad \mathcal{R}\{s\} > 0 \tag{46}$$

The pole-zero plot is shown in Figure 6CT.5, where the pole is indicated by overlapping crosses. This pole is called a *second-order* pole because the degree of the denominator s^2 is 2.

Figure 6CT.5
Pole-zero plot for the
unit ramp.

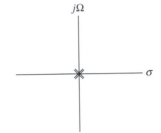

Additional Laplace transform properties are listed in Table 6CT.1. The differentiation property is established in Section 6CT.5. The proofs of the time-scaling, time-delay, and convolution properties are straightforward and are relegated to the problems. The properties we have listed are sufficient for most problems encountered in engineering.

Table 6CT.2 is a short list of Laplace transform pairs. We included several forms for the transform of a cosine to save time later.

Linearity	$a_1 x_1(t) + a_2 x_2(t)$	$a_1 \mathbf{X}_1(s) + a_2 \mathbf{X}_2(s)$
Differentiation (1)	$x^{(1)}(t)$	$s\mathbf{X}(s) - x(0^-)$
Differentiation (2)	$x^{(2)}(t)$	$s^2\mathbf{X}(s) - sx(0^-) - x^{(1)}(0^-)$
Differentiation (n)	$x^{(n)}(t)$	$s^n\mathbf{X}(s) - s^{n-1}x(0^-) - s^{n-2}x^{(1)}(0^-) - \cdots - x^{(n-1)}(0^-)$
Integration	$\displaystyle\int_0^\infty x(t)\, dt$	$\dfrac{1}{s}\mathbf{X}(s) + \displaystyle\int_{-\infty}^0 x(t)\, dt$
s shift	$x(t)e^{S_0 t}$	$\mathbf{X}(s - s_0)$
Convolution	$x_1(t) * x_2(t) = \displaystyle\int_0^t x_1(\lambda)x_2(t - \lambda)\, d\lambda$	$\mathbf{X}_1(s)\mathbf{X}_2(s)$
Time Delay	$x(t - \tau)u(t - \tau) \quad \tau \geq 0$	$\mathbf{X}(s)e^{-s\tau}$
Time Scale	$x(at) \quad a > 0$	$\dfrac{1}{a}\mathbf{X}(s/a)$

Table 6CT.1

Laplace Transform Properties

The region of convergence of the Laplace transform of $\delta(t)$ is the entire s-plane.
The region of convergence of every other entry is the half plane to the right of the rightmost pole.

	Waveform	**Transform**
6CT.LT1	$\delta(t)$	1
6CT.LT2	1	$\dfrac{1}{s}$
6CT.LT3	t	$\dfrac{1}{s^2}$
6CT.LT4	$\dfrac{1}{(n-1)!}t^{(n-1)}$	$\dfrac{1}{s^n}$
6CT.LT5	$e^{s_0 t}$	$\dfrac{1}{s-s_0}$
6CT.LT6	$t e^{s_0 t}$	$\dfrac{1}{(s-s_0)^2}$
6CT.LT7	$\dfrac{1}{(n-1)!}t^{(n-1)}e^{s_0 t}$	$\dfrac{1}{(s-s_0)^n}$
6CT.LT8	$e^{j\Omega_0 t}$	$\dfrac{1}{s-j\Omega_0}$
6CT.LT9	$e^{\sigma_0 t}$	$\dfrac{1}{s-\sigma_0}$
6CT.LT10	$\cos(\Omega_0 t)$	$\dfrac{s}{s^2+\Omega_0^2}=$ $$\dfrac{s}{(s-j\Omega_0)(s+j\Omega_0)}=\dfrac{\frac{1}{2}}{s-j\Omega_0}+\dfrac{\frac{1}{2}}{s+j\Omega_0}$$
6CT.LT11	$\sin(\Omega_0 t)$	$\dfrac{\Omega_0}{(s-j\Omega_0)(s+j\Omega_0)}$
6CT.LT12	$Ae^{\sigma_0 t}\cos(\Omega_0 t+\phi)=\mathcal{R}\{2Xe^{s_0 t}\}$	$\dfrac{X}{s-s_0}+\dfrac{X^*}{s-s_0^*}$ $X=\frac{1}{2}Ae^{j\phi},\, s_0=\sigma_0+j\Omega_0$
6CT.LT13	$\frac{1}{(n-1)!}t^{(n-1)}Ae^{\sigma_0 t}\cos(\Omega_0 t+\phi)$ $=\frac{1}{(n-1)!}t^{(n-1)}\mathcal{R}\{2Xe^{s_0 t}\}$	$\dfrac{X}{(s-s_0)^n}+\dfrac{X^*}{(s-s_0^*)^n}$ $X=\frac{1}{2}Ae^{j\phi},\, s_0=\sigma_0+j\Omega_0$

Table 6CT.2

Laplace Transform Pairs

6CT.5 Elementary Systems in the *s*-Domain

We can use the Laplace transform to create s-domain elementary system models. These elementary models include the effects of initial conditions (initial stored energy). Large-scale system models are obtained by interconnecting the s-domain elementary system models. Remember that the s-domain models refer to waveforms on the interval $t \geq 0$ because they are derived from the Laplace transform.

6CT.5.1
Ideal
Amplifiers and
Adders

Figure 6CT.6 illustrates the operations of amplification and addition in the time and s domains. The s domain operators are obtained directly from the linearity property of the Laplace transform, Eq. (26). Initial conditions are not a part of the s-domain systems because ideal amplification and addition are memoryless operations.

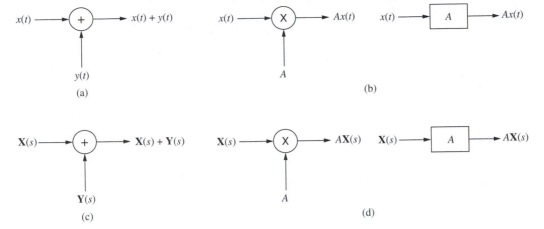

Figure 6CT.6
Amplification and addition in the time and s-domains: (a)–(b) addition and amplification in the time domain; (c)–(d) addition and amplification in the s domain.

6CT.5.2
Differentiators

We can obtain a recursion formula for the Laplace transform of the nth derivative of a waveform $x(t)$. To initialize the recursion, we denote the first derivative of $x(t)$ as

$$x^{(1)}(t) = \frac{d}{dt}x(t) \tag{47}$$

and the Laplace transform of $x^{(1)}(t)$ as

$$\mathbf{X}^{(1)}(s) = \int_0^\infty x^{(1)}(t)e^{-st}\,dt \tag{48}$$

To find $\mathbf{X}^{(1)}(s)$, we integrate Eq. (48) by parts. Integral calculus gives us the following rule for integration by parts.

$$\int_a^b u\,dv = uv\Big|_a^b - \int_a^b v\,du \tag{49}$$

If we set $u = e^{-st}$ and $dv = x^{(1)}(t)\,dt$, then the preceding equation becomes

$$\int_0^\infty x^{(1)}(t)e^{-st}\,dt = x(t)e^{-st}\Big|_0^\infty + s\int_0^\infty x(t)e^{-st}\,dt \tag{50}$$

which reduces to

$$\mathbf{X}^{(1)}(s) = s\mathbf{X}(s) - x(0^-) \tag{51}$$

The time and s domain representations of a differentiator are shown in Figure 6CT.7. Notice that the "initial condition" $x(0^-)$ appears explicitly in s-domain Eq. (51) but not in the t-domain Eq. (47).

Figure 6CT.8(a) shows two differentiators connected in series. The output of the second differentiator is the derivative of the output of the first. That is

$$x^{(2)}(t) \triangleq \frac{d^2}{dt^2}x(t) = \frac{d}{dt}x^{(1)}(t) \tag{52}$$

Figure 6CT.7
Differentiation in the
time and s-domains.

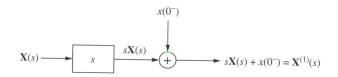

The s domain system corresponding to Figure 6CT.8(a) is shown in Figure 6CT.8(b). Each stage performs the operation shown in Figure 6CT.7(b) but with a different input. The output of the last differentiator of the s domain system is

$$\mathbf{X}^{(2)}(s) = s\mathbf{X}^{(1)}(s) - x^{(1)}(0^-) \tag{53}$$

By substituting Eq. (51) into (53), or by inspection of Figure 6CT.8(b), we see that

$$\mathbf{X}^{(2)}(s) = s^2\mathbf{X}(s) - sx(0^-) - x^{(1)}(0^-) \tag{54}$$

We can generalize Eq. (54) to n differentiators in series. We start with the definition

$$x^{(n)}(t) \triangleq \frac{d^n}{dt^n}x(t) = \frac{d}{dt}x^{(n-1)}(t) \tag{55}$$

The corresponding s-domain equation is

$$\mathbf{X}^{(n)}(s) = s\mathbf{X}^{(n-1)}(s) - x^{(n-1)}(0^-) \tag{56}$$

where $\mathbf{X}^{(n)}(s) = \mathcal{L}\{x^{(n)}(t)\}$. This equation is a recursion formula. Equations (51) and (53) can be viewed as special cases for $n = 1$ and 2, respectively. We can solve the general recursion by repeated substitution. The result is

$$\mathbf{X}^{(n)}(s) = s^n\mathbf{X}(s) - s^{n-1}x(0^-) - s^{n-2}x^{(1)}(0^-) - \cdots - x^{(n-1)}(0^-) \tag{57}$$

Figure 6CT.8
Two differentiators in
series.

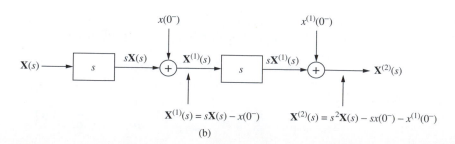

Figure 6CT.9
Integrator in time and
s-domains.

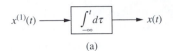

(a)

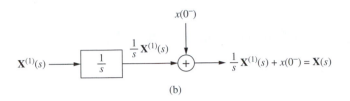

(b)

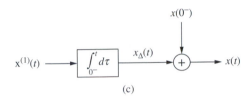

(c)

6CT.5.3
Integrators

Consider next the integrator of Figure 6CT.9(a). We can find the s-domain system by solving Eq. (51) for $\mathbf{X}(s)$. The result is

$$\mathbf{X}(s) = \frac{1}{s}\mathbf{X}^{(1)}(s) + \frac{1}{s}x(0^-) \qquad (58)$$

The s-domain version of the integrator is shown in Figure 6CT.9(b). The time domain equation corresponding Eq. (58) is

$$x(t) = x_\Delta(t) + x(0^-) \quad t \geq 0 \qquad (59)$$

The first term on right-hand side of Eq. (59) represents the part of the output of the integrator caused by $x^{(1)}(\lambda)$ in the time interval $0 \leq \lambda \leq t$.

$$x_\Delta(t) = \int_0^t x^{(1)}(\lambda) \, d\lambda \qquad (60)$$

The second term on the right-hand side of Eq. (59) represents the part of the output of the integrator caused by $x^{(1)}(\lambda)$ in the time interval $\lambda < 0$.

$$x(0^-) = \int_{-\infty}^0 x^{(1)}(\lambda) \, d\lambda \qquad (61)$$

Figure 6CT.9(c) shows the block diagram for the integrator based on (59). The block diagrams of Figures 6CT.9(a) and (c) are both commonly used. The difference is that the block diagram of Figure 6CT.9(a) applies for all t, and that of Figure 6CT.9(c) applies for $t \geq 0$.

It is easy to generalize the preceding results to n integrators connected in series. The main result can be obtained from Eq. (57), which rearranges to

$$\mathbf{X}(s) = \frac{1}{s^n}\mathbf{X}^{(n)}(s) + \frac{1}{s^n}x^{(n-1)}(0^-) + \frac{1}{s^{n-1}}x^{(n-1)}(0^-) + \cdots + \frac{1}{s}x(0^-) \qquad (62)$$

6CT.5.4
Delay Lines

We can also derive elementary system models for delay lines that contain initial conditions. For a delay line, the initial condition is the segment of a waveform that is stored in the delay line at time $t = 0$. The s domain model for a delay line having a stored waveform is developed in Problem 6CT.5.1.

6CT.6   Laplace Transform System Analysis

If we interconnect the elementary time domain systems described in the previous section, we obtain a system whose input and output are related by an ordinary linear integral-differential equation having constant coefficients. The s-domain equation of the system enables us to find the output for $t > 0$ using algebra and table lookup.

6CT.6.1
Basic
Procedure

The following four steps are used to solve most problems.

Basic Procedure

1. Obtain the s-domain I-O equation of the system. This step can be accomplished from either the system's differential I-O equation or the system's s-domain block diagram.
2. Solve the s-domain I-O equation for the s-domain output $\mathbf{Y}(s)$.
3. Use partial fraction expansion to put $\mathbf{Y}(s)$ is a form ready for table lookup.
4. Find $y(t)$ by table lookup.

The following examples demonstrate the four steps and serve as an introduction to the general development in the next subsection. Example 6CT.10 illustrate steps 1 and 2.

EXAMPLE 6CT.10 s-Domain I-O Equation of a First-Order System

Let's find the s-domain I-O equation of the first-order system of Example 2CT.3. The time and s domain system block diagrams, which address the time interval $t \geq 0$, are shown in Figure 6CT.10(a) and (b), respectively. The influence of the time interval $t < 0$ on the output for $t \geq 0$, is contained in the initial condition $y(0^-)$. The s-domain equation relating the LTI system's output to its input can be obtained either of two ways.

First, we can derive the s-domain I-O equation directly from the differential equation relating input and output of the time domain system. An inspection of Figure 6CT.10(a) reveals that

$$y(t) = \int_0^t y_\Delta^{(1)}(\lambda)\, d\lambda + y(0^-) \tag{63a}$$

and

$$y_\Delta^{(1)}(t) = \frac{1}{\tau_c}[x(t) - y(t)] \tag{63b}$$

If we eliminate $y_\Delta^{(1)}(t)$ from the preceding equations, we obtain

$$y(t) + \frac{1}{\tau_c}\int_0^t y(\lambda)\, d\lambda = \frac{1}{\tau_c}\int_0^t x(\lambda)\, d\lambda + y(0^-) \tag{64}$$

Figure 6CT.10
First-order system in
time and s-domains:
(a) time domain;
(b) s domain.

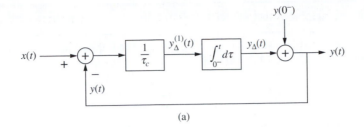

(a)

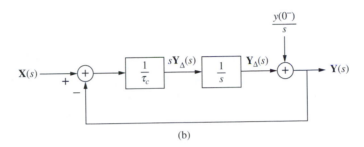

(b)

Equation (64) is an integral equation whose solution is $y(t)$. We prefer to work with a differential equation and therefore differentiate it to obtain

$$\tau_c y^{(1)}(t) + y(t) = x(t) \quad t > 0 \tag{65}$$

We take the Laplace transform of both sides to obtain the s-domain I-O equation

$$\tau_c[s\mathbf{Y}(s) - y(0^-)] + \mathbf{Y}(s) = \mathbf{X}(s) \tag{66}$$

We solve for $\mathbf{Y}(s)$

$$\mathbf{Y}(s) = \frac{1}{s\tau_c + 1}\mathbf{X}(s) + \frac{\tau_c}{s\tau_c + 1}y(0^-) \tag{67}$$

The effects of the input and the initial condition are represented by the first and second terms, respectively.

We can also obtain Eq. (67) from the s domain system shown in Figure 6CT.10(b). The s-domain system is obtained by replacing the time domain elements with their s-domain counterparts. An inspection of Figure 6CT.10(b) reveals that

$$\mathbf{Y}(s) = \mathbf{Y}_\Delta(s) + \frac{1}{s}y(0^-) \tag{68}$$

$$s\mathbf{Y}_\Delta(s) = \frac{1}{\tau_c}(\mathbf{X}(s) - \mathbf{Y}(s)) \tag{69}$$

By eliminating $\mathbf{Y}_\Delta(s)$ from the preceding equations, we obtain

$$\mathbf{Y}(s) + \frac{1}{s\tau_c}\mathbf{Y}(s) = \frac{1}{s\tau_c}\mathbf{X}(s) + \frac{1}{s}y(0^-) \tag{70}$$

which is a rearranged version of Eq. (66).

The following examples illustrate the third and fourth steps of the method.

EXAMPLE 6CT.11   Impulse Applied to a First-Order System with an Initial Condition

Assume that $x(t) = \delta(t)$ is applied to the system in Figure 6CT.10(a). The Laplace transform of $\delta(t)$ is 1. The substitution $X(s) = 1$ into Eq. (67) yields

$$Y(s) = \frac{1}{s\tau_c + 1} + \frac{\tau_c}{s\tau_c + 1} y(0^-) \tag{71}$$

Partial fraction expansion (step 3) is not necessary because both terms in the right-hand side can be found in Table 6CT.2, which gives the solution

$$y(t) = \frac{1}{\tau_c} e^{-\frac{t}{\tau_c}} + \frac{1}{\tau_c} y(0^-) e^{-\frac{t}{\tau_c}} \quad t > 0 \tag{72}$$

The first term in $y(t)$, due to the input, is called the *forced response*. The first term is the system's impulse $h(t)$ because it is the response to $x(t) = \delta(t)$ when the system is initially at rest ($y(0^-) = 0$). The second term, due to the initial condition, is called the *stored energy response*.

The following example introduces *partial fraction expansion*.

EXAMPLE 6CT.12   Step Applied to a First-Order System with an Initial Condition

Assume that $x(t) = u(t)$ is applied to the system in Figure 6CT.10(a). If we substitute $X(s) = 1/s$ from Table 6CT.2 into Eq. (67), we obtain

$$Y(s) = \frac{1}{s(s\tau_c + 1)} + \frac{\tau_c}{s\tau_c + 1} y(0^-) \tag{73}$$

Partial fraction expansion is necessary because the first term on the right-hand side does not appear in Table 6CT.2. We denote this term as $G(s)$. To make table lookup possible, we expand it into partial fractions.

$$G(s) = \frac{1}{s(s\tau_c + 1)} = \frac{K_1}{s} + \frac{K_2}{s\tau_c + 1} \tag{74}$$

The coefficients K_1 and K_2 are constants to be determined. There are several ways to find the values for the coefficients in a partial fraction expansion. One of the quickest, called *Heaviside's technique*, works as follows: To find K_1, we multiply both sides of Eq. (74) by s and then set $s = 0$. The multiplication step yields

$$sG(s) = \not{s}\frac{1}{\not{s}(s\tau_c + 1)} = \not{s}\frac{K_1}{\not{s}} + s\frac{K_2}{s\tau_c + 1} \tag{75}$$

The slashes show the common factors we eliminated. When we set $s = 0$, the preceding equation becomes

$$sG(s)\Big|_{s=0} = 1 = K_1 + 0 \tag{76}$$

so

$$K_1 = 1 \tag{77}$$

Similarly, to find K_2, we multiply Eq. (74) by $s\tau_c + 1$ and then set $s = -1/\tau_c$. The multiplication step yields

$$(s\tau_c + 1)G(s) = \frac{\cancel{(s\tau_c + 1)}1}{s\cancel{(s\tau_c + 1)}} = \frac{(s\tau_c + 1)K_1}{s} + \frac{\cancel{(s\tau_c + 1)}K_2}{\cancel{s\tau_c + 1}} \tag{78}$$

When we set $s = -1/\tau_c$, this becomes

$$(s\tau_c + 1)G(s)\Big|_{s=-\frac{1}{\tau_c}} = -\tau_c = 0 + K_2 \tag{79}$$

The substitution $K_1 = 1$ and $K_2 = -\tau_c$ into Eq. (54) yields

$$G(s) = \frac{1}{s(s\tau_c + 1)} = \frac{1}{s} - \frac{\tau_c}{s\tau_c + 1} \tag{80}$$

which, in turn, yields

$$\mathbf{Y}(s) = \frac{1}{s} - \frac{\tau_c}{s\tau_c + 1} + \frac{\tau_c}{s\tau_c + 1}y(0^-) \tag{81}$$

Each term in $\mathbf{Y}(s)$ now appears in Table 6CT.2. We can write the time domain output by inspection

$$y(t) = 1 - e^{-\frac{t}{\tau_c}} + y(0^-)e^{-\frac{t}{\tau_c}} \quad t > 0 \tag{82}$$

The first two terms are called the forced response. The last term is the stored energy response.

Now that we have seen the preliminary examples, let's develop the four basic steps formally and generally. The strength of the formal development is that it shows how to solve a large variety of problems. However, it takes time to understand the notation and analytic results in Section 6CT.6.2 and 6CT.6.3. We recommend that you read the developments the first time to gain a little familiarity with it. After that, focus on the examples and look back to the general development as needed.

**6CT.6.2
General Forms
of Time and
s-Domain
Equations of LTI
System Having
Initial Conditions**

In this section we turn our attention to arbitrary systems composed of adders, gains, integrators, and differentiators. These systems can be called *lumped* linear time-invariant systems.[4] Their I-O equations have the form

$$a_n y^{(n)}(t) + a_{n-1}y^{(n-1)}(t) + \cdots + a_1 y^{(1)}(t) + a_0 y(t)$$
$$= b_m x^{(m)}(t) + b_{m-1}x^{(m-1)}(t) + \cdots + b_1 x^{(1)}(t) + b_0 x(t) \tag{83}$$

where $x(t)$ and $y(t)$ are the input and output, respectively, and n is the order of the system. Equation (83) is an ordinary linear differential equation having constant coefficients. Our objective is to solve it for $y(t)$ subject to initial conditions $y^{(n-1)}(0^-), \ldots, y(0^-)$, $x^{m-1}(0^-), \ldots, x(0^-)$.

To obtain the solution, we transform Eq. (83) into the s-domain using the linearity and differentiation properties. The result is

$$a_n s^n \mathbf{Y}(s) + a_{n-1}s^{n-1}\mathbf{Y}(s) + \cdots + a_1 s\mathbf{Y}(s) + a_0\mathbf{Y}(s) + \mathbf{Y}_{\text{ic}}(s)$$
$$= b_m s^m \mathbf{X}(s) + b_{m-1}s^{m-1}\mathbf{X}(s) + \cdots + b_1 s\mathbf{X}(s) + b_0\mathbf{X}(s) + \mathbf{X}_{\text{ic}}(s) \tag{84}$$

---

[4]Systems that include ideal delay elements are called *distributed* systems because the waveforms they store must be distributed along some dimension. An example is the use of the recording tape to contain the audio performance of the singer in Figure 1.1. The audio waveform of the singer's voice is distributed along the tape. Distributed systems cannot be described by ordinary differential equations.

where $\mathbf{Y}_{ic}(s)$ and $\mathbf{X}_{ic}(s)$ are polynomials in s with degree $n - 1$ and $m - 1$, respectively. These polynomials arise from the differentiation property and are present only when the initial conditions $y^{(n-1)}(0^-), \ldots, y(0^-), x^{m-1}(0^-), \ldots, x(0^-)$ are not 0. An example is Eq. (66), which can be arranged into the form of (84).

It will be convenient to write Eq. (83) more compactly as

$$\mathcal{A}(p)y(t) = \mathcal{B}(p)x(t) \tag{85}$$

where

$$\mathcal{A}(p) = a_n p^n + a_{n-1} p^{n-1} + \cdots + a_1 p + a_0 \tag{86}$$

$$\mathcal{B}(p) = b_m p^m + b_{m-1} p^{m-1} + \cdots + b_1 p + b_0 \tag{87}$$

and

$$p^k = \frac{d^k}{dt^k} \tag{88}$$

for $k = 1, 2, \ldots$. The s domain Eq. (84) is then given by

$$\mathcal{A}(s)\mathbf{Y}(s) + \mathbf{Y}_{ic}(s) = \mathcal{B}(s)\mathbf{X}(s) + \mathbf{X}_{ic}(s) \tag{89}$$

The solution is

$$\mathbf{Y}(s) = \frac{\mathcal{B}(s)}{\mathcal{A}(s)}\mathbf{X}(s) + \frac{\mathbf{X}_{ic}(s) - \mathbf{Y}_{ic}(s)}{\mathcal{A}(s)} \tag{90}$$

where

$$\frac{\mathcal{B}(s)}{\mathcal{A}(s)} = \frac{b_m s^m + b_{m-1} s^{m-1} + \cdots + b_1 s + b_0}{a_n s + a_{n-1} s^{n-1} + \cdots + a_1 s + a_0} \tag{91}$$

The first term on the right hand side of corresponds to the system's forced response. It corresponds to the part of the output caused by $x(t)$ alone. The factor multiplying $\mathbf{X}(s)$ is called the system's *transfer function*.

$$\mathbf{H}(s) = \frac{\mathcal{B}(s)}{\mathcal{A}(s)} = \mathcal{L}\{h(t)\} \tag{92}$$

This equation states that $\mathbf{H}(s)$ is the Laplace transform of the system's impulse response $h(t)$. We can establish this relationship by recalling that $y(t) = h(t)$ when $x(t) = \delta(t)$ and the system is initially at rest (all initial conditions are 0). If we apply these conditions to Eq. (90), we find that $\mathbf{Y}(s) = \mathbf{H}(s)$ when $\mathbf{X}(s) = 1$ and $\mathbf{X}_{ic}(s) = \mathbf{Y}_{ic}(s) = 0$. The second term in Eq. (90) corresponds to the system's stored energy response. It corresponds to the part of the output caused by nonzero initial conditions.

A transfer function is specified by its poles and zeros. We show property this by expressing $\mathbf{H}(s)$ in factored form

$$\mathbf{H}(s) = \frac{\mathcal{B}(s)}{\mathcal{A}(s)} = \frac{b_m(s - z_{h1})(s - z_{h2}) \cdots (s - z_{hm})}{a_n(s - p_{h1})(s - p_{h2}) \cdots (s - p_{hn})} = \frac{b_m \Pi(s - z_{hi})}{a_n \Pi(s - p_{hi})} \tag{93}$$

where $p_{h1}, p_{h2}, \ldots, p_{hn}$, are the poles of $\mathbf{H}(s)$. The poles are the n roots to

$$\mathcal{A}(s) = a_n s^n + a s^{n-1} + \cdots + a_1 s + a_0 = 0 \tag{94}$$

Equation (94) is called the *characteristic equation associated with* Eq. (85). The poles of $\mathbf{H}(s)$ characterize the dynamic response of the system, independent of any applied input.

The poles are therefore also called the *complex natural frequencies* of the system. The expression in the denominator of Eq. (93) is the factored form of $A(s)$, i.e.

$$A(s) = a_n(s - p_{h1})(s - p_{h2}) \cdots (s - p_{hn}) \tag{95}$$

Similarly, the quantities $z_{h1}, z_{h2}, \ldots, z_{hm}$ are the zeros of $\mathbf{H}(s)$. They are the roots to

$$B(s) = b_m s^m + b_{m-1} s^{m-1} + \cdots + b_1 s + b_0 = 0 \tag{96}$$

The expression in the numerator of Eq. (93) is the factored form of $B(s)$.

$$B(s) = b_m(s - z_{h1})(s - z_{h2}) \cdots (s - z_{hm}) \tag{97}$$

The poles and zeros of $\mathbf{H}(s)$, along with the constant b_m/a_n, provide a complete description of the system. This means that we can obtain $\mathbf{H}(s)$, $h(t)$, and the differential equation of the system if we know b_m/a_n and the poles and zeros.

For many practical inputs, $\mathbf{X}(s)$ can also be written in factored form

$$\mathbf{X}(s) = \frac{\mathcal{C}(s)}{\mathcal{D}(s)} = k \frac{(s - z_{x1})(s - z_{x2}) \cdots (s - z_{xq})}{(s - p_{x1})(s - p_{x2}) \cdots (s - ps_{xp})} = k \frac{\Pi(s - z_{xi})}{\Pi(s - p_{xi})} \tag{98}$$

where the p_{xi} and z_{xi} are the poles and zeros of $\mathbf{X}(s)$. The poles of $\mathbf{X}(s)$ characterize the time constants and natural frequencies associated with the input (the forcing function) $x(t)$. The substitution of Eqs. (93), (95), and (98) into Eq. (90) yields

$$\mathbf{Y}(s) = c \frac{\Pi(s - z_{hi})\Pi(s - z_{xi})}{\Pi(s - p_{hi})\Pi(s - p_{xi})} + \frac{\mathbf{X}_{\text{ic}}(s) - \mathbf{Y}_{\text{ic}}(s)}{a_n \Pi(s - p_{hi})} \tag{99}$$

where $c = kb_m/a_n$. An example of Eq. (99) is given by Eq. (73).

6CT.6.3 Partial Fraction Expansions

Partial fraction expansion is used to put $\mathbf{Y}(s)$ in a form ready for table lookup. Let's consider how to find the partial fraction expansion of any term in $\mathbf{Y}(s)$ that has the form

$$G(s) = K \frac{(s - z_1)(s - z_2) \cdots (s - z_q)}{(s - p_1)(s - p_2) \cdots (s - p_l)} \tag{100}$$

The following terminology is used to describe $G(s)$.

Terminology

$G(s)$ is called a *proper rational* algebraic fraction if $q < l$, and an *improper rational* algebraic fraction if $q \geq l$. A pole p_i or a zero z_j has *order* k if its value repeats $k - 1$ times. (i.e., the value occurs k times). First-order poles or zeros are called *simple*.

The form of the partial fraction expansion of $G(s)$ depends on whether it is proper or improper and on the orders of the poles. In many practical problems, $G(s)$ is proper, and the poles are all different (all simple poles). This frequently occurring case is the simplest.

Proper G(s) and All Simple Poles

If $G(s)$ is proper and all the poles are simple, then its partial fraction expansion has the form

$$G(s) = \frac{K_1}{s - p_1} + \frac{K_2}{s - p_2} + \cdots + \frac{K_i}{s - p_i} + \cdots + \frac{K_l}{s - p_l} \tag{101}$$

The coefficients $K_1, K_2, \ldots,$ and K_l are called *residues*. Each residue can be evaluated using *Heaviside's formula*.

$$K_i = \mathcal{Res}\{G(s); p_i\} = (s - p_i)G(s)\Big|_{s=p_i} \quad i = 1, 2, \ldots, l \quad (102)$$

where $\mathcal{Res}\{G(s); s_i\}$ denotes the residue of $G(s)$ at the pole $s = s_i$. The time domain function associated with Eq. (101) is

$$g(t) = K_1 e^{p_1 t} + K_2 e^{p_2 t} + \cdots + K_l e^{p_l t} \quad t > 0 \quad (103)$$

In most applications, $g(t)$ represents the real valued output of a real system having a real valued input. For these applications, any complex poles and their associated residues occur in complex conjugate pairs. If there is a complex pole $p = \sigma + j\Omega$ with residue K, then there is also complex pole $p^* = \sigma - j\Omega$ with residue K^*. The partial fraction expansion (101) and the associated waveform Eq. (103), in turn, contain the terms

$$\frac{K}{s - p} + \frac{K^*}{s - p^*} \overset{LT}{\longleftrightarrow} K e^{pt} + K^* e^{p^* t} = 2|K| e^{\sigma t} \cos(\Omega t + \angle K) \quad (104)$$

We can prove this important statement by recognizing that, if the term $K^*/(s - p^*)$ did not appear along with $K/(s - p)$ in Eq. (104), then $g(t)$ could not be real. The result (104) implies that we need to evaluate only one residue of a complex pair of residues for real systems having real inputs. We can write down the real damped sinusoid on the right-hand side of Eq. (104) directly from knowledge of p and K.

EXAMPLE 6CT.13   Impulse Response of a Resonant Second-Order System

Second-order systems are described by second-order differential equations. The highest-order derivative on the output $y(t)$ equals 2. In this problem we find the impulse response of the second-order system characterized by

$$y^{(2)}(t) + 2\alpha y^{(1)}(t) + \Omega_o^2 y(t) = \Omega_o^2 x(t) \quad (105)$$

The constants α and Ω_o are called the *damping coefficient* and the *undamped natural frequency*. A block diagram of this system is shown in Figure 6CT.11(a). Circuit enthusiasts may recognize that this block diagram represents the *RLC* filter of Figure 6CT.11(b), where $\alpha = R/2L$ and $\Omega_o = \sqrt{1/LC}$.

Figure 6CT.11
Second-order system:
(a) block diagram;
(b) realization as a
series *RLC* circuit
where $\alpha = R/2L$ and
$\Omega_o = \sqrt{1/LC}$.

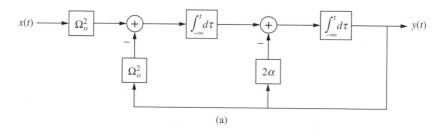

(a)

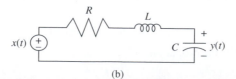

(b)

The s domain I-O equation is

$$s^2Y(s) - sy^{(1)}(0^-) - y(0^-) + 2\alpha sY(s) - 2\alpha y(0^-) + \Omega_o^2 Y(s) = \Omega_o^2 X(s) \tag{106}$$

The solution is

$$Y(s) = \frac{\Omega_o^2}{s^2 + 2\alpha s + \Omega_o^2}X(s) + \frac{sy^{(1)}(0^-) + (1 + 2\alpha)y(0^-)}{s^2 + 2\alpha s + \Omega_o^2} \tag{107}$$

The factor multiplying $X(s)$ is the system's transfer function.

$$H(s) = \mathcal{L}\{h(t)\} = \frac{\Omega_o^2}{s^2 + 2\alpha s + \Omega_o^2} \tag{108}$$

To expand $H(s)$ into partial fractions, we need to find the poles. The poles are the roots to

$$s^2 + 2\alpha s + \Omega_o^2 = 0 \tag{109}$$

We assume that $\Omega_o^2 > \alpha^2$ for which the roots are complex conjugates

$$p_{h1} = -\alpha + j\Omega_d \qquad p_{h2} = -\alpha - j\Omega_d \tag{110}$$

where

$$\Omega_d = \sqrt{\Omega_o^2 - \alpha^2} \tag{111}$$

The system is called *resonant* when $\Omega_o^2 > \alpha$. Ω_d is the *damped natural frequency* of the system. The pole-zero plot of the resonant system is shown in Figure 6CT.12. As shown in the figure, the poles are located on a circular path having radius Ω_o. For fixed Ω_o, the poles approach the $j\Omega$ axis of the s-plane as α decreases from Ω_o toward 0.

Figure 6CT.12
Pole-zero plot of
resonant
second-order system.

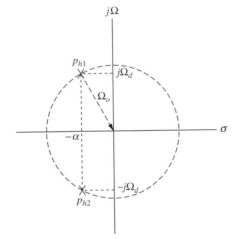

The partial fraction expansion is

$$H(s) = \frac{\Omega_o^2}{(s - p_{h1})(s - p_{h2})} = \frac{K_1}{s - p_{h1}} + \frac{K_2}{s - p_{h2}} \tag{112}$$

Heaviside' technique yields

$$K_1 = \frac{\Omega_o^2}{p_{h1} - p_{h2}} = \frac{\Omega_o^2}{2j\Omega_d} = \frac{\Omega_o^2}{2\Omega_d}\angle -90° \tag{113}$$

Figure 6CT.13
Impulse response of resonant second-order system.

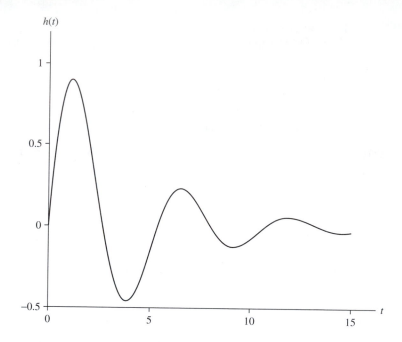

We know that $K_2 = K_1^*$ because the system and the input are real. (The reader is encouraged to use Heaviside's technique to confirm that $K_2 = K_1^*$.) Therefore Eq. (104) applies and we can write down

$$h(t) = \frac{\Omega_o^2}{\Omega_d} e^{-\alpha t} \cos(\Omega_d t - 90°) = \frac{\Omega_o^2}{\Omega_d} e^{-\alpha t} \sin(\Omega_d t) \tag{114}$$

A plot of $h(t)$ is given in Figure 6CT.13 for $\Omega_o = 1.2$, and $\alpha = 0.25$. The damped natural frequency $\Omega_d = 1.17367$.

G (s) with Second-Order Poles

Occasionally, we need to evaluate a partial fraction expansion involving one or more multiple-order poles in addition to simple poles. Let's consider first the most common case in which G(s) contains both simple poles and one or more second-order poles. The partial fraction expansion of G(s) includes terms of the form

$$\frac{K}{s - p} \overset{\text{LT}}{\longleftrightarrow} K e^{pt} \tag{115}$$

for each simple pole and two terms of the form

$$\frac{L_1}{s - p} + \frac{L_2}{(s - p)^2} \overset{\text{LT}}{\longleftrightarrow} L_1 e^{pt} + L_2 t e^{pt} \quad n \geq 0 \tag{116}$$

for each second-order pole. Only the coefficients K and L_1 are residues. We can find L_1 and L_2 from the formulas

$$L_2 = [(s - p)^2 G(s)]\big|_{s=p} \tag{117}$$

and

$$L_1 = \frac{d}{ds}[(s - p)^2 G(s)]\big|_{s=p} \tag{118}$$

As with simple poles, any complex poles and their associated partial fraction coefficients occur in complex conjugate pairs for real $g(t)$.

The following example illustrates partial fraction expansion involving a second-order real pole.

EXAMPLE 6CT.14   Ramp Response of a First-Order System

Consider the response of the first-order system of Example 6CT.10 to $x(t) = r(t)$ with $y(0^-) = 0$. If we substitute $y(0^-) = 0$ and $X(s) = 1/s^2$ from Table 6CT.2 into Eq. (67), we obtain

$$Y(s) = \frac{1}{s^2(\tau_c s + 1)} = \frac{\frac{1}{\tau_c}}{s^2\left(s + \frac{1}{\tau_c}\right)} \tag{119}$$

$Y(s)$ has a second-order pole at $s = 0$ and a first-order pole at $s = -1/\tau_c$. Therefore, the partial fraction expansion has the form

$$Y(s) = \frac{\frac{1}{\tau_c}}{s^2\left(s + \frac{1}{\tau_c}\right)} = \frac{K_1}{s + \frac{1}{\tau_c}} + \frac{L_1}{s} + \frac{L_2}{s^2} \tag{120}$$

We can use Heaviside's technique to evaluate K_1.

$$\left(s + \frac{1}{\tau_c}\right)Y(s)\bigg|_{s=-\frac{1}{\tau_c}} = \frac{\frac{1}{\tau_c}}{s^2}\bigg|_{s=-\frac{1}{\tau_c}} = K_1 + \left(s + \frac{1}{\tau_c}\right)\frac{L_1}{s^2}\bigg|_{s=-\frac{1}{\tau_c}} + \left(s + \frac{1}{\tau_c}\right)\frac{L_2}{s}\bigg|_{s=-\frac{1}{\tau_c}} \tag{121}$$

which yields

$$K_1 = \tau_c \tag{122}$$

We can use a modification of Heaviside's technique to evaluate L_2.

$$s^2 Y(s)\bigg|_{s=0} = \frac{\frac{1}{\tau_c}}{\left(s + \frac{1}{\tau_c}\right)}\bigg|_{s=0} = s^2\frac{K_1}{s + \frac{1}{\tau_c}}\bigg|_{s=0} + sL_1\bigg|_{s=0} + L_2 \tag{123}$$

which yields

$$L_2 = 1 \tag{124}$$

If we substitute $K_1 = \tau_c$ and $L_2 = 1$ into Eq. (120), we obtain

$$Y(s) = \frac{\frac{1}{\tau_c}}{s^2\left(s + \frac{1}{\tau_c}\right)} = \frac{\tau_c}{s + \frac{1}{\tau_c}} + \frac{L_1}{s} + \frac{1}{s^2} \tag{125}$$

The constant L_1 can be evaluated in several ways. One method is to substitute a convenient value for s into (125) and solve for L_1. For example, if we set $s = 1$, we obtain

$$\frac{\frac{1}{\tau_c}}{1 + \frac{1}{\tau_c}} = \frac{\tau_c}{1 + \frac{1}{\tau_c}} + L_1 + 1 \tag{126}$$

which yields $L_1 = -\tau_c$. We can also obtain this result by taking the limit $s \to \infty$ in Eq. (125). (We leave this method as an exercise.) A third way is to use Eq. (118)

$$L_1 = \frac{d}{ds}[s^2 Y(s)]\bigg|_{s=0} = \frac{d}{ds}\left(\frac{\frac{1}{\tau_c}}{s + \frac{1}{\tau_c}}\right)\bigg|_{s=0} = -\frac{\frac{1}{\tau_c}}{\left(s + \frac{1}{\tau_c}\right)^2}\bigg|_{s=0} = -\tau_c \tag{127}$$

which checks our previous result. Therefore

$$Y(s) = \frac{1}{s^2} - \frac{\tau_c}{s} + \frac{\tau_c}{s + \frac{1}{\tau_c}} \tag{128}$$

Table lookup gives the corresponding time domain response

$$y(t) = t - \tau_c + \tau_c e^{-t/\tau_c} \quad t > 0 \tag{129}$$

The stored energy response is 0 because we set $y(0^-) = 0$. All terms are part of the forced response.

$G(s)$ with mth-Order Poles

Here, the partial fraction expansion of $G(s)$ includes terms of the form

$$\frac{K}{s - p} \overset{\text{LT}}{\longleftrightarrow} K e^{pt} \tag{130}$$

for each simple pole and m terms of the form

$$\frac{L_1}{s - p} + \frac{L_2}{(s - p)^2} + \cdots + \frac{L_m}{(s - p)^m} \tag{131}$$

for each mth-order pole where $m = 2, 3, \ldots$. The coefficients K and L_1, associated with the first-degree denominators, are residues. The coefficient L_i is given by

$$L_i = \frac{1}{(m - i)!} \frac{d^{m-i}}{ds^{m-i}} [(s - p)^m G(s)] \Big|_{s=p} \tag{132}$$

for $i = 1, 2, ,\ldots, m$. The waveform corresponding to the ith term in Eq. (131) is

$$\frac{L_i}{(s - p)^i} \overset{\text{LT}}{\longleftrightarrow} \frac{L_i}{(i - 1)!} t^{i-1} e^{pt} \quad t \geq 0 \tag{133}$$

for $i = 1, 2, \ldots, m$.

$G(s)$ as an Improper Rational Algebraic Fraction ($q \geq l$)

If $G(s)$ is improper, we can use long division to put it in the form

$$G(s) = c_0 + c_1 s + \cdots + c_k s^{q-l} + G_1(s) \tag{134}$$

where $G_1(s)$ is a proper algebraic fraction. The method of long division is illustrated in the following example.

EXAMPLE 6CT.15   Impulse Response of a High-Pass Filter

Figure 6CT.14 shows the circuit diagram of a first-order high-pass filter. The circuit is the same as that considered in Example 2CT.3, except that the output is taken across the resistor instead of the capacitor. Let's find the impulse response $h(t)$ of this filter. The I-O equation is

$$\tau_c y^{(1)}(t) + y(t) = \tau_c x^{(1)}(t) \tag{135}$$

Figure 6CT.14
RC high-pass filter.

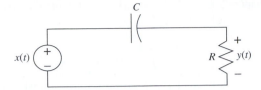

The corresponding s domain equation is

$$\tau_c[s\mathbf{Y}(s) - y(0^-)] + \mathbf{Y}(s) = \tau_c[s\mathbf{X}(s) - x(0^-)] \tag{136}$$

The solution is

$$\mathbf{Y}(s) = \frac{s\tau_c}{s\tau_c + 1}\mathbf{X}(s) + \frac{\tau_c}{s\tau_c + 1}[y(0^-) - x(0^-)] \tag{137}$$

The factor multiplying $\mathbf{X}(s)$ in the preceding equation is $\mathbf{H}(s)$.

$$\mathbf{H}(s) = \frac{s\tau_c}{s\tau_c + 1} \tag{138}$$

Since the degree of the numerator and denominator polynomials in $\mathbf{H}(s)$ are equal, the partial fraction expansion has the form

$$\mathbf{H}(s) = c_0 + \frac{K_1}{\tau_c s + 1} \tag{139}$$

One way to find c_0 is by long division. We divide the numerator in Eq. (138) by the denominator using the rules of long division as follows.

$$
\begin{array}{r}
1 \\
s\tau_c + 1 \overline{)s\tau_c } \\
\underline{s\tau_c + 1} \\
-1
\end{array}
$$

Therefore

$$\mathbf{H}(s) = 1 - \frac{1}{s\tau_c + 1} \tag{140}$$

and table lookup yields

$$h(t) = \delta(t) - \frac{1}{\tau_c}e^{-t/\tau_c} \tag{141}$$

We can explain this result by inspection of the circuit. The voltage impulse at the input produces a current impulse $(1/R)\delta(t)$. This current impulse produces a voltage impulse $R \times (1/R)\delta(t) = \delta(t)$ across the terminals of the resistor and charges the capacitor to $1/\tau_c$, as was shown in Example 2CT.3. The term $-(1/\tau_c)e^{-t/\tau_c}$ represents the discharge of the capacitor with time constant τ_c that occurs for $t > 0$.

The methods described in this subsection generalize easily to inputs and systems that include ideal delay (see Problems 6CT.6.18–6CT.6.23).

6CT.6.4
Stability

We showed in Chapter 2 that an LTI system is BIBO stable if and only if its impulse response $h(t)$ is absolutely integrable. For a causal system, this necessary and sufficient condition is written as

$$\int_0^\infty |h(t)|\, dt < \infty \tag{142}$$

A causal LTI system is characterized by the Laplace transform $\mathbf{H}(s) = \mathcal{L}\{h(t)\}$. For a lumped LTI system, $\mathbf{H}(s)$ and $h(t)$ can be expanded into terms having the forms

$$\underbrace{\frac{1}{(s - p_h)^k}}_{\text{Typical term in } H(s)} \sim \underbrace{\frac{1}{(k-1)!} t^{k-1} e^{p_h t}}_{\substack{\text{Corresponding} \\ \text{typical term in } h(t)}} \quad t > 0 \tag{143}$$

where p_h is a pole of $\mathbf{H}(s)$ having order k. It follows that a necessary and sufficient condition for a causal LTI system to be BIBO stable is that all the poles of $\mathbf{H}(s)$ lie in the left half of the s-plane.

A simple example is given by the familiar first-order system of Example 6CT.12 where $h(t) = \alpha e^{-\alpha t} u(t)$ and $\alpha = 1/\tau_c$. This system is BIBO stable if and only $\alpha > 1$, which means that the pole of $\mathbf{H}(s) = 1/(s + \alpha)$ is in the left half of the s-plane.

6CT.7   Response to Sinusoid Applied at $t = 0$

In this section we derive the response of a real LTI system to a sinusoid applied at $t = 0$. The result reveals that there is a fundamental relationship between a stable system's transient response and its steady-state sinusoidal response, as developed in Chapter 3CT. The relationship is best understood geometrically using the system's pole-zero plot. We assume throughout this section that all the poles of the system are simple and that the system is initially at rest.

Assume that the input to an LTI system at rest is a sinusoid

$$x(t) = A\cos(\Omega_x t + \phi) u(t) \tag{144}$$

The s domain response is

$$\mathbf{Y}(s) = \mathbf{H}(s)\mathbf{X}(s) \tag{145}$$

where

$$\mathbf{X}(s) = \left(\frac{X}{s - j\Omega_x} + \frac{X^*}{s + j\Omega_x} \right) \tag{146}$$

with $X = \tfrac{1}{2} A e^{j\phi}$. We can use Eqs. (93) and (146) to express $\mathbf{Y}(s)$ as

$$\mathbf{Y}(s) = \underbrace{\frac{b_m \, \Pi(s - z_{hi})}{a_n \, \Pi(s - p_{hi})}}_{\mathbf{H}(s)} \times \underbrace{\left[\frac{(s + j\Omega_x)X + (s - j\Omega_x)X^*}{(s - j\Omega_x)(s + j\Omega_x)} \right]}_{\mathbf{X}(s)} \tag{147}$$

where $p_{h1}, p_{h2}, \cdots, p_{hn}$ are the poles of $\mathbf{H}(s)$ and $\pm j\Omega_x$ are the poles of $\mathbf{X}(s)$. Let's assume that the poles of $\mathbf{Y}(s)$ are all simple. Because the poles are simple, the partial fraction expansion of $\mathbf{Y}(s)$ is

$$\mathbf{Y}(s) = \mathbf{H}(s)\mathbf{X}(s) = \frac{K_{h1}}{s - p_{h1}} + \frac{K_{h2}}{s - p_{h2}} + \cdots + \frac{K_{hn}}{s - p_{hn}} + \frac{K_{x1}}{s - j\Omega_x} + \frac{K_{x2}}{s + j\Omega_x} \tag{148}$$

We obtain the residues using Heaviside's technique.

$$K_{x1} = \mathcal{R}es\{\mathbf{H}(s)\mathbf{X}(s); s = j\Omega_x\} = (s - j\Omega_x)\mathbf{H}(s)\mathbf{X}(s)\big|_{s=j\Omega_x} = \mathbf{H}(j\Omega_x)X \quad (149)$$

and,

$$K_{x2} = \mathcal{R}es\{\mathbf{H}(s)\mathbf{X}(s); s = -j\Omega_x\} = (s + j\Omega_x)\mathbf{H}(s)\mathbf{X}(s)\big|_{s=-j\Omega_x} = \mathbf{H}^*(j\Omega_x)X^* \quad (150)$$

where we have used the fact that $\mathbf{H}(-j\Omega_x) = \mathbf{H}^*(j\Omega_x)$ for a real system. Notice that $K_{x2} = K_{x1}^*$. This result is to be expected because we are finding the response of a real system to a real input. Similarly

$$K_{hi} = \mathcal{R}es\{\mathbf{H}(s)\mathbf{X}(s); s = p_{hi}\} = (s - p_{hi})\mathbf{H}(s)\mathbf{X}(s)\big|_{s=p_{hi}} = \mathbf{X}(p_{hi})\mathcal{R}es\{\mathbf{H}(s); p_{hi}\} \quad (151)$$

for $i = 1, 2, \ldots, n$. Table lookup yields the output

$$y(t) = A|\mathbf{H}(j\Omega_x)|\cos(\Omega_x t + \phi + \angle\mathbf{H}(j\Omega_x)) + \sum_{i=1}^{n} \mathbf{X}(p_{hi})\mathcal{R}es\{\mathbf{H}(s); p_{hi}\}e^{p_{hi}t} \quad (152)$$

All the terms in $y(t)$ are part of the forced response. Any complex poles and coefficients in the summation occur in conjugate pairs and, for these poles, simplification (104) applies. If the system poles $p_{h1}, p_{h2}, \ldots p_{hl}$ are all in the left half of the s plane (stable system), then the terms in the summation are transients that become negligible for sufficiently large t. When all the transient terms are negligibly small, the response has then reached the sinusoidal steady state

$$\begin{aligned} y(t) &= A|\mathbf{H}(j\Omega_x)|\cos(\Omega_x t + \phi + \angle\mathbf{H}(j\Omega_x)) \\ &= A|H(F_x)|\cos(\Omega_x t + \phi + \angle H(F_x)) \end{aligned} \quad (153)$$

where we used the relation between the Laplace and Fourier transforms $\mathbf{H}(j2\pi F) = H(F)$ developed in Section 6CT.3.

The sinusoidal-steady-state response given by Eq. (153) is identical to the result we obtained in Chapter 3CT (Equation (3CT.19)).

EXAMPLE 6CT.16   Response of a First-Order System to a Cosine Applied at $t = 0$.

Consider the response of the first-order system

$$\mathbf{H}(s) = \frac{1}{s\tau_c + 1} = \frac{\frac{1}{\tau_c}}{s + \frac{1}{\tau_c}} \quad (154)$$

to

$$x(t) = A\cos(\Omega_x t)u(t) \overset{LT}{\longleftrightarrow} \mathbf{X}(s) = \frac{sA}{(s - j\Omega_x)(s + j\Omega_x)} \quad (155)$$

We can write down the expression for $y(t)$ by referring to Eq. (152). In this example, $n = 1$ and we can find $\mathbf{H}(j2\pi F)$, p_{h1}, K_{h1}, and $\mathbf{X}(p_{h1})$ easily from the preceding equations. The frequency response characteristic of the system is

$$\mathbf{H}(j2\pi F) = H(F) = \frac{1}{j2\pi F\tau_c + 1} = \frac{1}{\sqrt{(2\pi F\tau_c)^2 + 1}}\angle{-\arctan(2\pi F\tau_c)} \quad (156)$$

Figure 6CT.15
Response of a
first-order system to
a cosine applied at
$t = 0$. $A = 1$
$F_x = 1$ and $\tau_c = 1$.

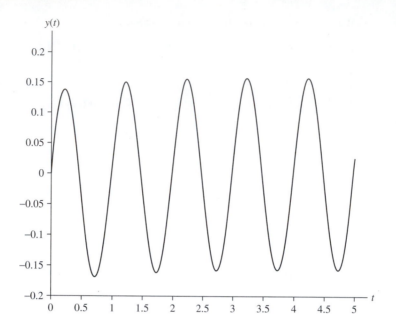

The system has one pole

$$p_{h1} = -\frac{1}{\tau_c} \tag{157}$$

The residue at that pole is

$$Res\{\mathbf{H}(s); p_{hi}\} = Res\left\{\mathbf{H}(s); -\frac{1}{\tau_c}\right\} = (s + 1/\tau_c)\frac{1/\tau_c}{s + 1/\tau_c}\Big|_{s=-\frac{1}{\tau_c}} = \frac{1}{\tau_c} \tag{158}$$

and the s-domain input evaluated at that pole is

$$\mathbf{X}(p_{h1}) = \frac{-\frac{1}{\tau_c}A}{\left(-\frac{1}{\tau_c} - j\Omega_x\right)\left(-\frac{1}{\tau_c} + j\Omega_x\right)} = \frac{-\tau_c A}{(\Omega_x \tau_c)^2 + 1} \tag{159}$$

Therefore, by Eq. (152)

$$y(t) = A\frac{1}{\sqrt{(\Omega_x \tau_c)^2 + 1}} \cos(\Omega_x t - \arctan(\Omega_x \tau_c)) - \frac{A}{(\Omega_x \tau_c)^2 + 1}e^{-t/\tau_c} \tag{160}$$

The first term in $y(t)$ is the sinusoidal steady-state response. This term can be obtained using the frequency domain methods of Chapter 3CT. There is no stored energy response because the initial conditions were 0. Both terms in $y(t)$ are part of the forced response. The second term is a transient that becomes negligible as t increases beyond a few time constants τ_c. A plot of $y(t)$ is given in Figure 6CT.15 for $A = 1$, $F_x = 1$, and $\tau_c = 1$.

Let's focus our attention on *the sinusoidal steady state response* term in Eq. (160). The amplitude and phase are determined by the frequency response characteristic of Eq. (156). A graphical interpretation of this characteristic is given in the pole-zero diagram Figure 6CT.16 where the denominator term in Eq. (154), $s + (1/\tau_c)$ is drawn as a vector that starts at the pole $s = p_h = -1/\tau_c$ and ends at the point $s = j2\pi F$. If we write $\mathbf{H}(j2\pi F)$ as

$$\mathbf{H}(j2\pi F) = H(F) = \frac{\frac{1}{\tau_c}}{j2\pi F - p_h} = \frac{\frac{1}{\tau_c}}{|j2\pi F - p_h|} \angle - (j2\pi F - p_h) \tag{161}$$

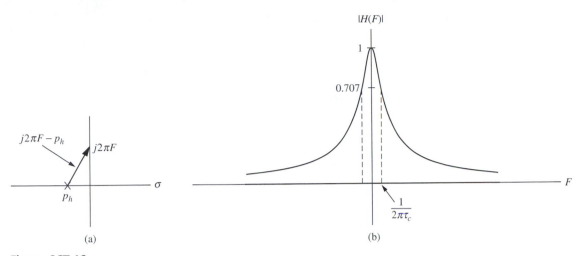

Figure 6CT.16
(a) Pole-zero plot; (b) amplitude response characteristic.

we see that the amplitude of the frequency response characteristic varies inversely as the length of this vector, and the phase of the frequency response characteristic varies as the negative of the angle of the vector. This geometric interpretation provides insight into the relation between the input frequency F and the system's own dynamics as represented by the pole. In the pole-zero plot, $j2\pi F$ describes the frequency of the input, and $p_h = -1/\tau_c$ describes the natural frequency of the system. Notice that $|H(F)|$ is relatively large when the point $j2\pi F$ is close to p_h and relatively small when $j2\pi F$ is far from p_h. Stated another way: $|H(F)|$ is large when the frequency of the input is close to the natural frequency of the system; $|H(F)|$ is small when the frequency of the input is far from the natural frequency of the system.

The next example further illustrates how a pole-zero diagram can provide insight into the relationship between a system's sinusoidal steady-state response and the system's poles.

EXAMPLE 6CT.17   Response of a Resonant Second-Order System to a Cosine Applied at $t = 0$

Consider the sinusoidal steady-state response of the second-order system

$$\mathbf{H}(s) = \frac{\Omega_o^2}{s^2 + 2\alpha s + \Omega_o^2} = \frac{\Omega_o^2}{(s - p_{h1})(s - p_{h2})} \tag{162}$$

with

$$p_{h1} = -\alpha + j\Omega_d \quad p_{h2} = p_{h1}^* \tag{163}$$

to

$$x(t) = A\cos(\Omega_x t)u(t) \overset{\text{LT}}{\longleftrightarrow} \mathbf{X}(s) = \frac{sA}{(s - j\Omega_x)(s + j\Omega_x)} \tag{164}$$

We can write down the expression for $y(t)$ by referring to Eq. (152). In this example, $n = 2$ but we only need $\mathbf{H}(j2\pi F)$, p_{h1} and K_{h1}, because the simplification Eq. (104) applies. The frequency response characteristic of the system is

$$\mathbf{H}(j2\pi F) = H(F) = \frac{\Omega_o^2}{(j2\pi F - p_{h1})(j2\pi F - p_{h2})} \tag{165}$$

Similarly

$$X(p_{h1}) = \frac{p_{h1} A}{(p_{h1} - j\Omega_x)(p_{h1} + j\Omega_x)} \tag{166}$$

and

$$\mathcal{R}es\{H(s); p_{h1}\} = (s - p_{h1})H(s)\big|_{s=p_{h1}} = \frac{\Omega_o^2}{(p_{h1} - p_{h2})} = \frac{\Omega_o^2}{2j\Omega_d} \tag{167}$$

K_{h1} is given by (151). $X(p_{h2})$ and $\mathcal{R}es\{H(s); p_{h1}\}$ are complex conjugates of $X(p_{h1})$ and $\mathcal{R}es\{H(s); p_{h1}\}$, respectively, we can write Eq. (152) as

$$y(t) = A|H(j\Omega_x)| \cos(\Omega_x t + \angle H(j\Omega_x)) + 2|K_{h1}|e^{-\alpha t} \cos(\Omega_d t + \angle K_{h1}) \tag{168}$$

If the damping coefficient α is positive, then the second term in Eq. (168) decreases as time increases and the system reaches the sinusoidal steady state when $t \gg \alpha^{-1}$. Figure 6CT.17 shows a plot of $y(t)$ where $F_x = 1$. As in the impulse response plot of Figure 6CT.13, $\Omega_o = 1.2$, $\alpha = 0.25$, and $\Omega_d = 1.17367$.

Figure 6CT.17
Response of a second-order resonant system to a cosine.

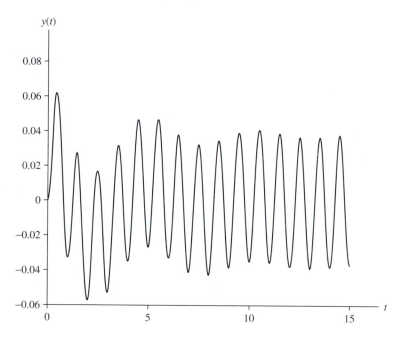

We can refer to the systems pole-zero diagram to show geometrically how the amplitude and phase of the sinusoidal steady-state response term in Eq. (168) depends on the system's internal dynamics as represented by its poles. The amplitude and phase are determined by the frequency response characteristic $H(j\Omega_x)$. Figure 6CT.18a illustrates the two factors in the denominator of $H(s)$ in Eq. (162) as vectors in the s-plane. The vectors start at the the poles p_{h1} and $p_{h2} = p_{h1}^*$, and they end on the $j\Omega$ axis at the "excitation point" $s = j\Omega_x$. Now if we write Eq. (165) in polar form

$$H(j2\pi F) = H(F) = \frac{\Omega_o^2}{|j2\pi F - p_{h1}||j2\pi F - p_{h2}|} \angle\{-(j2\pi F - p_{h1}) - (j2\pi F - p_{h2})\} \tag{169}$$

we see that the absolute value of $H(F)$ equals Ω_o^2 divided by the product of the lengths of the vectors and the angle of $H(F)$ equals the negative of the sum of the angles of the vectors. Notice that $|H(F)|$ becomes large when the point $j2\pi F$ is a close to either pole, p_{h1} or p_{h2}, and small when $j2\pi F$ is far from both poles.

Figure 6CT.18
Pole-zero plot;
(b) amplitude
response
characteristic.

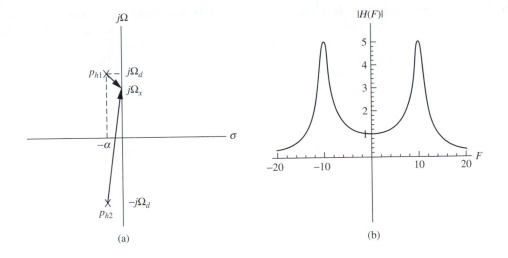

(a)

(b)

Therefore $|H(F)|$ is large when the input frequency is close to a natural frequency of the system, and small when the input frequency is far from a natural frequency of the system. The amplitude characteristic is plotted in Figure 6CT.18(b) for $p_{h1} = -2\pi + j20\pi$. Here $\alpha = 2\pi$, $\Omega_o = \pi\sqrt{404}$ and $\Omega_d = 20\pi$. The corresponding values in hertz are $\alpha/2\pi = 1$, $\Omega_o/2\pi \approx 10.05$ and $\Omega_d/2\pi = 10$. The peaks in the amplitude characteristic are called *resonance peaks*. Resonance occurs when a system is driven at a frequency close to the system's natural frequency.

Figure 6CT.19 shows another good way to visualize the amplitude characteristic. Here, $|H(s)|$ is plotted above the s-plane for $\mathcal{R}\{s\} \le 0$. As in Figure 6CT.18, $\alpha = 2\pi \approx 6.28$, $\Omega_o = \pi\sqrt{404} \approx 63.15$ and $\Omega_d = 20\pi \approx 62.83$. The amplitude characteristic $|H(j2\pi F)|$ is the intersection of $|H(s)|$ with the $\sigma = 0$ plane. The surface $|H(s)|$ is unbounded at the pole values $-\alpha \pm j\Omega_d \approx -6.28 \pm j62.83$. It is analogous to a rubber sheet supported by (physical) poles at the

Figure 6CT.19
Function $|H(s)|$
plotted above the
s-plane.

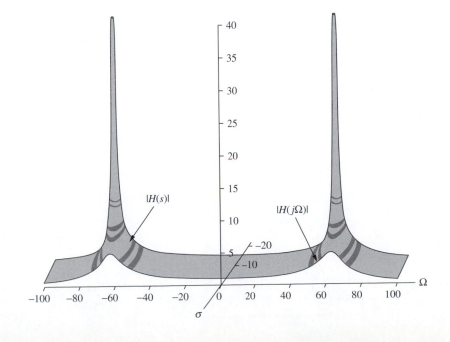

pole values. For an arbitrary $\mathbf{H}(s)$, the sheet would also be held down by tacks at the zero values. This rubber sheet analogy provides a quick way to visualize the shape of any amplitude characteristic. It has the merit of showing the importance of the poles and zeros of a transfer function in determining the shape of a system's amplitude characteristic. In general, poles close to the $j\Omega$ axis of the s-plane are associated with resonance peaks in $|\mathbf{H}(j\Omega)|$. Similarly, zeros close to the $j\Omega$ axis of the s-plane are associated with dips in $|\mathbf{H}(j\Omega)|$.

6CT.8 Initial and Final Value Theorems

The initial and final value theorems provide a way to find the initial and final values $x(0^+)$ and $x(\infty)$ of a waveform directly from $\mathbf{X}(s)$. These theorems are shown in Table 6CT.3.

The initial value theorem can be applied to any $x(t)$ that has an initial value. It does not work when $x(t)$ is unbounded or undefined at the origin. For rational algebraic $\mathbf{X}(s)$, the initial value theorem works if an only if $\mathbf{X}(s)$ is a proper algebraic fraction. Similarly, the final value theorem can be applied when $x(t)$ has a final value. It does not work if $x(t)$ becomes unbounded or oscillatory as $t \rightarrow \infty$. For rational algebraic $\mathbf{X}(s)$, the final value theorem works if an only if all the poles of $\mathbf{X}(s)$ are in the left half plane, except for possibly a pole with order one at the origin.

**6CT.8.1
Initial Value
Theorem**

The initial value theorem states that

$$x(0^+) = \lim_{s \to \infty} s\mathbf{X}(s) \tag{170}$$

provided the limit of $s\mathbf{X}(s)$ as $s \rightarrow \infty$ exists. $x(0^+)$ is defined as the limit

$$x(0^+) = \lim_{\substack{t \to 0 \\ t > 0}} x(t) \tag{171}$$

The proof is based on the differentiation property. Assume first that $x(t)$ is continuous at $t = 0$. Then

$$x(0^+) = x(0^-) \tag{172}$$

and Eq. (51) becomes

$$\int_0^\infty x^{(1)}(t)e^{-st}\,dt = s\mathbf{X}(s) - x(0^+) \tag{173}$$

The absolute value of the integral in the preceding equation is bounded by

$$\left| \int_0^\infty x^{(1)}(t)e^{-st}\,dt \right| \leq \int_0^\infty \left| x^{(1)}(t)e^{-st} \right| dt \leq \int_0^\infty |x^{(1)}(t)|e^{-\sigma t}\,dt \tag{174}$$

Table 6CT.3
Initial and Final Value
Theorems

Initial value	$x(0^+) = \lim_{s \to \infty} s\mathbf{X}(s)$
Final value	$\lim_{t \to \infty} x(t) = \lim_{s \to 0} s\,\mathbf{X}(s)$

The upper bound equals 0 in the limit $\sigma \to \infty$. Therefore if we take the limit $\sigma \to \infty$ (for which $s \to \infty$) of both sides of Eq. (173), we obtain

$$0 = \lim_{s \to \infty} s\mathbf{X}(s) - x(0^+) \tag{175}$$

which is the initial value theorem. If $x(t)$ is discontinuous at $t = 0$, then $x^{(1)}(t)$ contains an impulse with area $x(0^+) - x(0^-)$ at $t = 0$ and we can write Eq. (51) as

$$\int_0^\infty x^{(1)}(t)e^{-st}dt = \underbrace{\int_0^{0^+} x^{(1)}(t)e^{-st}dt}_{x(0^+)-x(0^-)} + \underbrace{\int_{0^+}^\infty x^{(1)}(t)e^{-st}dt}_{0 \text{ in the limit } s \to \infty} = s\mathbf{X}(s) - x(0^-) \tag{176}$$

As indicated by the braces, the first integral equals $x(0^+) - x(0^-)$ and the second integral equals 0 in the limit $s \to \infty$. We obtain the initial value theorem by rearranging Eq. (176).

EXAMPLE 6CT.18   Unit Step Applied to Second-Order System

Find the initial value of $y(t)$ when a unit step is applied to the second-order system of Example 6CT.13.

Solution

The s-domain response is

$$\mathbf{Y}(s) = \mathbf{H}(s)\mathbf{X}(s) = \left(\frac{\Omega_o^2}{s^2 + 2\alpha s + \Omega_o^2} \right) \left(\frac{1}{s} \right) \tag{177}$$

Therefore, by the initial value theorem

$$y(0^+) = \lim_{s \to \infty} s \left(\frac{\Omega_o^2}{s^2 + 2\alpha s + \Omega_o^2} \right) \left(\frac{1}{s} \right) = 0 \tag{178}$$

EXAMPLE 6CT.19   Ramp Applied to High-Pass Filter

Find the initial value of $y(t)$ when a ramp is applied to the high-pass filter of Example 6CT.15.

Solution

Here

$$\mathbf{Y}(s) = \left(\frac{s\tau_c}{s\tau_c + 1} \right) \left(\frac{1}{s^2} \right) \tag{179}$$

Application of the initial value theorem yields

$$y(0^+) = \lim_{s \to \infty} s \left(\frac{s\tau_c}{s\tau_c + 1} \right) \left(\frac{1}{s^2} \right) = 0 \tag{180}$$

EXAMPLE 6CT.20   Initial Value of $h(t)$ when $\mathbf{H}(s) = s/(s^2 + \Omega_0)^2$

Find the initial value of $y(t)$ when a unit impulse is applied to a system having transfer function.

$$\mathbf{H}(s) = \frac{s}{s^2 + \Omega_0^2} = \frac{s}{(s + j\Omega_0)(s - j\Omega_0)} \tag{181}$$

Solution

Application of the initial value theorem yields

$$h(0^+) = \lim_{s \to \infty} s\left(\frac{s}{s^2 + \Omega_0^2}\right) = 1 \tag{182}$$

6CT.8.2
Final Value
Theorem

The final value theorem states that

$$\lim_{t \to \infty} x(t) = \lim_{s \to 0} \mathbf{X}(s) \tag{183}$$

As with the initial value theorem, the proof is based on the differentiation property. We take the limit $s \to 0$ of both sides of Eq. (51). The result is

$$\int_0^\infty x^{(1)}(t)\, dt = \lim_{s \to 0} s\mathbf{X}(s) - x(0^-) \tag{184}$$

where the integral on the left-hand side is obtained by evaluating $\mathbf{X}^{(1)}(s) = \mathcal{L}\{x^{(1)}(t)\}$ at $s = 0$. We can obtain another expression for this integral from the definition of an integral.

$$\int_0^\infty x^{(1)}(t)\, dt = \lim_{t \to \infty} x(t) - x(0^-) \tag{185}$$

We obtain the final value theorem by combining the preceding two equations.

EXAMPLE 6CT.21   Unit Step Applied to Second-Order System

Find the final value of $y(t)$ when a unit step is applied to the second-order system of Example 6CT.13.

Solution

The s-domain response is

$$\mathbf{Y}(s) = \mathbf{H}(s)\mathbf{X}(s) = \left(\frac{\Omega_o^2}{s^2 + 2\alpha s + \Omega_o^2}\right)\left(\frac{1}{s}\right) \tag{186}$$

Application of the final value theorem yields

$$\lim_{t \to \infty} y(t) = \lim_{s \to 0} s\left(\frac{\Omega_o^2}{s^2 + 2\alpha s + \Omega_o^2}\right)\left(\frac{1}{s}\right) = \frac{1}{\Omega_o^2}. \tag{187}$$

EXAMPLE 6CT.22   Ramp Applied to High-Pass Filter

Find the final value of $y(t)$ when a ramp is applied to the high-pass filter of Example 6CT.15.

Solution

Here

$$\mathbf{Y}(s) = \left(\frac{s\tau_c}{s\tau_c + 1} \right) \left(\frac{1}{s^2} \right) \tag{188}$$

and the final value theorem yields

$$\lim_{t \to \infty} y(t) = \lim_{s \to 0} s \left(\frac{s\tau_c}{s\tau_c + 1} \right) \left(\frac{1}{s^2} \right) = \tau_c \tag{189}$$

EXAMPLE 6CT.23   Final Value of $h(t)$ where $\mathbf{H}(s) = s/(s^2 + \Omega_0)^2$

Find the final value of $y(t)$ when a unit impulse is applied to a system having the transfer function

$$\mathbf{H}(s) = \frac{s}{s^2 + \Omega_0^2} = \frac{s}{(s + j\Omega_0)(s - j\Omega_0)} \tag{190}$$

Solution

The system being considered is unstable because it has poles on the $j\Omega$ axis. We know from Table 6CT.2 that

$$h(t) = \cos(\Omega_0 t) \tag{191}$$

which is a function having no final value. That is, $\lim_{t \to \infty} h(t)$ does not exist. Application of the final value theorem gives the incorrect result

$$\lim_{t \to \infty} h(t) = \lim_{s \to 0} s \left(\frac{s}{s^2 + \Omega_0^2} \right) = 0 \tag{192}$$

6CT.9   ## Summary

The Laplace transform is the premier analytical aid in CT signals and systems theory for determining the response of an LTI system to a suddenly applied input. It applies to both stable and unstable systems and to signals such as ramps and raising exponentials that have no Fourier transform. The Fourier and Laplace transforms of a causal signal are equal on the $j\Omega$ axis of the s-plane. Therefore, the Fourier transform exists if and only if the region of convergence of the Laplace transform includes the $j\Omega$ axis.

Poles play a central role in Laplace transform system analysis. The poles of $\mathbf{H}(s)$ describe the damping coefficients and natural frequencies of the system. The are the foundation for partial fraction expansion of $\mathbf{H}(s)$, and in turn they determine the form of a system's impulse response $h(t)$. The influence of a system's poles and zeros on sinusoidal steady-state response becomes manifest by plotting $|\mathbf{H}(s)|$ as a surface above the s-plane. The intersection of this surface with the plane $\sigma = 0$ is the amplitude response characteristic $|\mathbf{H}(j\Omega)|$. Poles close to the $j\Omega$ axis of the s-plane are associated with resonance peaks in $|\mathbf{H}(j\Omega)|$. Zeros close to the $j\Omega$ axis are associated with dips in $|\mathbf{H}(j\Omega)|$

Summary of Definitions and Formulas

Definition of the Laplace Transform

$$\mathbf{X}(s) = \mathcal{L}\{x(t)\} = \int_0^\infty x(t)e^{-st}\,dt$$

where $s = \sigma + j\Omega$.

Waveforms that are different for $t < 0$ but equal for $t \geq 0$ have the same Laplace transform.

Existence of X(s)

Waveforms satisfying

$$|x(t)| < Me^{\sigma_a t}$$

where M and σ_a are finite, are said to have *exponential order*. If $x(t)$ has exponential order, then $\mathbf{X}(s)$ exists for all $\sigma > \sigma_a$. The region $\sigma > \sigma_a$ in the s-plane is called the *region of convergence*. The region of convergence of a rational algebraic fraction $\mathbf{X}(s)$ is the half plane lying to the right of the rightmost pole of $\mathbf{X}(s)$.

Relation of Laplace Transform to Fourier Transform

$$\mathbf{X}(s) = \mathcal{F}\{x(t)e^{-\sigma t}\} \qquad \text{for causal } x(t)$$

If the region of convergence of $\mathbf{X}(s)$ includes the $j\Omega \equiv j2\pi F$ axis, then the Fourier transform of $x(t)$ exists and is given by

$$\mathbf{X}(s)\big|_{s=j2\pi F} = \mathcal{F}\{x(t)\} = X(F) \qquad \text{for causal } x(t)$$

Differentiation Property

$$\mathbf{X}^{(1)}(s) = \mathcal{L}\left\{\frac{dx(t)}{dt}\right\} = s\mathbf{X}(s) - x(0^-)$$

This property is a key property: It makes it possible to transform differential equations having initial conditions into algebraic equations in s.

Extensions:

$$\mathbf{X}^{(2)}(s) = \mathcal{L}\left\{\frac{d^2x(t)}{dt^2}\right\} = s^2\mathbf{X}(s) - sx(0^-) - x^{(1)}(0^-)$$

$$\mathbf{X}^{(n)}(s) = \mathcal{L}\left\{\frac{d^nx(t)}{dt^n}\right\} = s^n\mathbf{X}(s) - s^{n-1}x(0^-) - s^{n-2}x^{(1)}(0^-) - \cdots - x^{(n-1)}(0^-)$$

See Tables 6CT.1 and 6CT.2 for Laplace transform properties and pairs.

To Solve Differential Equations Having Initial Conditions

1. Obtain the s-domain I-O equation of the system. Let

$$p = \frac{d}{dt}$$

$$A(p) = a_n p^n + a_{n-1} p^{n-1} + \cdots + a_1 p + a_0$$

$$B(p) = b_m p^m + b_{m-1} p^{m-1} + \cdots + b_1 p + b_0$$

Then write an ordinary differential equation having constant coefficients as

$$A(p)y(t) = B(p)x(t)$$

When we take the Laplace transform of the preceding equation, we obtain the s domain equation

$$A(s)\mathbf{Y}(s) + \mathbf{Y}_{ic}(s) = B(s)\mathbf{X}(s) + \mathbf{X}_{ic}(s)$$

where $\mathbf{X}_{ic}(s)$ and $\mathbf{Y}_{ic}(s)$ are due to initial conditions.

2. Solve the algebraic equation for the s-domain output $\mathbf{Y}(s)$. The solution is

$$\mathbf{Y}(s) = \frac{B(s)}{A(s)}\mathbf{X}(s) + \frac{\mathbf{X}_{ic}(s) - \mathbf{Y}_{ic}(s)}{A(s)}$$

In the above

$$\mathbf{H}(s) = \frac{B(s)}{A(s)} = \mathcal{L}\{h(t)\}$$

is the *transfer function* of the system. Often, we can write $\mathbf{Y}(s)$ as

$$\mathbf{Y}(s) = K\frac{\Pi(s - z_{hi})\Pi(s - z_{xi})}{\Pi(s - p_{hi})\Pi(s - p_{xi})} + \frac{\mathbf{X}_{ic}(s) - \mathbf{Y}_{ic}(s)}{a_n \Pi(s - p_{hi})}$$

where the p_{hi} are the roots to the characteristic equation $A(s) = 0$ and the p_{xi} are the poles of $\mathbf{X}(s)$.

3. Use partial fraction expansion to put $\mathbf{Y}(s)$ is a form ready for table lookup. For simple poles, the partial fraction expansion of $\mathbf{Y}(s)$ contains terms of the form

$$G(s) = \frac{K_1}{s - p_1} + \frac{K_2}{s - p_2} + \cdots + \frac{K_i}{s - p_i} + \cdots + \frac{K_l}{s - p_l}$$

where the residues $K_1, K_2, \ldots,$ and K_l can be evaluated using Heaviside's formula.

$$K_i = \mathcal{R}es\{G(s); p_i\} = (s - p_i)G(s)\Big|_{s=p_i} \quad i = 1, 2, \ldots, l$$

4. Find $y(t)$ by table lookup. For simples poles, as just considered, $y(t)$ contains terms of the form

$$g(t) = K_1 e^{p_1 t} + K_2 e^{p_2 t} + \cdots + K_l e^{p_l t} \quad t > 0$$

Complex Poles

Complex poles always occur in conjugate pairs for a real system having a real input. The preceding partial fraction for $G(s)$ and its associated waveform $g(t)$ then contain terms having the form

$$G(s) = \frac{K}{s - p} + \frac{K^*}{s - p^*}$$

and

$$g(t) = K e^{pt} + K^* e^{p^* t} = 2|K| e^{\sigma t} \cos(\Omega t + \angle K)$$

Stability

For a causal LTI system having a rational algebraic transfer function, a necessary and sufficient condition for BIBO stability is that all the poles of $\mathbf{H}(s)$ lie in the left half of the s-plane.

Response of LTI System to Sinusoid Applied at $t = 0$

When the poles of $\mathbf{Y}(s)$ are distinct

$$A\cos(\Omega_x t + \phi)u(t) \xrightarrow{h(t)} A|\mathbf{H}(j\Omega_x)|\cos(\Omega_x t + \phi + \angle\mathbf{H}(j\Omega_x))$$

$$+ \sum_{i=1}^{n} \mathbf{X}(p_{hi})\mathcal{R}es\{\mathbf{H}(s);\ p_{hi}\}e^{p_{hi}t}$$

For a stable LTI system, the terms in the summation $\sum_{i=1}^{n}$ on the right-hand side are transients, which die out as t increases. When the transient terms are all negligible, $y(t)$ has reached the *sinusoidal steady state*

$$y(t) = A|\mathbf{H}(j\Omega_x)|\cos(\Omega_x t + \phi + \angle\mathbf{H}(j\Omega_x))$$

The sinusoidal steady-state response was studied extensively in Chapter 3CT.

Initial and Final Value Theorems

$$x(0^+) = \lim_{s\to\infty} s\mathbf{X}(s)$$

$$\lim_{t\to\infty} x(t) = \lim_{s\to 0} s\mathbf{X}(s)$$

6CT.10 Problems

6CT.2 Definition of the Laplace Transform

6CT.2.1 Find the Laplace transforms, the pole-zero plots, and the regions of convergence for the following waveforms:
a) 3
b) $7\delta(t)$
c) e^{5t}
d) e^{-2t}
e) e^{j8t}

6CT.2.2 Find the Laplace transforms, the pole-zero plots, and the regions of convergence for the following waveforms:
a) $\delta(t-6)$
b) $\delta(t+3) + \delta(t-6)$
c) $u(t-6)$
d) $u(t+3)$

6CT.2.3 The *bilateral Laplace transform* is defined as

$$\mathbf{X}_B(s) = \int_{-\infty}^{\infty} x(t)e^{-st}dt = \int_{-\infty}^{0} x(t)e^{-st}dt + \int_{0}^{\infty} x(t)e^{-st}dt \qquad (193)$$

where $s = \sigma + j\Omega$. Find the bilateral Laplace transform and its region of convergence for:
a) $x(t) = e^{-t}u(t)$.
b) $x(t) = e^{t}u(-t)$.
c) $x(t) = e^{-t}u(t) + e^{t}u(-t)$.

6CT.2.4 Repeat Problem 6CT.2.3 for:
  a) $x(t) = e^t u(t)$.
  b) $x(t) = e^{-t} u(-t)$.
  c) $x(t) = e^t u(t) + e^{-t} u(-t)$.

6CT.3   Relationship Between the Laplace and Fourier Transforms of Causal Waveforms

6CT.3.1 **a)** Find the Laplace transform and region of convergence of $x(t) = \sqcap((1 - 0.5\tau)/\tau)$, where $\tau > 0$.
  b) Use your answer from part (a) to write down the Fourier transform of $\sqcap((1 - 0.5\tau)/\tau)$, where $\tau > 0$.
  Express your answer in terms of the sinc function.

6CT.3.2 Explain why the Laplace transform of $u(t)$ exists, but the Fourier transform does not. Then explain why we can define the Fourier transform of $u(t)$ as $\frac{1}{2}\delta(F) + 1/j2\pi F$. Does $\frac{1}{2}\delta(F) + 1/j2\pi F$ "exist"?

6CT.3.3 **a)** Find the Laplace transform and region of convergence for $x(t) = e^{5t}$.
  b) Use your answer to part (a) to state whether or not $e^{5t} u(t)$ has a Fourier transform.

6CT.4   Properties of the Laplace Transform

6CT.4.1 State and prove the time-delay property of the Laplace transform.

6CT.4.2 State and prove the time-scale property of the Laplace transform.

6CT.4.3 State and prove the convolution property of the Laplace transform.

6CT.4.4 Does the convolution property of the Laplace transform work if one or both of the waveforms are noncausal? Justify your answer.

6CT.4.5 Derive the Laplace transform of $x(t) = Ate^{\sigma_0 t} \cos(\Omega_0 t + \phi)$ and confirm that your answer agrees with that in Table 6CT.2.

6CT.4.6 Sketch the pole-zero plot for the sinusoid of Example 6CT.7.

6CT.4.7 Sketch the pole-zero plot for the damped sinusoid of Example 6CT.8.

6CT.5   Elementary Systems in the s-Domain

6CT.5.1 **a)** Derive the s-domain representation for an ideal delay line. Hint: Let $x(t)$ and $w(t)$ denote the input and the output to the delay line, respectively. Start with the equation $w(t) = x(t - \tau)$ and find $W(s)$. Do not assume that $x(t)$ is a causal waveform.
  b) Draw the s-domain system.
  c) What information is needed to describe the initial condition of the ideal delay line at $t = 0$?
  d) Assume that at $t = 0$ a delay line is connected to a feedback system described by the equation

$$y(t) = a\frac{d}{dt}y(t) + w(t) \tag{194}$$

  Assume that the input to a delay line is $x(t) = 0$ for all time before $t = 0$ and that $x(t) = 1$ for $t \geq 0$. Find $y(t)$ for $t > 0$ subject to the initial condition $y(0^-) = 0$.
  e) Redo part (d) but assume that the input to a delay line is $x(t) = 1$ for all time before $t = 0$ and that $x(t) = 0$ for $t \geq 0$. Find $y(t)$ for $t > 0$ subject to the initial condition $y(0^-) = 0$.

6CT.6   Laplace Transform System Analysis

6CT.6.1 Confirm Equations (68), (69), and (70).

6CT.6.2 Invent two ways to find K_1 and K_2 in Example 6CT.12 other than that invented by Heaviside. Apply your methods to confirm the answers given in the Example 12.

6CT.6.3 Show that it is possible to chose the initial condition $y(0^-)$ in Example 6CT.12 so that the system immediately reaches the steady-state output $y(t) = 1$ when a step is applied.

6CT.6.4 Find the functions $\mathcal{A}(s)$, $\mathcal{B}(s)$, and $\mathbf{H}(s)$ for Example 6CT.12.

6CT.6.5 a) Find $\mathbf{Y}(s)$ for the second-order system of Example 6CT.13 if $x(t) = u(t)$. Put your answer in partial fraction form. Assume that the system poles are complex and that the initial conditions are all 0.
b) Use your answer to part (a) to find $y(t)$.

6CT.6.6 Use the Laplace transform to find the step response of a system described by $\tau y^{(1)}(t) + y(t) = \tau x^{(1)}(t)$.

6CT.6.7 Use the Laplace transform to find the ramp response of a system described by $\tau y^{(1)}(t) + y(t) = \tau x^{(1)}(t)$.

6CT.6.8 The response of the second-order system of Example 6CT.13 is called *overdamped* if $\alpha^2 > \Omega_o^2$.
a) Find $\mathbf{H}(s)$ for $\alpha^2 > \Omega_o^2$.
b) Sketch the pole-zero plot.
c) Find $h(t)$.

6CT.6.9 The response of the second-order system of Example 6CT.13 is called *critically damped* if $\alpha^2 = \Omega_o^2$.
a) Find $\mathbf{H}(s)$ for $\alpha^2 = \Omega_o^2$.
b) Sketch the pole-zero plot.
c) Find $h(t)$

6CT.6.10 Use a method of circuit theory to confirm Eq. (105) for the *RLC* filter of Figure 6CT.11(b). Hint: You can obtain the result by inspection if you recognize that this circuit is a voltage divider.

6CT.6.11 The circuit of Figure 6CT.11(b) is often referred to as an *RLC band-pass filter* when the output voltage is taken across the resistor instead of the capacitor.
a) Find the expressions for $\mathbf{H}(s)$ and $h(t)$ for the series *RLC* circuit of Figure 6CT.11(b), where the output voltage is taken across the resistor instead of the capacitor. Assume that the circuit is underdamped.
b) Sketch the pole-zero diagram for $\mathbf{H}(s)$.
c) Write a computer program to plot $|H(F)| = |\mathbf{H}(j2\pi F)|$ for $\Omega_o = 1.2$ and $\alpha = 0.25$. Compare your plot with that in the example.
d) Write a computer program to plot $h(t)$ for $\Omega_o = 1.2$ and $\alpha = 0.25$. Compare your plot with that in the example.

6CT.6.12 a) Find the expressions for $\mathbf{H}(s)$ and $h(t)$ for the series *RLC* circuit of Figure 6CT.11(b), where the output voltage is taken across the inductor instead of the capacitor. Assume that the circuit is underdamped.
b) Sketch the pole-zero diagram for $\mathbf{H}(s)$.
c) Write a computer program to plot $|H(F)| = |\mathbf{H}(j2\pi F)|$ for $\Omega_o = 1.2$ and $\alpha = 0.25$. Compare your plot with that in the example.
d) Write a computer program to plot $h(t)$ for $\Omega_o = 1.2$ and $\alpha = 0.25$. Compare your plot with that in the example.

6CT.6.13 Find the partial fraction expansion of

$$G(s) = 3\frac{s + 2.1}{(s + 2)(s + 1 + j0.25)(s + 1 - j0.25)}$$

6CT.6.14 Find the partial fraction expansion of

$$G(s) = \frac{s^3 + 3}{s^2 + 3s + 2} = \frac{s^3 + 3}{(s + 1)(s + 2)}$$

6CT.6.15 The MATLAB function *residue* can be used to calculate the partial fraction expansion of $G(s)$. Read the MATLAB help for *residue* and use this function to verify the your answers to Problems 6CT.6.13 and 6CT.6.14.

6CT.6.16 The MATLAB function `tf2zp` can be used convert $\mathbf{H}(s)$ from polynomial form to pole-zero form. Read the MATLAB help for `tf2zp` and use this function to find the pole-zero form of $\mathbf{H}(s) = (s^3 + 1)/(s^2 + 2s + 2)$. Plot the pole-zero diagram.

6CT.6.17 We can use the Laplace transform in combination with other techniques, such as those in Chapter 2CT, to find an output. For example, to find the output of an LTI system to a rectangular pulse, $x(t) = u(t) - u(t - T)$, we can use the Laplace transform to find the system's step response $h_{-1}(t)$ and then write down the pulse response $y(t) = h_{-1}(t) - h_{-1}(t - T)$.
a) Use this combined approach to find the response of Example 6CT.12 to
$x(t) = u(t) - u(t - T)$ if $y(0^-) = 0$.
b) Use this combined approach to find the response of Example 6CT.12 to
$x(t) = u(t) - u(t - T)$ if $y(0^-) \neq 0$.
Problems 6CT.18–6CT.23 involve systems or inputs that contain delays. As you can see by doing these problems, the methods of Section 6CT.6 work for these systems.

6CT.6.18 Set the initial condition to 0 in Example 6CT.12 and use the Laplace transform to find the response of the system to each of the following and check your answer using some other method.
a) $x(t) = u(t - T)$
b) $x(t) = u(t) - u(t - T)$
c) $x(t) = u(t) - u(t - T) + u(t + 2T) - u(t - 3T) + \cdots$

6CT.6.19 The input $x(t) = u(t - T)$ is applied to a system having the transfer function $\mathbf{H}(s) = (1/s)e^{-sT}$. Use Laplace transforms to find $h(t)$ and $y(t)$. Check your answer using some other method.

6CT.6.20 The input $x(t) = u(t)$ is applied to a system having the transfer function $\mathbf{H}(s) = (1/s)(1 - e^{-sT})$. Find $h(t)$ and $y(t)$. Check your answer using some other method.

6CT.6.21 The input $x(t) = u(t) - u(t - T)$ is applied to a system having the transfer function $\mathbf{H}(s) = (1/s)(1 - e^{-sT})$. Use Laplace transforms to find $y(t)$. Check your answer using some other method.

6CT.6.22 Find the I-O equation of a system having transfer function $\mathbf{H}(s) = (1/s)(1 - e^{-sT})$.

6CT.6.23 **a)** Find the transfer function and its region of convergence for the acoustic feedback problem introduced in Introduction. Hint: Recall from Chapter 2 that
$h(t) = \alpha \sum_{n=0}^{\infty} (\alpha\beta)^n \delta(t - n\tau)$.
b) Find the locations of the poles and plot the pole-zero diagram.
c) Does your answer to part (b) require that $\alpha\beta < 1$? Justify your answer.

6CT.7   Response to Sinusoid Applied at $t = 0$

6CT.7.1 Is it possible to set the initial condition on the first-order system of Example 6CT.16 so that the sinusoidal steady state is attained immediately? If so, find the formula for that initial condition.

6CT.7.2 Find expressions for the absolute values and angles in Eq. (168). Express your answer in terms of $A, \alpha, \Omega_o, \Omega_d$ and Ω_x.

6CT.7.3 Example 6CT.17 assumed zero initial conditions. Find the output $y(t)$ for $x(t) = A\cos(\Omega_x t)u(t)$ if the initial conditions are nonzero.

6CT.7.4 a) Find the response of the first-order system of Example 6CT.16 to
$x(t) = A\cos(\Omega_x t + \phi)u(t)$, where ϕ is an arbitrary phase angle. Assume that the system is initially at rest.
b) Is there a ϕ for which the sinusoidal steady state is attained immediately for $x(t) = A\cos(\Omega_x t + \phi)u(t)$? If so, find the formula for that ϕ.

6CT.7.5 a) Find the output of the second-order system of Example 6CT.17 for an input
$x(t) = A\cos(\Omega_x t + \phi)u(t)$, where ϕ is an arbitrary phase angle. Assume that the system is initially at rest.
b) Is there a ϕ for which the sinusoidal steady state is attained immediately for $x(t) = A\cos(\Omega_x t + \phi)u(t)$? If so, find the formula for that ϕ.

6CT.7.6 Consider Figure 6CT.18 of Example 6CT.17. Assume that the system is highly resonant. For a highly resonant system the poles are very close to the $j\Omega$ axis: $|\alpha| \ll \Omega_d$.
a) Use simple trigonometric approximations to show that the resonance peak in $|H(j\Omega)|$ occurs at $\Omega \approx \Omega_d$ for a highly resonant system.
b) The half-power frequencies of the system in radians per sample are defined as the values of Ω for which $|H(j\Omega)| = 1/\sqrt{2}|H(j\Omega_{\text{peak}})|$, where Ω_{peak} is the frequency at which the amplitude characteristic is a maximum. The half-power bandwidth of a resonance peak is defined as $\Delta\Omega = \Omega_+ - \Omega_-$, where Ω_+ and Ω_- are the half-power frequencies on the high- and low-frequency sides of the peak, respectively. Use simple trigonometric approximations to show that the half-power bandwidth of the resonance peak is approximately $\Delta\Omega \approx 2\alpha$.

6CT.8   Initial and Final Value Theorems

6CT.8.1 A unit step is applied to a system having the transfer function $H(s)$.
a) Show that, if the initial and final values of the output $y(t)$ exist, they are given by

$$y(0^+) = \lim_{\sigma \to \infty} H(s)$$

respectively

$$\lim_{t \to \infty} y(t) = \lim_{s \to 0} H(s)$$

b) Give an example to illustrate each of these results.

6CT.8.2 A unit ramp is applied to a system having the transfer function $H(s)$.
a) Assume that the initial and final values of the output $y(t)$ exist and find the formulas for the initial and final values.
b) Give an example to illustrate you answers to part (a).

6CT.8.3 The input $\cos(\Omega t)u(t)$ is applied to a system having the transfer function $H(s)$.
a) Assume that the initial and final values of the output $y(t)$ exist and find the formulas for the initial and final values.
b) Give an example to illustrate you answers to part (a).

6CT.8.4 A ramp $r(t)$ is applied to a first-order system having transfer function $H(s) = 1/(\tau_c s + 1)$. The difference between output and input is $d(t) = y(t) - r(t)$. Assume that the system is initially at rest.

a) Use the final value theorem to find the final value of $d(t)$.

b) Use the initial value theorem to find the initial value of $d(t)$.

c) Use the result from Example 6CT.14 to check your answers to parts (a) and (b).

d) Sketch $d(t)$.

6CT.8.5 There are n first-order systems placed in series. Each system has transfer function $H(s) = 1/(\tau_c s + 1)$ and is initially at rest. A ramp $r(t)$ is applied to the input. The difference between output and input is $d(t) = y(t) - r(t)$.

a) Assume that $n = 3$ and show that the initial and final values of $d(t)$ are 0 and $3\tau_c$, respectively.

b) Show that for arbitrary n the initial and final values of $d(t)$ are 0 and $n\tau_c$, respectively.

c) Use your answer to part (b) to find the (approximate) equation for $y(t)$ for large t. Use this equation and your engineering insight to make an approximate sketch of $y(t)$ for $t \geq 0$.

6CT.8.6 The second-order system of Example 6CT.13 and the high-pass filter of Example 6CT.15 are placed in series. Both systems are initially at rest. A unit step is applied to input to the series connection. Find the initial and final values of the output.

6CT.8.7 Repeat Problem 6CT.8.6 for the parallel connection of the second-order and high-pass systems.

The z-Transform

6DT

6DT.1 Introduction

The z transform is the premier analytical aid in DT signals and systems theory for determining the response of an LTI system to a suddenly applied input. It is an analytical extension of the discrete-time Fourier transform based on the idea of complex frequency. The concept of a signal's spectrum (DTFT) $X(e^{j2\pi f})$, so important in previous chapters, is included in the framework of the z-transform. The z-transform, however, also introduces entirely new concepts. These new concepts (*poles, zeros,* and *regions of convergence*) enable us to solve for the response of both stable and unstable LTI systems to suddenly applied inputs using just algebra and table lookup. The new concepts provide us with a deeper understanding of (1) a signal's spectrum $X(e^{j2\pi f})$, (2) a system's frequency response characteristic $H(e^{j2\pi f})$, and (3) how the sinusoidal steady-state response and transient response are related. Thus, the z transform not only simplifies determining the transient and steady-state responses of LTI systems, but also provides intuition and insight into DT signals and systems concepts.

6DT.2 Definition of the z-Transform

The z-transform[1] of a sequence $x[n]$ is defined as

---

z-Transform

$$X(z) = \mathcal{Z}\{x[n]\} = \sum_{n=0}^{\infty} x[n]z^{-n} \tag{1}$$

where z is a complex variable having absolute value ρ and angle θ.

$$z = \rho e^{j\theta} \tag{2}$$

The sum defining $X(z)$ ignores the index interval $n = -1, -2, \ldots$. Therefore, sequences that are different for $n < 0$ but equal for $n \geq 0$ have the same z-transform. If we

---

[1] There are two versions of the z-transform: the *unilateral* z-transform defined in Eq. (1) and the *bilateral* z-transform introduced in Problems 6DT.2.3 and 6DT.2.4. We follow popular terminology when we refer to the unilateral z-transform as "the" z-transform.

like, we can assume for convenience that $x[n] = 0$ for $n = -1, -2, \ldots$. This or any other assumption about $x[n]$ for $n < 0$ has no effect on $X(z)$.

The class of sequences that are z-transformable is considerably larger than the class of Fourier-transformable sequences. We can establish this fact by upper-bounding the absolute value of $X(z)$:

$$|X(z)| = \left| \sum_{n=0}^{\infty} x[n] z^{-n} \right| \leq \sum_{n=0}^{\infty} |x[n] z^{-n}| = \sum_{n=0}^{\infty} |x[n]| \rho^{-n} \tag{3}$$

If finite numbers M and ρ_a exist such that

$$|x[n]| < M \rho_a^n \tag{4}$$

then

$$\sum_{n=0}^{\infty} |x[n]| \rho^{-n} < M \sum_{n=0}^{\infty} \left(\frac{\rho_a}{\rho} \right)^n \tag{5a}$$

The sum on the right-hand side converges for all ρ greater than ρ_a. Therefore, if $|x[n]|$ satisfies Eq. (4), then

$$\sum_{n=0}^{\infty} |x[n]| \rho^{-n} < \infty \qquad \text{for } |z| > \rho_a \tag{5b}$$

and $X(z)$ exists for values of z outside a circle having radius ρ_a in the z-plane. The term ρ_a stands for the *radius of convergence*. Sequences satisfying (4) are said to have *exponential order*. If we compare (5b) with Eq. (4DT.104), we see that the factor ρ^{-n} can allow the z-transform of a sequence to exist when the DTFT does not.

In the following examples we derive the z-transforms of some familiar sequences.

EXAMPLE 6DT.1 z-Transform of a DC Sequence

If $x[n]$ is the DC sequence

$$x[n] = 1 \tag{6}$$

then Eq. (1) becomes

$$X(z) = \sum_{n=0}^{\infty} z^{-n} = \frac{1}{1 - z^{-1}} = \frac{z}{z - 1} \quad |z| > 1 \tag{7}$$

where we have used Eq. (B.4) of Appendix B. We can display $X(z)$ using the *pole-zero diagram* shown in Figure 6DT.1. In this diagram, a circle is drawn at $z = 0$ to indicate the value of z where $X(z) = 0$. This value of z is called a *zero*

Figure 6DT.1
Pole-zero plot for DC sequence: The region of convergence is the unshaded region $|z| > 1$.

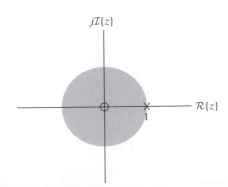

of $X(z)$. A cross is drawn at $z = 1$ to indicate where the denominator of $X(z)$ is 0. This value of z is called a *pole*. We see from (7) that $X(z)$ is defined only for $|z| > 1$, i.e., for points lying outside the unit circle. This region, called the *region of convergence* of $X(z)$, is shown unshaded in the figure. Notice that the region of convergence is outside the circle on which the pole lies.

Before leaving this example, it is a good idea to show that the z-transform of a unit step $u[n]$ is also given by (7).

The complex exponential sequence in the following example provides a building block for many other sequences.

EXAMPLE 6DT.2   z-Transform of a Complex Exponential

Consider the complex exponential

$$x[n] = z_0^n \tag{8}$$

where z_0 is a complex constant

$$z_0 = \rho_0 e^{j\theta_0} \tag{9}$$

The substitution of (8) into (1) yields

$$\mathcal{Z}\{z_0^n\} = \sum_{n=0}^{\infty} z_0^n z^{-n} = \sum_{n=0}^{\infty} \left(\frac{z}{z_0}\right)^{-n} = \frac{z}{z - z_0} \quad |z| > |z_0| \tag{10}$$

The pole-zero plot is shown in Figure 6DT.2. The region of convergence is outside the circle having radius ρ_0 on which the pole lies.

Figure 6DT.2
Pole-zero plot for a complex exponential: The region of convergence is the unshaded region $|z| > |z_0|$.

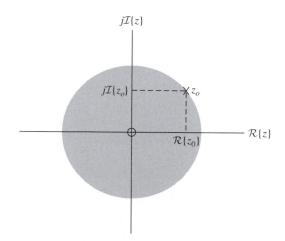

The real exponential $x[n] = \rho_0^n$ and the complex sinusoid $x[n] = e^{j\theta_0 n}$ are special cases of Eq. (10), obtained by setting $\theta_0 = 0$ and $\rho_0 = 1$, respectively. The sequence $x[n] = 1$ that was considered in the first example is a special case obtained by setting $\theta_0 = 0$ and $\rho_0 = 1$.

EXAMPLE 6DT.3 z-Transform of a Unit Impulse

If $x[n]$ is an impulse

$$x[n] = \delta[n] \tag{11}$$

then Eq. (1) becomes

$$X(z) = \sum_{n=0}^{\infty} \delta[n] z^{-n} \tag{12}$$

which simplifies to

$$X(z) = 1 \tag{13}$$

Here the region of convergence is the entire z-plane.

6DT.3 Relationship Between the *z*-Transform and DTFT of Causal Sequences

A sequence is called *causal* if it equals 0 for $n < 0$. The z- and Fourier transforms of causal sequences are related in a fundamental way. We can establish the relationship if we set $z = \rho e^{j2\pi f}$ and assume that $x[n]$ is causal. Equation (1) then becomes

$$X(z)\big|_{z=\rho e^{j2\pi f}} = \sum_{n=0}^{\infty} x[n](\rho e^{j2\pi f})^{-n} = \sum_{n=0}^{\infty} (x[n]\rho^{-n}) e^{-j2\pi f n} \tag{14}$$

Equation (14) shows that the z-transform of $x[n]$ equals the DTFT of the exponentially weighted version $x[n]\rho^{-n}$ of $x[n]$.

$$X(z)\big|_{z=\rho e^{j2\pi f}} = \mathcal{DTFT}\{x[n]\rho^{-n}\} \qquad \text{for causal } x[n] \tag{15}$$

We noted earlier that the factor ρ^{-n} makes it possible for the z-transform of a sequence to exist when the DTFT does not. If we set $\rho = 1$, then (15) becomes

$$X(z)\big|_{z=e^{j2\pi f}} = X(e^{j2\pi f}) \qquad \text{for causal } x[n] \tag{16}$$

where $X(e^{j2\pi f})$ is the DTFT of $x[n]$. Therefore, the z-transform of $x[n]$ evaluated on the unit circle of the z-plane is the DTFT of $x[n]$. It follows from (16) that the DTFT (spectrum) of $x[n]$ exists if and only if the z-transform of $x[n]$ converges on the unit circle.

EXAMPLE 6DT.4 $x[n] = a^n u[n]$

Consider the sequence $a^n u[n]$. It follows from Example 6DT.2 that

$$X(z) = \frac{z}{z - a} \qquad |z| > |a| \tag{17}$$

As we see in Figure 6DT.3, the region of convergence, $|z| > |a|$, includes the unit circle if $|a| < 1$.

Figure 6DT.3
Pole-zero plot for
$a^n u[n]$: The region of
convergence is the
unshaded region
$|z| > |a|$.

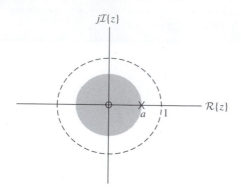

Therefore, if $|a| < 1$, the DTFT of $a^n u[n]$ exists and can set $z = e^{j2\pi f}$ in Eq. (17) to obtain the spectrum of $a^n u[n]$.

$$X(e^{j2\pi f}) = \frac{e^{j2\pi f}}{e^{j2\pi f} - a} = \frac{1}{1 - ae^{-j2\pi f}} \qquad \text{for } |a| < 1 \tag{18}$$

If $|a| \geq 1$, then the region of convergence $|z| > |a|$ does not include the unit circle and the DTFT of $a^n u[n]$ does not exist.

EXAMPLE 6DT.5 $x[n] = u[n]$

Consider the unit step $u[n]$. It follows from the preceding example with $a = 1$ that

$$X(z) = \frac{z}{z - 1} \qquad \text{for } |z| > 1 \tag{19}$$

The region of convergence does not include the unit circle. Therefore, DTFT of a unit step sequence does not exist. This conclusion agrees with Appendix 3DT where we showed that the DTFT does not converge at the point $f = 0$ on the interval $-\frac{1}{2} \leq f < \frac{1}{2}$. Although $X(e^{j2\pi f})$ does not exist $f = 0$, we showed that we can define $X(e^{j2\pi f})$ as

$$X(e^{j2\pi f}) = \frac{1}{2}\delta(f) + \frac{1}{1 - e^{-j2\pi f}} \tag{20}$$

6DT.4   Properties of the z-Transform

We can obtain many additional z-transforms by applying the properties of the z-transform to the result of Examples 6DT.1–6DT.3. One of the most important properties is linearity.

Linearity Property

$$\mathcal{Z}\{ax[n] + by[n]\} = a\mathcal{Z}\{x[n]\} + b\mathcal{Z}\{y[n]\} \tag{21}$$

The linearity property follows directly from Eq. (1). As shown in the following example, the region of convergence of $\mathcal{Z}\{ax[n] + by[n]\}$ is located just outside a circle in the z-plane defined by pole in $\mathcal{Z}\{ax[n] + by[n]\}$ whose absolute value is largest.

EXAMPLE 6DT.6   z-Transform of a Sum of Two Exponentials

Consider

$$w[n] = 5\left(\frac{2}{3}\right)^n + 6\left(-\frac{1}{3}\right)^n \tag{22}$$

Because the z-transform is a linear operation

$$W(z) = \mathcal{Z}\left\{5\left(\frac{2}{3}\right)^n + 6\left(-\frac{1}{3}\right)^n\right\} = 5\mathcal{Z}\left\{\left(\frac{2}{3}\right)^n\right\} + 6\mathcal{Z}\left\{\left(-\frac{1}{3}\right)^n\right\} \tag{23}$$

The z-transforms of the terms on the right-hand side of the preceding equation follow by setting $z_0 = 2/3$ and in $z_0 = -1/3$ in Eq. (10). The results are

$$\mathcal{Z}\left\{\left(\frac{2}{3}\right)^n\right\} = \frac{z}{z - 2/3} \qquad \text{for } |z| > 2/3 \tag{24}$$

and

$$\mathcal{Z}\left\{\left(-\frac{1}{3}\right)^n\right\} = \frac{z}{z + 1/3} \qquad \text{for } |z| > 1/3 \tag{25}$$

$W(z)$ exists if and only if *both* $\mathcal{Z}\{(2/3)^n\}$ and $\mathcal{Z}\{(-1/3)^n\}$ exist. Therefore, the region of convergence for $W(z)$ is $|z| > 2/3$ and we have

$$W(z) = 5\frac{z}{z - \frac{2}{3}} + 6\frac{z}{z + \frac{1}{3}} = \frac{11z\left(z - \frac{7}{33}\right)}{\left(z - \frac{2}{3}\right)\left(z + \frac{1}{3}\right)} \qquad |z| > \frac{2}{3} \tag{26}$$

The pole-zero plot of $W(z)$ is shown in Figure 6DT.4. Notice that the region of convergence lies just outside of the largest circle containing a pole. A region of convergence can never include a pole because the z-transform does not converge at a pole.

Figure 6DT.4
Pole-zero plot for $5(2/3)^n + 6(-1/3)^n$: The region of convergence is the unshaded region just outside the largest circle containing a pole.

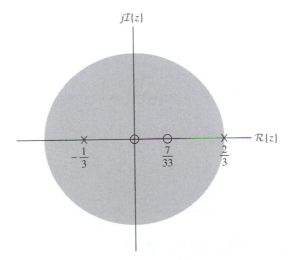

We saw in the above example that the region of convergence of the sum of two scaled exponential waveforms lies outside the circle whose radius is defined by the pole having the largest absolute value. The z-transform of this sum was a *rational algebraic fraction*, $\mathbf{W}(z)$, given by the right-hand-side of (26). We can generalize this result as follows: Whenever a

z-transform is a rational algebraic fraction, the region of convergence lies outside the circle having radius defined by the pole having the largest absolute value. The general form of a rational algebraic fraction is given by (83) of Section 6DT.6.3.

In the following example, we apply the linearity property and the result of Example 6DT.2 to find the z-transform of a sinusoid.

EXAMPLE 6DT.7 z-Transform of a Sinusoid

Consider

$$x[n] = A\cos(\theta_0 n + \phi) \tag{27}$$

We can write the preceding

$$x[n] = Xe^{j\theta_0 n} + X^* e^{-j\theta_0 n} \tag{28}$$

where

$$X = \tfrac{1}{2}Ae^{j\phi} \tag{29}$$

By the linearity property and the result of Example 6DT.2 we obtain

$$\mathcal{Z}\{A\cos(\omega_0 n + \phi)\} = X\mathcal{Z}\{e^{j\theta_0 n}\} + X^*\mathcal{Z}\{e^{-j\theta_0 n}\}$$

$$= \frac{zX}{z - e^{j\theta_0}} + \frac{zX^*}{z - e^{-j\theta_0}} \quad |z| > 1 \tag{30}$$

A second useful important property is the *z-scale property*. It states that if we multiply $x[n]$ by z_0^n, then we scale the argument of $X(z)$ by $1/z_0$.

z-Scale Property

$$x[n]z_0^n \overset{\text{ZT}}{\longleftrightarrow} X(z/z_0) \tag{31}$$

We can easily prove the z-scale property from the definition of the z-transform of $x[n]z_0^n$.

$$\mathcal{Z}\{x[n]e^{z_0 n}\} = \sum_{n=0}^{\infty} x[n]z_0^n z^{-n} = \sum_{n=0}^{\infty} x[n](z/z_0)^{-n} = X(z/z_0) \tag{32}$$

EXAMPLE 6DT.8 z-Transform of a Damped Sinusoid

Consider the damped sinusoid

$$x[n] = A\rho_0^n \cos(\theta_0 n + \phi) \tag{33}$$

If we apply the z-scale property with $z_0 = \rho_0$ to the result of Example 6DT.7, we obtain

$$X(z) = \frac{\frac{z}{\rho_0}X}{\frac{z}{\rho_0} - e^{j\theta_0}} + \frac{\frac{z}{\rho_0}X^*}{\frac{z}{\rho_0} - e^{-j\theta_0}} = \frac{zX}{z - \rho_0 e^{j\theta_0}} + \frac{zX^*}{z - \rho_0 e^{-j\theta_0}} \quad |z| > |\rho_0| \tag{34}$$

A third useful property is the differentiation property,

Differentiation Property

$$nx[n] \xleftrightarrow{ZT} -z\frac{d}{dz}X(z) \qquad (35)$$

The proof follows directly from (1) for if

$$X(z) = \sum_{n=0}^{\infty} x[n]z^{-n} \qquad (36)$$

then

$$-z\frac{d}{dz}X(z) = -z\sum_{n=0}^{\infty}(-n)x[n]z^{-n-1} = \sum_{n=0}^{\infty} nx[n]z^{-n} = \mathcal{Z}\{nx[n]\} \qquad (37)$$

In the following example, we use the differentiation property to find the z-transform of a unit ramp.

EXAMPLE 6DT.9   z-Transform of a Ramp

Consider

$$x[n] = r[n] = nu[n] \qquad (38)$$

We showed in Example 6DT.5 that

$$u[n] \xleftrightarrow{ZT} \frac{z}{z-1} \quad |z| > 1 \qquad (39)$$

The differentiation property yields

$$nu[n] \xleftrightarrow{ZT} -z\frac{d}{dz}\frac{z}{z-1} = \frac{z}{(z-1)^2} \quad |z| > 1 \qquad (40)$$

The pole-zero plot is shown in Figure 6DT.5, where the pole is indicated by overlapping crosses. This pole is called a *second-order* pole because the degree of the denominator factor $(z-1)^2$ is 2.

Figure 6DT.5
Pole zero plot for unit ramp.

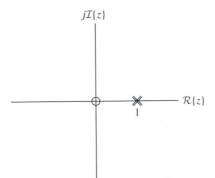

Additional z-transform properties are listed in Table 6DT.1. The sequence shift property is established in Section 6DT.5. The proofs of the convolution and accumulation properties are straightforward and are relegated to the problems. The properties we have listed are sufficient for most engineering problems.

Table 6DT.2 is a list of z-transform pairs. We included several forms for the transform of a cosine to save time later.

Table 6DT.1
Operational
Properties of the
One-Sided
z-Transform

Linearity	$a_1 x_1[n] + a_2 x_2[n]$	$a_1 X_1(z) + a_2 X_2(z)$
Shift right (delay) by 1	$x[n-1]$	$z^{-1} X(z) + x[-1]$
Shift right (delay) by 2	$x[n-2]$	$z^{-2} X(z) + z^{-1} x[-1] + x[-2]$
Shift right (delay) by k	$x[n-k]$	$z^{-k} X(z) + \displaystyle\sum_{m=1}^{k} x[-m] z^{-(k-m)}$
Shift left by 1	$x[n+1]$	$z X(z) - z x[0]$
Shift left by 2	$x[n+2]$	$z^2 X(z) - z^2 x[0] - z x[-1]$
Shift left by k	$x[n+k]$	$z^k X(z) - \displaystyle\sum_{m=0}^{k-1} x[m] z^{k-m}$
Accumulate	$\displaystyle\sum_{m=0}^{n} x[m]$	$\dfrac{z}{z-1} X(z)$
z-scale	$x[n] z_0^n$	$X(z/z_0)$
Convolution	$x_1[n] * x_2[n]$	$X_1(z) X_2(z)$
Differentiation	$n x[n]$	$-z \dfrac{d}{dz} X(z)$

The region of convergence of the z-transform of $\delta[n]$ is the entire z-plane.
The region of convergence of every other entry is the region just outside of the largest circle containing a pole.

	Sequence	Transform
Impulse	$\delta[n]$	1
DC or step	1	$\dfrac{z}{z-1}$
Ramp	n	$\dfrac{z}{(z-1)^2}$
Parabola	n^2	$\dfrac{z(z+1)}{(z-1)^3}$
Exponential	p^n	$\dfrac{z}{z-p}$
Second-order pole	$(n+1)p^n$	$\left(\dfrac{z}{z-p}\right)^2$
kth order pole	$\dfrac{(n+1)(n+2)\cdots(n+k-1)}{(k-1)!} p^n$	$\left(\dfrac{z}{z-p}\right)^k$
Cosine	$\cos(\omega_0 n)$	$\dfrac{1}{2}\dfrac{z}{z-e^{j\omega_0}} + \dfrac{1}{2}\dfrac{z}{z-e^{-j\omega_0}}$
Cosine	$\cos(\omega_0 n)$	$\dfrac{z(z-\cos\omega_0)}{(z-e^{j\omega_0})(z-e^{-j\omega_0})} = \dfrac{z(z-\cos\omega_0)}{z^2 - 2z\cos\omega_0 + 1}$
Sine	$\sin(\omega_0 n)$	$\dfrac{1}{2j}\dfrac{z}{z-e^{j\omega_0}} - \dfrac{1}{2j}\dfrac{z}{z-e^{-j\omega_0}}$
Sine	$\sin(\omega_0 n)$	$\dfrac{z\sin\omega_0}{(z-e^{j\omega_0})(z-e^{-j\omega_0})} = \dfrac{z\sin\omega_0}{z^2 - 2z\cos\omega_0 + 1}$
Complex conjugate poles	$2\lvert K\rvert \rho^n \cos(\theta n + \angle K)$	$\dfrac{Kz}{z-p} + \dfrac{K^*z}{z-p^*}; \; p = \rho e^{j\theta}$
kth-order complex pole	$\frac{2(n+1)(n+2)\cdots(n+k-1)}{(k-1)!}\lvert K\rvert \rho^n$ $\times \cos(\theta n + \angle K); \quad \rho = \lvert p\rvert, \; \theta = \angle K$	$K\left(\dfrac{z}{z-p}\right)^k + K^*\left(\dfrac{z}{z-p^*}\right)^k$

Table 6DT.2

One-Sided z-Transforms

6DT.5   Elementary Systems in the *z*-Domain

We can use the *z*-transform to create a set of *z*-domain elementary system models. These elementary models include the effects of initial conditions (initial stored energy). Large-scale system models are obtained by interconnecting the *z*-domain elementary system models. Remember that the *z*-domain models refer to sequences on the interval $n = 0, 1, 2, \dots$ because they are derived from the *z*-transform.

6DT.5.1
Ideal Amplifiers and Adders

Figure 6DT.6 illustrates the operations of amplification and addition in the sequence and *z*-domains. The *z*-domain operators are obtained directly from the linearity property of the *z*-transform (21). Initial conditions are not a part of the *z*-domain systems because ideal amplification and addition are memoryless operations.

6DT.5.2
Ideal Delay (Right Shift)

To describe the *z*-domain system for right shift, it is convenient to denote a unit delay operation as

$$x^{(1)}[n] = x[n-1] \tag{41}$$

and the *z*-transform of $x^{(1)}[n]$ as

$$X^{(1)}(z) = \sum_{n=0}^{\infty} x[n-1]z^{-n} = x[-1] + \sum_{n=1}^{\infty} x[n-1]z^{-n} \tag{42}$$

If we set $n = m + 1$, we obtain:

$$X^{(1)}(z) = x[-1] + \sum_{m=0}^{\infty} x[m]z^{-(m+1)} = x[-1] + z^{-1}X(z) \tag{43}$$

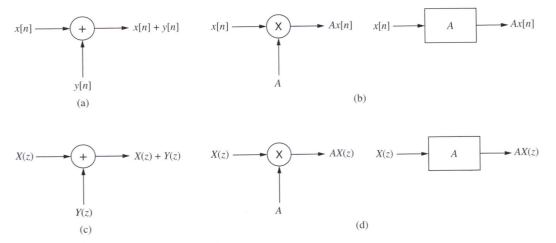

Figure 6DT.6
Amplification and addition in the sequence and *z*-domains. (a)–(b) Addition and Amplification in Sequence-Domain (c)–(d) Addition and Amplification in *z*-Domain.

Figure 6DT.7
Unit delay element in
the sequence and
z-domains:
(a) sequence domain;
(b) z-domain showing
initial conditions.

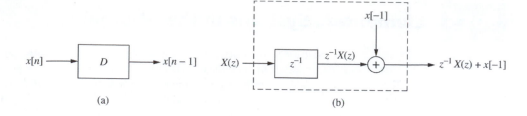

(a)

(b)

Therefore

$$X^{(1)}(z) = z^{-1}X(z) + x[-1] \tag{44}$$

The sequence and z-domain representations of the unit delay element is shown in Figure 6DT.7

Figure 6DT.8(a) show two unit unit delay elements connected in series. The overall output is

$$x^{(2)}[n] = x[n-2] \tag{45}$$

The z-domain system corresponding to Figure 6DT.8(a) is shown in Figure 6DT.8(b). Each stage performs the operation shown in Figure 6DT.7(b) but with a different input. The output of the last unit delay of the z-domain system is

$$X^{(2)}(z) = z^{-1}X^{(1)}(z) + x^{(1)}[-2] \tag{46}$$

By substituting Eq. (44) into Eq. (46), or by inspection of Figure 6DT.8(b), we see that

$$X^{(2)}(z) = z^{-2}X(z) + z^{-1}x[-1] + x[-2] \tag{47}$$

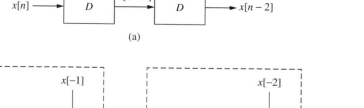

(a)

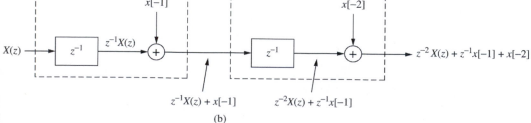

(b)

Figure 6DT.8
Two unit delay elements in series: (a) sequence domain; (b) z-domain showing initial conditions.

We can generalize these results to l unit delays in series. We start with the definition

$$x^{(l)}[n] = x^{(l-1)}[n-1] \tag{48}$$

The corresponding the z-domain equation is

$$X^{(l)}(z) = z^{-1}X^{(l-1)}(z) + x^{(l-1)}[n-1] \tag{49}$$

where $X^{(l)}(z) = \mathcal{Z}\{x^{(l)}[n]\}$. This equation is a recursion formula. Equations (44) and (46) can be viewed as special cases of Eq. (49) for $l = 1$ and 2, respectively. We can solve the general recursion by repeated substitution. The result is

$$X^{(l)}(z) = z^{-l}X(z) + z^{-l+1}x[-1] + z^{-l+2}x[-2] + \cdots + x[-l] \tag{50}$$

These results are summarized in Table 6DT.1. The left-shift operators listed in Table 6DT.1 can be derived in a similar way (Problem 6DT.5.2)

6DT.6   z-Transform System Analysis

If we interconnect the elementary systems described in the preceding section, we obtain a system whose input and output are related by a linear difference equation. We can find the output of this system for $n \geq 0$ for arbitrary initial conditions by deriving and then solving the I-O equation of the system.

**6DT.6.1
Basic
Procedure**

For most engineering problems, the output can be obtained using the following four steps:

Basic Procedure

1. Obtain the z-domain I-O equation of the system. This step can be accomplished from either the system's difference I-O equation or from the system's z-domain block diagram.
2. Solve the z-domain I-O equation for the z-domain output $Y(z)$.
3. Use partial fraction expansion to put $Y(z)$ in a form ready for table lookup.
4. Look up the terms of $Y(z)$ in a table to find $y[n]$.

The following examples demonstrate the four steps and serve as an introduction to the general development in the next subsection. Example 6DT.10 illustrates the first two steps of the basic procedure.

EXAMPLE 6DT.10   *z*-domain I-O Equation of a First-Order System

Let's find the z-domain I-O equation of the first-order system of Example 2DT.3. The system block diagram is redrawn in Figure 6DT.9(a). The z-domain system, which addresses the index interval $n \geq 0$, is shown in Figure 6DT.9(b). The influence of the interval $n < 0$ on the output for $n \geq 0$ is contained in the *initial condition* $y[-1]$. The z-domain equation relating the LTI system's output to its input can be obtained either of two ways.

Figure 6DT.9
First-order system in
sequence and
z-domains:
(a) sequence domain;
(b) z-domain.

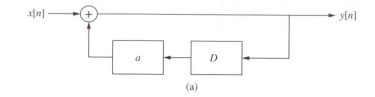

(a)

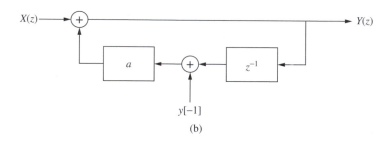

$y[-1]$

(b)

First, we can derive the z-domain I-O equation directly from the difference equation relating input and output of the sequence domain system. An inspection of Figure 6DT.9(a) reveals that

$$y[n] = ay[n-1] + x[n] \quad n \geq 0 \tag{51}$$

We take the z-transform of both sides to obtain the z-domain I-O equation. The result is

$$Y(z) = az^{-1}Y(z) + ay[-1] + X(z) \tag{52}$$

We can solve for $Y(z)$.

$$Y(z) = \frac{z}{z-a}X(z) + \frac{z}{z-a}ay[-1] \tag{53}$$

The effects of the input and the initial condition are represented by the first and second terms, respectively.

We can also obtain Eq. (53) from the z-domain system shown in Figure 6DT.9(b). The z-domain system is obtained by replacing the sequence domain elements by their z-domain counterparts. An inspection of Figure 6DT.9(b) reveals that

$$Y(z) = az^{-1}Y(z) + ay[-1] + X(z) \tag{54}$$

which agrees with Eq. (52).

The following examples illustrate the last two steps of the method.

EXAMPLE 6DT.11   Response to Impulse and Initial Condition

Let's find $y[n]$ for the system of Figure 6DT.9 if $x[n] = \delta[n]$ and there is a some arbitrary value $y[-1]$ stored in the delay at $n = 0$. The z-transform of $\delta[n]$ is 1. The substitution $X(z) = 1$ into Eq. (53) yields

$$Y(z) = \frac{z}{z-a} + \frac{z}{z-a}ay[-1] \tag{55}$$

Partial fraction expansion (step 3) is not necessary because both terms in the right-hand side can be found in Table 6DT.2, which gives the solution.

$$y[n] = a^n + y[-1]a^{n+1} \quad n \geq 0 \tag{56}$$

The first term in $y[n]$, due to the input, is called the *forced response*. The first term is the system's impulse $h[n]$ because it is the response to $x[n] = \delta[n]$ when the system is initially at rest $(y[-1] = 0)$. The second term, due to the initial condition, is sometimes referred to as the *stored energy response*.

The following example introduces *partial fraction expansion*.

EXAMPLE 6DT.12   Response to Step and Initial Condition

Consider the system of Figure 6DT.9 with input $x[n] = u[n]$. If we substitute $X(z) = z/(z-1)$ from Table 6DT.2 into Eq. (53), we obtain

$$Y(z) = \frac{z^2}{(z-a)(z-1)} + \frac{z}{z-a}ay[-1] \tag{57}$$

Partial fraction expansion is necessary because the first term on the right-hand side does not appear in Table 6DT.2. We denote this term as $G(z)$. To make table lookup possible, we expand it into partial fractions.

$$G(z) = \frac{z^2}{(z-a)(z-1)} = \frac{K_1 z}{z-a} + \frac{K_2 z}{z-1} \tag{58}$$

where we assume that $a \neq 1$. The coefficients K_1 and K_2 are constants to be determined. There are several ways to find the values for the coefficients in a partial fraction expansion. One of the quickest, called *Heaviside's technique*, works as follows: To find K_1, multiply both sides of Eq. (58) by $(z-a)/z$ and then set $z = a$. The multiplication step yields

$$\left(\frac{z-a}{z}\right)G(z) = \left(\frac{z-a}{z}\right)\frac{z^2}{(z-a)(z-1)} = \left(\frac{z-a}{z}\right)\frac{K_1 z}{z-a} + \left(\frac{z-a}{z}\right)\frac{K_2 z}{(z-1)} \tag{59}$$

The slashes show some common factors we eliminated. When we set $z = a$, the preceding equation becomes

$$\left(\frac{z-a}{z}\right)G(z)\bigg|_{z=a} = \frac{a}{a-1} = K_1 + 0 \tag{60}$$

so

$$K_1 = -\frac{a}{1-a} \tag{61}$$

Similarly, to find K_2, we multiply Eq. (58) by $(z-1)/z$ and then set $z = 1$. The multiplication step yields

$$\left(\frac{z-1}{z}\right)G(z) = \left(\frac{z-1}{z}\right)\frac{z^2}{(z-a)(z-1)} = \left(\frac{z-1}{z}\right)\frac{K_1 z}{z-a} + \left(\frac{z-1}{z}\right)\frac{K_2 z}{(z-1)} \tag{62}$$

When we set $z = 1$, this becomes

$$\left(\frac{z-1}{z}\right)G(z)\bigg|_{z=1} = \frac{1}{1-a} = K_2 \tag{63}$$

The substitution $K_1 = -a/(1-a)$ and $K_2 = 1/(1-a)$ into Eq. (58) yields

$$G(z) = -\left(\frac{a}{1-a}\right)\frac{z}{z-a} + \left(\frac{1}{1-a}\right)\frac{z}{z-1} \tag{64}$$

which, in turn, yields

$$Y(z) = -\left(\frac{a}{1-a}\right)\frac{z}{z-a} + \left(\frac{1}{1-a}\right)\frac{z}{z-1} + \frac{z}{z-a}ay[-1] \tag{65}$$

Each term in $Y(z)$ now appears in Table 6DT.2. We can write down the sequence domain output by inspection.

$$y[n] = -\left(\frac{a}{1-a}\right)a^n + \frac{1}{1-a} + y[-1]a^{n+1} \quad n \geq 0 \tag{66}$$

which rearranges to

$$y[n] = \frac{1}{1-a}(1 - a^{n+1}) + y[-1]a^{n+1} \quad n \geq 0 \tag{67}$$

The first term is the forced response. The last term is the stored energy response.

Now that we have seen the preliminary examples, let's develop the four basic steps formally and generally. The strength of the formal development is that it shows how to solve a large variety of problems. However, it takes time to understand the notation and analytic results in Sections 6DT.6.2 and 6DT.6.3. We recommend that you read the developments the first time to gain a little familiarity with it. After that, focus on the examples and look back to the general development as needed.

6DT.6.2 General Forms of Sequence and z-Domain Equations of LTI Systems

The difference I-O equation of a system composed of adders, gains, and unit delays has the form

$$a_l y[n-l] + a_{l-1} y[n-(l-1)] + \cdots + a_1 y[n-1] + a_0 y[n]$$
$$= b_m x[n-m] + b_{m-1} x[n-(m-1)] + \cdots + b_1 x[n-1] + b_0 x[n] \tag{68}$$

where $x[n]$ and $y[n]$ are the input and output, respectively, and l is the order of the system. Equation 68 is an ordinary linear difference equation having constant coefficients. Our objective is to solve Eq. (68) for $y[n]$ subject to initial conditions $y[-l], \ldots, y[-1]$, $x[-m], \ldots, x[-1]$.

To obtain the solution, we transform Eq. (68) into the z-domain using the linearity and sequence shift properties. The result is

$$a_l z^{-l} Y(z) + a_{l-1} z^{-(l-1)} Y(z) + \cdots + a_1 z^{-1} Y(z) + a_0 Y(z) + Y_{ic}(z)$$
$$= b_m z^{-m} X(z) + b_{m-1} z^{-(m-1)} X(z) + \cdots + b_1 z^{-1} X(z) + b_0 X(z) + X_{ic}(z) \tag{69}$$

where $Y_{ic}(z)$ and $X_{ic}(z)$ are polynomials in z^{-1} with degree $l - 1$ and $m - 1$, respectively. These polynomials arise from the delay property and are present only when the initial conditions $y[-l], \ldots, y[-1], x[-m], \ldots, x[-1]$ are not 0. An example is Eq. (52), which can be arranged into the form (68).

It will be convenient to write Eq. (68) compactly as

$$\mathcal{A}(D)y[n] = \mathcal{B}(D)x[n] \tag{70}$$

where

$$\mathcal{A}(D) = a_l D^l + a_{l-1} D^{l-1} + \cdots + a_1 D + a_0 \tag{71}$$
$$\mathcal{B}(D) = b_m D^m + b_{m-1} D^{m-1} + \cdots + b_1 D + b_0 \tag{72}$$

and D is the *shift operator*, defined by

$$D^k w[n] \triangleq w[n-k] \tag{73}$$

for any sequence $w[n]$ and integer $k \geq 0$. The z-domain Eq. (69) is then given by

$$\mathcal{A}(z^{-1})Y(z) + Y_{ic}(z) = \mathcal{B}(z^{-1})X(z^{-1}) + X_{ic}(z) \tag{74}$$

The solution is

$$Y(z) = \frac{\mathcal{B}(z^{-1})}{\mathcal{A}(z^{-1})}X(z) + \frac{X_{ic}(z) - Y_{ic}(z)}{\mathcal{A}(z^{-1})} \tag{75}$$

where

$$\frac{\mathcal{B}(z^{-1})}{\mathcal{A}(z^{-1})} = \frac{b_m z^{-m} + b_{m-1} z^{-(m-1)} + \cdots + b_1 z^{-1} + b_0}{a_l z^{-l} + a_{l-1} z^{-(l-1)} + \cdots + a_1 z^{-1} + a_0}$$

$$= z^{l-m}\left(\frac{b_0 z^m + b_1 z^{m-1} p + \cdots + b_{m-1} z + \cdots + b_m}{a_0 z^l + a_1 z^{l-1} + \cdots + a_{l-1} z + a_l}\right) \tag{76}$$

The first term in Eq. (75) describes the system's forced response. It corresponds to the part of the output caused by $x[n]$ acting alone. The factor multiplying $X(z)$ is the system's *transfer function*.

$$H(z) = \frac{\mathcal{B}(z^{-1})}{\mathcal{A}(z^{-1})} = \mathcal{Z}\{h[n]\} \tag{77}$$

Equation (77) states that $H(z)$ is the z-transform of the system's impulse response $h[n]$. We can establish this relationship by recalling that $y[n] = h[n]$ when $x[n] = \delta[n]$ and the system is initially at rest (all initial conditions are 0). If we apply these conditions to Eq. (75), we find that $Y(z) = H(z)$ when $X(z) = 1$ and $X_{ic}(z) = Y_{ic}(z) = 0$. The second term in Eq. (75) describes the system's stored energy response. It corresponds to the part of the output caused by nonzero initial conditions.

A transfer function is specified by its poles and zeros. We show property this by referring to Eq. (76) to write $H(z)$ in factored form

$$H(z) = \left(\frac{z^{-m}}{z^{-l}}\right)\frac{b_0(z - z_{h1})(z - z_{h2})\cdots(z - z_{hm})}{a_0(z - p_{h1})(z - p_{h2})\cdots(z - p_{hl})} = \frac{b_0 z^{-m}\prod_{i=1}^{m}(z - z_{hi})}{a_0 z^{-l}\prod_{i=1}^{l}(z - p_{hi})} \tag{78}$$

where $p_{h1}, p_{h2}, \ldots, p_{hl}$ are poles of $H(z)$. The poles include the n roots to the denominator polynomial of Eq. (76).

$$a_0 z^l + a_1 z^{l-1} + \cdots + a_{l-1} z + a_l = 0 \tag{79}$$

plus an additional $m - l$ poles at $z = 0$ if $m > l$. The poles of $H(z)$ characterize the dynamic response of the system, independent of any applied input. The quantities $z_{h1}, z_{h2}, \ldots, z_{hm}$ are called the *zeros* of $H(z)$. The zeros include the roots to the numerator polynomial of Eq. (76)

$$b_0 z^m + b_1 z^{m-1} + \cdots + b_{m-1} z + \cdots + b_m = 0 \tag{80}$$

plus an additional $l - m$ zeros, $z = 0$, if $l > m$. A count of the number of poles and zeros of $H(z)$, including those at the origin, shows that the number of poles equals the number of zeros. This number is the maximum of l and m. The poles and zeros, along with the constant b_0/a_0, are a complete description of the system. This means that we can obtain $H(z)$, $h[n]$, and the difference equation of the system if we know b_0/a_0 and the poles and zeros.

For many practical inputs, $X(z)$ can also be written in factored form:

$$X(z) = k \frac{z^{-v} \prod_{i=1}^{v}(z - z_{xi})}{z^{-u} \prod_{i=1}^{u}(z - p_{xi})} \tag{81}$$

where the p_{Xi} and z_{Xi} are poles and zeros of $X(z)$, respectively. The substitution of Eqs. (78) and (81) into Eq. (75) yields

$$Y(z) = K \frac{z^{-(m+v)} \prod_{i=1}^{m}(z - z_{hi}) \prod_{i=1}^{v}(z - z_{xi})}{z^{-(l+u)} \prod_{i=1}^{l}(z - p_{hi}) \prod_{i=1}^{u}(z - p_{xi})} + \frac{X_{ic}(z) - Y_{ic}(z)}{a_0 z^{-l} \prod_{i=1}^{l}(z - p_{hi})} \tag{82}$$

where $K = kb_0/a_0$. An example of Eq. (82) is given by (53).

6DT.6.3 Partial Fraction Expansions

Partial fraction expansion is used to put $Y(z)$ in a form ready for table lookup. Let's consider how to find the partial fraction expansion of any term in $Y(z)$ that can be put in any one of the following forms.

$$G(z) = \frac{c_q z^{-q} + c_{q-1} z^{-(q-1)} + \cdots + c_1 z^{-1} + c_0}{d_r z^{-r} + d_{r-1} z^{-(r-1)} + \cdots + d_1 z^{-1} + a_0} = \frac{c_0}{a_0} \frac{(1 - z_1 z^{-1})(1 - z_2 z^{-1}) \cdots (1 - z_q z^{-1})}{(1 - p_1 z^{-1})(1 - p_2 z^{-1}) \cdots (1 - p_r z^{-1})}$$

$$= z^{r-q} \left(\frac{c_0 z^q + c_1 z^{q-1} p + \cdots + c_{q-1} z + c_q}{d_0 z^r + d_1 z^{r-1} + \cdots + d_{r-1} z + d_r} \right) = \frac{c_0}{a_0} z^{r-q} \frac{(z - z_1)(z - z_2) \cdots (z - z_q)}{(z - p_1)(z - p_2) \cdots (z - p_r)} \tag{83}$$

We organize our description of partial fraction expansions using the following terminology.

Terminology

$G(z)$ is called a *proper rational* algebraic fraction if $q < r$, and an *improper rational* algebraic fraction if $q \geq r$. A pole p_i or a zero z_j has *order k* if its value repeats $k - 1$ times (i.e., the value occurs k times). First-order poles or zeros are called *simple*.

The form of the partial fraction expansion of $G(z)$ depends on whether it is proper or improper and on the orders of the poles. In many practical problems, $G(z)$ is proper and the poles are all first order (all simple poles). This frequently occuring case is the simplest.

Proper G(z) and All Simple Poles

If $G(z)$ is proper and all the poles are simple, then its partial fraction expansion has the form

$$G(z) = \frac{K_1 z}{z - p_1} + \frac{K_2 z}{z - p_2} + \cdots + \frac{K_i z}{z - p_i} + \cdots + \frac{K_r z}{z - p_r} \tag{84}$$

The coefficients $K_1, K_2, \ldots$, and K_r are called *residues*. Each residue can be evaluated using Heaviside's formula.

$$K_i = \mathcal{Res}\{G(z); p_i\} = \left(\frac{z - p_i}{z}\right) G(z)\Big|_{z=p_i} \quad i = 1, 2, \ldots, r \tag{85}$$

where $\mathcal{Res}\{G(z); p_i\}$ denotes "the residue of $G(z)$ at the pole $z = p_i$." The sequence associated with Eq. (84) is

$$g[n] = K_1 p_1^n + K_2 p_2^n + \cdots + K_r p_r^n \quad n \geq 0 \tag{86}$$

In most applied work, $g[n]$ represents the real valued output of a real system having a real valued input. Whenever $g[n]$ is real, any complex poles and their associated residues must occur in complex conjugate pairs: If $G(z)$ has a complex pole $p = \rho\angle\theta$ with residue K, then $G(z)$ also has a complex pole $p^* = \rho\angle -\theta$ with residue K^*. The partial fraction expansion (84) and the associated sequence (86), in turn, must contain pairs of terms

$$\frac{Kz}{z - p} + \frac{K^*z}{z - p^*} \xrightarrow{ZT} Kp^n + K^* p^{*n} = 2|K|\rho^n \cos(\theta n + \angle K) \tag{87}$$

We can prove this statement simply by noticing that if the term $K_i^* z/(z - p_i^*)$ did not appear along with $K_i z(z - p_i)$ in Eq. (87), then $g[n]$ could not be real. The result Eq. (87) implies that we only need to find one residue K of a complex pair of residues for real systems having real inputs. We can write down the real damped sinusoid on the right-hand side of Eq. (87) directly from knowledge of p and K.

EXAMPLE 6DT.13   Impulse Response of a Resonant Second-Order System

Second-order systems are described by second-order difference equations. The highest-order difference on the output $y[n]$ equals 2. In this problem we find the impulse response of the second-order system characterized by

$$cy[n - 2] + by[n - 1] + y[n] = x[n] \tag{88}$$

A block diagram of this system is shown in Figure 6DT.10

Figure 6DT.10
Second-order system.

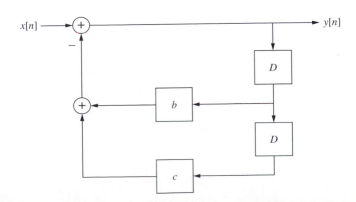

The z-domain I-O equation is

$$cz^{-2}Y(z) + cz^{-1}y[-1] + cy[-2] + bz^{-1}Y(z) + by[-1] + Y(z) = X(z) \tag{89}$$

The solution is

$$Y(z) = \frac{1}{cz^{-2} + bz^{-1} + 1} X(z) - \frac{cz^{-1}y[-1] + cy[-2] + by[-1]}{cz^{-2} + bz^{-1} + 1} \tag{90}$$

The factor multiplying $X(z)$ is the system's transfer function.

$$H(z) = \frac{1}{cz^{-2} + bz^{-1} + 1} = \frac{z^2}{z^2 + bz + c} \tag{91}$$

To expand $H(z)$ into partial fractions, we need to find the poles. The poles are the roots to

$$z^2 + bz + c = 0 \tag{92}$$

We assume that $c > (b/2)^2$ for which the poles are complex conjugates.

$$p_{h1} = -\frac{b}{2} + j\sqrt{c - \left(\frac{b}{2}\right)^2} \qquad p_{h2} = p_{h1}^* = -\frac{b}{2} - j\sqrt{c - \left(\frac{b}{2}\right)^2} \tag{93}$$

The pole-zero plot is shown in Figure 6DT.11.

Figure 6DT.11
Pole-zero plot of
resonant
second-order system.

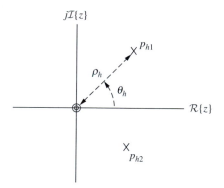

We see from this figure that the poles are located at polar coordinates.

$$p_{h1} = \rho_h \angle \theta_h \qquad p_{h2} = \rho_h \angle -\theta_h \tag{94}$$

where $\rho_h = \sqrt{c}$ and $\theta_h = (\pi/2) + \cos^{-1}(b/2)$. The partial fraction expansion is

$$H(z) = \frac{z^2}{(z - p_{h1})(z - p_{h1}^*)} = \frac{K_1 z}{z - p_{h1}} + \frac{K_2 z}{z - p_{h1}^*} \tag{95}$$

Heaviside's technique yields

$$K_1 = \frac{p_{h1}}{p_{h1} - p_{h2}} \tag{96}$$

Figure 6DT.12
Impulse response of resonant second-order system.

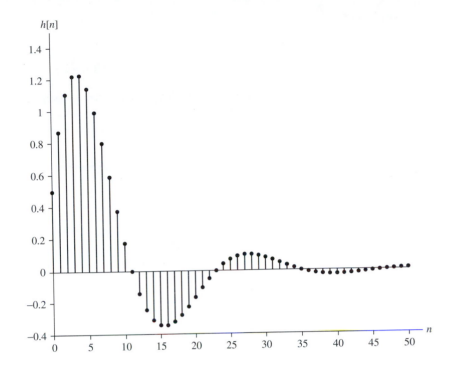

We know that $K_2 = K_1^*$ because the system and the input are real. (The reader is encouraged to use Heaviside's technique to confirm that $K_2 = K_1^*$.) Therefore Eq. (87) applies and we can write down

$$h[n] = 2|K_1|\rho_h^n \cos(\theta_h n + \angle K_1) \tag{97}$$

where $|K_1| = c/b^2$ and $\angle K_1 = \tan^{-1}\{-\sqrt{(c/(2b)^2) - 1}\}$. A plot of $h[n]$ is given in Figure 6DT.12 for $\rho_h = 0.9$ and $\theta_h = 15°$.

G(z) *with Second-Order Poles*

Occasionally, we need to evaluate a partial fraction expansion involving one or more multiple-order poles in addition to simple poles. Let's consider first the most common case in which $G(z)$ contains both simple poles and one or more second-order poles. The partial fraction expansion of $G(z)$ includes terms of the form

$$\frac{Kz}{z - p} \xleftrightarrow{\text{ZT}} Kp^n \tag{98}$$

for each simple pole, and two terms of the form

$$\frac{L_1 z}{z - p} + \frac{L_2 z^2}{(z - p)^2} \xleftrightarrow{\text{ZT}} L_1 p^n + L_2(n + 1)p^n \quad n \geq 0 \tag{99}$$

for each second-order pole. Only the coefficients K and L_1 are residues. We can find L_2 and L_1 from the formulas

$$L_2 = [(1 - pz^{-1})^2 G(z)]\Big|_{z=p} = \left[\left(\frac{z - p}{z}\right)^2 G(z)\right]\Bigg|_{z=p} \tag{100}$$

and

$$L_1 = -\frac{1}{p}\frac{d}{d(z^{-1})}[(1 - pz^{-1})^2 G(z)]\Big|_{z=p} = \frac{1}{p}z^2\frac{d}{dz}\left[\left(\frac{z-p}{z}\right)^2 G(z)\right]\Big|_{z=p} \tag{101}$$

As with simple poles, any complex poles and their associated partial fraction coefficients occur in complex conjugate pairs for real $g[n]$.

The following example illustrates partial fraction expansion involving a second-order real pole.

EXAMPLE 6DT.14   Ramp Response of a First-Order System

Consider the response of the first-order system defined by $h[n] = (1 - a)a^n u[n]$ to $x[n] = r[n]$ with $y[-1] = 0$. This system was shown in Figure 2DT.20. Here, the transfer function is

$$H(z) = (1 - a)\frac{z}{(z - a)} \tag{102}$$

and the z-domain input is (Table 6DT.2)

$$X(z) = \frac{z}{(z - 1)^2} \tag{103}$$

Because the initial condition is 0, the z-domain output is

$$Y(z) = H(z)X(z) = (1 - a)\frac{z}{(z - a)}\frac{z}{(z - 1)^2} \tag{104}$$

$Y(z)$ has a second-order pole at $z = 1$ and a first-order pole at $z = a$. Therefore, the partial fraction expansion has the form

$$Y(z) = \frac{(1 - a)z^2}{(z - a)(z - 1)^2} = \frac{Kz}{z - a} + \frac{L_1 z}{z - 1} + \frac{L_2 z^2}{(z - 1)^2} \tag{105}$$

We can use Heaviside's technique to evaluate K.

$$\left(\frac{z-a}{z}\right)Y(z)\Big|_{z=a} = (1 - a)\frac{z}{(z-1)^2}\Big|_{z=a} = K + \left(\frac{z-a}{z}\right)\frac{L_1 z}{z-1}\Big|_{z=a} + \left(\frac{z-a}{z}\right)\frac{L_2 z^2}{(z-1)^2}\Big|_{z=a} \tag{106}$$

This yields

$$K = \frac{a}{1 - a} \tag{107}$$

We can use Eq. (100) with $G(z) \equiv Y(z)$ to evaluate L_2

$$\left(\frac{z-1}{z}\right)^2 Y(z)\Big|_{z=1} =$$

$$\left(\frac{z-1}{z}\right)^2\frac{(1 - a)z^2}{(z - a)(z - 1)^2}\Big|_{z=1} = \left(\frac{z-1}{z}\right)^2\frac{Kz}{z-a}\Big|_{z=1} + \left(\frac{z-1}{z}\right)^2\frac{L_1 z}{z-1}\Big|_{z=1} + L_2 \tag{108}$$

which yields

$$L_2 = 1 \tag{109}$$

If we substitute Eqs. (107) and (109) into Eq. (105), we obtain

$$\frac{(1 - a)z^2}{(z - a)(z - 1)^2} = \frac{a}{1 - a}\frac{z}{z - a} + \frac{L_1 z}{z - 1} + \frac{z^2}{(z - 1)^2} \tag{110}$$

L_1 can be evaluated in several ways. One method is to substitute a convenient value for z and solve for L_1. If we divide the preceding equation by z and set $z = 0$, we find that

$$0 = -\frac{1}{1-a} - L_1 \tag{111}$$

which yields

$$L_1 = \frac{-1}{1-a} \tag{112}$$

We can also obtain this result by taking the limit $z \to \infty$ in Eq. (110). Another way to find L_1 is to use Eq. (101). We set $p = 1$ in Eq. (101) to obtain

$$L_1 = z^2 \frac{d}{dz}\left[\left(\frac{z-1}{z}\right)^2 Y(z)\right]\Bigg|_{z=1} = z^2 \frac{d}{dz}\frac{(1-a)}{z-a}\Bigg|_{z=1} = z^2 \frac{-(1-a)}{(z-a)^2}\Bigg|_{z=1} = \frac{-1}{1-a} \tag{113}$$

which checks our previous result. Therefore

$$Y(z) = \frac{a}{1-a}\frac{z}{z-a} - \frac{1}{1-a}\frac{z}{z-1} + \frac{z^2}{(z-1)^2} \tag{114}$$

Table lookup gives the corresponding sequence domain response.

$$y[n] = \frac{a^{n+1}}{1-a} - \frac{1}{1-a} + n + 1$$
$$= n + 1 - \frac{a}{1-a} + \frac{a^{n+1}}{1-a} \quad n \geq 0 \tag{115}$$

Figure 6DT.13 shows a plot of $y[n]$ for $a = 0.9$.

Figure 6DT.13
Ramp response of
a first-order system.

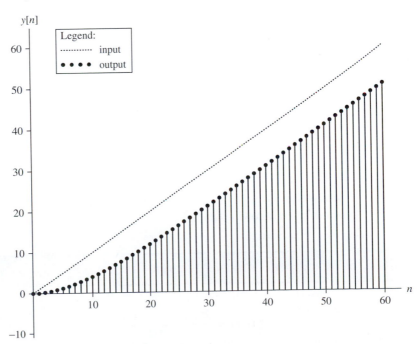

G(z) with Multiple-Order Poles

The partial fraction expansion of $G(z)$ includes terms of the form

$$\frac{Kz}{z-p} \xleftrightarrow{\text{ZT}} Kp^n \tag{116}$$

for each simple pole and m terms of the form

$$\frac{L_1 z}{z-p} + \frac{L_2 z^2}{(z-p)^2} \cdots + \frac{L_m z^m}{(z-p)^m} \tag{117a}$$

for each mth-order pole, where $m = 2, 3, \ldots$. An equivalent form for Eq. (117a) is

$$\frac{L_1}{1 - pz^{-1}} + \frac{L_2}{(1 - pz^{-1})^2} + \cdots + \frac{L_m}{(1 - pz^{-1})^m} \tag{117b}$$

The coefficients K and L_1 associated with the first-degree denominators are residues. The coefficient L_i is given by

$$L_i = \frac{1}{(m-i)!(-p)^{m-i}} \frac{d^{m-i}}{d(z^{-1})^{m-i}} [(1 - pz^{-1})^m G(z)] \Big|_{z=p} \tag{118}$$

for $i = 1, 2, \ldots, m$. The sequence corresponding to the ith term in Eq. (117) is

$$\frac{L_i z^i}{(z-p)^i} \xleftrightarrow{\text{ZT}} \frac{L_i}{(i-1)!} (n+1)(n+2) \cdots (n+i-1) p^n \quad n \geq 0 \tag{119}$$

for $i = 1, 2, \ldots, m$.

G(z) as an Improper Rational Algebraic Fraction (q ≥ r)

If $G(z)$ is improper, we can use long division to put it in the form

$$G(z) = e_0 + e_1 z^{-1} + \cdots + e_{q-r} z^{-(q-r)} + G_1(z) \tag{120}$$

where $G_1(z)$ is a proper rational algebraic fraction. The method of long division is illustrated in the following example.

EXAMPLE 6DT.15 Improper Rational Algebraic Fraction

Consider

$$G(z) = \frac{2 + 0.5z^{-1} - 0.5z^{-2}}{1 - 0.5z^{-1}} \tag{121}$$

We divide the numerator by the denominator using the ordinary rules of long division:

$$
\begin{array}{r}
z^{-1} \qquad 1 \\
-0.5z^{-1} \quad 1 \overline{)\,-0.5z^{-2} \quad 0.5z^{-1} \quad 2} \\
-0.5z^{-2} \quad z^{-1} \\
\hline
-0.5z^{-1} \quad 2 \\
-0.5z^{-1} \quad 1 \\
\hline
1
\end{array}
$$

The result is

$$G(z) = 1 + z^{-1} + \underbrace{\frac{1}{1 - 0.5z^{-1}}}_{G_1(z)} \tag{122}$$

Therefore

$$g[n] = \delta[n] + \delta[n-1] + (0.5)^n \quad n \geq 0 \tag{123}$$

6DT.6.4 Stability

We showed in Chapter 2DT that an LTI system is BIBO stable if and only if its impulse response $h[n]$ is absolutely summable. For a causal system, this necessary and sufficient condition is written as

$$\sum_{n=0}^{\infty} |h[n]| < \infty \tag{124}$$

A causal LTI system is characterized by $H(z) = \mathcal{Z}\{h[n]\}$. We know from our work in the preceding section that if $H(z)$ has the form (78), then $h[n]$ is a sum of impulses and weighted exponentials. If every pole of $H(z)$ lies inside the unit circle, then the weighted exponentials decay to 0 as n increases and Eq. (124) is satisfied. If one or more poles lies on or outside the unit circle, then the weighted exponentials do not decay to zero as n increases and Eq. (124) is not satisfied. This shows that, for every causal LTI system having a rational algebraic transfer function, a necessary and sufficient condition for stability is that all the poles of $H(z)$ lie inside the unit circle.

A simple example is given by the familiar first-order system $h[n] = a^n u[n]$. We know from Chapter 2DT that this system is BIBO stable if and only if $|a| < 1$. This is the same condition that the pole of $H(z) = z/(z - a)$ is inside the unit circle.

6DT.7 Response to Sinusoid Applied at $n = 0$

In this section we derive the response of a real LTI system to a sinusoid that is applied at $n = 0$. The result reveals that there is a fundamental relationship between a stable system's transient response and its steady-state sinusoidal response, as developed in Chapter 3DT. The relationship is best understood geometrically using the system's pole-zero plot. We assume throughout this section that all the poles of the system are simple and that the system is initially at rest.

Assume that the input to an LTI system at rest is a sinusoid.

$$x[n] = A\cos(\omega_x n + \phi)u[n] \tag{125}$$

The z-domain response is

$$Y(z) = H(z)X(z) \tag{126}$$

where

$$X(z) = \frac{Xz}{z - e^{j\omega_x}} + \frac{X^*z}{z - e^{-j\omega_x}} = \frac{z(z - e^{-j\omega_x})X + z(z - e^{j\omega_x})X^*}{(z - e^{j\omega_x})(z - e^{-j\omega_x})} \tag{127}$$

with $X = \frac{1}{2}Ae^{j\phi}$. We can use Eqs. (78) and (127) to express $Y(z)$ as

$$Y(z) = \underbrace{z^{l-m}\frac{b_0 \prod_{i=1}^{m}(z - z_{hi})}{a_0 \prod_{i=1}^{l}(z - p_{hi})}}_{H(z)} \times \underbrace{\left[\frac{z(z - e^{-j\omega_x})X + z(z - e^{j\omega_x})X^*}{(z - e^{j\omega_x})(z - e^{-j\omega_x})}\right]}_{X(z)} \tag{128}$$

where $p_{h1}, p_{h2}, \ldots, p_{hl}$ are the poles of $H(z)$ and $e^{\pm j\omega_x}$ are the poles of $X(z)$. (We assume that $l \geq m$.) For simple poles, the partial fraction expansion of $Y(z)$ is:

$$Y(z) = H(z)X(z) = \frac{K_{h1}z}{z - p_{h1}} + \frac{K_{h2}z}{z - p_{h2}} + \cdots + \frac{K_{hn}z}{z - p_{hl}} + \frac{K_{x1}z}{z - e^{j\omega_x}} + \frac{K_{x1}^*z}{z - e^{-j\omega_x}} \tag{129}$$

where the residues can be found using Heaviside's method.

$$K_{x1} = \mathcal{R}es\{H(z)X(z); e^{j\omega_x}\} = \left(\frac{z - e^{j\omega_x}}{z}\right)H(z)X(z)\bigg|_{z=e^{j\omega_x}} = H(e^{j\omega_x})X \tag{130}$$

$$K_{x2} = \mathcal{R}es\{H(z)X(z); e^{-j\omega_x}\} = \left(\frac{z - e^{-j\omega_x}}{z}\right)H(z)X(z)\bigg|_{z=e^{-j\omega_x}} = H^*(e^{j\omega_x})X^* \tag{131}$$

and

$$K_{hi} = \mathcal{R}es\{H(z)X(z); p_{hi}\} = \left(\frac{z - p_{hi}}{z}\right)H(z)X(z)\bigg|_{z=p_{hi}} = X(p_{hi})\mathcal{R}es\{H(z); p_{hi}\} \tag{132}$$

for $i = 1, 2, \ldots$. Because the system is real, $H(e^{-j\omega_x}) = H^*(e^{j\omega_x})$. The sequence domain response follows by table lookup:

$$y[n] = A|H(e^{j\omega_x})|\cos(\omega_x n + \phi + \angle H(e^{j\omega_x})) + \sum_{i=1}^{l} K_{hi}p_{hi}^n \tag{133}$$

All complex system poles occur in conjugate pairs. The conjugate pairs of terms in the sum Eq. (133), if any, combine to real damped sinusoids, as in Eq. (87).

For a stable system, $|p_{hi}| < 1$, for $i = 1, 2, \ldots, l$, and all the exponential terms in the sum are transients. Eventually, as n increases, they are negligible compared to the sinusoidal term $A|H(e^{j\omega_x})|\cos(\omega_x n + \phi + \angle H(e^{j\omega_x}))$. The steady-state response of a stable LTI system driven by a suddenly applied sinusoid is sinusoidal with the same frequency as the input.

The following example shows how Eq. (133) provides insight into the relationship between a system's transient response and its steady-state sinusoidal response.

EXAMPLE 6DT.16   Response of a Resonant Second-Order System to a Cosine Applied at $n = 0$

---

1. Find the response of the resonant second-order system of Example 6DT.13 (Figure 6DT.10) to

$$x[n] = \cos(\omega_x n) u[n] \tag{134}$$

This input has the form (125) where $A = 1$ and $\phi = 0$. Therefore, $X = 0.5$ in (127).

2. Describe the system's frequency response characteristic in terms of the system's pole-zero plot.

Solution

1. The response is given by Eq. (133) with $l = 2$. The system's poles p_{h1} and p_{h2} are given by Eq. (93). The factor $H(e^{j\omega_x})$ in Eq. (133) is obtained by setting $z = e^{j\omega_x}$ in Eq. (95).

$$H(e^{j\omega_x}) = \frac{e^{j2\omega_x}}{(e^{j\omega_x} - p_{h1})(e^{j\omega_x} - p_{h1}^*)} \tag{135}$$

The factor $X(p_{h1})$ in K_{h1} of (132) is obtained by setting $z = p_{h1}$.

$$X(p_{h1}) = \frac{p_{h1}(p_{h1} - \cos\omega_x)}{(p_{h1} - e^{j\omega_x})(p_{h1} - e^{-j\omega_x})} \tag{136}$$

The residue K_{h1} is given by (132) where $X(p_{h1})$ is given by (136) and

$$Res\{H(z); p_{h1}\} = \left(\frac{z - p_{h1}}{z}\right) H(z)\Big|_{s=p_{h1}} = \frac{p_{h1}}{p_{h1} - p_{h1}^*} \tag{137}$$

The $l = 2$ term in the sum Eq. (129) is the complex conjugate of the $l = 1$ term because the system is real and it has a real input. Therefore, the $l = 1$ and $l = 2$ terms combine to yield the result

$$y[n] = |H(e^{j\omega_x})| \cos(\omega_x n + \angle H(e^{j\omega_x})) + 2|K_{h1}|\rho_h^n \cos(\theta_h n + \angle K_{h1}) \tag{138}$$

We can see from the preceding equation that if $\rho_h < 1$, $y[n]$ approaches the sinusoidal steady state

$$y[n] = |H(e^{j\omega_x})| \cos(\omega_x n + \angle H(e^{j\omega_x})) \tag{139}$$

as n increases. This sinusoidal steady-state term is identical to that described in Chapter 3DT. $y[n]$ is plotted in Figure 6DT.14 for $\rho_h = 0.9$, $\theta_h = 15°$, and $\omega_x = 20°$.

2. Figure 6DT.15 illustrates the numerator and the two factors in the denominator of Eq. (135) as vectors in the z-plane. The numerator vector $e^{j2\omega_x}$ is a unit vector that starts at the origin and has angle $2\omega_x$. The denominator vectors start at the the pole locations p_{h1} and $p_{h2} = p_{h1}^*$ and end on the unit circle at the point $e^{j\omega_x}$. The amplitude characteristic $|H(e^{j2\pi f})|$ equals the inverse of the product of the lengths of these vectors because

$$|H(e^{j2\pi f})| = \frac{1}{|e^{j2\pi f} - p_{h1}||e^{j2\pi f} - p_{h1}^*|} \tag{140}$$

Similarly, the phase characteristic

$$\angle H(e^{j2\pi f}) = 2 \times 2\pi f - \angle(e^{j2\pi f} - p_{h1}) - \angle(e^{j2\pi f} - p_{h1}^*) \tag{141}$$

equals the difference between the vectors associated with the numerator and denominator. This figure enables us to visualize the effect of the system poles and zeros on the amplitude and phase characteristics. These characteristics

Figure 6DT.14
Response of a
second-order system
to a cosine applied at
$n = 0$.

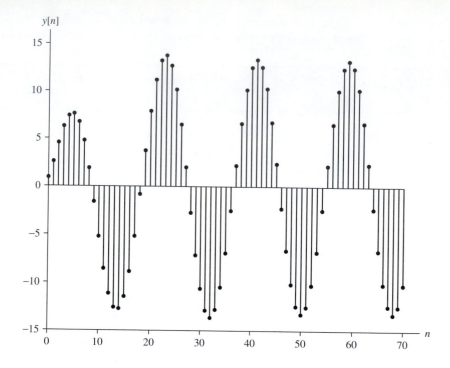

Figure 6DT.15
Interpretation of
frequency response
characteristic.

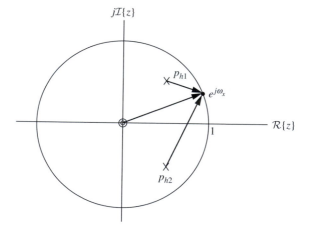

are plotted in Figure 6DT.16 for $\rho_h = 0.9$, $\theta_h = 15°$. We see that the amplitude characteristic is large when the input frequency is close to a natural frequency of the system, and it is small when the input frequency is far from a natural frequency of the system. The peaks in the amplitude characteristic shown in Figure 6DT.16(b) are called *resonance peaks*. Resonance occurs when a system is driven at a frequency close to the system's natural frequency.

Figure 6DT.17 shows a good way to visualize the amplitude characteristic. Here $|H(z)|$ is plotted above the *z*-plane for $|z| \le 1$. The amplitude characteristic $|H(e^{j2\pi f})|$ is the intersection of $|H(z)|$ with the $|z| = 1$ cylinder. The surface $|H(z)|$ is unbounded at the pole values $0.9 < \pm 15°$. It is analogous to a rubber sheet supported by

Figure 6DT.16
Frequency response
characteristic of
second-order system:
(a) amplitude;
(b) phase.

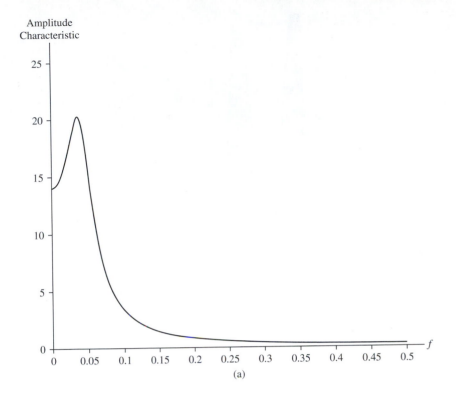

Amplitude
Characteristic

(a)

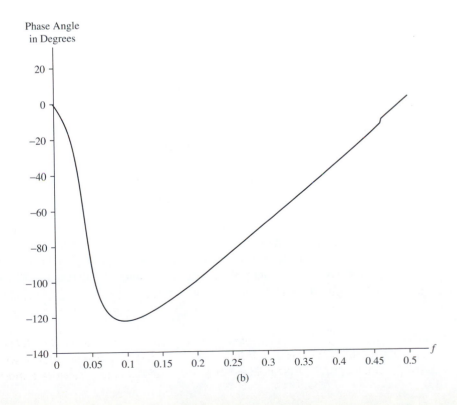

Phase Angle
in Degrees

(b)

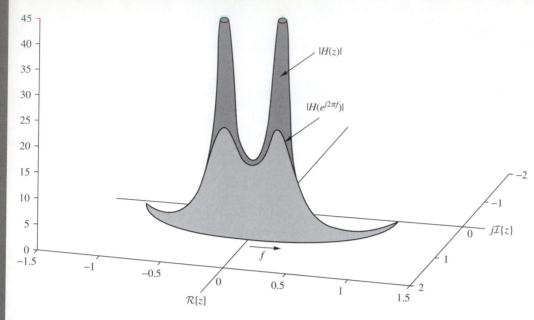

Figure 6DT.17
Function $|H(z)|$ plotted above the z-plane.

(physical) poles at the pole values. For an arbitrary $H(z)$, the sheet is held down by tacks at the zero values. This rubber sheet analogy provides a quick way to visualize the shape of any amplitude characteristic. It has the merit of showing the importance of the poles and zeros of a transfer function in determining the shape of a system's amplitude characteristic.

6DT.8   Initial and Final Value Theorems

The initial and final value theorems provide a way to find the initial and final values $x[0]$ and $x[\infty]$ of a sequence directly from $X(z)$. These theorems are shown in Table 6DT.3.

The initial value theorem can be applied to any real or complex valued sequence. The final value theorem can be applied only when $x[n]$ has a final value. It does not work if $x[n]$ becomes unbounded or oscillatory as $n \to \infty$. For rational algebraic $X(z)$, the final value theorem works if and only if $X(z)$ all the poles of $X(z)$ are inside the unit circle, except for possibly a pole with order one at $z = 1$.

Table 6DT.3
Initial and Final Value
Theorems

Initial value	$x[0] = \lim\limits_{z \to \infty} X(z)$
Final value	$\lim\limits_{n \to \infty} x[n] = \lim\limits_{z \to 1}(z-1)X(z)$

6DT.8.1
Initial Value
Theorem

The initial value theorem states that

$$x[0] = \lim_{z \to \infty} X(z) \qquad (142)$$

The proof follows immediately from the definition of the z-transform (1).

EXAMPLE 6DT.17   Initial Value of Step Response of Second-Order System

Find the initial value of the output when a unit step is applied to the second-order system of Example 6DT.14.

Solution

The z-domain response is

$$Y(z) = H(z)X(z) = \frac{z^2}{z^2 + bz + c} \frac{z}{z - 1} \qquad (143)$$

Therefore, the initial value theorem gives the result

$$y[0] = \lim_{z \to \infty} Y(z) = 1 \qquad (144)$$

EXAMPLE 6DT.18   Initial Value of $h[n]$

Find the initial value of the response when a unit impulse is applied to a system having the transfer function

$$H(z) = \frac{z(z - \cos \omega_x)}{z^2 - 2z \cos \omega_x + 1} \qquad (145)$$

Solution

The initial value theorem states that

$$h[0] = \lim_{z \to \infty} H(z) = 1 \qquad (146)$$

6DT.8.2
Final Value
Theorem

The final value theorem states that

$$\lim_{n \to \infty} x[n] = \lim_{z \to 1} (z - 1) X(z) \qquad (147)$$

To prove the final value theorem, we use the linearity and delay properties to write

$$\mathcal{Z}\{x[n+1] - x[n]\} = zX(z) - zx[0] - X(z) = (z - 1)X(z) - zx[0] \qquad (148)$$

By definition

$$\mathcal{Z}\{x[n+1] - x[n]\}\Big|_{z=1} = \lim_{n \to \infty} \sum_{i=0}^{n} (x[i+1] - x[i])z^{-i}\Big|_{z=1} \qquad (149)$$

The right-hand side of the preceding equation is

$$\lim_{n \to \infty} (x[1] - x[0]) + (x[2] - x[1]) + (x[3] - x[2]) + \cdots + (x[n+1] - x[n])$$

$$= \lim_{n \to \infty} x[n+1] - x[0] \qquad (150)$$

and therefore

$$\mathcal{Z}\{x[n+1] - x[n]\}\big|_{z=1} = \lim_{n \to \infty} x[n+1] - x[0] \tag{151}$$

If we combine (151) and (148) we obtain

$$(z-1)X(z)\big|_{z=1} - x[0] = \lim_{n \to \infty} x[n+1] - x[0] \tag{152}$$

which yields the final value theorem.

EXAMPLE 6DT.19   Final Value of Step Response of Second-Order System

Find the final value of the response when a unit step is applied to the second-order system of Example 6DT.13.

Solution

The z-domain response is

$$Y(z) = H(z)X(z) = \frac{z^2}{z^2 + bz + c} \frac{z}{z - 1} \tag{153}$$

The final value theorem gives the result

$$\lim_{n \to \infty} y[n] = \lim_{z \to 1}(z-1)Y(z) = \frac{1}{1 + b + c} \tag{154}$$

EXAMPLE 6DT.20   Final Value of $h[n]$

Find the final value of the response when a unit impulse is applied to a system having the transfer function

$$H(z) = \frac{z(z - \cos \omega_x)}{z^2 - 2z \cos \omega_x + 1} \tag{155}$$

Solution

The system being considered has poles on the unit circle and is therefore unstable. We know from Table 6DT.2 that

$$h[n] = \cos(\omega_x n) \tag{156}$$

which is a function having no final value. Application of the final value theorem gives the incorrect result

$$\lim_{n \to \infty} h[n] = \lim_{z \to 1}(z-1)H(z) = 0 \tag{157}$$

6DT.9   Summary

The z-transform is the premier analytical aid in DT signals and systems theory for determining the response of an LTI system to a suddenly applied input. It applies to both stable and unstable systems and to infinite energy signals such as ramps and raising exponentials.

The discrete-time Fourier and z-transforms of a causal signal are equal on the unit circle of the z-plane. Therefore, the DTFT exists if and only if the region of convergence of the z transform includes the unit circle.

Poles and zeros play a central role in z transform systems analysis. We saw that the terms in a partial fraction expansion were based on poles. We showed how we can visualize the influence of the poles and zeros on the sinusoidal steady-state response of a system by plotting $|H(z)|$ as a surface above the z-plane. The intersection of this surface with the unit cylinder is the amplitude response characteristic $|H(e^{j2\pi f})|$. The influence of the poles and zeros on both the transient and sinusoidal steady-state response was illustrated by first- and second-order system examples.

Summary of Definitions and Formulas

Definition of the *z*-Transform

$$X(z) = \mathcal{Z}\{x[n]\} = \sum_{n=0}^{\infty} x[n]z^{-n}$$

where $z = \rho e^{j\theta}$.

Sequences that are different for $n < 0$ but equal for $n \geq 0$ have the same z-transform.

Existence of X(z)

Sequences satisfying

$$|x[n]| < M\rho_{\mathrm{a}}^n$$

where M and ρ_{a} are finite, are said to have *exponential order*. If $x[n]$ has exponential order, then $X(z)$ exists for all $|z| > \rho_{\mathrm{a}}$. The region $|z| > \rho_{\mathrm{a}}$ in the z-plane is called the *region of convergence*. The region of convergence of a rational algebraic fraction $X(z)$ is the region outside the circle defined by the pole of $X(z)$ having the largest absolute value.

Relation of *z*-Transform to the DTFT

$$X(z)|_{z=\rho e^{j2\pi f}} = \mathcal{DTFT}\{x[n]\rho^n\} \qquad \text{for causal } x[n]$$

If the region of convergence of $X(z)$ includes the unit circle, then

$$X(z)|_{z=e^{j2\pi f}} = \mathcal{DTFT}\{x[n]\} \qquad \text{for causal } x[n].$$

Unit Delay Property

$$\mathcal{Z}\{x[n-1]\} = z^{-1}X(z) + x[0]$$

This property is a key property. It makes it possible to transform difference equations having initial conditions into algebraic equations in z.

Extensions:

$$\mathcal{Z}\{x[n-2]\} = z^{-2}X(z) + z^{-1}x[0] + x[-1]$$

$$\mathcal{Z}\{x[n-k]\} = z^{-k}X(z) + \sum_{m=0}^{k} x[-m]z^{-(k-m)}$$

See Tables 6DT.1 and 6DT.2 for z-transform properties and pairs.

To Solve Difference Equations Having Initial Conditions

1. Obtain the z-domain I-O equation of the system. With

$$D^k w[n] = w[n - k]$$

$$\mathcal{A}(D) = a_l D^l y[n] + a_{l-1} D^{l-1} y[n] + \cdots + a_1 D y[n] + a_0 y[n]$$

$$\mathcal{B}(D) = b_m D^m x[n] + b_{m-1} D^{m-1} x[n] + \cdots + b_1 D x[n] + b_0 x[n]$$

we can write a difference equation as

$$\mathcal{A}(D) y[n] = \mathcal{B}(D) x[n]$$

When we take the z-transform of this equation, we obtain the z-domain equation:

$$\mathcal{A}(z^{-1}) Y(z) + Y_{ic}(z) = \mathcal{B}(z^{-1}) X(z^{-1}) + X_{ic}(z)$$

where $Y_{ic}(z)$ and $X_{ic}(z)$ are due to initial conditions.

2. Solve the algebraic equation for the z-domain output $Y(z)$. The solution is

$$Y(z) = \frac{\mathcal{B}(z^{-1})}{\mathcal{A}(z^{-1})} X(z) + \frac{X_{ic}(z) - Y_{ic}(z)}{\mathcal{A}(z^{-1})}$$

In the preceding equation

$$H(z) = \frac{\mathcal{B}(z^{-1})}{\mathcal{A}(z^{-1})} = \mathcal{Z}\{h[n]\}$$

is the *transfer function* of the system. Often, we can write $Y(z)$ as

$$Y(z) = K \frac{z^{-(m+v)} \prod_{i=1}^{m}(z - z_{hi}) \prod_{i=1}^{v}(z - z_{xi})}{z^{-(l+u)} \prod_{i=1}^{l}(z - p_{hi}) \prod_{i=1}^{u}(z - p_{xi})} + \frac{X_{ic}(z) - Y_{ic}(z)}{a_0 z^{-l} \prod_{i=1}^{l}(z - p_{hi})}$$

where the p_{hi} are the roots to the characteristic equation, $\mathcal{A}(z) = 0$, and the p_{xi} are the poles of $X(z)$.

3. Use partial fraction expansion to put $Y(z)$ in a form ready for table lookup. For simple poles, the partial fraction expansion of $Y(z)$ contains terms of the form

$$G(z) = \frac{K_1 z}{z - p_1} + \frac{K_2 z}{z - p_2} + \cdots + \frac{K_i z}{z - p_i} + \cdots + \frac{K_r z}{z - p_r}$$

where the residues, $K_1, K_2, \ldots$, and K_r can be evaluated using Heaviside's formula.

$$K_i = \mathcal{R}es\{G(z); p_i\} = \left(\frac{z - p_i}{z} \right) G(z) \Big|_{z = p_i} \quad i = 1, 2, \ldots r$$

4. Look up the terms of $Y(z)$ in a table to find $y[n]$. For simple poles, as jsut considered, $y(t)$ contains terms of the form

$$g[n] = K_1 p_1^n + K_2 p_2^n + \cdots + K_r p_r^n \quad n \geq 0$$

Complex Poles

Complex poles always occur in conjugate pairs for a real system having a real input. The preceding partial fraction for $G(z)$ and its associated waveform $g[n]$ then contain terms having the form

$$G(z) = \frac{K z}{z - p} + \frac{K^* z}{z - p^*}$$

and

$$g[n] = Kp^n + K^*p^{*n} = 2|K|\rho^n \cos(\theta n + \angle K)$$

where $p = \rho < \theta$.

Stability

A necessary and sufficient condition for a causal LTI system to be BIBO stable is that all the poles of $H(z)$ lie inside the unit circle.

Response of LTI System to Sinusoid Applied at $n = 0$

$$A \cos(\omega_x n + \phi)u[n] \xrightarrow{h[n]} A|H(e^{j\omega_x})| \cos(\omega_x n + \phi + \angle H(e^{j\omega_x}))$$

$$+ \sum_{i=1}^{l} X(p_{hi})\mathcal{R}es\{H(z); p_{hi}\}p_{hi}^n$$

For a stable LTI system, the terms in the summation $\sum_{i=1}^{l}$ on the right-hand side are transients, which die out as n increases. When these transients are all negligible, $y[n]$ has reached the *sinusoidal steady state*

$$A|H(e^{j\omega_x})| \cos(\omega_x n + \phi + \angle H(e^{j\omega_x}))$$

The sinusoidal steady-state response was studied extensively in Chapter 3DT.

Initial and Final Value Theorems

$$x[0] = \lim_{z \to \infty} X(z)$$

$$\lim_{n \to \infty} x[n] = \lim_{z \to 1}(z - 1)X(z).$$

6DT.10   Problems

6DT.2   Definition of the z-Transform

6DT.2.1 Find the z-transforms, the pole-zero plots, and the regions of convergence for the following sequences:

a) 3
b) $7\delta[n]$
c) $(0.9)^n$
d) $(0.9)^{-n}$
e) e^{j2n}

6DT.2.2 Find the z-transforms of the following sequences:

a) $\delta[n - 6]$
b) $\delta[n + 3] + \delta[n - 6]$
c) $u[n - 6]$
d) $u[n + 3]$

6DT.2.3 The *bilateral z-transform* is defined as

$$X_B(z) = \sum_{n=-\infty}^{\infty} x[n]z^{-n} = \sum_{n=-\infty}^{-1} x[n]z^{-n} + \sum_{n=0}^{\infty} x[n]z^{-n} \qquad (158)$$

where $z = \rho e^{j\theta}$. Find the bilateral z-transform and its region of convergence for the following:

a) $x[n] = (0.9)^n u[n]$

b) $x[n] = (0.9)^{-n} u[-n]$

c) $x[n] = (0.9)^n u[n] + (0.9)^{-n} u[-n]$

6DT.2.4 Repeat Problem 6DT.2.3 for the following:

a) $x[n] = (0.9)^{-n} u[n]$

b) $x[n] = (0.9)^n u[-n]$

c) $x[n] = (0.9)^{-n} u[n] + (0.9)^n u[-n]$

6DT.3   Relationship Between the *z*-Transform and DTFT of Causal Sequences

6DT.3.1 a) Find the z-transform and region of convergence for $\sqcap[n-2; 5]$.

b) Use your answer from part (a) to write down the DTFT of $\sqcap[n-2; 5]$. Express your answer in terms of the cinc function.

6DT.3.2 Explain why the z-transform of $u[n]$ exists, but the DTFT does not. Then explain why we can define the DTFT of $u[n]$ as $H(e^{j2\pi f}) = \frac{1}{2}\delta(f) + 1/(1 - e^{-j2\frac{1}{2}f})$ for $-\frac{1}{2} < f < \frac{1}{2}$. Does $\frac{1}{2}\delta(f) + 1/(1 - e^{-j2\frac{1}{2}f})$ "not exist"?

6DT.3.3 a) Find the z-transform and region of convergence for $x[n] = 10^n$.

b) Use your answer to part (a) to state whether or not $10^n u[n]$ has a DTFT.

6DT.4   Properties of the *z*-Transform

6DT.4.1 State and prove the convolution property of the z-transform.

6DT.4.2 State and prove the accumulation property of the z-transform.

6DT.4.3 Does the convolution property of the z-transform work if one or both of the sequences are noncausal? Justify your answer.

6DT.4.4 Derive the z-transform of $x[n] = 2|K|n\rho^n \cos(\theta n + \angle K)$ and confirm that your answer agrees with that in Table 6DT.2.

6DT.4.5 Sketch the pole-zero plot for the sinusoid of Example 6DT.7.

6DT.4.6 Sketch the pole-zero plot for the damped sinusoid of Example 6DT.8.

6DT.4.7 An LTI system has input $x[n] = \delta[n] + \delta[n-1]$ and output $y[n] = \delta[n] + \delta[n-1] + \delta[n-2]$. Find:

a) $H(e^{j2\pi f})$.

b) $h[n]$.

6DT.5   Elementary Systems in the *z*-Domain

6DT.5.1 A system has impulse response $h[n] = \delta[n-3]$.

a) Draw the block diagram of the system.

b) Write down the I-O equation of the system.

c) Write down the z-transform of your answer to part (b).

d) Draw the z-domain system.

e) What information is needed to describe the "initial condition" at $n = 0$?

6DT.5.2 Derive the left-shift operator properties in Table 6DT.1.

6DT.6   z-Transform System Analysis

6DT.6.1 Invent two ways to find K_1 and K_2 in Example 6DT.12 other than that invented by Heaviside. Apply your methods to confirm the answers given in the example.

6DT.6.2 Show that it is possible to chose the initial condition in Example 6DT.12 so that the system immediately reaches the steady-state output when a step is applied.

6DT.6.3 Find $\mathcal{A}(z)$ and $\mathcal{B}(z)$ and $H(z)$ for Example 6DT.12.

6DT.6.4 **a)** Find $Y(z)$ for the second-order system of Example 6DT.13 if $x[n] = u[n]$. Put your answer in partial fraction form. Assume that the system poles are complex and that the initial conditions are all 0.
b) Use your answer to part (a) to find $y[n]$.

6DT.6.5 Confirm Eq. (101).

6DT.6.6 Find L_1 in Eq. (110) by taking the limit $z \to \infty$.

6DT.6.7 Use the z-transform to find the step response of a system described by $\Delta y[n] + y[n] = \Delta x[n]$.

6DT.6.8 Use the z-transform to find the ramp response of a system described by $\Delta y[n] + y[n] = b\Delta x[n]$.

6DT.6.9 Consider the second-order system of Example 6DT.13. The system response is *overdamped* if $c < (b/2)^2$.
a) Find $H(z)$ for $c < (b/2)^2$.
b) Sketch the pole-zero plot.
c) Find $h[n]$.

6DT.6.10 Consider the second-order system of Example 6DT.13. The system response is *critically damped* if $c = (b/2)^2$.
a) Find $H(z)$ for $c = (b/2)^2$.
b) Sketch the pole-zero plot.
c) Find $h[n]$.

6DT.6.11 Find the partial fraction expansion of

$$G(z) = \frac{z^2}{(z+2)(z-0.1)} \tag{159}$$

6DT.6.12 Find the partial fraction expansion of

$$G(z) = \frac{z^2}{z^2 + 3z + 2} \tag{160}$$

6DT.6.13 We can use the z-transform in combination with other techniques, including those in Chapter 2DT, to find an output. For example, to find the output $y[n]$ of an LTI system to $x[n] = u[n] - u[n-N]$, we can use the Laplace transform to find the system's step response $h_{-1}[n]$ and then write down $y[n] = h_{-1}[n] - h_{-1}[n - N]$. Illustrate this method using the second-order system of Examples 6DT.13 and 6DT.16. Hint: You can obtain the step response of this system as a special case of the response to a sinusoid (Example 6DT.16).

6DT.6.14 We saw in Example 6DT.16 how to interpret the frequency response characteristic $H(e^{j2\pi f})$, using vectors in the z-plane. Find and discuss a similar interpretation of the factors $X(p_{h1})$ and $Res\{H(z); p_{h1}\}$ appearing in (132).

6DT.6.15 The result (133) assumed that $l > m$ in Eq. (78). Consider the other case for which $l \le m$ by finding the response of an LTI system to a sinusoid applied at $n = 0$, where $l = 0$ and $m \ge 1$.

6DT.6.16 The MATLAB function `residuez` can be used to calculate the partial fraction expansion of $G(z)$. Read the MATLAB help for `residuez` and use this function to verify your answers to Problems 6DT.6.11 and 6DT.12.

6DT.6.17 The MATLAB function `tf2zpk` can be used convert $H(z)$ from polynomial form to pole-zero form. Read the MATLAB help for `tf2zpk` and use this function to find the pole-zero form of $H(s) = (z^3 + 1)/(z^3 + 2z + 2)$.

6DT.6.18 Use the z-transform to find the step response of the first-order system of Example 6DT.14 to the following, and check your answer using some other method.
a) $x[n] = u[n - 3]$
b) $x[n] = u[n] - u[n - 3]$
c) $x[n] = u[n] - u[n - 3] + u[n - 6] - u[n - 9] + \cdots$

6DT.6.19 The input $x[n] = u[n - N]$ is applied to a system having $H(z) = z^{-N}[z/(z - 1)]$. Use z-transforms to find $h[n]$ and $y[n]$. Check your answer using some other method.

6DT.6.20 The input $x[n] = u[n]$ is applied to a system having $H(z) = (1 - z^{-N})z/(z - 1)$. Find $h[n]$ and $y[n]$. Check your answer using some other method.

6DT.6.21 The input $x[n] = u[n] - u[n - N]$ is applied to a system having $H(z) = (1 - z^{-N})z/(z - 1)$. Find $y[n]$. Check your answer using some other method.

6DT.6.22 Find the I-O equation if $H(z) = z/(z - 1)(1 - z^{-N})$.

6DT.7   Response to Sinusoid Applied At $n = 0$

6DT.7.1 Example 6DT.16 assumed zero initial conditions. Find the output $y[n]$ for $x[n] = A\cos(\omega_x n)u[n]$ if the initial conditions are nonzero.

6DT.7.2 **a)** Find the output of the second-order system of Example 6DT.16 for an input $x[n] = A\cos(\omega_x n + \phi)u[n]$ where ϕ is an arbitrary phase angle. Assume that the system is initially at rest.
b) Is there a ϕ for which the sinusoidal steady state is attained immediately, such that. $y[n]$ is purely sinusoidal for all $n \geq 0$, if $x[n] = A\cos(\omega_x n + \phi)u[n]$? If so, find the formula for that ϕ.

6DT.7.3 Consider Figure 6DT.15 of Example 6DT.16. Assume that the system is highly resonant. For a highly resonant system the poles will be very close to the unit circle: $|p_{h1}| = \rho_h = \sqrt{c} \approx 1$.
a) Use simple trigonometric approximations to show that the resonance peak in $|H(e^{j\omega})|$ occurs at $\omega \approx \angle p_{h1}$ for a highly resonant system.
b) The half-power frequencies of the system in radians per sample are defined as the values of ω for which $|H(e^{j\omega})| = (1/\sqrt{2})|H(e^{j\omega_{peak}})|$ where ω_{peak} is the frequency at which the amplitude characteristic is a maximum. The half-power bandwidth of a resonance peak is defined as $\Delta\omega = \omega_+ - \omega_-$, where ω_+ and ω_- are the half-power frequencies on the high- and low-frequency sides of the peak, respectively. Use simple trigonometric approximations to show that $\Delta\omega \approx 2(1 - \sqrt{c})$ for a highly resonant system.

6DT.7.4 The Matlab function `freqz` can be used to plot the frequency response characteristic $H(e^{j2\pi f})$. Read the MATLAB help for `freqz` and use this function to plot $H(e^{j2\pi f})$ for the second-order system of Example 6DT.17.

6DT.8   Initial and Final Value Theorems

6DT.8.1 A unit step is applied to a system having the transfer function $H(z)$.

  a) Show that, if the initial and final values of the output $y[n]$ exist, then they are given by

$$y[0] = \lim_{z \to \infty} \frac{z}{z-1} H(z) \qquad (161)$$

  and

$$\lim_{n \to \infty} y[n] = \lim_{z \to 1} z H(z) \qquad (162)$$

  respectively.

  b) Give an example to illustrate each of the preceding results.

6DT.8.2 A unit ramp is applied to a system having the transfer function $H(z)$.

  a) Assume that the initial and final values of the output $y[n]$ exist and find the formulas for the initial and final values.

  b) Give an example to illustrate your answers to part (a).

6DT.8.3 The input $\cos(2\pi f n) u[n]$ is applied to a system having transfer function $H(z)$.

  a) Assume that the initial and final values of the output $y[n]$ exist and find the formulas for the initial and final values.

  b) Give an example to illustrate your answers to part (a).

6DT.8.4 A ramp $r[n]$ is applied to a first-order system having the transfer function $H(z) = (1-a)z/(z-a)$. The difference between input and output is $d[n] = r[n] - y[n]$. Assume that the system is initially at rest.

  a) Use the final value theorem to find the final value of $d[n]$.

  b) Use the initial value theorem to find the initial value of $d[n]$.

  c) Use the result from Example 6DT.14 to check your answers to parts (a) and (b).

  d) Sketch $d[n]$.

6DT.8.5 These are n first-order systems placed in series. Each system has the transfer function $H(z) = (1-a)z/(z-a)$ and is initially at rest. A ramp $r[n]$ is applied to the input. The difference between output and input is $d[n] = y[n] - r[n]$.

  a) Assume that $n = 3$ and find the initial and final values of $d[n]$.

  b) Find the initial and final values of $d[n]$ for arbitrary n.

  c) Use your answer to part (b) find the (approximate) equation for $y[n]$ for large n. Use this equation and your engineering insight to make an approximate sketch of $y[n]$ for $n \geq 0$.

6DT.8.6 A certain high-pass filter has the transfer function

$$H(z) = \frac{\beta}{1+\beta} \frac{z-1}{z-p_1}$$

  where $\beta > 1$ and $p_1 = (\beta - 1)/(\beta + 1)$.

  a) Sketch the pole-zero diagram of the filter.

  b) Sketch the amplitude characteristic of the filter.

  c) Assume that the filter is initially at rest and find the initial and the final values of the output when the input is a unit step.

6DT.8.7 The second-order system of Example 6DT.13 and the high-pass filter of Problem 6DT.8.6 are placed in series. Both systems are initially at rest. A unit step is applied to input to the series connection. Find the initial and final values of the output.

6DT.8.8 Repeat Problem 6DT.8.7 for the paralled connection of the first-order system of Example 6DT.10 and high-pass filter of Problem 6DT.8.6.

7 DT Processing of CT Signals

7.1 Introduction

It is common practice to process a waveform using a hybrid system composed of an analog-to-digital converter, a digital computer, and a digital-to-analog converter. The basic model for the hybrid processor is shown in Figure 7.1(a). It includes a CT-to-DT convertor, a DT processor, and a DT-to-CT convertor. This model neglects quantization errors that occur when a continuous-amplitude variable is represented by a finite-length binary number. Quantization errors are negligible in a well designed hybrid processor. Because the input and output of the hybrid system are both waveforms, it is natural to compare the performance of the hybrid system to that of a purely analog system, as illustrated in Figure 7.1b.

The hybrid system has several advantages over a purely analog system. First, the I-O relation of the hybrid system can be changed simply by reprogramming the computer. Second, the hybrid system can achieve levels of performance that would be impossible for the analog system using realistic component values. Finally, the digital components of the hybrid system are less susceptible to the degrading influences of time and environment than the analog system.

We show in this chapter how to design the hybrid system model so that it has the same or approximately the same performance as a hypothetical analog system. We also introduce the standard architectures for implementing DT processors.

7.2 Conversion from CT to DT

Our first step in understanding the hybrid system of Figure 7.1(a) is to analyze the CT-to-DT converter. The input to the converter is the waveform $x(t)$ and the output is the sequence

$$x[n] = x(nT_s) \tag{1}$$

We are interested in the relationship between the spectrum $X(e^{j2\pi f})$ of the sequence of samples and the spectrum $X(F)$ of the CT waveform $x(t)$. We need to know this relationship to design the DT processor. Recall that, in general, $X(e^{j2\pi f})$ and $X(F)$ are fundamentally different.[1]

$$X(e^{j2\pi f}) = \sum_{n=-\infty}^{\infty} x[n]e^{-jn2\pi f} \tag{2}$$

---

[1] The use of the common symbol X will not confuse you if you remember that the the argument $e^{j2\pi f}$ always identifies $X(e^{j2\pi f})$ as a DTFT and the argument F always identifies $X(F)$ as an FT.

Figure 7.1
Hybrid and CT system
models: (a) hybrid
system; (b) CT system.
The frequency
domain description
for the CT system is
depicted in the
figure. The frequency
domain description
for the DT system is
developed in the text.

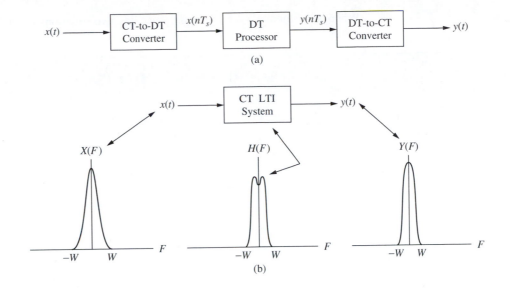

is a periodic function of f with period 1 and

$$X(F) = \int_{-\infty}^{\infty} x(t)e^{-j2\pi Ft}dt \tag{3}$$

is not a periodic function of F. To find the relationship between $X(e^{j2\pi f})$ and $X(F)$, we model the sampled waveform as we did in Eq. (4CT.65).

$$x_s(t) = x(t) \sum_{k=-\infty}^{\infty} \delta(t - kT_s) = \sum_{n=-\infty}^{\infty} x(nT_s)\delta(t - nT_s) \tag{4}$$

But now, instead of applying the multiplication property of the Fourier transform as we did in Eq. (4CT.67), we substitute Eq. (4) directly into the Fourier integral.

$$X_s(F) = \int_{-\infty}^{\infty} x_s(t)e^{-j2\pi Ft}dt = \int_{-\infty}^{\infty} \sum_{n=-\infty}^{\infty} x(nT_s)\,\delta(t - nT_s)\,e^{-j2\pi Ft}dt$$

$$= \sum_{n=-\infty}^{\infty} x(nT_s) \int_{-\infty}^{\infty} \delta(t - nT_s)\,e^{-j2\pi Ft}dt = \sum_{n=-\infty}^{\infty} x(nT_s)\,e^{-jn2\pi FT_s} \tag{5}$$

If we compare the right-hand side of Eq. (5), where $x(nT_s) = x[n]$, with Eq. (2), we see that

$$X_s(F) = X(e^{j2\pi FT_s}) \tag{6}$$

Therefore, the Fourier transform of $x_s(t)$ equals the DTFT of $x[n]$. We know from our work in Section 5.2 of Chapter 4CT that the Fourier transform of $x_s(t)$ is also given by

$$X_s(F) = F_s \sum_{n=-\infty}^{\infty} X(F - nF_s) \tag{7}$$

When we eliminate $X_s(F)$ from the preceding two equations we obtain the following basic result.

Basic Relationship Between DTFT and FT

$$X(e^{j2\pi FT_s}) = F_s \sum_{n=-\infty}^{\infty} X(F - nF_s) \qquad (8a)$$

which is

$$X(e^{j2\pi f}) = F_s \sum_{n=-\infty}^{\infty} X((f - n)F_s) \qquad (8b)$$

where

$$f = FT_s = \frac{F}{F_s} \qquad (9)$$

Equation (8b) states that $X(e^{j2\pi f})$ is the sum of translated versions of $X(F)$, where $F = fF_s$. The translations make the sum on the right-hand side of Eq. (8b) a periodic function of f with period 1. Equation (8) is illustrated in Figure 7.2. Figure 7.2(a) shows

Figure 7.2
Relationship between $X(e^{j2\pi f})$ and $X(F)$: (a) $X(F)$; (b) $X(e^{j2\pi f})$ for $F_s \leq 2W$; (c) $X(e^{j2\pi f})$ for $F_s > 2W$. When $F_s > 2W$ as, in (c), $X(e^{j2\pi f})$ is a scaled version of $X(F)$ for $-\frac{1}{2} \leq f < \frac{1}{2}$ $X(e^{j2\pi f}) = F_s X(fF_s)$, $\frac{1}{2} \leq f < \frac{1}{2}$.

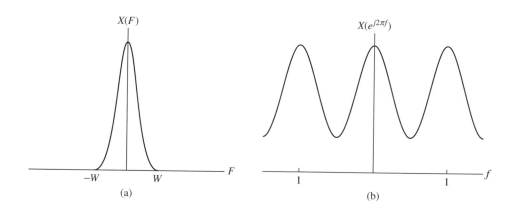

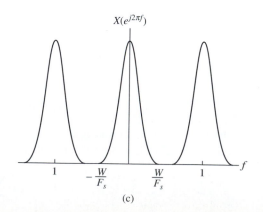

Figure 7.3
CT-to-DT converter:
(a) time domain;
(b) frequency
domain. If $F_s > 2W$,
then, for $\frac{1}{2} \le f < \frac{1}{2}$,
the spectrum of the
output $X(e^{j2\pi f})$ is a
scaled version of the
spectrum of the input
$X(e^{j2\pi f}) = F_s X(fF_s)$,
$\frac{1}{2} \le f < \frac{1}{2}$.

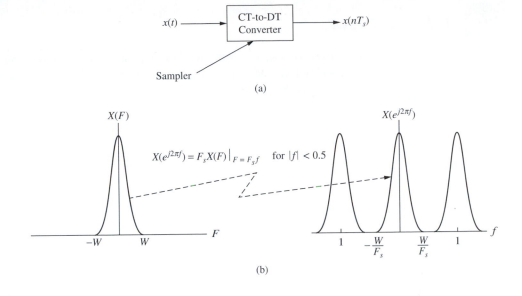

(a)

(b)

a hypothetical spectrum $X(F)$ of the input waveform $x(t)$. Figures 7.2(b) and 7.2(c) illustrate the spectrum $X(e^{j2\pi f})$ of the sequence $x[n] = x(nT_s)$. In Figure 7.2(b), it is assumed that $F_s \le 2W$, where W is the highest significant frequency in $x(t)$. Here the sampling frequency $F_s = 1/T_s$ is below the Nyquist rate $2W$, and aliasing occurs. In Figure 7.2(c), it is assumed that $F_s > 2W$. Here the sampling frequency is above the Nyquist rate, and aliasing does not occur.

These results are consistent with the analysis in Section 4CT.6, where we found that a signal can be reconstructed from its samples if $F_s > 2W$. Reconstruction is generally not possible if $F_s < 2W$ because information about the signal is lost when aliasing occurs. The conclusion is that we must sample $x(t)$ at a rate $F_s > 2W$ if we want the performance of the hybrid system to match that of a purely analog system.[2] If we look at the region $-\frac{1}{2} \le f < \frac{1}{2}$ in Figure 7.2(c) and refer to Eq. (8), we obtain the most important result of this section.

Relationship Between DTFT and FT for $F_s > 2W$

When there is no aliasing, the DTFT of $x[n]$ is a scaled replica of the Fourier transform of $x(t)$ for $-\frac{1}{2} \le f < \frac{1}{2}$.

$$X(e^{j2\pi f}) = F_s X(fF_s) \quad -\frac{1}{2} \le f < \frac{1}{2} \tag{10}$$

The preceding result is illustrated in Figure 7.3. Because our goal in designing the hybrid system is to match the performance of the CT system, we assume for the remainder of this chapter that the condition $F_s > 2W$ is satisfied.

---

[2] If the sampling rate falls below the Nyquist rate, then the degrading effects of spectrum foldover can be decreased by means of an antialiasing filter. Antialiasing filters trade spectrum foldover for loss in high-frequency signal components. (See Section 4CT.6.2.)

DT Processor

Our next task in analyzing the hybrid system of Figure 7.1(a) is to consider the DT processor. Like all DT LTI processors, it operates under the principles described in Chapters 2DT–5DT: It performs a convolution

$$y[n] = \sum_{m=-\infty}^{\infty} x[m]h[n-m] = x[n] * h[n] \tag{11}$$

or, equivalently, a frequency domain multiplication

$$Y(e^{j2\pi f}) = H(e^{j2\pi f})X(e^{j2\pi f}) \tag{12}$$

where

$$H(e^{j2\pi f}) = \sum_{n=-\infty}^{\infty} h[n]e^{-jn2\pi f} \tag{13}$$

If we substitute Eq. (10) and into Eq. (12), we obtain

$$Y(e^{j2\pi f}) = H(e^{j2\pi f})F_s X(fF_s) \quad -\tfrac{1}{2} \le f < \tfrac{1}{2} \tag{14}$$

We can now answer the question of how to select $H(e^{j2\pi f})$ so that the I-O relation of the hybrid system matches that of the purely analog system.

For Matched Performance of the Hybrid and Analog System

Set

$$H(e^{j2\pi f}) = H_m(e^{j2\pi f})$$

where

$$H_m(e^{j2\pi f}) = H(fF_s) \quad -\tfrac{1}{2} \le f < \tfrac{1}{2} \tag{15}$$

and $H(F)$ is the frequency response characteristic of the analog system.

The following example illustrates the matched DT system frequency response characteristic $H_m(e^{j2\pi f})$ for a differentiator.

EXAMPLE 7.1   A DT Differentiator

Consider the objective of making the hybrid system have the same output as a CT differentiator. The frequency response characteristic of a CT differentiator is

$$H(F) = j2\pi F \tag{16}$$

The frequency response characteristic of the matched DT differentiator is, by Eq. (15),

$$H_m(e^{j2\pi f}) = j2\pi f F_s \quad -\tfrac{1}{2} \le f < \tfrac{1}{2} \tag{17}$$

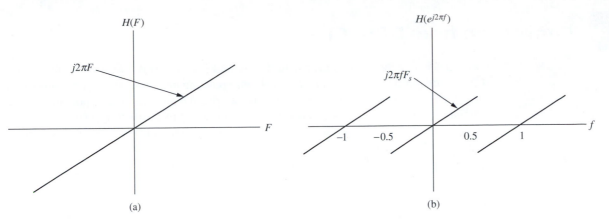

Figure 7.4
Frequency response characteristics of ideal differentiators: (a) CT; (b) DT (for use in the hybrid system of Figure 7.1(a)).

This frequency response characteristic is plotted in Figure 7.4 for $-1.5 < f < 1.5$. Notice that like all DT systems, the frequency response of the DT differentiator is periodic with period 1.

If we substitute $H_\mathrm{m}(e^{j2\pi f})$ from Eq. (15) for $H(e^{j2\pi f})$ into Eq. (14), we obtain

$$Y(e^{j2\pi f}) = H(f F_s) F_s X(f F_s) \quad -\tfrac{1}{2} \le f < \tfrac{1}{2} \tag{18}$$

This basic result is summarized in Figure 7.5.

The problem of selecting the elements of a DT filter so that its frequency response characteristic equals $H_\mathrm{m}(e^{j2\pi f})$ is considered in Section 7.5. In the next section we consider the final stage of the hybrid system.

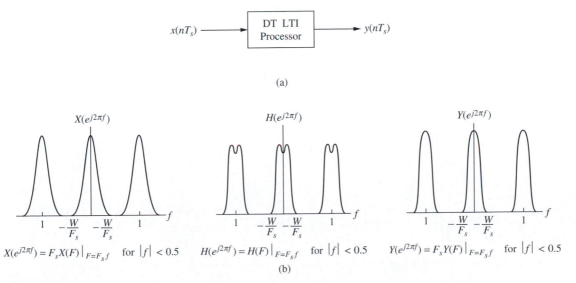

Figure 7.5
Matched DT processor: (a) system in sequence domain; (b) frequency domain description.

7.4   Conversion from DT to CT

Our final step in the design of the hybrid system of Figure 7.1(a) is to perform the DT-to-CT conversion. This conversion amounts to interpolating among samples of the processed data $y[n]$ to produce a CT waveform $y(t)$. Recall that the problem of interpolating among samples was considered in Section 4CT.6 in the context of reconstructing a waveform from its samples. The current problem is to design the interpolation filter so that the systems in Figure 7.1 have the same outputs.

The spectrum at the output of the CT processor (Figure 7.1(b)) is

$$Y(F) = H(F)X(F) \tag{19}$$

The spectrum of the DT sequence $y(nT_s)$, applied to the DT-to-CT converter is, from Eq. (18)

$$Y(e^{j2\pi f}) = H(fF_s)F_sX(fF_s) \quad -\tfrac{1}{2} \le f < \tfrac{1}{2} \tag{20}$$

We can use the two-stage model shown in Figure 7.6(a) to mathematically relate input and output of the DT to CT. The first stage is a modulator that converts the sequence $y(nT_s)$ to a CT waveform.

$$y_s(t) = \sum_{n=-\infty}^{\infty} y(nT_s)\,\delta(t - nT_s) \tag{21}$$

The second stage is a CT filter having the impulse response $h_i(t)$. The response of the filter to $y_s(t)$ is

$$\hat{y}(t) = \sum_{n=-\infty}^{\infty} y(nT_s)h_i(t - nT_s) \tag{22}$$

We see from Eq. (22) that the filter interpolates among the samples $y(nT_s)$. (The subscript i in $h_i(t)$ denotes interpolation.) We use the notation $\hat{y}(t)$ to emphasize that we want the filter output to equal $y(t)$, the output of the CT system of Figure 7.1(b). To select $h_i(t)$, we need to know the spectrum of $\hat{y}(t)$. We can find this spectrum with the use of the linearity

Figure 7.6
DT-to-CT conversion:
(a) modeled as a
modulator and filter
in series;
(b) frequency domain
with ideal
interpolator
frequency response
characteristic.

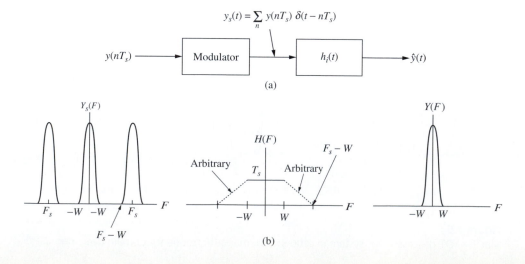

and time shift properties of the Fourier transform.

$$\hat{Y}(F) = \mathcal{F}\left\{\sum_{n=-\infty}^{\infty} y(nT_s)h_i(t - nT_s)\right\} = \sum_{n=-\infty}^{\infty} y(nT_s)\mathcal{F}\{h_i(t - nT_s)\}$$

$$= \sum_{n=-\infty}^{\infty} y(nT_s)H_i(F)\,e^{-2\pi FT_s} = Y(e^{j2\pi FT_s})H_i(F) \tag{23}$$

The substitution of Eq. (18) into Eq. (23) then yields, with the aid of Eq. (9),

$$\hat{Y}(F) = H(F)F_s X(F)H_i(F) = Y(F)F_s H_i(F) \qquad -\frac{F_s}{2} \leq F < \frac{F_s}{2} \tag{24}$$

where we have used Eq. (19) in the last equality. The factors in the preceding equation are illustrated in Figure 7.6(b). We can see from the figure that if we chose

$$H_i(F) = \begin{cases} T_s & |F| < W \\ \text{arbitrary} & W < |F| < F_s - W \\ 0 & |F| \geq F_s - W \end{cases} \tag{25}$$

then Eq. (24) becomes

$$\hat{Y}(F) = Y(F) \tag{26}$$

Therefore, the ideal interpolation filter has a frequency response characteristic given by Eq. (25). Notice that the ideal interpolation filter (25) is the same as the one obtained in Section 6 of Chapter 4CT in connection with reconstructing a signal from its samples. The time domain version of Eq. (26) states that

$$\hat{y}(t) = y(t) \tag{27}$$

which is the bottom-line result that the hybrid system has the same output as the purely analog system.

7.5   Filter Design

So far, we have established that the performance of the hybrid system matches that of the analog system if the sampling frequency exceeds the Nyquist rate and if we use the matched processor and interpolation filter of Eqs. (15) and (25), respectively. We now ask (1) how do we design the transfer function $H(z)$ of the DT filter of Figure 7.1 so that Eq. (15) is satisfied? (2) What are the various filter structures that we can use to implement the DT filter? We shall see that Eq. (15) can be *exactly* satisfied, in general, only if the DT filter is composed of an infinite number of DT elements (adders, gains, and delays) and if the processing delay is arbitrarily long. We shall also see that Eq. (15) can be closely approximated using a finite number of DT elements along with a finite delay.

The word "approximated" suggests that we start with filter design specifications ("specs") that state acceptable limits for the approximation error. Design specs occur in many varieties. They can be given in either the sequence or the frequency domains. Figure 7.7 illustrates a typical frequency domain design spec for a low-pass filter: In this figure, the frequency axis is divided into three regions and the amplitude specs are stated in decibels. The decibel is a logarithmic unit used to describe a ratio. The ratio here is $|H(e^{j2\pi f})|$, which is the ratio of the output amplitude at frequency f to the input amplitude at frequency f. Decibel values are related to the corresponding ratio values by

$$|H(e^{j2\pi f})|_{\text{dB}} = 20\log_{10}|H(e^{j2\pi f})| \tag{28}$$

Figure 7.7
Typical design spec
for a digital low-pass
filter.

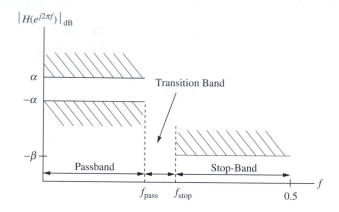

The low-pass filter design specs shown in the figure are summarized as follows.

Low-Pass Filter Design Specifications (Figure 7.7)

Passband:

$$-\alpha \le |H(e^{j2\pi f})|_{dB} \le \alpha \qquad \text{for } 0 \le f \le f_{pass} \tag{29}$$

where α is maximum allowable passband deviation in dB ($\alpha \ge 0$).
Transition band:

$$f_{pass} < f < f_{stop} \tag{30}$$

Stop-band:

$$|H(e^{j2\pi f})|_{dB} \le -\beta \qquad \text{for } f_{stop} \le f < 0.5 \tag{31}$$

where β is the minimum allowable stop-band attenuation in dB ($\beta \ge 0$).

Low-pass filter design specs are often normalized for a maximum amplitude ratio of unity, or 0 dB, in the passband. A typical normalized design specs is shown in Figure 7.8. The normalized low-pass filter design specs shown in the figure are summarized as follows.

Figure 7.8
Typical normalized
design spec for a
digital low-pass filter.

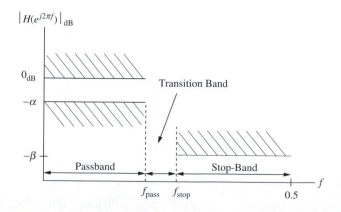

Normalized Low-Pass Filter Design Specifications (Figure 7.8)

Passband:

$$-\alpha \leq |H(e^{j2\pi f})|_{dB} \leq 0 \qquad \text{for } 0 \leq f \leq f_{\text{pass}} \tag{32}$$

where α is maximum allowable passband attenuation in dB ($\alpha \geq 0$).

Transition band:

$$f_{\text{pass}} < f < f_{\text{stop}} \tag{33}$$

Stop-band:

$$|H(e^{j2\pi f})|_{dB} \leq -\beta \qquad \text{for } f_{\text{stop}} \leq f < 0.5 \tag{34}$$

where β is the minimum allowable stop-band attenuation in dB ($\beta \geq 0$).

Specifications similar to those in Figures 7.8 and 7.9 are used to design high-pass and band-pass filters.

Figure 7.9 illustrates a possible design spec for the amplitude characteristic of a differentiator. Here the specification applies only to the region $0 \leq f \leq f_p < 0.5$, where it is required that

$$\left. \begin{array}{l} 0 \ \ \text{for } f < \epsilon \\ 2\pi(f - \epsilon)F_s \qquad \text{for } \epsilon \leq f \leq f_p \end{array} \right\} \leq |H(e^{j2\pi f})| \leq 2\pi(f + \epsilon)F_s \tag{35}$$

A specification like this makes sense if it is known that the input sequence does not have significant frequency components in the range $f_p < |f| < 0.5$.

Considerable care should be devoted to the determination of a design specification because, in general, as we make a design spec more stringent we need more DT elements and processing delay to satisfy it. Once a spec has been established, the filter designer starts by considering the type of filter that is best able to satisfy it. There are two basic filter types: FIR filters and IIR filters. FIR filters are generally preferred in applications where a linear phase characteristic is important. These applications include video signal processing because human vision is sensitive to phase distortion in an image. FIR filters are also preferred in advanced applications in which the filter must adapt to different operating conditions. IIR filters are generally preferred in applications where phase distortion is not an important consideration. They are often used for audio signal processing because human hearing is relatively insensitive to the phase distortion in a sound. IIR filters are also preferred when computational efficiency is an important consideration.

Figure 7.9
Possible design spec for the amplitude characteristic of a digital differentiator: Notice that this spec applies only to the frequency range $0 \leq |f| \leq f_p$.

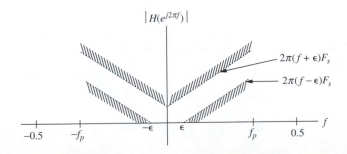

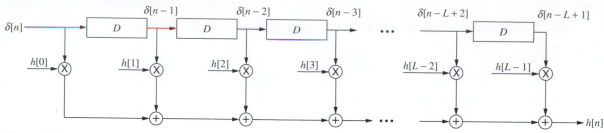

Figure 7.10
FIR filter.

7.5.1
FIR Filters

An *FIR filter* is a DT filter that has a finite-length impulse response.

FIR Filter Impulse Response

$$h[n] = h[0]\delta[n] + h[1]\delta[n-1] + h[2]\delta[n-2] + \cdots + h[L-1]\delta[n-L+1] \tag{36}$$

In the preceding equation, L is the finite length of $h[n]$. Because L is finite, a FIR filter impulse response is always absolutely summable.

$$\sum_{n=0}^{L-1} |h[n]| < \infty \tag{37}$$

Therefore, every FIR filter is BIBO stable. This guaranteed stability is an important property of FIR filters. We can implement an FIR filter with the system shown in Figure 7.10. We can see by inspection that this implementation works because if we set $x[n] = \delta[n]$, we obtain $h[n]$ of Eq. (36). Other structures for FIR filters exist, but the one shown in the figure is the standard form (Problems 7.5.1, 7.5.2 and 7.5.16.)

The I-O equation of the FIR filter is easy to find by inspection of the figure or by convolving $h[n]$ of Eq. (36) with $x[n]$.

$$y[n] = h[0]x[n] + h[1]x[n-1] + h[2]x[n-2] + \cdots + h[L-1]x[n-L+1] \tag{38}$$

This equation has the form

$$\mathcal{A}\{D\}y[n] = \mathcal{B}\{D\}x[n] \tag{39}$$

where

$$\mathcal{A}\{D\} = 1 \tag{40}$$

and

$$\mathcal{B}\{D\} = h[0] + h[1]D + h[2]D^2 + \cdots h[L-1]D^{L-1} \tag{41}$$

We can find the transfer function by assuming zero initial conditions and applying the shift property of the z-transform. The result is

---

Transfer Function of FIR Filter

$$H(z) = \frac{\mathcal{B}(z^{-1})}{\mathcal{A}(z^{-1})} = h[0] + h[1]z^{-1} + h[2]z^{-2} + \cdots + h[L-1]z^{-(L-1)} \tag{42}$$

Because $H(z)$ contains an $L - 1$th-order pole at $z = 0$, an FIR filter having length L is said to have order $L - 1$. For most applications, L is an odd integer:

$$L = 2K + 1 \tag{43}$$

where K is a nonnegative integer.

Noncausal FIR Design Filter

We can see from both its impulse response Eq. (36) and its structure (Figure 7.10) that an FIR filter is causal. When we design an FIR filter, however, we often want the FIR filter frequency response characteristic to match or approximate that of a hypothetical *noncausal* filter such as an ideal low-pass filter. We stated in Chapter 1 that it is possible for a causal system to mimic the performance of a noncausal one if we allow an overall delay. To design the causal FIR filter $h[n]$, it is helpful to work with an associated *noncausal version* $h_{nc}[n]$ of $h[n]$.

$$h_{nc}[n] \triangleq h_{nc}[-K]\delta[n + K] + h_{nc}[-K + 1]\delta[n + K - 1] + \cdots + h_{nc}[0]\delta[n] + \cdots$$
$$+ h_{nc}[K]\delta[n - K] \tag{44}$$

The impulse response of the causal FIR filter of Figure 7.10 is a delayed version of $h_{nc}[n]$.

$$h[n] = h_{nc}[n - K] \tag{45}$$

Figure 7.11 shows an example of $h[n]$ and $h_{nc}[n]$ for $K = 10$. The transfer functions of $h[n]$ and $h_{nc}[n]$ are related by

$$H(z) = z^{-K} H_{nc}(z) \tag{46}$$

where

$$H_{nc}(z) \triangleq h_{nc}[-K]z^K + h_{nc}[-K + 1]z^{K-1} + \cdots + h_{nc}[0] + \cdots + h_{nc}[K]z^{-K} \tag{47}$$

In designing the FIR filter $h[n]$, we design the noncausal filter $h_{nc}[n]$ first. Then we shift $h_{nc}[n]$ as in Eq. (45) to obtain $h[n]$. We shall see that the performance of an FIR filter generally improves as the filter length, $L = 2K + 1$, is increased. When K increases, so does the K sample processing delay associated with the shift from $h_{nc}[n]$ to $h[n]$.

The frequency response characteristics corresponding to Eq. (46) is

$$H(e^{j2\pi f}) = e^{-jK2\pi f} H_{nc}(e^{j2\pi f}) \tag{48}$$

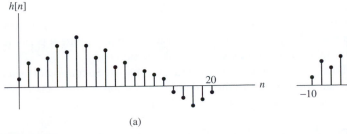

(a)

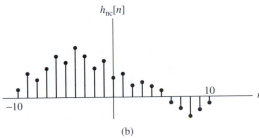

(b)

Figure 7.11
FIR impulse response $h[n]$ and associated noncausal version $h_{nc}[n]$.

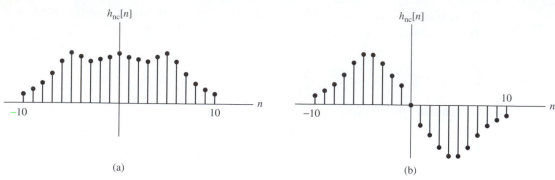

Figure 7.12
Symmetry for $h_{nc}[n]$: (a) even symmetry; (b) odd symmetry.

where

$$H_{nc}(e^{j2\pi f}) = \sum_{k=-K}^{K} h_{nc}[k]e^{jk2\pi f} \qquad (49)$$

In most applications, $h_{nc}[n]$ is real and has either even or odd symmetry, as illustrated in Figure 7.12. When $h_{nc}[n]$ is real and has even symmetry, Eq. (49) simplifies to

$$H_{nc}(e^{j2\pi f}) = h_{nc}[0] + 2\sum_{k=1}^{K} h_{nc}[k]\cos(k2\pi f) \qquad (50)$$

When $h_{nc}[n]$ is real and has even symmetry, Eq. (49) simplifies to

$$H_{nc}(e^{j2\pi f}) = 2j\sum_{k=1}^{K} h_{nc}[k]\sin(k2\pi f) \qquad (51)$$

Notice that $H_{nc}(e^{j2\pi f})$ is real and even when $h_{nc}[n]$ is real and even, and $H_{nc}(e^{j2\pi f})$ is imaginary and odd when $h_{nc}[n]$ is real and odd.

FIR Filter Design

The term "FIR filter design" means to find numerical values for $h_{nc}[-K]$, $h_{nc}[-K+1]$, ..., $h_{nc}[K]$ so that $H_{nc}(e^{j2\pi f})$ approximates the design goal $H_m(e^{j2\pi f})$ of Eq. (15) within given design specs. The technical problem is to approximate the periodic function $H_m(e^{j2\pi f})$ with a sum of sinusoidal functions (49), (50), or (51). This problem is very much like the one we considered in Chapter 5CT in connection with the partial Fourier series of a periodic waveform. There, we approximated a periodic function of time with a sum of complex sinusoidal functions of time. In the present problem, we want to approximate a periodic function of f, $H_m(e^{j2\pi f})$ with a sum of complex sinusoidal functions of f.

In solving the partial Fourier series problem in Chapter 5CT, we used a minimum mean square error criterion. We can similarly solve the FIR filter design problem by choosing $h_{nc}[n]$ to minimize the mean square approximation error (MMSE).

$$\mathcal{E} = \int_{-0.5}^{0.5} \left| H_{nc}(e^{j2\pi f}) - H_m(e^{j2\pi f}) \right|^2 df \qquad (52)$$

The solution can be obtained using the same steps we used in Chapter 5CT. The result is as follows.

MMSE FIR Filter Design Equation

The $h_{nc}[n]$ that minimize mean square approximation error $\mathcal{E}$ in Eq. (52) equal the Fourier coefficients of the Fourier series expansion of $H_m(e^{j2\pi f})$. This Fourier series expansion of $H_m(e^{j2\pi f})$ is

$$H_m(e^{j2\pi f}) = \sum_{k=-\infty}^{\infty} H_m[k]e^{jk2\pi f} \tag{53}$$

and the Fourier coefficients are given by

$$h_m[n] = \int_{-0.5}^{0.5} H_m(e^{j2\pi f})\, e^{-j2\pi nf}\, df \tag{54}$$

For a length $L = 2K + 1$ FIR filter, we take $h_{nc}[n] \equiv h_m[n]$ for $-K \leq n \leq K$.

$$h_{nc}[n] = h_m[n] \qquad \text{for } -K \leq n \leq K \tag{55}$$

It follows that, with MMSE design, $H_{nc}(e^{j\omega})$ of Eq. (49) is the partial Fourier series for $H_m(e^{j2\pi f})$ and $H_{nc}(e^{j\omega})$ converges to $H_m(e^{j2\pi f})$ in the MMSE sense as K increases.

Often the design goal $H_m(e^{j2\pi f})$ is either real and even or imaginary and odd. When $H_m(e^{j2\pi f})$ is real and even, Eq. (54) simplifies to

$$h_m[n] = \int_{-0.5}^{0.5} H_m(e^{j2\pi f}) \cos(2\pi f n)\, df \tag{56}$$

When $H_m(e^{j2\pi f})$ is imaginary and odd Eq. (54) simplifies to

$$h_m[n] = -j \int_{-0.5}^{0.5} H_m(e^{j2\pi f}) \sin(2\pi f n)\, df \tag{57}$$

EXAMPLE 7.2   FIR Low-Pass Filter Based on MMSE

Here our design goal $H_m(e^{j2\pi f})$ is an ideal low-pass filter having the bandwidth $w = 0.2$. Our design specs are

Passband: $-1 \leq |H(e^{j2\pi f})|_{dB} \leq 1$ for $0 \leq f \leq 0.15$

Transition band: $0.15 < f < 0.25$

Stop-band: $|H(e^{j2\pi f})|_{dB} \leq -30$ for $0.25 \leq f < 0.5$

Because $H_m(e^{j\omega})$ is real and even, we can find the FIR filter impulse response using Eq. (56), which yields

$$h_{nc}[n] \equiv h_m[n] = \int_{-w}^{w} \cos(2\pi f n)\, df = 2w\, \text{sinc}(2wn) \tag{58}$$

To conclude the design, we need to find the minimum value of K for which the design specs are met. This is a trial-and-error process. For each trial K, we plot $H_{nc}(e^{j\omega})$ of Eq. (50) and compare the plot with the design spec. We can write our own program to make the plots, or use the MATLAB function `fir 1` as outlined in Problem 7.5.5. Trial and error reveals that we can satisfy the design spec only if $K \geq 22$. Figure 7.13 shows the amplitude characteristic and the

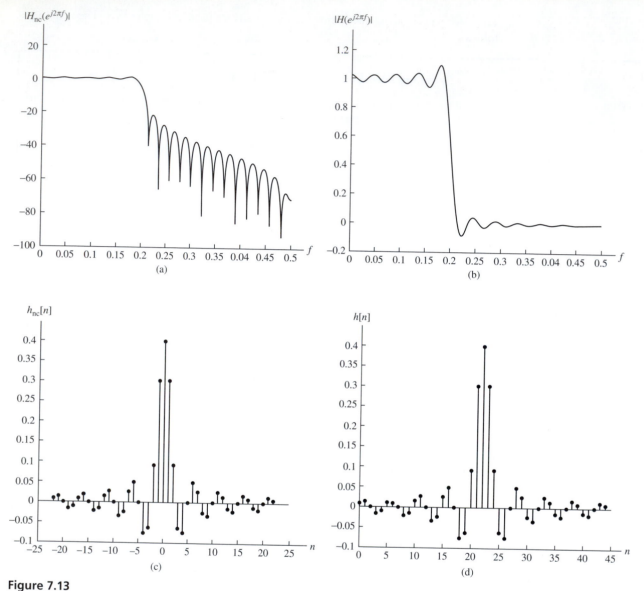

Figure 7.13
FIR low-pass filter based on MMSE design with $L = 45$: (a) $|H_{nc}(e^{j2\pi f})|_{dB}$; (b) $H_{nc}(e^{j2\pi f})$, (c) $h_{nc}[n]$; (d) $h[n]$.

impulse responses $h_{nc}[n]$ and $h[n]$ for $K = 22$ ($L = 45$). The amplitude characteristic is shown in decibels in Figure 7.13(a), and it is shown with a linear scale in Figure 7.13(b). Notice from Figure 7.13(b) that there is an overshoot of approximately 9 percent at either side of the discontinuity at $f = 0.2$ of the design goal $H_m(e^{j\omega})$. This overshoot and the oscillatory behavior near $f = 0.2$ is the Gibbs phenomenon we observed in connection with the MMSE convergence Fourier series in Example 5CT.12. We know from that example that the ripples compress closer to the discontinuity as K is increased but the 9 percent overshoot does not decrease in amplitude.

In conclusion, we can satisfy the design spec using MMSE FIR design with $L = 45$. The processing delay is $K = 22$ samples.

EXAMPLE 7.3   FIR Differentiator Based on MMSE Design

Here our design goal is the ideal differentiator characteristic shown in Figure 7.4. We use the design spec of Figure 7.9, with $f_p = 0.4$, $\epsilon = 1/8\pi$, and $F_s = 1$. Because the design goal $H_m(e^{j\omega})$ is imaginary and odd, we find the filter impulse response values using Eq. (57), with $H_m(e^{j2\pi f}) = j2\pi f$. The result is

$$h_{nc}[n] \equiv h_m[n] = \int_{-0.5}^{0.5} 2\pi f \sin(2\pi f n)df = \begin{cases} 0 & n = 0 \\ \frac{\cos(\pi n)}{n} & n \neq 0 \end{cases} \qquad (59)$$

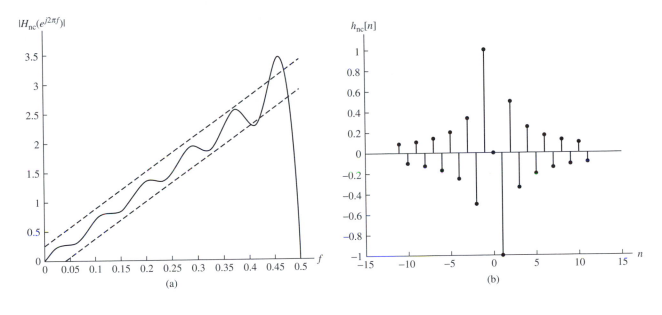

(a)

(b)

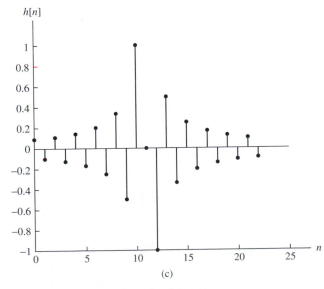

(c)

Figure 7.14
FIR differentiator filter-based on MMSE design with $L = 23$: (a) $|H_{nc}(e^{j2\pi f})|$; (b) $h_{nc}[n]$; (c) $h[n]$.

We substitute values for K into Eq. (51) and compare $H_{nc}(e^{j\omega})$ with $H_m(e^{j\omega})$. We find by trial and error that we can satisfy the design spec if and only if $K \geq 11 \Rightarrow L \geq 23$. The amplitude characteristic $|H_{nc}(e^{j2\pi f})|$ is shown in Figure 7.14 for $L = 23$. Figure 7.14(b) and (c) shows plots of $h_{nc}[n]$ and $h[n]$ for $L = 23$.

In conclusion, we have met the design spec using MMSE design with $L = 23$ and an overall processing delay of $K = 11$ samples.

Windowing

The windowing technique substantially decreases the Gibbs overshoot and ripple associated with MMSE design. By the use of windowing, we can usually obtain an FIR filter that satisfies the design specs with fewer elements and less processing delay than the MMSE filter.

The impulse response of a FIR filter based on windowed design is

$$h_w[n] = h_{ncw}[n - K] \tag{60}$$

where

Windowed FIR Filter Design Equation

$$h_{ncw}[n] = w[n]h_m[n] \tag{61}$$

where $w[n]$ is a length $L = 2K + 1$ window.

$$w[n] = 0 \quad |n| > K \tag{62}$$

and the $h_m[n]$ are given by Eq. (54).

In general, windows are positive even sequences that decrease smoothly toward 0 as $|n|$ increases to K. They are normalized for

$$w[0] = 1 \tag{63}$$

The term "window" is used because the windowed sequence $h_{ncw}[n]$ includes only the Fourier coefficients $h_m[-K], h_m[-K+1], \ldots, h_m[K-1]$, and $h_m[K]$, defined by Eq. (54).

Typical examples of windows include the *triangular window*

$$w_T[n] = \wedge[n; 2K] \tag{64}$$

the *raised cosine* or *Hann window*

$$w_H[n] = \frac{1}{2}\left\{1 + \cos\left(\frac{\pi n}{K+1}\right)\right\} \sqcap (n, K) \tag{65}$$

and the *Nuttal window*

$$w_N[n] = 0.3635819 - 0.4891775 \cos\left(\frac{\pi n}{K+1}\right)$$

$$+ 0.1365995 \cos\left(\frac{2\pi n}{K+1}\right) - 0.0106411 \cos\left(\frac{2\pi n}{K+1}\right) \tag{66}$$

Notice that a Hann window is the raised cosine sequence we considered in Example 4DT.5. The MMSE FIR filter design we developed earlier can be viewed as a special case of windowed filter design where $w[n]$ is the *rectangular window*

$$w_R[n] = \sqcap[n; K] \tag{67}$$

The rectangular window simply truncates $h_m[n]$; $n = 0, \pm 1, \pm 2, \ldots$ to the index region $-K \leq n \leq K$.

The frequency response characteristic corresponding to a causal windowed FIR filter Eq. (60) is

$$H_w(e^{j2\pi f}) = e^{-jK2\pi f} H_{ncw}(e^{j2\pi f}) \tag{68}$$

where

$$H_{ncw}(e^{j2\pi f}) = \sum_{k=-K}^{K} h_{ncw}[k] e^{jk2\pi f} \tag{69}$$

When $h_{ncw}[n]$ is real and has even symmetry, Eq. (69) simplifies to

$$H_{ncw}(e^{j2\pi f}) = h_{ncw}[0] + 2 \sum_{k=1}^{K} h_{ncw}[k] \cos(k2\pi f) \tag{70}$$

When $h[n]$ is real and has even symmetry, Eq. (69) simplifies to

$$H_{ncw}(e^{j2\pi f}) = 2j \sum_{k=1}^{K} h_{ncw}[k] \sin(k2\pi f) \tag{71}$$

EXAMPLE 7.4   FIR Low-Pass Filter Designed with Hann Window

Let's reconsider the ideal low-pass filter design problem of Example 7.2. In an effort to satisfy the design spec with an FIR filter having a length shorter than 45, let's try the *Hann* window. The windowed impulse response is obtained by substituting Eq. (58) into Eq. (61). The result is

$$h_{ncw}[n] = w_H[n] 2w \ \text{sinc}(2wn) \tag{72}$$

where $w_H[n]$ is given by Eq. (65). Because $h_{ncw}[n]$ is a real, even sequence, the frequency response characteristic corresponding to $h_{ncw}[n]$ is

$$H_{ncw}(e^{j\omega}) = h_{ncw}[K] + 2 \sum_{k=1}^{K} h_{ncw}[k] \cos(k\omega) \tag{73}$$

Again, we make plots of $H_{ncw}(e^{j\omega})$ to find the minimum value of K for which the design specs are met. Trial and error reveals that the specs are met for $K = 13$ but not for $K = 12$. Figure 7.15 shows the plots of $|H_{ncw}(e^{j2\pi f})|_{dB}$, $H_{ncw}(e^{j2\pi f})$, $h_{ncw}[n]$, and) $h_w[n]$ for $K = 12$ ($L = 27$).

We see that the Hann window design has resulted in a low-pass filter that satisfies the design spec of Example 7.2 with $L = 27$. This compares very favorably to the length $L = 45$ FIR filter we obtained using MMSE design.

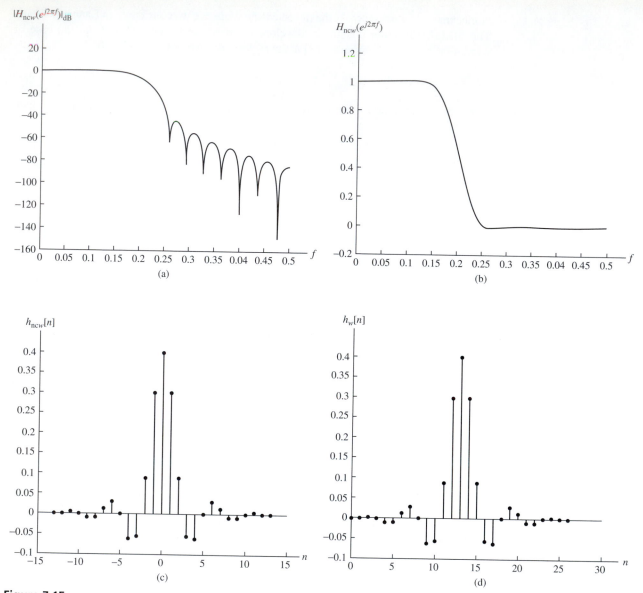

Figure 7.15
FIR Low-pass filter based on Hann window design with $L = 27$: (a) $|H_{ncw}(e^{j2\pi f})|_{dB}$; (b) $H_{ncw}(e^{j2\pi f})$; (c) $h_{ncw}[n]$, d) $h_w[n]$.

For our final example, we apply the windowing technique to the design of an ideal differentiator.

EXAMPLE 7.5   FIR Differentiator Designed with a Hann Window

Let's consider once more the ideal differentiator characteristic shown in Figure 7.4. We use the same design spec we used in Example 7.3, but this time try a windowed FIR filter design using a *Hann* window. The windowed impulse response is given by Eq. (61), where $h_{nc}[n]$ is given by Eq. (59) and $w[n]$ is given by Eq. (65).

$$h_{ncw}[n] = \begin{cases} 0 & n = 0 \\ w_H[n]\frac{\cos(\pi n)}{n} & n \neq 0 \end{cases} \tag{74}$$

Using trial and error we find that we can satisfy the design specs, provided that the FIR filter length $L = 2K + 1$ is a least 13. Plots of $\left|H_{nc}(e^{j\omega})\right|$, $h_{nc}[n]$ and $h[n]$ are shown in Figure 7.16. In conclusion, we can satisfy the design spec with $L = 13$ and an overall processing delay of $K = 6$ samples. The windowed design compares very favorably to the MMSE design of Example 7.3, where $L = 23$ and $K = 11$.

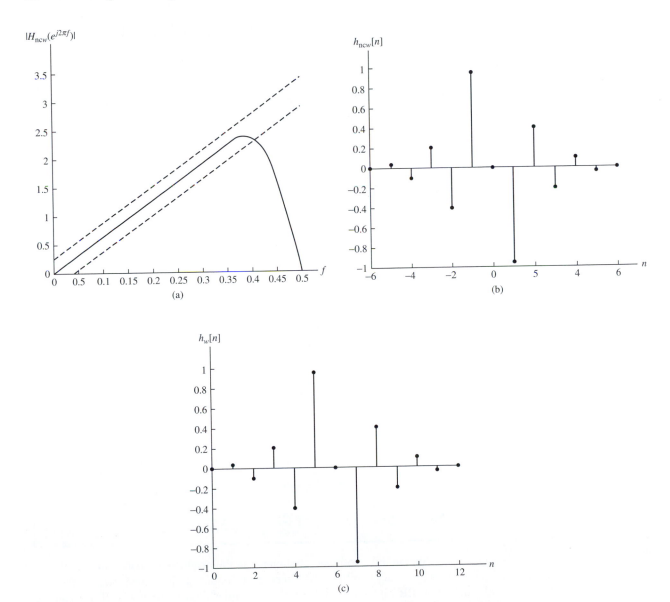

Figure 7.16
FIR differentiator based on a Hann window design with $L = 13$: (a) $H_{ncw}(e^{j2\pi f})$; (b) $h_{ncw}[n]$; (c) $h_w[n]$.

Theoretical insight into how windowing works can be obtained by considering the frequency domain version of Eq. (61).

$$H_{\text{ncw}}(e^{j2\pi f}) = W(e^{j2\pi f}) \circledast H_{\text{m}}(e^{j2\pi f}) \tag{75}$$

In Eq. (75) $W(e^{j2\pi f})$ is the DTFT of $w[n]$ and $\circledast$ denotes circular convolution. The convolution has the effect of smoothing edges in $H_{\text{m}}(e^{j2\pi f})$. For a rectangular window (MMSE design), $W(e^{j2\pi f})$ equals

$$W_R(e^{j2\pi f}) = \mathcal{DTFT}\{\sqcap[n; K]\} = (2K + 1)\,\text{cinc}(f; K) \tag{76}$$

which was plotted and described in Example 4DT.2. (It was called $X(e^{j2\pi f})$ there.) When $W_R(e^{j2\pi f})$ is convolved with an edge in $H_{\text{m}}(e^{j2\pi f})$, the result is the 9 percent overshoot and gradually decreasing oscillations near the edge seen in Example 7.2. For the Hann window of Example 7.4, $W(e^{j2\pi f})$ equals

$$
\begin{aligned}
W_H(e^{j2\pi f}) &= \mathcal{DTFT}\left\{\frac{1}{2}\left\{1 + \cos\left(\frac{\pi n}{K+1}\right)\right\}\sqcap(n, K)\right\} \\
&= \frac{1}{2}(2K+1)\left\{\text{cinc}(f; K) + \frac{1}{2}\text{cinc}\left(f + \frac{1}{2(K+1)}; K\right)\right. \\
&\quad \left. + \frac{1}{2}\text{cinc}\left(f - \frac{1}{2(K+1)}; K\right)\right\}
\end{aligned}
\tag{77}
$$

which was plotted and described in Example 4DT.5. (It was called $X(e^{j2\pi f})$ there.) We saw in that example that $W_H(e^{j2\pi f})$ has substantially smaller sidelobes than those in $W_R(e^{j2\pi f})$ and that the sidelobes of $W_H(e^{j2\pi f})$ also attenuate much faster than those in $W_R(e^{j2\pi f})$. For this reason, the convolution $W_H(e^{j2\pi f}) \circledast H_{\text{m}}(e^{j2\pi f})$ results in much less overshoot and oscillations that decrease faster near edges of $H_{\text{m}}(e^{j2\pi f})$ than the convolution $W_R(e^{j2\pi f}) \circledast H_{\text{m}}(e^{j2\pi f})$.

**7.5.2
IIR Filters**

IIR (infinite-length impulse response) filters have infinite-length impulse responses.

IIR Filter Impulse Response

$$h[n] = h[0]\delta[n] + h[1]\delta[n-1] + h[2]\delta[n-2] + \cdots \tag{78}$$

Because an IIR filter impulse response continues indefinitely, there is no guarantee that $h[n]$ is absolutely summable. Therefore, IIR filters can be either stable or unstable.

Examples of IIR filters are given by the first-order filter of Example 2DT.3 and the second-order filter of Examples 6DT.13 and 6DT.16. These filters were specified by first- and second-order difference equations, respectively. More generally, an lth-order IIR filter is specified by an lth order difference equation having the form

$$y[n] + a_1 y[n-1] + a_2 y[n-2] + \cdots + a_l y[n-l]$$
$$= b_0 x[n] + b_1 x[n-1] + \cdots + b_m x[n-m] \tag{79}$$

We can use the operator notation of Chapter 6DT to write this as

$$\mathcal{A}\{D\}y[n] = \mathcal{B}\{D\}x[n] \tag{80}$$

where

$$\mathcal{A}\{D\} = 1 + a_1 D + a_2 D^2 + \cdots a_l D^l \tag{81}$$

with $a_l \neq 0$, and

$$\mathcal{B}\{D\} = b_0 + b_1 D + b_2 D^2 + \cdots b_m D^m \tag{82}$$

The following transfer function of an IIR filter follows directly from the shift property of the z-transform (or from Eq. 6DT.77 with $a_0 = 1$).

Transfer Function of IIR Filter

$$H(z) = \frac{\mathcal{B}(z^{-1})}{\mathcal{A}(z^{-1})} = \frac{b_0 + b_1 z^{-1} + b_2 z^{-2} + \cdots b_m z^{-m}}{1 + a_1 z^{-1} + a_2 z^{-2} + \cdots a_l z^{-l}} \tag{83}$$

A straightforward implementation for an IIR filter can be obtained by rearranging Eq. (79) to

$$y[n] = b_0 x[n] + b_1 x[n-1] + \cdots + b_m x[n-m]$$
$$- a_1 y[n-1] - a_2 y[n-2] - \cdots - a_l y[n-l] \tag{84}$$

The operations shown in Eq. (84) are shown in the block diagram of Figure 7.17 for $l = m$. This implementation is called *Direct Form 1*. Notice that this implementation contains $2m$ memory elements. We can derive an implementation that contains only m memory elements by commuting the two subsystems in Direct Form 1, as shown in Figure 7.18. An inspection of Figure 7.18 reveals that the two m-element vertical shift registers are redundant because they have the same input. We can eliminate one shift register to obtain the system shown in Figure 7.19. This system, *Direct Form 2*, is called a *canonic form* because it contains the minimum possible number of memory elements. Figure 7.20 shows the Direct Form 2 realizations of first- and second-order systems.

Additional implementations can be obtained by manipulating $H(z)$ algebraically, as illustrated in the following example.

Figure 7.17
Direct Form 1
realization of IIR filter:
The figure assumes
that $l = m$.

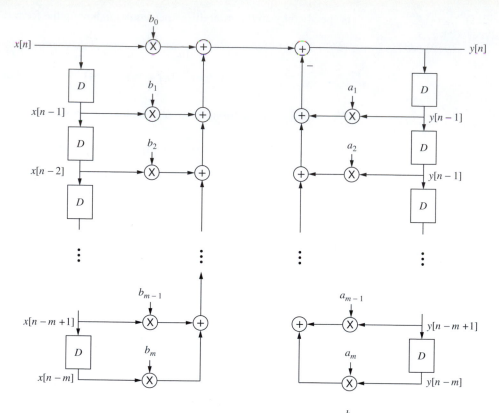

Figure 7.18
Transition from Direct
Form 1 to Direct
Form 2.

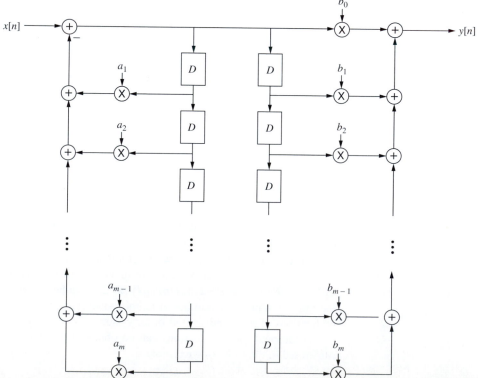

Figure 7.19
Direct Form 2, a
canonic form because
it contains the
minimum possible
number of unit delay
elements.

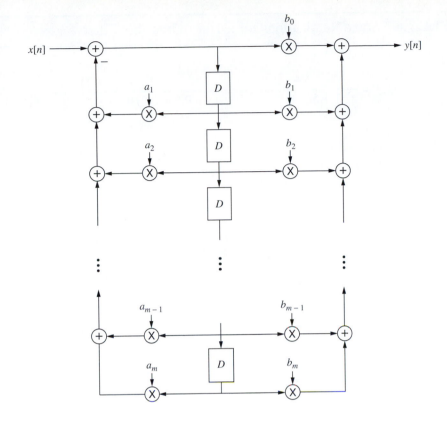

Figure 7.20
Direct Form 2
realizations for first-
and second-order
systems:
(a) first-order: $m = 1$;
(b) second-order
system: $m = 2$.

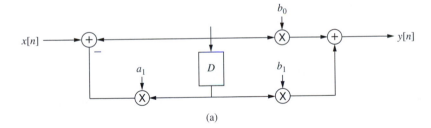

(a)

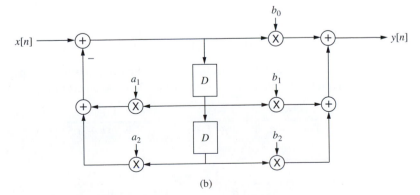

(b)

EXAMPLE 7.6   Some Realizations for a Third-Order IIR Filter

Consider the transfer function

$$H(z) = \frac{z^3 + z^2}{z^3 + 1.5z^2 + z + 0.25} = \frac{z^2(z+1)}{(z + 0.5 + 0.5j)(z + 0.5 - 0.5j)(z + 0.5)} \tag{85}$$

a) Draw the Direct Form 1 realization.
b) Draw the Direct Form 2 realization.
c) Find a realization composed of two systems in series.
d) Find a realization composed of two systems in parallel.

Solution

a)–b) We can draw the Direct Form realizations by inspection if we write $H(z)$ in the form

$$H(z) = \frac{1 + z^{-1}}{1 + 1.5z^{-1} + z^{-2} + 0.25z^{-3}} \tag{86}$$

Figures 7.21(a) and 7.21(b) follow directly from the preceding equation and Figures 7.17 and 7.19.
c) We can write $H(z)$ as

$$H(z) = \frac{z^2(z+1)}{(z^2 + z + 0.5)(z + 0.5)} = H_1(z)H_2(z) \tag{87}$$

where

$$H_1(z) = \frac{z(z+1)}{z^2 + z + 0.5} = \frac{1 + z^{-1}}{1 + z^{-1} + 0.5z^{-2}} \tag{88}$$

$$H_2(z) = \frac{z}{z + 0.5} = \frac{1}{1 + 0.5z^{-1}} \tag{89}$$

The series realization corresponding to the preceding equations is shown in Figure 7.21(c), where we have used Direct Form 2 for $H_1(z)$ and $H_2(z)$.
d) We can use the techniques of Chapter 6DT to write $H(z)$ in partial fraction form. The result is

$$H(z) = \frac{5}{1 + 0.5z^{-1}} + \frac{-3 + j}{1 - j + z^{-1}} + \frac{-3 - j}{1 + j + z^{-1}} \tag{90}$$

This expression represents $H(z)$ as a parallel connection of three first-order systems. The systems corresponding to the last two terms contain complex multipliers. If we combine the last two terms in Eq. (90)

$$\frac{-3 + j}{1 - j + z^{-1}} + \frac{-3 - j}{1 + j + z^{-1}} = -\frac{4 + 3z^{-1}}{1 + z^{-1} + 0.5z^{-2}} \tag{91}$$

we can write $H(z)$ as

$$H(z) = \frac{5}{1 + 0.5z^{-1}} - \frac{4 + 3z^{-1}}{1 + z^{-1} + 0.5z^{-2}} \tag{92}$$

This expression can be implemented by the parallel connection of first- and second-order real systems, as shown in Figure 7.21(d).

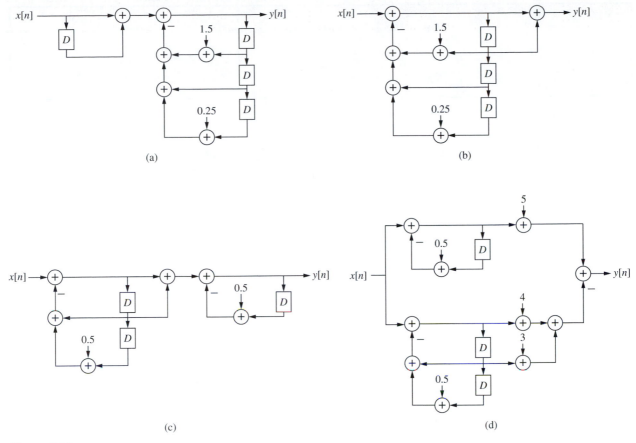

Figure 7.21
Possible realizations for third-order system example: (a) Direct Form 1; (b) Direct Form 2; (c) series; (d) parallel.

IIR filters are typically designed using analog filter counterparts called analog *prototypes*. We now turn to two important IIR filter design techniques called *impulse invariant design* and *bilinear transform design*. Each of these techniques comes with a stability guarantee: If the analog prototype filter is stable, then the digital filter is also stable.

Impulse-Invariant IIR Filter Design

The goal of impulse invariant design is to make the digital filter impulse $h[n]$ proportional to samples of the impulse response $h(t)$ of some analog filter prototype.

Impulse-Invariant Design Goal

$$h[n] = T_s h(nT_s) \tag{93}$$

It is possible to design an IIR filter so that Eq. (93) is satisfied exactly. Because $h(t) = \mathcal{L}^{-1}\{\mathbf{H}(s)\}$ and $h(nT_s) = \mathcal{L}^{-1}\{\mathbf{H}(s)\}|_{t=nT_s}$, we can write Eq. (93) as

$$h[n] = T_s \mathcal{L}^{-1}\{\mathbf{H}(s)\}\Big|_{t=nT_s} \tag{94}$$

The z-transform of the preceding equation give us $H(z)$.

Impulse-Invariant Design Equation

$$H(z) = T_s \mathcal{Z} \left\{ \mathcal{L}^{-1}\{H(s)\} \Big|_{t=nT_s} \right\} \qquad (95)$$

EXAMPLE 7.7   First-Order Low-Pass Filter by Impulse-Invariant Design

Consider the use of impulse-invariant design based on an analog low-pass filter prototype having the transfer function

$$\mathbf{H}(s) = \frac{1}{1 + s\tau} \qquad (96)$$

a) Find the transfer function $H(z)$ of the corresponding impulse-invariant digital filter.
b) Show that the digital filter is stable for all $\tau > 0$.

Solution

a) We can find $H(z)$ using the steps indicated in Eq. (95). The impulse response of the analog filter is

$$h(t) = \mathcal{L}^{-1}\left\{ \frac{1}{1 + s\tau} \right\} = \frac{1}{\tau} e^{-\frac{t}{\tau}} u(t) \qquad (97)$$

and the impulse response of the digital filter is

$$h[n] = T_s h(nT_s) = \frac{T_s}{\tau} \left(e^{-\frac{T_s}{\tau}} \right)^n u[n] \qquad (98)$$

The z-transform of $h[n]$ gives us

$$H(z) = \mathcal{Z}\{h[n]\} = \frac{T_s}{\tau} \frac{z}{z - e^{-\frac{T_s}{\tau}}} = \frac{T_s}{\tau} \frac{1}{1 - e^{-\frac{T_s}{\tau}} z^{-1}} \qquad (99)$$

Figure 7.22 shows the Direct Form 2 realization.

b) The digital filter has a single simple pole at $z = e^{-T_s/\tau}$. This pole is inside the unit circle for all $\tau > 0$. Therefore, The filter is stable. A more direct proof is to note that $h[n]$ as given by Eq. (98) is absolutely summable because $|e^{-\frac{T_s}{\tau}}| < 1$.

Figure 7.22
Realization of
impulse-invariant
low-pass filter.

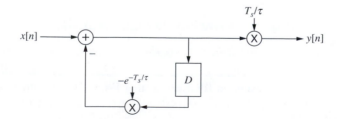

The preceding example demonstrates the stability guarantee of impulse-invariant filters. A general proof of the stability of IIR filters can be based on partial fraction expansions. Problem 7.5.19 develops the proof for systems having transfer functions $\mathbf{H}(s)$ that are proper algebraic fractions containing only simple poles.

A theoretical analysis of the frequency response characteristic of an impulse-invariant filter can be performed using a derivation similar to that in Section 7.2. The following expression comes from Eq. (8), where $X(e^{j2\pi f})$ is replaced with $H(e^{j2\pi f})$ and $X(F)$ is replaced with $T_s H(f F_s)$.

$$H(e^{j2\pi f}) = \sum_{m=-\infty}^{\infty} H((f-m)F_s)$$

$$= H(f F_s) + \sum_{\substack{m=-\infty \\ m\neq 0}}^{\infty} H((f-m)F_s) \tag{100}$$

This expression show us that the frequency response characteristic $H(e^{j2\pi f})$ of an impulse-invariant filter is composed of periodic repetitions of the analog filter frequency response $H(f F_s)$. Notice that the term $H(f F_s)$ (the $m = 0$ term in the sum) satisfies the matched processor goal (15) exactly. The remaining terms in Eq. (100) introduce aliasing. Thus, in impulse-invariant design, the difference between $H(e^{j2\pi f})$ and $H(f F_s)$ is caused by aliasing. When we use the impulse-invariant method to design a low-pass filter, we can decrease the aliasing error by increasing the sampling rate F_s.

EXAMPLE 7.8 The Frequency Response Characteristic of the Digital Filter of Example 7.7

Find and plot the amplitude characteristics of the digital filter of Example 7.7 and the matched processor $H_m(e^{j2\pi f}) = H(f F_s)$ of Eq. (15) .

Solution

$H(e^{j2\pi f})$ is obtained from $H(z)$, Eq. (100), by setting $z = e^{j2\pi f}$. The result is

$$H(e^{j2\pi f}) = \mathcal{DTFT}\{h[n]\} = \frac{T_s}{\tau} \frac{e^{j2\pi f}}{e^{j2\pi f} - e^{-\frac{T_s}{\tau}}} = \frac{T_s}{\tau} \frac{1}{1 - e^{-\frac{T_s}{\tau}}e^{-j2\pi f}} \tag{101}$$

$H(F)$ is obtained from $\mathbf{H}(s)$, Eq. (96), by setting $s = j2\pi F$. The result is

$$H(F) = \mathbf{H}(j2\pi F) = \frac{1}{1 + j2\pi F\tau} = \frac{1}{1 + jF/F_c} \tag{102}$$

where $F_c = 1/2\pi\tau$ is the analog filter's half-power frequency. It follows that

$$H(f F_s) = \frac{1}{1 + jf F_s/F_c} \tag{103}$$

Plots of the amplitude characteristics of $H(e^{j2\pi f})$ and $H(f F_s)$ are shown in Figure 7.23 for $F_s = 20F_c$ and $F_s = 100F_c$. The plots illustrate the property that aliasing error decreases as F_s increases.

Figure 7.23
Amplitude
characteristics
resulting from
impulse invariant
design: (a) $F_s = 20F_c$;
(b) $F_s = 100F_c$.

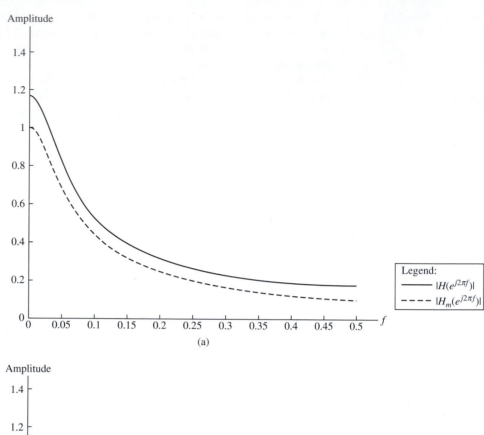

(a)

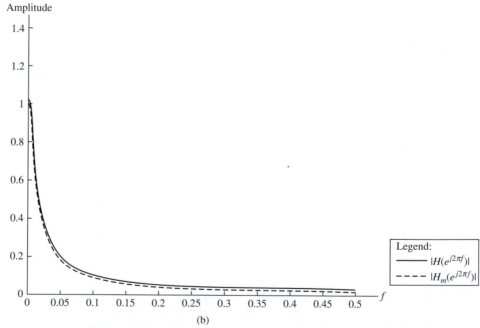

(b)

The notion of impulse-invariant design can be extended to include invariance to other
signals such as steps and ramps. This extension is developed in Appendix 7.A.

Figure 7.24
Frequency warping
from F to f: The
entire CT frequency
axis is mapped onto
the DT frequency
interval
$-0.5 \le f < 0.5$ The
mapping $F \to f$
repeats periodically
with period 1; only
the interval
$-0.5 \le f < 0.5$ is
shown.

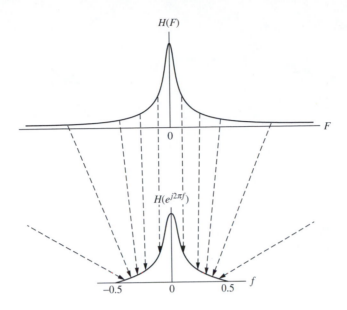

The Bilinear Transform

We saw in the previous section that the impulse-invariant design technique produces a digital filter frequency response characteristic that is an aliased version of the analog filter frequency response characteristic. The bilinear transform design technique avoids aliasing error by mapping the entire analog frequency axis $-\infty < F < \infty$ onto the digital frequency interval $-0.5 \le f < 0.5$, as illustrated in Figure 7.24.

The mapping from F to f performed by the bilinear transform is shown geometrically in Figure 7.25. In the figure, a unit circle has been drawn above the point $F = 0$ on the frequency line $-\infty < F < \infty$. The apex of the circle (point D) is a distance F_s/π above this frequency line. A line segment, DA, connects the apex of the circle to the frequency F (point A) on the frequency axis. This line segment intersects the unit circle at point B. A second line segment CB connects the center of the circle to point B and forms an angle $2\pi f$ with the vertical. The bilinear transform maps F to f. As F varies from $-\infty$ to ∞, $f = f(F)$ varies from -0.5 to $+0.5$.

Figure 7.25
Geometry of bilinear
transform map
$F \to f$.

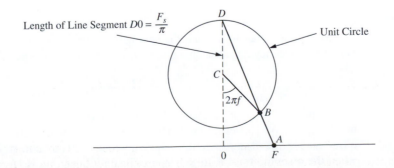

Figure 7.26
Frequency warping
between analog and
digital filter amplitude
characteristics: A
similar figure applies
for phase
characteristics.

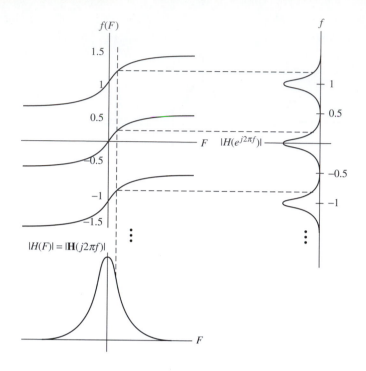

Trigonometry yields the following formula.

Frequency Mapping of the Bilinear Transform

$$F = \frac{F_s}{\pi} \tan(\pi f) \tag{104a}$$

$$f(F) = \frac{1}{\pi} \arctan\left(\pi \frac{F}{F_s}\right) \tag{104b}$$

The term $f(F)$ is plotted in Figure 7.26. An inspection of Figure 7.26 or Eq. (104) reveals that the transformation $F \rightarrow f$ is approximately linear for small F

$$f \approx \frac{F}{F_s} \qquad \text{for } F \ll \frac{1}{\pi} F_s \tag{105}$$

and becomes increasingly nonlinear as F increases beyond $1/\pi F_s$.

The bilinear transform maps the analog frequency response characteristic $H(F)$ to the DT filter frequency response characteristic $H(e^{j2\pi f})$ by the equation

$$H(e^{j2\pi f}) = \mathbf{H}(j2\pi F)\Big|_{F = \frac{F_s}{\pi} \tan(\pi f)} \tag{106}$$

The mapping (106) is illustrated in Figure 7.26. You can see from this figure that there is no aliasing error. However, another form of error, known as *frequency warping* error, is introduced by the nonlinear compression of F onto f. The frequency warping error is small when the mapping from F to f is approximately linear but is large when the mapping is

7.5 Filter Design 457

highly nonlinear. Frequency warping generally decreases as F_s increases because the linear approximation (105) holds for $F \ll \frac{1}{\pi}F_s$.

We can extend the mapping Eq. (106) to the s- and z-planes by writing Eq. (104) as

$$2\pi F = 2F_s \frac{\sin(\pi f)}{\cos(\pi f)} = \frac{\frac{1}{2j}(e^{j\pi f} - e^{-j\pi f})}{\frac{1}{2}(e^{j\pi f} + e^{-j\pi f})} 2F_s = \frac{1}{j}\left(\frac{1 - e^{-j2\pi f}}{1 + e^{-j2\pi f}}\right) 2F_s \qquad (107)$$

If we replace $j2\pi F$ by s and $e^{j2\pi f}$ by z as we did in Chapters 6CT and 6DT, we obtain the bilinear transform.

Bilinear Transform

$$s = \frac{1 - z^{-1}}{1 + z^{-1}} 2F_s \qquad (108)$$

The digital filter transfer function $H(z)$ is given by

Bilinear Transform Filter Design Equation

$$H(z) = \mathbf{H}(s)\Big|_{s=\frac{1-z^{-1}}{1+z^{-1}}2F_s} \qquad (109)$$

It can be shown that Eq. (108) maps the left half of the s-plane onto the inside of the unit circle in the z-plane. Thus it maps all the left-half plane poles of $\mathbf{H}(s)$ to poles of $H(z)$ that are inside the unit circle of the z-plane. This means that the digital filter defined by $H(z)$ in Eq. (109) is stable if the analog prototype filter is stable.

EXAMPLE 7.9 Bilinear Transform Design with a Fixed $\mathbf{H}(s)$

Consider the analog filter transfer function

$$\mathbf{H}(s) = \frac{1}{1 + s\tau} \qquad (110)$$

where τ is a fixed number. Recall that the half-power frequency of this filter is $F_c = 1/(2\pi\tau)$.

a) Use the bilinear transform to find the transfer function $H(z)$ of a corresponding digital filter.

b) Plot the amplitude characteristics of both the digital filter and the matched processor $H_m(e^{j2\pi f}) = H(fF_s)$ of Eq. (15).

c) Show that the DT filter is stable for all $\tau > 0$.

Solution

a) Equation (109) yields

$$H(z) = \frac{1}{1 + \left(\frac{1-z^{-1}}{1+z^{-1}}\right)2F_s\tau} = \frac{z+1}{z(1 + 2F_s\tau) + 1 - 2F_s\tau} \qquad (111)$$

Figure 7.27
Amplitude
characteristics
resulting from
bilinear transform
design: (a) $F_s = 20F_c$;
(b) $F_s = 100F_c$.

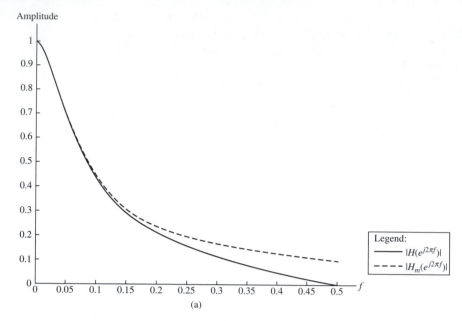

(a)

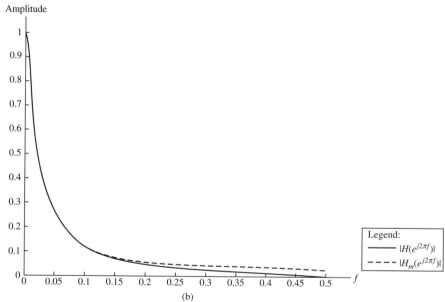

(b)

b) The amplitude characteristics $|H(e^{j2\pi f})|$ and $|H(f F_s)|$ are shown in Figure 7.27(a) for $F_s/F_c = 20$ and Figure 7.27(b) for $F_s/F_c = 100$, respectively. Notice that the $|H(e^{j2\pi f})|$ becomes a better approximation to $|H(f F_s)|$ as F_s is increased.

c) The bilinear transform has mapped the analog filter pole at $s = -1/\tau$ to the point

$$z = \frac{1 - 2F_s\tau}{1 + 2F_s\tau} \tag{112}$$

This pole lies inside the unit circle in the z-plane for all $\tau > 0$. Therefore, the DT filter is stable.

Figure 7.28
Design spec mapping: The bilinear transform maps DT filter design specs to CT filter design specs and vice versa.

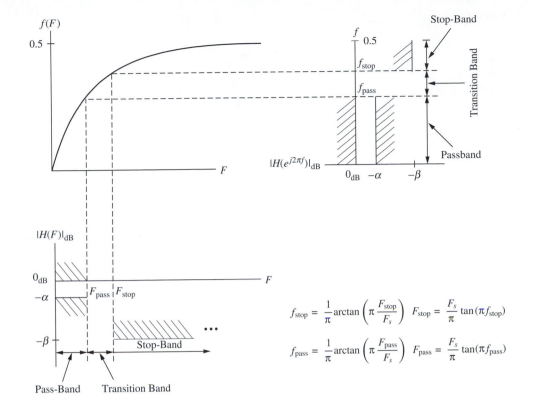

$$f_{\text{stop}} = \frac{1}{\pi}\arctan\left(\pi\frac{F_{\text{stop}}}{F_s}\right) \qquad F_{\text{stop}} = \frac{F_s}{\pi}\tan(\pi f_{\text{stop}})$$

$$f_{\text{pass}} = \frac{1}{\pi}\arctan\left(\pi\frac{F_{\text{pass}}}{F_s}\right) \qquad F_{\text{pass}} = \frac{F_s}{\pi}\tan(\pi f_{\text{pass}})$$

In the preceding example, the prototype transfer function $\mathbf{H}(s)$ was completely specified in the statement of the problem. This specification allowed us to obtain $H(z)$ in one step. More often, the statement of the problem includes a digital filter design spec, not a completely specified $\mathbf{H}(s)$. We then work with an analog prototype with variable parameters that we can chose to meet the design spec. We use the bilinear transform to map the digital filter design spec to an analog filter spec, as illustrated in Figure 7.28. The frequencies F_{pass}, f_{pass}, F_{stop}, and f_{stop} are the analog and digital filters' passband and stop-band *edge frequencies*, defined by

$$|H_{\text{B}n}(e^{j2\pi f_{\text{pass}}})|_{dB} = |H_{\text{B}n}(F_{\text{pass}})|_{dB} = -\alpha \tag{113}$$

$$|H_{\text{B}n}(e^{j2\pi f_{\text{stop}}})|_{dB} = |H_{\text{B}n}(F_{\text{stop}})|_{dB} = -\beta \tag{114}$$

If effect, the analog and the digital filters are designed simultaneously. This design procedure is illustrated in Examples 7.10 and 7.11 for two commonly used analog prototypes.

Let us now turn our attention to the two analog prototypes considered in the examples. Each prototype is referred to as an *approximation* because it provides a causal approximation to an ideal low-pass filter.

The Butterworth Approximation

Amplitude Characteristic. The low-pass *nth-order Butterworth filter* is defined by its amplitude characteristic.

Figure 7.29
Amplitude
characteristic of
nth-order
Butterworth filter for
$n = 1, 2, 3,$ and 4.

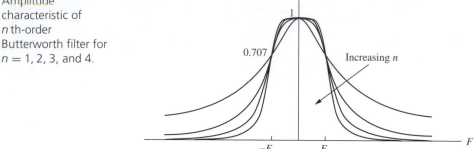

Amplitude Characteristic of nth-Order Low-Pass Butterworth Filter

$$|H_{Bn}(F)| = |\mathbf{H}_{Bn}(j2\pi F)| = \frac{1}{\sqrt{1 + \left(\frac{F}{F_c}\right)^{2n}}} \tag{115}$$

In Eq. (115), F_c is the *half-power frequency* of the filter in hertz. An important property of the Butterworth amplitude characteristic is that it is *maximally flat*. This term means that the first $2n - 1$ derivatives of $|H_{Bn}(F)|$ equal 0 at $F = 0$.

$$\left. \frac{d^i}{dF^i} |H_{Bn}(F)| \right|_{F=0} = 0 \qquad i = 1, 2, \ldots, 2n - 1 \tag{116}$$

Figure 7.29 shows plots of $|H_{Bn}(F)|$ for $n = 1, 2, 3,$ and 4. The plots illustrate the following properties of Butterworth filter amplitude characteristics.

Properties of Butterworth Filter Amplitude Characteristics

- $|H_{Bn}(F)|$ is maximally flat at $F = 0$.
- For every n, $|H_{Bn}(F)|$ decreases monotonically towards 0 as $|F|$ increases.
- The DC gain is unity.

$$|H_{Bn}(F)| = 1 \qquad \text{for } F = 0 \tag{117}$$

- The gain at the half-power frequency $F = F_c$ is

$$|H_{Bn}(F_c)| = \frac{1}{\sqrt{2}} \tag{118}$$

- As n increases, $|H_{Bn}(F)|$ becomes an increasingly better approximation to the amplitude characteristic of an ideal low-pass filter having bandwidth $W = F_c$.

Transfer Function.   We can derive the transfer function of a filter from the filter's amplitude characteristic. To find $\mathbf{H}_{Bn}(s)$, we square both sides of Eq. (115) and replace $j2\pi F$

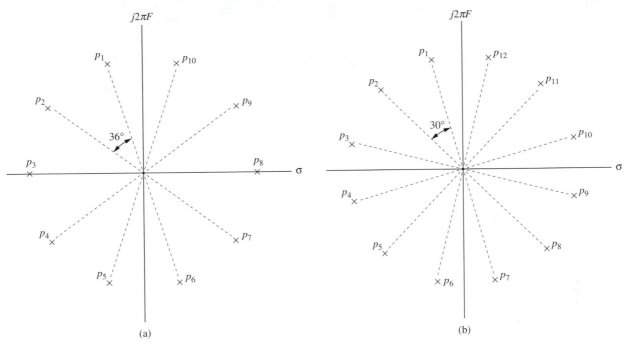

Figure 7.30
Pole-zero plots of $|H_{Bn}(s)|^2$: (a) $n = 5$; (b) $n = 6$.

with s. The result is

$$|\mathbf{H}_{Bn}(s)|^2 = \mathbf{H}_{Bn}(s)\mathbf{H}^*_{Bln}(s) = \frac{\Omega_{lc}^{2n}}{\Omega_c^{2n} + (-s^2)^n} \qquad (119)$$

where Ω_c is the half-power frequency in radians per second

$$\Omega_c = 2\pi F_c \qquad (120)$$

The poles of $|\mathbf{H}_{Bn}(s)|^2$ are the $2n$ roots to the equation

$$\Omega_c^{2n} + (-s^2)^n = 0 \qquad (121)$$

These roots are

$$p_i = \Omega_c e^{j\pi\left(\frac{1}{2} + \frac{2i-1}{2n}\right)} \quad i = 1, 2, \ldots, 2n \qquad (122)$$

Figure 7.30 shows the pole-zero diagram of $|\mathbf{H}_{Bn}(s)|^2$ for $n = 5$ and 6. The poles are equally spaced around a circle in the s-plane having radius Ω_c in the s-plane.

To obtain $\mathbf{H}_{Bn}(s)$, we need to determine which n of the $2n$ poles of $|\mathbf{H}_{Bn}(s)|^2$ belong to $\mathbf{H}_{Bn}(s)$ and which n belong to $\mathbf{H}^*_{Bn}(s)$. This step is called *spectrum factorization*: We can factor $|\mathbf{H}_{Bn}(s)|^2$ by recalling from Chapter 6CT that the poles of every stable, causal filter lie in the left half of the s-plane. Therefore, we chose the left-half poles of $|\mathbf{H}_{Bn}(s)|^2$ for $\mathbf{H}_{Bn}(s)$ and the right-half poles for $\mathbf{H}^*_{Bln}(s)$. The result follows.

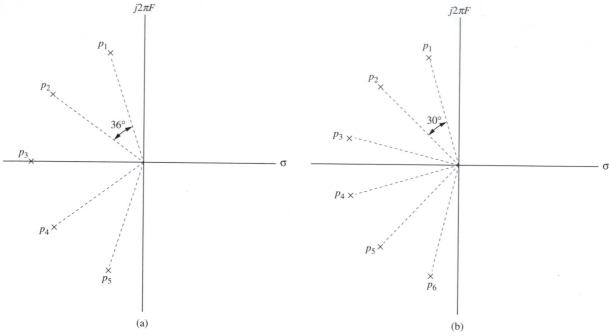

Figure 7.31
Pole-zero diagrams of Butterworth filters: (a) $n = 5$; (b) $n = 6$.

Transfer Function of nth-Order Low-Pass Butterworth Filter in Factored Form

$$\mathbf{H}_{\mathrm{B}n}(s) = \frac{\Omega_c^n}{(s - p_1)(s - p_2) \cdots (s - p_n)} \tag{123}$$

where the p_i are given by Eq. (122) for $i = 1, 2, \ldots, n$.

Figure 7.31 shows the pole-zero diagrams of a Butterworth filter for $n = 5$ and 6. Figure 7.32 illustrates the relation between $|\mathbf{H}_{\mathrm{B}n}(s)|$ and the amplitude characteristic $|\mathbf{H}_{\mathrm{B}n}(j2\pi F)|$, where $n = 5$. If we expand the factors in the denominator of Eq. (123), we obtain

---

Transfer function of nth-Order Low-Pass Butterworth Filter in Expanded Form

$$\mathbf{H}_{\mathrm{B}n}(s) = \frac{\Omega_c^n}{s^n + a_{n-1}s^{n-1} + a_{n-2}s^{n-2} + \cdots + a_1 s + a_0} \tag{124}$$

---

where $a_0 = \Omega_c^n$ and the values $a_1, \ldots, a_{n-1}$ are obtained from the algebraic expansion from Eqs. (123) to (124).

The work involved in the hand calculation of Eq. (122), (123), and (124) is routine, but time-consuming. You can use MATLAB functions to eliminate the need for hand calculation. This software is introduced in the problems at the end of this chapter.

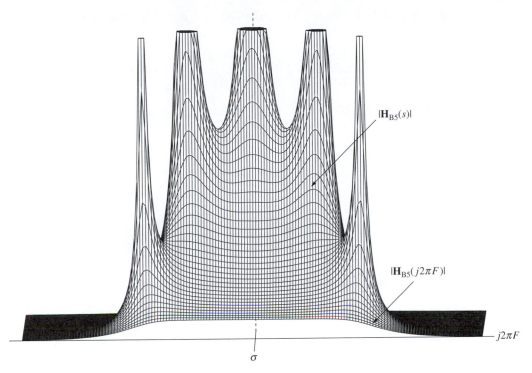

Figure 7.32
Butterworth amplitude characteristic $|H_{B5}(2\pi F)|$, plotted as intersection of $|H_{B5}(s)|$ and $\sigma = 0$ Planes. The poles of a Butterworth Filter are equally spaced on a semicircle circle, resulting in a maximally flat amplitude characteristic.

Order Specification.   When we design a digital filter using a Butterworth analog filter, we use the design specs to find the analog filter order n. A formula for the minimum order for which a Butterworth filter satisfies the normalized analog design specs of Figure 7.28 can be obtained by substituting Eq. (115) into

$$|H_{Bn}(F_{\text{pass}})|_{dB} = -\alpha \tag{125}$$

and

$$|H_{Bn}(F_{\text{stop}})|_{dB} = -\beta \tag{126}$$

and solving for n. The result is

$$n = \frac{\log_{10}(10^{0.1\alpha} - 1) - \log_{10}(10^{0.1\beta} - 1)}{2(\log_{10} F_{\text{pass}} - \log_{10} F_{\text{stop}})} \tag{127}$$

The right-and side of Eq. (127) is not necessarily an integer. Since n must be an integer, we set

$$n^* = \text{ceil}\left\{\frac{\log_{10}(10^{0.1\alpha} - 1) - \log_{10}(10^{0.1\beta} - 1)}{2(\log_{10} F_{\text{pass}} - \log_{10} F_{\text{stop}})}\right\} \tag{128}$$

where ceil$[\lambda]$ is MATLAB notation for the smallest integer greater than or equal to λ. The notation n^* is a reminder that the value of n given by Eq. (128) can be larger than the theoretical noninteger value needed to meet the specs.

One or both of the design specs are exceeded if the right-hand side of Eq. (128) is not an integer. We can set

$$|H_{Bn}(F_{\text{pass}})|_{dB} = \left|\frac{1}{\sqrt{1+\left(\frac{F_{\text{pass}}}{F_c}\right)^{2n^*}}}\right|_{dB} = -\alpha \tag{129}$$

to obtain the following half-power frequency of the filter that meets the passband spec Eq. (125) but exceeds the stop-band spec Eq. (126).

$$F_c = \frac{F_{\text{pass}}}{(10^{0.1\alpha}-1)^{\frac{1}{2n^*}}} \tag{130}$$

Alternatively, we can set

$$|H_{Bn}(F_{\text{stop}})|_{dB} = \left|\frac{1}{\sqrt{1+\left(\frac{F_{\text{stop}}}{F_c}\right)^{2n^*}}}\right|_{dB} = -\beta \tag{131}$$

we obtain the following half-power frequency that meets the stop-band spec Eq. (126) but exceeds the pass band spec Eq. (125).

$$F_c = \frac{F_{\text{stop}}}{(10^{0.1\beta}-1)^{\frac{1}{2n^*}}} \tag{132}$$

Butterworth filter design is illustrated in Example 7.10. You can skip directly to this example without loss in continuity. The following section introduces a second analog prototype used in digital filter design.

Chebyshev Approximation

Amplitude Characteristic.   The low-pass *nth-order Chebyshev filter* is defined by its amplitude characteristic.

Amplitude Characteristic of *n*th-Order Low-Pass Chebyshev Filter

$$|H_{Cn}(F)| = \sqrt{\frac{1}{1+\epsilon^2 T_n^2\left(\frac{F}{F_c}\right)}} \tag{133}$$

In Eq. (133), $T_n(\xi)$ is the *nth-order Chebyshev polynomial*, defined by

$$T_n(\xi) = \begin{cases} \cos(n\cos^{-1}\xi) & |\xi| \leq 1 \\ \cosh(n\cosh^{-1}\xi) & |\xi| > 1 \end{cases} \tag{134}$$

The Chebyshev polynomials vary between -1 and $+1$ for $|\xi| \leq 1$, where

$$|T_n(0)| = \begin{cases} 1 & \text{for } n = 0, 2, 4, \ldots \\ 0 & \text{for } n = 1, 3, 5, \ldots \end{cases} \tag{135}$$

and $|T_n(1)| = 1$. Illustrative plots of $T_n(\xi)$ are shown in Figure 7.33 for $n = 2, 3, 4,$ and 5. Notice that $|T_n(\xi)|$ increases monotonically for $|\xi| > 1$ and that this increase becomes more rapid as n increases.

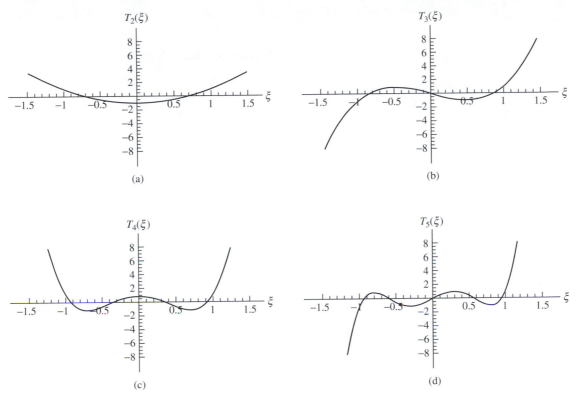

Figure 7.33
Chebyshev polynomials $T_n(\xi)$: (a) $n = 2$; (b) $n = 3$; (c) $n = 4$; (d) $n = 5$.

It can be shown from Eq. (134) that

$$T_n(\xi) = 2\xi T_{n-1}(\xi) - T_{n-2}(\xi) \qquad n = 3, 4, \ldots \tag{136}$$

where $T_0(\xi) = 1$ and $T_1(\xi) = \xi$. We can use this recurrence relation to express the $T_n(\xi)$ as polynomials. The first nine Chebyshev polynomials are listed in Table 7.1.

Figure 7.34 shows the amplitude characteristics of Chebyshev filters having orders $n = 2, 3, 4$, and 5. Notice that the properties of the Chebyshev amplitude characteristics in Figure 7.34 correspond to the properties of the Chebyshev polynomials in Figure 7.33. The plots illustrates the following properties of Chebyshev filters.

Table 7.1
Chebyshev
Polynomials

n	$T_n(\xi)$
0	1
1	ξ
2	$-1 + 2\xi^2$
3	$-3\xi + 4\xi^3$
4	$1 - 8\xi^2 + 8\xi^4$
5	$5\xi - 20\xi^3 + 16\xi^5$
6	$-1 + 18\xi^2 - 48\xi^4 + 32\xi^6$
7	$-7\xi + 56\xi^3 - 112\xi^5 + 64\xi^7$
8	$1 - 32\xi^2 + 160\xi^4 - 256\xi^6 + 128\xi^8$

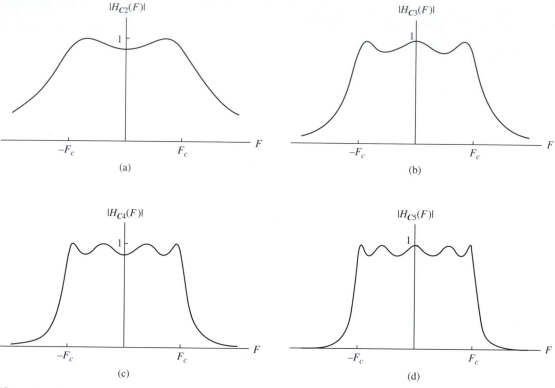

Figure 7.34
Amplitude characteristics of chebyshev filters: (a) $n = 2$; (b) $n = 3$; (c) $n = 4$; (d) $n = 5$.

Properties of Chebyshev Filter Amplitude Characteristics

$|H_{Cn}(F)|$ is equiripple in the passband:

$$\sqrt{\frac{1}{1+\epsilon^2}} \leq |H_{Cn}(F)| \leq 1 \qquad \text{for } 0 \leq F \leq F_c \tag{137}$$

DC gain

$$|H_{Cn}(0)| = \begin{cases} \sqrt{\frac{1}{1+\epsilon^2}} & \text{for } n = 0, 2, 4, \ldots \\ 1 & \text{for } n = 1, 3, 5, \ldots \end{cases} \tag{138}$$

The gain at the *passband corner frequency* $F = F_c$ is

$$|H_{Cn}(F_c)| = \sqrt{\frac{1}{1+\epsilon^2}} \tag{139}$$

F_c is the highest frequency for which $|H_{Cn}(F)| = \sqrt{1/(1+\epsilon^2)}$.
$|H_{Cn}(F)|$ decreases monotonically toward 0 in the stop band.
$|H_{Cn}(F)|$ decreases faster in the stop band as the filter order n is increased.

The value of ϵ^2 satisfying the normalized analog design specs of Figure 7.28 can be found with the use of Eq. (139).

$$20 \log_{10} \sqrt{\frac{1}{1+\epsilon^2}} = -\alpha \tag{140}$$

which yields

$$\epsilon^2 = 10^{0.1\alpha} - 1 \tag{141}$$

Transfer Function.   We can obtain $H_{Cn}(s)$ using spectrum factorization. We replace F by $s/2\pi j$ and square both sides of Eq. (133) to obtain

$$|\mathbf{H}_{cln}(s)|^2 = \mathbf{H}_{cln}(s)\mathbf{H}^*_{cln}(s) = \frac{1}{1+\epsilon^2 T_n^2\left(\frac{s}{2\pi j F_c}\right)} \tag{142}$$

To perform spectrum factorization, we express $T_n(s/2\pi j F_c)$ as a polynomial and solve for the poles of $|\mathbf{H}_{cln}(s)|^2$. As with the Butterworth filter, we associate the left-hand plane poles of $|\mathbf{H}_{cln}(s)|^2$ with $\mathbf{H}_{cn}(s)$ and the right-half plane poles with $\mathbf{H}^*_{cn}(s)$. When we write $\mathbf{H}_{cn}(s)$ in expanded form, the result is

---

Transfer Function of nth-Order Low-Pass Chebyshev Filter

$$\mathbf{H}_{Cn}(s) = \frac{b}{s^n + a_{n-1}s^{n-1} + a_{n-2}s^{n-2} + \cdots + a_1 s + a_0} \tag{143}$$

---

where

$$b = \begin{cases} \frac{a_0}{1+\epsilon^2} & \text{for } n = 0, 2, 4, \ldots \\ a_0 & \text{for } n = 1, 3, 5, \ldots \end{cases}$$

It can be shown that the poles of a Chebyshev filter are equally spaced on a semiellipse. The eccentricity of the ellipse brings the poles closer to the $j2\pi F$ axis. The result is equal amplitude ripples in the passband, and a steeper falloff in the stopband than for a Butterworth filter having the same order. Figure 7.35 illustrates the relation between $|\mathbf{H}_{Cn}(s)|$ and the amplitude characteristic $|\mathbf{H}_{Cn}(2\pi F)|$ where $n = 5$. As with the Butterworth filter, hand calculation of the Chebyshev filter poles and coefficients b, $a_0, a_1, \ldots, a_{n-1}$ is a straightforward, but tedious task. MATLAB functions, introduced in the problems, eliminate the need for hand calculation.[3]

Order Specification.   We can find the minimum order for which the stop-band design spec shown in Figure 7.28 is satisfied by equating $|H_{Cn}(F_{\text{stop}})|_{dB}$ to $-\beta$.

$$|H_{Cn}(F_{\text{stop}})|_{dB} = \sqrt{\frac{1}{1+\epsilon^2[\cosh(n \cosh^{-1}\{F_{\text{stop}}/F_c\})]^2}}\Bigg|_{dB} = -\beta \tag{144}$$

and solving for n. The result is

$$n = \frac{\cosh^{-1}\{\sqrt{(10^{0.1\beta}-1)/\epsilon^2}\}}{\cosh^{-1}\{F_{\text{stop}}/F_c\}} \tag{145}$$

---

[3] Analytical expressions for the poles are given in [Mitra, Sanjit K., 1998].

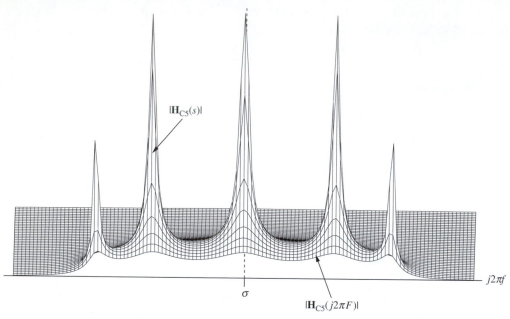

Figure 7.35
Chebyshev filter amplitude characteristic $|H_{C5}(2\pi F)|$, plotted as intersection of $|H_{C5}(s)|$ and $\sigma = 0$ plane: The poles of a Chebyshev filter are placed on an ellipse, whose eccentricity brings the poles closer to the $j2\pi F$ axis than the poles of a Butterworth filter, resulting in equiripples in the passband and a sharper falloff in the stop-band.

The right-and side of Eq. (145) is not necessarily an integer. Since n must be an integer, we set

$$n^* = \text{ceil}\left\{\frac{\cosh^{-1}\{\sqrt{(10^{0.1\beta} - 1)/\epsilon^2}\}}{\cosh^{-1}(F_{\text{stop}}/F_c)}\right\} \tag{146}$$

Either the passband or the stop-band design spec is exceeded when the quantity in the brackets is not an integer. If use the value for ϵ^2 determined by the passband spec (141), we obtain a filter that meets the passband spec and exceeds the stop-band spec. Alternatively, we can solve Eq. (146) for ϵ^2 to obtain

$$\epsilon^2 = \frac{10^{0.1\beta} - 1}{[\cosh(n^* \cosh^{-1}(F_{\text{stop}}/F_c))]^2} \tag{147}$$

If we use the preceding value for ϵ^2 and the value for β given in the design spec, then we obtain a filter that meets the stop-band spec with passband ripple that is less than that given in the design spec.

Chebyshev filter design is illustrated in Example 7.10.

Low-Pass Filter Design Using the Bilinear Transform

Now that we have introduced the Butterworth and Chebyshev prototype filters, we can commence with digital filter design using the bilinear transform. We can start with design specs for either the digital filter $H(e^{j2\pi f})$ and/or the corresponding analog filter $H(F)$ (Figure 7.1). For a low-pass DT filter, the design can then be completed using the steps shown in Figure 7.36. In step (a) of the flow graph, we use Eq. (104) to convert the digital filter design specs into corresponding analog filter design specs. (Skip step (a) if you are

Figure 7.36
Flow diagram 1.

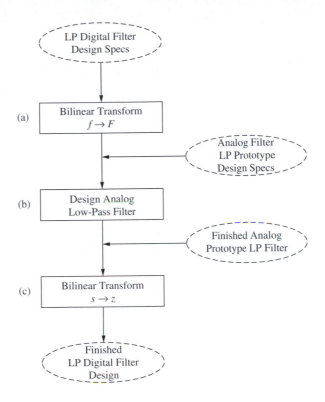

already given the analog filter design specs.) Step (b) includes the use of Eq. (128) or Eq. (146) to determine the order n. Step (c) is to use the bilinear transform Eq. (108) to convert the finished analog filter to the finished digital filter.

The following examples illustrate the procedure.

EXAMPLE 7.10   Low-Pass Butterworth Filter Design

Use the bilinear transform to design a digital Butterworth filter that satisfies the following specs:

$$\text{Passband: } -1 \le |H(e^{j2\pi f})|_{dB} \le 0 \qquad \text{for } 0 \le f \le 0.15 \tag{148}$$

$$\text{Transition band: } 0.15 < f < 0.25 \tag{149}$$

$$\text{Stop-band: } |H(e^{j2\pi f})|_{dB} \le -30 \qquad \text{for } 0.25 \le f < 0.5 \tag{150}$$

The sampling rate is $F_s = 1\text{kHz}$.

Solution

Flow diagram 1 indicates the steps involved in the design. Here we are given the specs for a digital filter. Therefore, the first step is to use the bilinear transform to obtain the corresponding analog filter design spec (see Figure 7.28 and its caption). The result is

$$\text{Passband: } -1 \le |H(F)|_{dB} \le 0 \qquad \text{for } 0 \le F \le 0.1622 \text{ kHz} \tag{151}$$

$$\text{Transition band: } 0.1622 \text{ kHz} < F < 0.3183 \text{ kHz} \tag{152}$$

$$\text{Stop-band: } |H(e^{j2\pi f})|_{dB} \le -30 \qquad \text{for } 0.3183 \text{ kHz} \le F \tag{153}$$

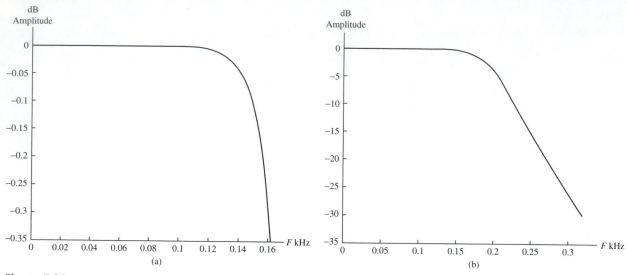

Figure 7.37
Amplitude characteristics of 7th-order Butterworth analog and digital filters: (a)–(b) analog filter, (c)–(e) digital filters.

The second step is to design the analog prototype low-pass filter. Equation (127) leads to

$$n = \frac{\log_{10}(10^{0.1\alpha} - 1) - \log_{10}(10^{0.1\beta} - 1)}{2(\log_{10} F_{\text{pass}} - \log_{10} F_{\text{stop}})}$$

$$= \frac{\log_{10}(10^{0.1 \times 1} - 1) - \log_{10}(10^{0.1 \times 30} - 1)}{2(\log_{10} 0.1622 - \log_{10} 0.318)} = 6.1236 \tag{154}$$

The minimum order Butterworth filter satisfying the specs is, by Eq. (128)

$$n^* = 7 \tag{155}$$

Because $n^* > n$, we can obtain a filter that exceeds either the passband spec or the stop-band spec. We here chose to design a filter that exceeds the passband spec by solving Eq. (132) for F_c.

$$F_c = \frac{F_{\text{stop}}}{(10^{0.1\beta} - 1)^{\frac{1}{2n}}} = \frac{0.3183 \text{ kHz}}{(10^{0.1 \times 30} - 1)^{\frac{1}{2 \times 7}}} = 0.1944 \text{ kHz} \tag{156}$$

We can now find $\mathbf{H}_{\text{B}n}(s)$ in pole-zero form using Eqs. (122) and (123), and expand it algebraically to obtain the following coefficient form (124).

$$\mathbf{H}_{\text{B}n}(s) = \frac{4.0563}{s^7 + 5.4892s^6 + 15.0654s^5 + 26.5911s^4 + 32.4798s^3 + 27.4542s^2 + 14.9240s + 4.0563} \tag{157}$$

Because the unit of F_c was kilohertz, the units of s are kiloradians (or kilonepers) per second.

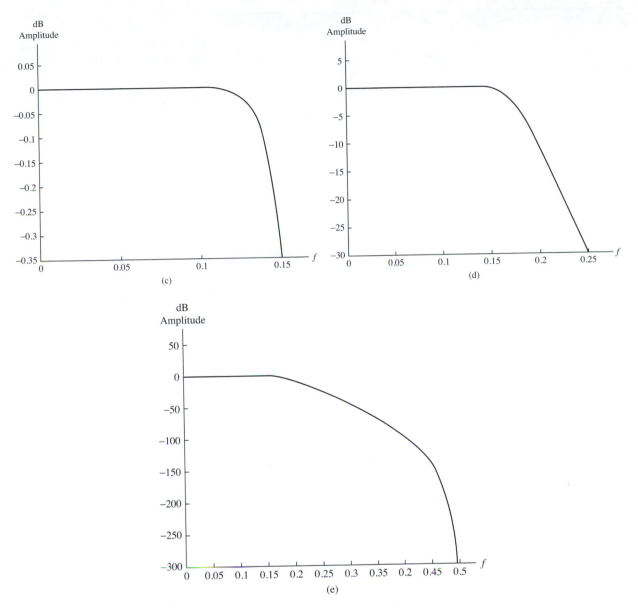

Figure 7.37
(*Cont.*).

The final step is to use the bilinear transform as in Eq. (109) to obtain the digital filter.

$$H(z) = \frac{0.0023 + 0.0159z^{-1} + 0.0476z^{-2} + 0.0793z^{-3} + 0.0793z^{-4} + 0.0476z^{-5} + 0.0159z^{-6} + 0.0023z^{-7}}{1 - 2.0986z^{-1} + 2.6472z^{-2} - 2.0035z^{-3} + 1.0124z^{-4} - 0.3233z^{-5} + 0.0609z^{-6} - 0.0051z^{-7}}$$

(158)

The amplitude characteristics of the analog and digital filters are shown in Figure 7.37. The half-power frequency can be found from by using the bilinear transform to map $F_c = 0.1944$ kHz to f_c. The result is $f_c = 0.1745$.

EXAMPLE 7.11   Low-Pass Chebyshev Filter Design

Use the bilinear transform and a Chebyshev analog prototype to design a low-pass digital filter that satisfies the following spec:

$$\text{Passband: } -1 \le |H(e^{j2\pi f})|_{\text{dB}} \le 0 \qquad \text{for } 0 \le f \le 0.15 \tag{159}$$

$$\text{Transition band: } 0.15 < f < 0.25 \tag{160}$$

$$\text{Stop-band: } |H(e^{j2\pi f})|_{\text{dB}} \le -30 \qquad \text{for } 0.25 \le f < 0.5 \tag{161}$$

The sampling rate is $F_s = 1\text{kHz}$.

Solution

Again, Flow diagram 1 indicates the steps involved in the design. Because we were given the design specs for a digital filter, the first step is to use the bilinear transform to obtain the corresponding analog filter design spec (see Figure 7.28 and its caption). The result is:

$$\text{Passband: } -1 \le |H(F)|_{\text{dB}} \le 0 \qquad \text{for } 0 \le F \le 0.1622 \text{ kHz} \tag{162}$$

$$\text{Transition band: } 0.1622 \text{ kHz} < F < 0.3183 \text{ kHz} \tag{163}$$

$$\text{Stop-band: } |H(e^{j2\pi f})|_{\text{dB}} \le -30 \qquad \text{for } 0.3183 \text{ kHz} \le F \tag{164}$$

The original value for ϵ^2 is, by Eq. (141),

$$\epsilon^2 = 10^{0.1\alpha} - 1 = 10^{0.1\times1} - 1 = 0.2589 \tag{165}$$

The minimum-order analog Chebyshev filter satisfying these specs is, from Eq. (145),

$$n = \frac{\cosh^{-1}\{\sqrt{(10^{0.1\beta} - 1)/\epsilon^2}\}}{\cosh^{-1}\{F_{\text{stop}}/F_c\}}$$

$$= \frac{\cosh^{-1}\{\sqrt{(10^{0.1\times30} - 1)/0.2589}\}}{\cosh^{-1}\{0.3183/0.1622\}} = 3.7233 \tag{166}$$

Equation (146) yields the integer

$$n^* = 4 \tag{167}$$

Because $n^* > n$, a fourth-order Chebyshev filter with $F_c = 0.1622$ and $\epsilon^2 = 0.2589$ meets the passband spec and exceeds the stop-band spec. We chose instead to obtain a filter that meets the stop-band spec with equality and exceeds the passband spec. To obtain this filter, we substitute $n^* = 4$ and $\beta = 30$ into Eq. (147) to obtain

$$\epsilon^2 = \frac{10^{0.1\beta} - 1}{[\cosh(n^*\cosh^{-1}(F_{\text{stop}}/F_c))]^2}$$

$$= \frac{10^{0.1\times30} - 1}{[\cosh(4\cosh^{-1}(0.318/0.1622))]^2} = 0.1265 \tag{168}$$

which, by Eq. (140), corresponds to a maximum passband attenuation of 0.5172 dB.
The amplitude characteristics of the analog and digital filters are shown in Figure 7.38.

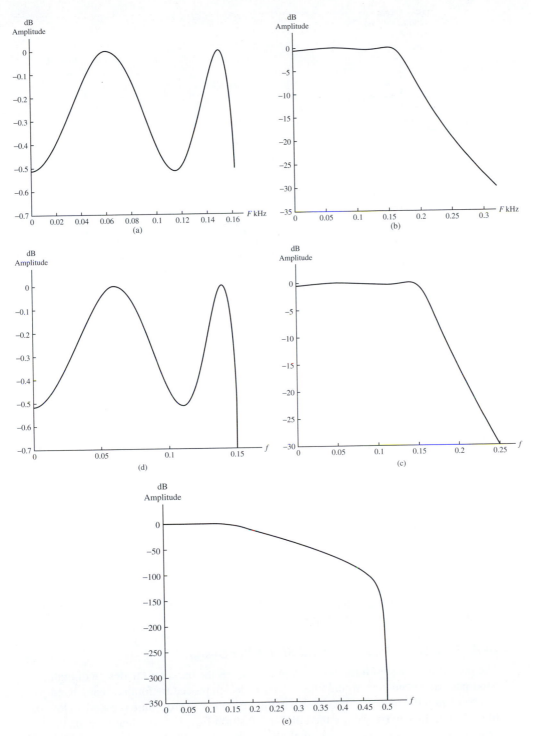

Figure 7.38
Amplitude characteristics of fourth-order Chebyshev analog and digital filters: (a)–(b) analog filter; (c)–(e) digital filter.

Figure 7.39
Flow diagram 2.

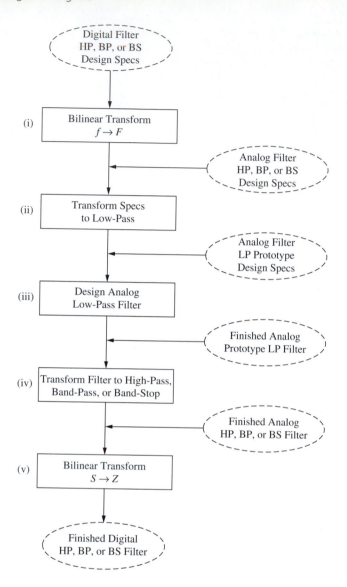

The end of chapter problems introduce MATLAB functions that avoid the labor of hand calculations involved in filter design.

High-Pass, Band-Pass, and Band-Stop Filter Design

The preceding low-pass filter design technique forms the basis for the design of high-pass, band-pass, and band-stop digital filters. A basic design procedure for high-pass, band-pass, and band-stop IIR filter design is shown in Figure 7.39. The procedure is based on low-pass filter design. In step (i), we use the bilinear transform Eq. (104) to obtain the analog filter design spec corresponding to the digital filter design spec. (Skip this step if you are already given the analog filter specs.) In step (ii), we use a frequency transformation to convert the analog high-pass, band-pass, or band-stop spec to a corresponding analog low-pass spec. In step (iii), we design the analog low-pass filter using the design specs obtained from step (ii).

Table 7.2

Frequency Transformations Used in IIR Filter Design

High-pass, band-pass, and band-stop analog filter prototypes can be obtained from low-pass analog filters by the use of frequency transformations listed in this table. The third column can be obtained from the second column by setting $s = j2\pi F$ and $s_l = j2\pi F_l$.

Low-pass to high-pass	$F_l = -\dfrac{F_{lp} F_{hp}}{F}$	$s_l = \dfrac{\Omega_{lp}\Omega_{hp}}{s}$
Low-pass to band-pass	$F_l = -F_{lp}\dfrac{F_o^2 - F^2}{F(F_{bp2} - F_{bp1})}$	$s_l = \Omega_{lp}\dfrac{s^2 + \Omega_o^2}{s(\Omega_{bp2} - \Omega_{bp1})}$
Low-pass to band-stop	$F_l = -F_{ls}\dfrac{F(F_{bs2} - F_{bs1})}{F_o^2 - F^2}$	$s_l = \Omega_{ls}\dfrac{s(\Omega_{bs2} - \Omega_{bs1})}{s^2 + \Omega_0^2}$

This step is similar to step (b) in Figure 7.36. We then use a frequency transformation to convert the finished (completely specified) low-pass analog prototype to a corresponding finished analog high-pass, band-pass, or band-stop prototype. The final step is use Eq. (108) to convert the analog filter obtained in step (iv) to the finished digital filter.

Let us now focus on the essential difference between flow graphs 1 and 2. This difference involves the frequency transformations indicated in steps (ii) and (iv) of flow graph 2. These frequency transformations are listed in Table 7.2 and are shown graphically in Figures 7.40–7.42.

Figure 7.40 shows how the low-pass to high-pass transformation works. In the lower left-hand corner of the figure, the amplitude characteristic $|H_{lp}(F_l)|$ of a typical low-pass filter is plotted versus F_l. From Table 7.2, the relationship between the coordinates F_l

Figure 7.40

Low-pass to high-pass filter conversion.

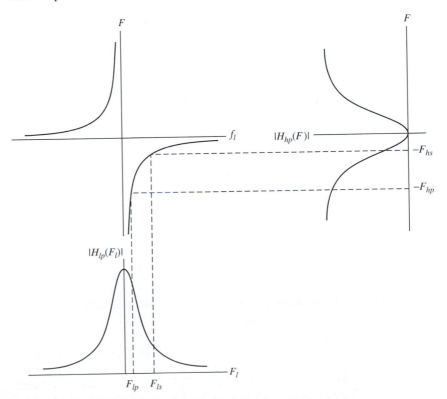

and F is

$$F_l = -\frac{F_{lp} F_{hp}}{F} \tag{169}$$

This relationship is plotted in the upper left-hand corner of the figure. The corresponding high-pass filter amplitude characteristic $|H_{lp}(F)|$ is plotted versus F in the upper right-hand corner. The dotted lines show how values of $|H_{lp}(F_l)|$ are mapped to the corresponding values of $|H_{lp}(F)|$. Notice, in particular, that the low-pass filter passband edge frequency F_{lp} is mapped to the high-pass filter passband edge frequency $-F_{hp}$. Similarly, $-F_{lp}$ is mapped to F_{hp}. A plot similar to that of Figure 7.40 applies for phase characteristics.

The expression for the frequency response of the high-pass filter is given by

$$H_{hp}(F) = H_{lp}\left(-\frac{F_{lp} F_{hp}}{F}\right) \tag{170}$$

where $H_{lp}(F)$ is the frequency response of the low-pass filter. F_{lp} and F_{hp} are, respectively, the passband edge frequencies of the low-pass and high-pass filters. Similarly, the expression for the transfer function of the high-pass filter is given by

$$\mathbf{H}_{hp}(s) = \mathbf{H}_{lp}\left(\frac{\Omega_{lp}\Omega_{hp}}{s}\right) \tag{171}$$

where $\Omega_{lp} = 2\pi F_{lp}$ and $\Omega_{hp} = 2\pi F_{hp}$.

EXAMPLE 7.12   High-Pass First-Order Butterworth Filter

Use the low-pass to high-pass frequency transformation to obtain a first-order high-pass Butterworth filter. We start with a low-pass Butterworth filter having the half-power frequency Ω_{lc}. The transfer function of the low-pass Butterworth filter is

$$\mathbf{H}_{B1l}(s) = \frac{\Omega_{lc}}{s + \Omega_{lc}} \tag{172}$$

a) Find the corresponding high-pass Butterworth filter transfer function by means of Table 7.2. Assume that the design spec stipulates that the half-power frequency Ω_{lc} of the low-pass filter is to be mapped to the half-power frequency Ω_{hc} of the high-pass filter.

b) Find the frequency response characteristic of the high-pass Butterworth filter.

Solution

a) To map Ω_{lc} to Ω_{hc}, we set $\Omega_{lp} = \Omega_{lc}$ and $\Omega_{hp} = \Omega_{hc}$ in the low-pass to high-pass frequency transformation (171). The transfer function of the first-order Butterworth high-pass filter is obtained by replacing s by $\Omega_{lc}\Omega_{hc}/s$ in Eq. (172). The result is

$$\mathbf{H}_{B1h}(s) = \frac{\Omega_{lc}}{\frac{\Omega_{lc}\Omega_{hc}}{s} + \Omega_{lc}} = \frac{s}{\Omega_{hc} + s} \tag{173}$$

b) The frequency response characteristic $H_{B1h}(F)$ can be obtained directly from Eq. (173) by replacing s with $j2\pi F$. The result is

$$H_{B1h}(F) = \mathbf{H}_{B1l}(j2\pi F) = \frac{j2\pi F}{\Omega_{hc} + j2\pi F} = \frac{jF}{F_{hc} + jF} \tag{174}$$

Figure 7.41
Low-pass to
band-pass filter
conversion.

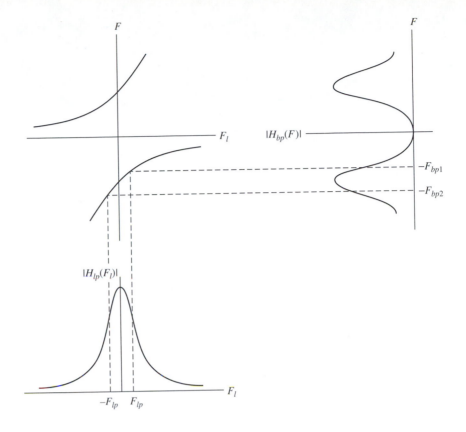

Figure 7.41 illustrates the low-pass to band-pass filter transformation. The frequency mapping is

$$F_l = -F_{lp} \frac{F_o^2 - F^2}{F(F_{bp2} - F_{bp1})} \tag{175}$$

where F_{bp2} and F_{bp1} are the upper and lower passband edge frequencies of the band-pass pass filter. F_o is determined by the equation

$$F_o^2 = F_{bp2} F_{bp1} \tag{176}$$

The difference

$$F_{bp2} - F_{bp1} = W \tag{177}$$

is the bandwidth of the filter. A high-pass filter design spec ordinarily specifies either W and F_o, or F_{bp2} and F_{bp1}. Notice that Eq. (175) maps $F = 0$ to $F = \pm F_o$. Therefore, if the low-pass amplitude characteristic has a peak at $F = 0$, the band-pass amplitude characteristic has peaks at $F = \pm F_o$. Similarly, Eq. (175) maps the low-pass edge frequency F_{lp} to the bandpass edge frequencies. The band-stop filter transformation, illustrated in Figure 7.42, works in a similar way.

As with low-pass filter design, the design procedure for the high-pass, band-pass, and band-stop filters is often too laborious to be done by hand. The end-of-chapter problems introduce MATLAB programs that avoid the labor of hand calculation.

Figure 7.42
Low-pass to
band-stop filter
conversion.

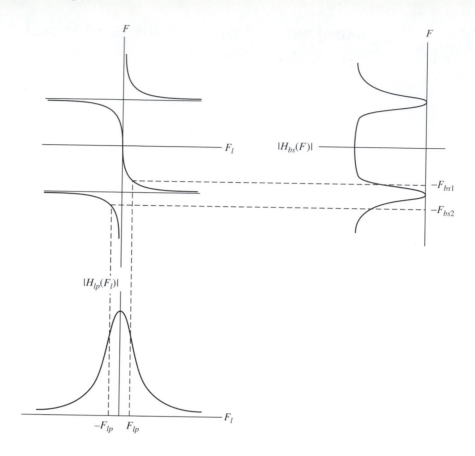

7.6 Summary

This chapter considered the design of a hybrid system, composed of an analog-to-digital converter, a digital computer, and a digital-to-analog converter. We noted that a hybrid system has the following advantages over a purely analog system: (1) The I/O relation of the hybrid system can be changed by reprogramming its computer; (2) the hybrid system can achieve levels of performance that would be impossible for a practical analog system using realistic component values; and (3) the digital components of the hybrid system are less susceptible to the degrading influences of time and environment than a theoretically equivalent analog system.

We modeled the hybrid system as a sampler, a DT processor, and an interpolator connected in series. We showed that the hybrid system model can, in theory, perform as well as any hypothetical LTI analog system provided that (1) the input signal $x(t)$ is sampled above the Nyquist sampling rate, (2) the frequency response characteristic of the DT processor matches that of the analog system Eq. (15), (3) the interpolation filter in the D/A converter is the ideal of Eq. (25). Much of the chapter focused on design techniques and structures for implementing the DT processor. Examples demonstrated that the matched processor design goal can be better approximated by allowing more processing delay and more DT processor elements.

Summary of Definitions and Formulas

Conversion from CT to DT

$$x[n] = x(nT_s)$$

$$X(e^{j2\pi FT_s}) = F_s \sum_{n=-\infty}^{\infty} X(F - nF_s)$$

$$X(e^{j2\pi f}) = F_s \sum_{n=-\infty}^{\infty} X((f - n)F_s)$$

where $f = FT_s = F/F_s$.

When there is no aliasing, $X(e^{j2\pi f}) = F_s X(fF_s)$, $-\frac{1}{2} < f < \frac{1}{2}$.

DT Processor

$$y[n] = \sum_{m=-\infty}^{\infty} x[m]h[n - m] = x[n] * h[n]$$

$$Y(e^{j2\pi f}) = H(e^{j2\pi f})X(e^{j2\pi f})$$

To match the performance of an analog system having the frequency response characteristics $H(F)$, set

$$H(e^{j2\pi f}) = H_m(e^{j2\pi f})$$

where $H_m(e^{j2\pi f}) = H(fF_s)$, $-\frac{1}{2} \le f < \frac{1}{2}$.

Conversion from DT to CT
Ideal Interpolation

$$\hat{y}(t) = \sum_{n=-\infty}^{\infty} y(nT_s)h_i(t - nT_s)$$

If we chose

$$H_i(F) = \begin{cases} T_s & |F| < W \\ \text{arbitrary} & W < |F| < F_s - W \\ 0 & |F| \ge F_s - W \end{cases}$$

then

$$\hat{y}(t) = y(t)$$

FIR Filters

$$h[n] = h[0]\delta[n] + h[1]\delta[n - 1] + h[2]\delta[n - 2] + \cdots + h[L - 1]\delta[n - L + 1]$$

$$H(z) = \frac{B(z^{-1})}{A(z^{-1})} = h[0] + h[1]z^{-1} + h[2]z^{-2} + \cdots + h[L - 1]z^{-(L-1)}$$

For most applications, L is an odd integer $L = 2K + 1$, where K is a nonnegative integer.

Every FIR filter is BIBO stable.

Noncausal FIR Design Filter

$$h[n] = h_{nc}[n - K]$$

$$H(z) = z^{-K} H_{nc}(z)$$

$$H(e^{j2\pi f}) = e^{-jK2\pi f} H_{nc}(e^{j2\pi f})$$

$$H_{nc}(e^{j2\pi f}) = \sum_{k=-K}^{K} h_{nc}[k] e^{jk2\pi f}$$

FIR Filter Design

The $h_{nc}[n]$ that minimize mean square approximation error

$$\mathcal{E} = \int_{-0.5}^{0.5} \left| H_{nc}(e^{j2\pi f}) - H_m(e^{j2\pi f}) \right|^2 df$$

are given by

$$h_m[n] = \int_{-0.5}^{0.5} H_m(e^{j2\pi f}) e^{-j2\pi nf} df$$

For a length $L = 2K + 1$ FIR filter, set $h_{nc}[n] \equiv h_m[n]$ for $-K \le n \le K$.

Windowed FIR Filter Design

$$h_w[n] = h_{ncw}[n - K]$$

$$h_{ncw}[n] = w[n] h_m[n]$$

where $w[n] = 0, |n| > K$.

IIR Filters

$$h[n] = h[0]\delta[n] + h[1]\delta[n - 1] + h[2]\delta[n - 2] + \cdots$$

$$H(z) = \frac{b_0 + b_1 z^{-1} + b_2 z^{-2} + \cdots + b_m z^{-m}}{a_0 + a_1 z^{-1} + a_2 z^{-2} + \cdots + a_l z^{-l}}$$

$$a_0 y[n] + a_1 y[n - 1] + a_2 y[n - 2] + \cdots + a_l y[n - l]$$
$$= b_0 x[n] + b_1 x[n - 1] + \cdots + b_m x[n - m]$$

IIR filters can be unstable.

Impulse Invariant IIR Filter Design

$$h[n] = T_s h(nT_s)$$

$$H(z) = T_s \mathcal{Z} \left\{ \mathcal{L}^{-1}\{\mathbf{H}(s)\} \Big|_{t=nT_s} \right\}$$

$$H(e^{j2\pi f}) = H(f F_s) + \sum_{\substack{m=-\infty \\ m \ne 0}}^{\infty} H((f - m) F_s)$$

The Bilinear Transform

$$s = \frac{1 - z^{-1}}{1 + z^{-1}} 2F_s$$

$$H(z) = H(s)\Big|_{s = \frac{1-z^{-1}}{1+z^{-1}} 2F_s}$$

$$H(e^{j2\pi f}) = \mathbf{H}(j2\pi F)\Big|_{F = \frac{F_s}{\pi} \tan(\pi f)}$$

Analog Butterworth Filters

$$|H_{Bn}(F)| = |\mathbf{H}_{Bn}(j2\pi F)| = \frac{1}{\sqrt{1 + \left(\frac{F}{F_c}\right)^{2n}}}$$

where F_c is the half-power frequency.
$|H_{Bn}(F)|$ is maximally flat at $F = 0$.
For every n, $|H_{Bn}(F)|$ decreases monotonically toward 0 as F increases.

$$|H_{Bn}(0)| = 1 \qquad \text{and} \qquad |H_{Bn}(F_c)| = \frac{1}{\sqrt{2}}$$

As n increases, $|H_{Bn}(F)|$ becomes an increasingly better approximation to the amplitude characteristic of an ideal low-pass filter having the cutoff frequency F_c.

Analog Chebyshev Filters

$$|H_{Cn}(F)| = \sqrt{\frac{1}{1 + \epsilon^2 T_n^2\left(\frac{F}{F_c}\right)}}$$

where F_c is the *passband corner frequency*.
$|H_{Cn}(F)|$ is equiripple in the passband:

$$\sqrt{\frac{1}{1 + \epsilon^2}} \le |H_{Cn}(F)| \le 1 \qquad \text{for } 0 \le F \le F_c$$

$$\text{DC gain } |H_{Cn}(0)| = \begin{cases} \sqrt{\frac{1}{1+\epsilon^2}} & \text{for } n = 0, 2, 4, \ldots \\ 1 & \text{for } n = 1, 3, 5, \ldots \end{cases}$$

$$\text{Gain at } F = F_c \ |H_{Cn}(F_c)| = \sqrt{\frac{1}{1 + \epsilon^2}}$$

$|H_{Cn}(F)|$ decreases monotonically towards 0 as F increases beyond F_c.
$|H_{Cn}(F)|$ decreases faster in the stop-band as n increases.
Filter Design Using the Bilinear Transform
See Figures 7.36 and 7.39 and Figures 7.40–7.42.

7.7   Problems

7.2   Conversion from CT to DT

7.2.1 The waveform $x(t) = \text{sinc}^2(t/\tau)$ is sampled at rate F_s to produce a data sequence $x[n] = x(nT_s)$.
a) Find and plot $X(F)$.
b) plot $F_s X(f F_s)$ versus f.
c) Find and plot $X(e^{j2\pi f})$ for various sampling rates. Find the minimum sampling rate for which aliasing does not occur.

7.3   DT Processor

7.3.1 Consider designing a hybrid system (Figure 7.1) to match the performance of an ideal low-pass filter having bandwidth W. Assume that the input waveform $x(t)$ is band limited so that $X(F) = 0$ for $|F| > W_x$, where $W_x > W$.

a) What is the minimum sampling rate for which the performance of the hybrid system can theoretically match that of the analog system?

b) Find and plot the frequency response characteristic of the digital filter $H_m(e^{j2\pi f})$ in the hybrid system that provides a match to the performance of the analog system.

7.3.2 An analog system having

$$H(F) = \begin{cases} -j & F \geq 0 \\ j & F < 0 \end{cases} \tag{178}$$

is called a *Hilbert transformer* or a *90-degree phase shifter*. The input and the output of a Hilbert transformer are often denoted by $x(t)$ and $\hat{x}(t)$, respectively.

a) Plot the amplitude and the phase characteristics of a Hilbert transformer. Refer to your plots to explain why the term *90-degree phase shifter* is appropriate.

b) Find the Hilbert transform of $x(t) = \cos(2\pi Ft)$.

c) Find the Hilbert transform of $\delta(t)$.

7.3.3 Refer to the preceding problem.

a) Find the frequency response characteristic Eq. (15) $H_m(e^{j2\pi f})$ for a digital Hilbert transformer. Plot the amplitude and phase characteristics of $H_m(e^{j2\pi f})$ for $-1 \leq f < 1$.

b) Find the output of $H_m(e^{j2\pi f})$ for an input $x[n] = \cos(2\pi fn)$.

c) Show that the impulse response of a digital Hilbert transformer is

$$h_m[n] = \mathcal{IDTFT}\{H_m(e^{j2\pi f})\} = \begin{cases} 0 & \text{for } n \text{ even} \\ \frac{2}{\pi n} & \text{for } n \text{ odd} \end{cases} \tag{179}$$

7.5   Filter Design

7.5.1 A particular FIR filter has the impulse response $h[n] = h_1[n] * h_2[n] * \cdots * h_I[n]$, where $h_i[n] = \frac{1}{2}\delta[n] + \frac{1}{2}\delta[n-1], i = 1, 2, \ldots, I$.

a) Draw a realization for $h[n]$ for $I = 5$. Your realization should consist of a cascade of five FIR filters.

b) Show that $H[z] = (\frac{1}{2} + \frac{1}{2}z^{-1})^I$.

c) Use the result of part (b) to show that $H(e^{j2\pi f}) = e^{-jI\pi f}\cos^I(\pi f)$. Plot the amplitude and characteristics for $I = 8$ and $-\frac{1}{2} \leq f < \frac{1}{2}$.

d) Use the binomial theorem $(1 + x)^I = \sum_{k=0}^{I}\{I!/[k!(I-k)!]\}x^k$ to find $h[n]$. Plot $h[n]$ for $I = 8$.

7.5.2 Repeat Problem 7.5.1 with $h_i[n] = \frac{1}{2}\delta[n] - \frac{1}{2}\delta[n-1], i = 1, 2, \ldots, I$.

Noncausal FIR Design Filter

7.5.3 Refer to Problems 7.3.2 and 7.3.3. We showed in Problem 7.3.3 that the impulse response of a digital Hilbert transformer is

$$h_m[n] = \mathcal{IDTFT}\{H_m(e^{j2\pi f})\} = \begin{cases} 0 & \text{for } n \text{ even} \\ \frac{2}{\pi n} & \text{for } n \text{ odd} \end{cases} \tag{180}$$

a) Plot $h_m[n]$.

b) Let $h_{nc}[n] = h_m[n]$ for $n = \pm1, \pm2, \ldots, \pm K$ and $h_{nc}[n] = 0$ for $|n| > K$. Assume that $K = 5$

and plot $h_{nc}[n]$.

c) Plot $h[n] = h_{nc}[n - K]$, where $K = 5$.

d) Use a computer to evaluate $H_{nc}(e^{j2\pi f}) = \sum_{k=-K}^{K} h_{nc}[k]e^{jk2\pi f}$. Plot the amplitude and phase spectrums of $H_{nc}(e^{j2\pi f})$ for $K = 5$ and for $K = 25$. Comment on your results.

7.5.4 Refer to problem 7.5.3 and use windowing to design a length $L = 11$ and a length $L = 51$ digital Hilbert transformer. Use any window of your choice and comment on your results.

FIR Low-Pass Filter Design Using MATLAB

You can use the MATLAB functions `fir1(n,Wn,window)`, `freqz(b,1,npts)`, and `plot(x,y)` to compute and plot the frequency response characteristics of a low-pass FIR filter.

- `b=fir1(n,Wn,window)` computes the impulse response of an approximation to an ideal lowpass filter having the cutoff frequency F_c Hz.
  - n is the order of the FIR filter (n $= L - 1$).
  - b is a length $L = $ n $+ 1$ vector representing the impulse response.
  - Wn is defined as Wn $= F_c/(F_s/2)$. It is the ideal low-pass filter's cutoff frequency normalized to $F_s/2$. (Wn $= 2f_c$, where $f_c = F_c/F_s$ is the normalized frequency defined in the text).
  - window is a length $L = $ n $+ 1$ vector representing the window.

The MATLAB Signal Processing toolbox contains an extensive description of commonly used windows, along with two tools for visualizing them. You can see the list of windows by typing `windows` in the MATLAB help menu. You can obtain a detailed description of each window by clicking the name in the list. You can type `wintool` in the MATLAB workspace to display the sequence and frequency domain characteristics of windows entered from a menu. You can also display these characteristics using the function `wvtool(...)`. For example, to display the characteristics of 65-point rectangular, hann, and nuttall windows, type `wvtool(rectwin(65),hann(65), nuttallwin(65))` in the workspace.

- `[h,w]=freqz(b,1,npts)` computes the frequency response characteristic of the FIR filter.
  - h is a vector representing the frequency response characteristic.
  - w is a vector representing the corresponding frequencies in h.
  - w is related to normalized frequency f by w $= 2\pi f$.
  - h and w both have length `npts`.
  - b is the impulse response vector computed from `b=fir1(n,Wn,window)`.
- `plot(x,y)` plots y versus x where x and y are vectors having the same length.
  - `plot(w/(2*pi),abs(h))` plots the amplitude characteristic versus f for $0 \leq f < \frac{1}{2}$.
  - `plot(w/(2*pi),angle(h))` plots the phase characteristic versus f for $0 \leq f < \frac{1}{2}$.
  - `plot(w/(2*pi),20*log10(abs(h)))` plots the amplitude characteristic in decibels versus f for $0 \leq f < \frac{1}{2}$.

Example

You can use the following script to generate the low-pass filter characteristics described in Example 7.2

```
b=fir1(44,0.4,rectwin(45));
[h,w]=freqz(b,1,128);
plot(w/(2*pi),20*log10(abs(h)))
```

7.5.5 Read the MATLAB help file for the functions just described. Then run the script in the preceding example and comment on the result.

7.5.6 Use MATLAB to help you design an FIR low-pass filter like the one in Example 7.2 but with -30 dB replaced by -28dB. Base your design on the MMSE criterion. Use trial and error to find the minimum possible filter length L. Plot the resulting amplitude characteristic in decibels.

7.5.7 Use MATLAB to help you design a low-pass FIR filter like the one in Example 7.2 but with ± 1 dB replaced with ± 1.5 dB and -30 dB replaced with -20 dB. Base your design on the MMSE criterion. Use trial and error to find the minimum possible filter length L. Plot the resulting amplitude characteristic in decibels.

7.5.8 Use MATLAB to help you design a low-pass FIR filter like the one in Example 7.2 but with no restrictions on the pass-band deviation. Base your design on the MMSE criterion. Use trial and error to find the minimum possible filter length L. Plot the resulting amplitude characteristic in decibels.

7.5.9 Use the MATLAB function `hann(n)` and script analogous to that in the preceding MATLAB example to make a plot similar to that in Example 7.4. (Read the MATLAB help file for the function `hann(n)`.)

7.5.10 Experiment with alternative windows to see if you can reduce the length of the FIR filter needed to meet the design specs of Examples 7.2 and 7.4. Plot the resulting amplitude characteristic in decibels.

7.5.11 Use MATLAB to help you design a windowed FIR filter like the one in Example 7.4 but with -30 dB replaced by -28 dB. Use trial and error and any alternative window of your choice to find the minimum possible filter length L. Plot the resulting amplitude characteristic in decibels.

7.5.12 Write a MATLAB script and program your computer to duplicate the plots shown in Example 7.3.

7.5.13 Revise your MATLAB script in the preceding problem to include windowing. Use your program to duplicate the plots shown in Example 7.5. Then experiment with two additional windows to see if you can reduce the required filter length.

7.5.14 Write a MATLAB script and program your computer to generate the frequency response characteristics of an FIR Hilbert transformer. (See Problem 7.3.0.). Base the design on MMSE. Plot the amplitude and phase characteristics for various filter lengths L and comment on your results.

7.5.15 Revise your MATLAB script in the preceding problem to include windowing. Experiment with two alternative windows of your choice and comment on the results.

IIR Filters

7.5.16 Consider the FIR filter $h[n] = \frac{1}{50}(u[n] - u[n-50])$.
   a) How many delay elements are needed to implement the filter in the form shown in Figure 7.10?
   b) What operation does this filter perform on its input?
   c) Use the
   z-transform $\mathcal{Z}\{u[n]\} = 1/(1 - z^{-1})$, the delay property, and superposition to find $H(z)$.
   d) Use your result from part (c) to show that the FIR filter can be realized by an IIR first-order Direct Form 2 implementation. Do you see any potential problems using this structure? Hint: Consider the stability of the feedback loop.

Impulse-Invariant IIR Filter Design

7.5.17 a) Find the impulse-invariant DT filter corresponding to the analog filter $h(t) = te^{-t/\tau}u(t)$.
   b) Draw the Direct Form 2 realization.
   c) Plot the amplitude response of the digital filter and compare it to $H_m(e^{j2\pi f})$ for $0 \le f < \frac{1}{2}$ with $\tau = 1$ and $F_s = 2$.

7.5.18 Repeat the preceding problem for $h(t) = (1 - e^{-t/\tau})u(t)$. Use values $\tau = 0.1$ and $F_s = 3$ for the plots.

7.5.19 Consider the partial fraction expansion of a CT system for which $\mathbf{H}(s)$ is a proper algebraic fraction with only simple poles [See Eq. (6DT.80)]. Find the transfer function of the corresponding impulse-invariant filter. Assume that the CT filter is stable and prove that the corresponding impulse-invariant DT filter is also stable.

7.5.20 Find the impulse-invariant DT filter corresponding to a band-pass analog filter having the impulse response

$$h(t) = \frac{1}{\Omega_d} e^{-\alpha t} \sin(\Omega_d t) u(t) \qquad (181)$$

This system was described in Example 6CT.13. Draw a realization for the digital filter. Plot the amplitude response characteristics of the digital filter with $\alpha = 0.2$, $\Omega_d = 12$ and $F_s = 20$.

7.5.21 Someone proposes to use the technique of impulse in variance to design a high-pass filter. Point out the problems in attempting to design a digital high-pass filter based on $\mathbf{H}(s) = s\tau/(s\tau + 1)$.

7.5.22 Read Appendix 7.A.
a) Draw a filter realization for a step-invariant DT filter corresponding to
$h(t) = (1/\tau)e^{-t/\tau}u(t)$.
b) Plot the amplitude response characteristic of
the digital filter and compare it to that obtained using impulse invariance (Example 7.7).

7.5.23 Read Appendix 7.A.
a) Find $H(z)$ for a ramp-invariant DT filter corresponding to $h(t) = (1/\tau)e^{-t/\tau}u(t)$.
b) Draw a filter realization.

7.5.24 Read Appendix 7.A.
a) Show that the frequency response of an IIR filter designed on the basis of step invariance is given by

$$H(e^{j2\pi f}) = \left(\frac{1 - e^{-j2\pi f}}{j2\pi f F_s}\right) \mathbf{H}(j2\pi f F_s) + (1 - e^{-j2\pi f}) \sum_{\substack{m=-\infty \\ m \neq 0}}^{\infty} \frac{\mathbf{H}(j2\pi(f-m)F_s)}{j2\pi(f-m)F_s} \qquad (182)$$

b) Show that the factor multiplying $\mathbf{H}(j2\pi f F_s)$ is approximately unity for $|f| \ll 0.5$.

The Bilinear Transform

7.5.25 Consider the use of the bilinear transform to design a digital differentiator. The analog filter on which the design is based has transfer function $\mathbf{H}(s) = s$.
a) Find $H(z)$.
b) Draw the Direct Form 2 realization.
c) Find $H(e^{j2\pi f})$.
d) Plot the amplitude and phase characteristics
of the digital differentiator and compare them to those of the matched system $H_m(e^{j2\pi f})$.

7.5.26 Assume that we sample a waveform $x(t)$ and use straight-line interpolation between the samples. If we compute the area under the interpolated values of $x(t)$ from $t = 0$ to $t = nT_s$, we obtain an approximation to the integral

$$y(nT_s) = \int_0^{nT_s} x(t)dt \qquad (183)$$

The approximation is given by

$$\hat{y}(nT_s) = \sum_{m=-\infty}^{n} \frac{T_s}{2}\{x(mT_s) + x((m-1)T_s)\} \qquad (184)$$

which we can write as

$$\hat{y}[n] = \sum_{m=-\infty}^{n} \frac{T_s}{2}(x[m] + x[m-1]) \tag{185}$$

This numerical method of approximating an integral is called *trapezoidal integration*.
a) Use figures to illustrate the difference between exact integration and a trapezoidal approximation.
b) Show that $\hat{y}[n]$ can be computed using the recursion

$$\hat{y}[n] = \hat{y}[n-1] + \frac{T_s}{2}(x[n] + x[n-1]) \tag{186}$$

c) Show that the transfer function $H(z)$ of a trapezoidal integrator is

$$H(z) = \frac{T_s}{2}\frac{1+z^{-1}}{1-z^{-1}} \tag{187}$$

d) Recall that an analog integrator has the transfer function $\mathbf{H}(s) = 1/s$. The bilinear approximation to an analog integrator is therefore

$$H(z) = \mathbf{H}(s)\Big|_{s=\frac{1-z^{-1}}{1+z^{-1}}2F_s} = \frac{T_s}{2}\frac{1+z^{-1}}{1-z^{-1}} \tag{188}$$

Compare the preceding two equations and comment.
e) Would it be correct to say that, when you use a bilinear transform to obtain a digital system from an analog system, you are in effect approximating the integrators in the analog system with trapezoidal integrators? Give an example using a first-order analog system.
f) Plot the amplitude and phase characteristics of a trapezoidal integrator for $0 \leq f < \frac{1}{2}$, and compare them to the amplitude and phase characteristics of the DT system $H_m(e^{j2\pi f})$ matched to an analog integrator.

Low-Pass IIR Filter Design Using MATLAB

Several MATLAB functions are available for designing IIR filters. These programs, or others that you yourself program, are important because they make it possible for you to design IIR filters that would be too laborious to design by hand. You can type `iir` in the MATLAB search menu to open the iirfilter design help window. The functions `buttord`, `butter`, and `cheby1ord`, `cheby1` listed there can be used to design both the digital Butterworth and Chebyshev filters introduced in this book. You can use the MATLAB functions `freqz` and `plot(x,y)` to compute and plot the frequency response characteristics. The following descriptions of these functions help you get started.
For a low-pass filter, use the script:
■ `[n,Wn] = buttord(Wp,Ws,Rp,Rs)` to compute the order and normalized half-power frequency of a low-pass digital Butterworth filter.
■ `[b,a] = butter(n,Wn)` to compute the coefficient vectors of a low-lass Butterworth filter.
In these functions:
   ■ `Rp` is the filter's maximum passband attenuation in decibels ($\mathtt{Rp} = \alpha$ in the text).
   ■ `Rs` is the filter's minimum stop band attenuation in decibels ($\mathtt{Rp} = \beta$ in the text).
   ■ `Wp` is the filter's passband edge frequency normalized to the Nyquist frequency ($\mathtt{Wp} = F_{pass}/(F_s/2) = 2f_{pass}$).
   ■ `Ws` is the filter's stop-band edge frequency normalized to $F_s/2$ ($\mathtt{Ws} = F_{stop}/(F_s/2) = 2f_{stop}$).
   ■ `n` is the order of the filter ($n = L - 1$).

■ Wn is the Butterworth filter's half-power frequency normalized to $F_s/2$
  ($\texttt{Wn} = 2f_c$).
■ b and a are the numerator and denominator coefficient vectors, respectively.
MATLAB's indexing of coefficient vectors differs from that of the text. The correspondence
between the MATLAB indexing and that in this book (Eq. (79) with $m = n$ and $a_0 = 1$) is

$$\text{MATLAB} \longleftrightarrow \text{This book}$$

$$\texttt{b=[b}_1\texttt{,b}_2\texttt{, ...,b}_{n+1}\texttt{]} \longleftrightarrow [b_0, b_1, \ldots, b_n] \tag{189}$$

$$\texttt{a=[1,a}_2\texttt{,...,a}_{n+1}\texttt{]} \longleftrightarrow [1, a_1, \ldots, a_n] \tag{190}$$

Similarly, use the script:
■ $\texttt{[n,Wn]} = \texttt{cheb1ord(Wp,Ws,Rp,Rs)}$ to compute the order and normalized
  passband edge frequency of a low-pass digital Chebyshev filter.
■ $\texttt{[b,a]} = \texttt{cheby1(n,Rp, Wn)}$ to compute the coefficient vectors of the low-pass
  IIR Chebyshev filter described in this book.
In these functions:
■ Wn is the Chebyshev filter's corner frequency normalized to the Nyquist frequency
  ($\texttt{Wn}= 2f_c$).
■ Wp, Ws, Rp, Rs, n, b and a are defined as in the low-pass Butterworth filter.
Use the script:
■ $\texttt{h,w]=freqz(b,a,npts)}$ to compute the frequency response characteristic of
  the IIR filter.
Where:
■ h is a vector representing the frequency response characteristic.
■ w is a vector representing the corresponding frequencies in h.
■ w is related to normalized frequency f by $\texttt{w}= 2\pi f$.
■ h and w both have length npts.
■ b and a are coefficient vectors computed by $\texttt{[b,a]} = \texttt{cheby1(n,Wn)}$.
The script:
■ $\texttt{plot(x,y)}$ plots y versus x, where x and y are vectors having the same length.
Use:
■ $\texttt{plot(w/(2*pi),abs(h))}$ to plot the amplitude characteristic versus f for
  $0 \le f < \frac{1}{2}$.
■ $\texttt{plot(w/(2*pi),angle(h))}$ to plot the phase characteristic versus f for
  $0 \le f < \frac{1}{2}$.
■ $\texttt{plot(w/(2*pi),20*log10(abs(h)))}$ to plot the amplitude characteristic
  in decibels versus f for $0 \le f < \frac{1}{2}$.

Example

You can use the following script to generate and plot the low-pass filter
characteristics described in Example 7.10.

```
[n,Wn]=buttord(2*0.15,2*0.25,1,30)
[b,a] = butter(n,Wn)
[h,w]=freqz(b,a,128);
plot(w/(2*pi),20*log10(abs(h)))
```

 7.5.27 Read the MATLAB help file for the preceding functions. Then run the script in the example
and comment on the result.

7.5.28 Use MATLAB to help you design a low-pass Butterworth filter like the one in Example 7.10
but with the 1-dB passband attenuation spec replaced with 2 dB. Plot the resulting amplitude
characteristic in decibels.

7.5.29 Use MATLAB to help you design a low-pass Butterworth filter like the one in Example 7.10 but with the 1-dB passband attenuation spec replaced with 3 dB. Plot the resulting amplitude characteristic in decibels.

7.5.30 Use MATLAB to help you design a low-pass Chebyshev filter like the one in Example 7.11 but with the 1-dB passband attenuation spec replaced with 2 dB. Plot the resulting amplitude characteristic in decibels.

7.5.31 Use MATLAB to help you design a low-pass Chebyshev filter like the one in Example 7.11 but with the 1-dB passband attenuation spec replaced with 3 dB. Plot the resulting amplitude characteristic in decibels.

High-Pass, Band-Pass, and Band-Stop Filter Design

7.5.32 In Example 7.11 we used a low-pass to high-pass frequency transformation to transform analytically a first-order Butterworth filter to a first-order high-pass Butterworth filter.
a) Use a similar method to transform the first-order Butterworth filter to a band-pass filter.
b) What is the resulting order of the band-pass filter?
c) Find the analytical expressions for the poles and zeros of the band-pass filter.
d) Sketch the amplitude characteristic of the band-pass filter.

7.5.33 In Example 7.12 we used a low-pass to high-pass frequency transformation to transform analytically a first-order Butterworth filter to a first-order high-pass Butterworth filter.
a) Use a similar method to transform the first-order Butterworth filter to a band-stop filter.
b) What is the resulting order of the band-stop filter?
c) Find the analytical expressions for the poles and zeros of the band-stop filter.
d) Sketch the amplitude characteristic of the band-stop filter.

IIR High-Pass, Band-Pass, and Band-Stop Filter Design Using MATLAB

The MATLAB functions `buttord`, `butter`, `cheby1ord`, and `cheby1` just described can also be used to design high-pass, band-pass, and band-stop Butterworth and Chebyshev filters. The frequency transformations described in the text are built into these functions. Again, you can use the MATLAB functions `freqz`, and `plot(x,y)` to compute and plot the frequency response characteristics. The MATLAB help menu shows you how to use these functions. The following problem helps you get started.

7.5.34 Consider the following MATLAB script.

```
    Wp=[2*0.15 2*0.2];
    Ws=[2*0.1 2*0.25];
 [n,Wn]= buttord(Wp,Ws,2,40)
  [b,a]= butter(n,Wn);
  [h,w]= freqz(b,a,256);
plot(w/(2*pi), 20*log10(abs(h)))
```

a) Read the appropriate MATLAB help files and explain the significance of each numerical value given in the script.
b) What kind of filter does the script generate?
c) Run the script and comment on the results.
d) What is the order of the filter corresponding to your plot?

7.5.35 a) Modify the script in the preceding example to generate and plot a Chebyshev filter meeting the same specs.
b) Run your script and comment on the results.
c) What is the order of the filter corresponding to your plot?

Appendix 7.A Signal Invariant Design

This appendix generalizes the concept of impulse-invariant design to include inputs other than impulses. Consider a CT system having an input $x(t)$ and a DT system having input $x[n] = x(nT_s)$. Our goal is to design the DT system so that its output $y[n]$ equals samples of the output of the CT system.

$$y[n] = y(nT_s) \tag{A1}$$

We can derive the transfer function of the DT system by noting that

$$y(t) = x(t) * h(t) = \mathcal{L}^{-1}\{\mathbf{H}(s)\mathbf{X}(s)\} \tag{A2}$$

where $h(t)$ is the impulse response and $H(s)$ is the transfer function of the CT system. It follows from these equations that

$$y[n] = \mathcal{L}^{-1}\{\mathbf{H}(s)\mathbf{X}(s)\}\big|_{t=nT_s} \tag{A3}$$

The z-transform of $y[n]$ is

$$Y(z) = \mathcal{Z}\left\{\mathcal{L}^{-1}\{\mathbf{H}(s)\mathbf{X}(s)\}\big|_{t=nT_s}\right\} \tag{A4}$$

We find the transfer function of the DT filter $H(z)$ by dividing $Y(z)$ by $X(z)$.

$$H(z) = \frac{Y(z)}{X(z)} = \frac{\mathcal{Z}\left\{\mathcal{L}^{-1}\{\mathbf{H}(s)\mathbf{X}(s)\}\big|_{t=nT_s}\right\}}{X(z)} \tag{A5}$$

EXAMPLE 7.13 Step Invariant Filter Design

a) Find the design equation for the step-invariant design.

b) Find $H(z)$ for the step-invariant DT filter that corresponds to the CT filter

$$\mathbf{H}(s) = \frac{1}{1+s\tau} \tag{A6}$$

a) Here $x(t) = u(t)$ and $x[n] = u[n]$. Therefore

$$\mathbf{X}(s) = \frac{1}{s} \tag{A7}$$

and

$$X(z) = \frac{z}{z-1} = \frac{1}{1-z^{-1}} \tag{A8}$$

The substitution of Eq. (A7) and Eq. (A8) into Eq. (A5) yields the following design equation for the step-invariant filter.

$$H(z) = (1 - z^{-1})\mathcal{Z}\left\{\mathcal{L}^{-1}\left\{\frac{1}{s}\mathbf{H}(s)\right\}\big|_{t=nT_s}\right\} \tag{A9}$$

b) Here

$$\mathbf{H}(s)\mathbf{X}(s) = \frac{1}{s(1+s\tau)} = \frac{1}{s} - \frac{1}{1+s\tau} \tag{A10}$$

where we have used partial fraction expansion. The step response of the CT filter is

$$y(t) = \mathcal{L}^{-1}\{\mathbf{H}(s)\mathbf{X}(s)\} = (1 - e^{-\frac{t}{\tau}})u(t) \tag{A11}$$

The corresponding output for the DT filter is

$$y[n] = (1 - e^{-\frac{nT_s}{\tau}})u[n] \tag{A12}$$

The z-transform of $y[n]$ is

$$Y(z) = \mathcal{Z}\{y[n]\} = \frac{z}{z-1} - \frac{z}{z - e^{-\frac{T_s}{\tau}}} = \frac{z(1 - e^{-\frac{T_s}{\tau}})}{(z-1)(z - e^{-\frac{T_s}{\tau}})} \tag{A13}$$

Therefore, Eq. (9) yields

$$H(z) = (1 - z^{-1})\frac{z(1 - e^{-\frac{T_s}{\tau}})}{(z-1)(z - e^{-\frac{T_s}{\tau}})} = \frac{1 - e^{-\frac{T_s}{\tau}}}{z - e^{-\frac{T_s}{\tau}}} \tag{A14}$$

Problem 7.5.22 asks you to draw the structure of this filter and compare its amplitude characteristic with that of Example 7.7.

Introduction to Random Signals

8.1 Introduction

This chapter is a significant point of departure from all that has preceded it in this book: In earlier chapters, the sequences and waveforms we considered were described by known mathematical formulas and our work focused on the problem of determining the response of an LTI system driven by the known input. In the present chapter we are concerned with input sequences and waveforms that are characterized by probabilities and statistical averages. These signals are called random, or stochastic, signals or stochastic processes. We are interested in determining the probabilities and statistical averages that characterize the response.

The incorporation of probability into signal and system theory was an inevitable step in its development.[1] The step was inevitable because most real-life signals are not known until they are actually observed. They are, in principle, more like the sequence of numbers you get when you repeatedly toss a pair of dice or like the temperature variations in your city than like the sinusoid $\cos(2\pi 60t)$. Instead of deterministic formula, we need the tools of probability theory to describe these signals and to design systems that process them.

The topic of random signals is traditionally offered in courses that follow signals and systems. This chapter is intended as an introduction. The material is presented without the formality and mathematical depth found in later courses. The goal is to give you an *intuitive* understanding of the main concepts and results.

8.2 Probability

8.2.1 Definitions of Probability

Three definitions of probability have been proposed over time. Each is still used. All three definitions address the fundamental idea that probability is a mathematical model for relative frequency.

Relative Frequency Definition

The relative frequency definition of probability directly captures the idea that probability is a mathematical model for relative frequency. Here, the probability of an event $\mathcal{A}$ is defined

---

[1] The pioneering work of twentieth-century mathematicians Norbert Weiner, Claude Shannon, and Andrei Kolmogorov provided the mathematical foundation for the modern communication and control theories.

as the relative number of times that the event $\mathcal{A}$ occurs in N trials of an experiment when N is arbitrarily large.

$$P(\mathcal{A}) = \lim_{N \to \infty} \frac{n(\mathcal{A})}{N} \tag{1}$$

where $n(\mathcal{A})$ is the number of times that event $\mathcal{A}$ occurs in N trials. For example, suppose we are tossing a die having faces f_i, $i = 1, 2, 3, 4, 5, 6$. The probability of the event "face 2 shows" is defined as

$$P(f_2) = \lim_{N \to \infty} \frac{n\{f_2\}}{N} \tag{2}$$

where $n\{f_2\}$ is the number of times face 2 shows in N trials. Of course, if the die is fair, we intuitively expect that the limit on the right-hand side of Eq. (2) is 1/6.

Although the relative frequency definition of probability is intuitively satisfying, a moment's thought reveals that it has a fundamental weakness: It is not possible to physically repeat any trial an infinite number of times. Therefore, its definition of probability is *only* conceptual. In spite of this weakness, the relative frequency does help us appreciate many results of probability theory. Also, experience has shown that we can use the relative frequency definition with a sufficiently large but *finite* value for N to obtain reliable estimates of probability.[2]

Classical Definition

The classical definition of probability was motivated by games of chance like cards and dice. In this definition, the probability of an event $\mathcal{A}$, $P(\mathcal{A})$, is defined as the ratio of favorable outcomes to the total possible number of outcomes. Consider, for example, tossing a six-sided die. The probability that face 2 shows $P\{f_2\}$ is 1/6 because there is one favorable outcome and six possible outcomes. Similarly, the probability that *either* a face 2 or a face 5 shows $P\{f_2, f_5\}$ equals 2/6. The obvious weakness in this definition is that it does not work for an unfair die. In spite of this shortcoming, practical experience has proven the value of the classical definition in many applications.

Axiomatic Definition

In the *axiomatic* theory, no effort is made to specify numerical values for probabilities. Instead, the probabilities of events are simply defined as real numbers that obey certain rules. The rules describe how the probabilities of some events are related to the probabilities of other events. Thus, the axiomatic theory of probability is analogous to Newton's theory of motion in which the laws of motion are stated without specifying actual values for mass or force.

In the axiomatic theory, events are sets of elements. An event, a set $\mathcal{A}$, is said to occur if one of its elements in $\mathcal{A}$ is the outcome of the experiment. For example, in a die toss experiment, the event $\mathcal{A} = \{f_1, f_5\}$ occurs if the outcome is either an f_1 or an f_5.

The probability of an event $\mathcal{A}$, $P(\mathcal{A})$, is defined as a real number that satisfies the following three axioms.

---

[2] Simulation experiments that demonstrate this property are found in Problem 8.3.5.

Axioms of Probability Theory

Axiom 1: $P(\mathcal{A}) \geq 0$.

Axiom 2: $P(\mathcal{S}) = 1$, where $\mathcal{S}$ is the certain event, $\mathcal{S} = \{\text{all possible outcomes}\}$.

Axiom 3: If $\mathcal{A}$ and $\mathcal{B}$ are mutually exclusive events, then

$$P(\mathcal{A} \cup \mathcal{B}) = P(\mathcal{A}) + P(\mathcal{B}) \tag{3}$$

An example of Axiom 2 is given by the die toss experiment, where $\mathcal{S} = \{f_1, f_2, f_3, f_4, f_5, f_6\}$ which means that face 1 *or* face 2 or face 3 or face 4 *or* face 5 *or* face 6 shows. Certainly some face *must* show, so $P(\mathcal{S}) = 1$.

In axiom 3, The symbol $\cup$ means set union. $\mathcal{A} \cup \mathcal{B} \equiv \mathcal{B} \cup \mathcal{A}$ is the set having elements belonging to $\mathcal{A}$ or $\mathcal{B}$. For example, if $\mathcal{A} = \{f_2\}$ and $\mathcal{B} = \{f_5\}$, then $\mathcal{A} \cup \mathcal{B} = \{f_2, f_5\}$. $\mathcal{A}$ and $\mathcal{B}$ are *mutually exclusive* events if they do not contain a common element, i.e., if their *intersection* $\mathcal{A} \cap \mathcal{B}$ is empty. The set intersection. $\mathcal{A} \cap \mathcal{B} \equiv \mathcal{B} \cap \mathcal{A}$ is the set of elements that $\mathcal{A}$ and $\mathcal{B}$ have in common. $\mathcal{A} \cap \mathcal{B}$ should be read "$\mathcal{A}$ and $\mathcal{B}$". Clearly $\{f_2\}$ and $\{f_5\}$ do not contain a common element. Axiom 3 says that the probability that either a face 2 or a face 5 appears equals the probability that face 2 appears plus the probability that face 5 appears.

The axioms agree with the relative frequency definition of probability: Relative frequency is nonnegative, hence axiom 1. The relative frequency that *some* outcome occurs is unity, hence axiom 2. To consider axiom 3, consider again the die toss experiment. The number of times a face 2 or a face 5 appears, $n(\{f_2, f_5\})$, is the sum $n(\{f_2, f_5\}) = n(\{f_2\}) + n(\{f_5\})$. This leads to axiom 3. Similarly, we can show that the axioms agree with the classical definition of probability.

Because events are sets in the axiomatic theory, definitions and results from set theory are often helpful in manipulating probabilities. We have already seen the definitions for set union $\cup$ and intersection $\cap$. Another concept is the *empty set* ϕ defined as the set containing no elements. The empty set is defined to have the following properties.

$$\phi \cup \mathcal{A} = \mathcal{A} \tag{4}$$

and

$$\phi \cap \mathcal{A} = \phi \tag{5}$$

for every set $\mathcal{A}$. Using the empty set, we can express the idea that two sets $\mathcal{A}$ and $\mathcal{B}$ are mutually exclusive as

$$\mathcal{A} \cap \mathcal{B} = \phi \tag{6}$$

You can verify from using axiom 3 that the probability of ϕ equals 0.

$$P(\phi) = 0 \tag{7}$$

The empty set ϕ is called the *impossible event* because *some* outcome always occurs in a trial of an experiment. It is impossible that no outcome occurs.

Another idea from set theory is set complementation. The *complement* $\bar{\mathcal{A}}$ of the set $\mathcal{A}$ is defined as the set containing all the elements of $\mathcal{S}$ that are not in $\mathcal{A}$. It follows from this definition that

$$\mathcal{A} \cup \bar{\mathcal{A}} = \mathcal{S} \tag{8}$$

and

$$\mathcal{A} \cap \bar{\mathcal{A}} = \phi \tag{9}$$

Figure 8.1
Venn diagrams:
(a) union $A \cup B$
(shaded);
(b) intersection $A \cap B$
(shaded);
(c) complement $\bar{A}$
(shaded);
(d) inclusion $A \subset B$.

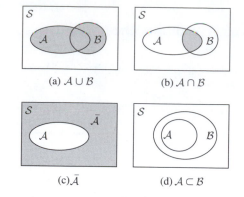

(a) $A \cup B$ (b) $A \cap B$

(c) $\bar{A}$ (d) $A \subset B$

We are often interested in $P(\bar{A})$, which is the probability that some event A does *not* occur. It follows from the preceding two equations and axioms 1 and 3 that

$$P(A) + P(\bar{A}) = 1 \qquad (10)$$

Set relationships and operations can be visualized using a picture representation of abstract sets called a Venn diagram. Venn diagrams are often used to motivate or describe a proof involving sets visually.[3] Figures 8.1(a), (b), and (c) show Venn diagrams that represent set a union, an intersection, and a complement respectively. Figure 8.1(d) shows the Venn diagram for set inclusion, $A \subset B$. We can see from this diagram that a set A is included in a set B, $A \subset B$, if and only if every element of A is also an element of B. We can also see that $A \subset B$ if and only if $A \cap B = A$.

In the following example, we use Venn diagrams to help find $P(A \cup B)$, where A and B are not necessarily mutually exclusive.

EXAMPLE 8.1   Some Venn Diagrams

As an example of the usefulness of Venn diagrams, let's find the probability that either event A or event B occurs when A and B are not mutually exclusive. You can see by drawing a Venn diagram that the set $A \cup B$ can be broken into two mutually exclusive parts, A and $\bar{A} \cap B$. The set $\bar{A} \cap B$ consists of the elements that are in B but not in A.

$$A \cup B = A \cup (\bar{A} \cap B) \qquad (11)$$

Axiom 3 then yields the result

$$P(A \cup B) = P(A) + P(\bar{A} \cap B) \qquad (12)$$

You can see by drawing another Venn diagram that the set B can be broken into two mutually exclusive parts, $A \cap B$ and $\bar{A} \cap B$. The set $A \cap B$ consists of the elements that are in both B and A.

$$B = (\bar{A} \cap B) \cup (A \cap B) \qquad (13)$$

Axiom 3 then yields

$$P(B) = P(\bar{A} \cap B) + P(A \cap B) \qquad (14)$$

Now if we eliminate $P(\bar{A} \cap B)$ from Eqs. (12) and (14), we obtain the final result.

$$P(A \cup B) = P(A) + P(B) - P(A \cap B) \qquad (15)$$

---

[3] Venn diagrams, however, do not in themselves provide a rigorous proof.

8.3 **Statistical Dependence and Independence**

If you toss a die repeatedly, the outcome of one toss does not ordinarily effect the probabilities of another toss. For example, if you get a f_2 on toss 1, the probabilities for the various faces showing on the second toss are the same as they were before you made the first toss. We say that the first and second tosses are *statistically independent*.

Statistically Independent events

Two events A and B are statistically independent if the probability that they both occur, $P(A \cap B)$, equals the product of the probabilities of each event.

$$P(A \cap B) = P(A)P(B) \tag{16}$$

The symbol $\cap$ denotes set intersection; so $A \cap B \equiv B \cap A$ is the set of elements that A and B have in common. $A \cap B$ should be read "A and B."

EXAMPLE 8.2   Statistically Independent Events

Consider tossing a fair die twice. In the case of the die toss, there are $6 \times 6 = 36$ possible outcomes for two tosses. Only one of these 36 outcomes has f_2 on both the first and the second toss. Thus, by the classical definition (which assumes equally likely outcomes)

$$P\{f_2 \text{ on first toss } and \text{ } f_2 \text{ on second toss}\}$$

$$= P(\{f_2 \text{ on first toss}\} \cap \{f_2 \text{ on second toss}\})$$

$$= \frac{1}{36} = \frac{1}{6} \times \frac{1}{6} = P\{f_2 \text{ on first toss}\}P\{f_2 \text{ on second toss}\} \tag{17}$$

which has the form $P(A \cap B) = P(A)P(B)$.

A formal treatment of this example is to set up the sample space S as illustrated in Figure 8.2. The set of all possible experimental outcomes is S. There are 36 possible outcomes when you toss a die twice. Because the die is fair and because the two tosses are statistically independent, each of the 36 outcomes has the same probability: 1/36. We see

Figure 8.2
Die toss sample space with events A and B.

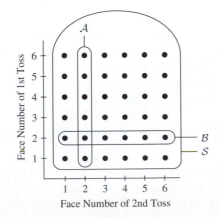

Face Number of 2nd Toss

in Figure 8.2 the events or sets $A = \{f_2$ on first toss$\}$ and $B = \{f_2$ on second toss$\}$. Note that A and B each contain six elements, each element with a probability of 1/36. Thus $P(A) = P(B) = 6/36 = 1/6$ and $P(A)P(B) = 1/36$. The event $A \cap B$ contains only one element, and $P(A \cap B) = 1/36$. Thus, $P(A \cap B) = P(A)P(B)$.

EXAMPLE 8.3   Previous Example with Different A and B

As a second example of statistically independent events, consider the event $\{f_2$ or f_4 on first toss *and* f_2 or f_6 on second toss$\}$ for the experiment of the preceding example. This event has four equally likely members, which are $\{f_2f_2, f_2f_6, f_4f_2, f_4f_6\}$. Thus

$$P\{f_2 \text{ or } f_4 \text{ on first toss and } f_2 \text{ or } f_6 \text{ on second toss}\}$$

$$= P(\{f_2 \text{ or } f_4 \text{ on first toss}\} \cap \{f_2 \text{ or } f_6 \text{ on second toss}\})$$

$$= \frac{4}{36} = \frac{2}{6} \times \frac{2}{6} = P\{f_2 \text{ or } f_4 \text{ on first toss } P\{f_2 \text{ or } f_6 \text{ on second toss}\} \tag{18}$$

which again has the form $P(A \cap B) = P(A)P(B)$. The events S, $A = \{f_2$ or f_4 on first toss$\}$ and $B = \{f_2$ or f_6 on second toss$\}$ are illustrated in Figure 8.3. As we have just shown, A and B are statistically independent.

Figure 8.3
Die toss sample space with different events A and B.

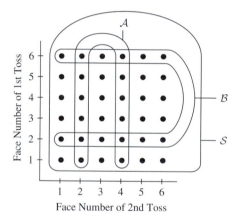

In the preceding two examples, the probabilities for the second toss did not depend on the outcome of the first toss. The two tosses were therefore statistically independent. Consider what would happen if the table was designed to magnetize the die. The direction of the magnetic field induced in the die would depend on which face of the die showed on the first toss. This would, for example, make $P\{f_2$ on second toss *given we have* f_2 on the first toss$\}$ different from $P\{f_2$ on second toss$\}$ (where we are not given the results of the first toss). Thus, the use of a magnetic table (and a die that can be magnetized) would introduce statistical dependence between the two tosses.

EXAMPLE 8.4   Statistically Dependent Events

Consider tossing a fair die twice over a table that is designed to magnetize the die. For the first toss, we have $P\{f_i\} = 1/6$, $i = 1, 2, \ldots, 6$ because the die is fair and has not been magnetized yet by the table. Suppose that the magnetic field of the table is so strong that it fully magnetizes the die so that the outcome of the second toss is certain to be the same as the outcome of the first. For example, if you get a face 2 on the first toss, it is certain that you will get a face 2 on the second toss; and if you got a face 3 on the first toss, it is certain that you will get a face 3 on the second toss, and so on. We can define the sets $\mathcal{S}$, $\mathcal{A}$, and $\mathcal{B}$ just as we did before. But the probabilities of the 36 elements are no longer 1/36 each. Instead, we have all the probabilities concentrated along the diagonal, i.e.

$$P\{f_i \text{ on first toss and } f_j \text{ on second toss}\} = \begin{cases} 0 & \text{for } i \neq j \\ 1/6 & \text{for } i = j \end{cases} \tag{19}$$

$i, j = 1, 2, \ldots, 6$. Here $P(\mathcal{A}) = P(\mathcal{B}) = 1/6$ as before, but $P(\mathcal{A} \cap \mathcal{B}) = 1/6$ is not as before. For this example of statistically dependent events, we see that $P(\mathcal{A} \cap \mathcal{B}) \neq P(\mathcal{A})P(\mathcal{B})$. Of course, this example is an extreme case. For a somewhat less powerful magnetic field, we might have

$$P\{f_i \text{ on first toss and } f_j \text{ on second toss}\} = \begin{cases} 1/180 & \text{for } i \neq j \\ 5/36 & \text{for } i = j \end{cases} \tag{20}$$

Three events, $\mathcal{A}_1$, $\mathcal{A}_2$, and $\mathcal{A}_3$ are statistically independent if they are independent in pairs

$$P(\mathcal{A}_i \cap \mathcal{A}_j) = P(\mathcal{A}_i)P(\mathcal{A}_j) \quad i < j = 1, 2, 3 \tag{21}$$

and if

$$P(\mathcal{A}_1 \cap \mathcal{A}_2 \cap \mathcal{A}_3) = P(\mathcal{A}_1)P(\mathcal{A}_2)P(\mathcal{A}_3) \tag{22}$$

For N independent events, $\mathcal{A}_1 \mathcal{A}_2, \ldots, \mathcal{A}_N$ the probabilities of all combinations of intersections of the events similarly factor into products of probabilities of individual events.

8.3.1 Conditional Probability

We said at the start of Section 8.2 that the use of a magnetic table makes $P\{f_2$ on the second toss *given we have* f_2 on the first toss$\}$ different from $P\{f_2$ on the second toss$\}$ (where we do not have the condition that f_2 occurred on the first toss). The probability $P\{f_2$ on the second toss *given we have* f_2 on the first toss$\}$ is a *conditional probability* and is denoted by $P\{f_2$ on the second toss$|f_2$ on the first toss$\}$, where the condition goes to the right of the vertical bar ($|$). In general, the probability of an event $\mathcal{B}$ conditioned on an event $\mathcal{A}$ is denoted as $P\{\mathcal{B}|\mathcal{A}\}$. We can motivate the definition of $P\{\mathcal{B}|\mathcal{A}\}$ using the relative frequency definition of probability.

$$P\{\mathcal{B}|\mathcal{A}\} = \lim_{N \to \infty} \frac{n(\mathcal{A} \cap \mathcal{B})}{n(\mathcal{A})} \tag{23}$$

In Eq. (23), $n(\mathcal{A})$ is the number of times event $\mathcal{A}$ occurred and $n(\mathcal{A} \cap \mathcal{B})$ is the number of times both events $\mathcal{A}$ and $\mathcal{B}$ occurred in N trials of an experiment. Thus, $n(\mathcal{A} \cap \mathcal{B})/n(\mathcal{A})$ is the number of times both $\mathcal{A}$ and $\mathcal{B}$ occurred, divided by the number of times $\mathcal{A}$ occurred in the N trials of an experiment. The following example increases your understanding of Eq. (23).

EXAMPLE 8.5   Relative Frequency and Conditional Probability

Consider the experiment of tossing a die twice, i.e., *one trial* of this experiment is defined as tossing a die *twice*. Suppose that we perform this experiment $N = 1{,}200{,}000$ times, and find that f_2 has shown on the first toss $n(\mathcal{A}) = n\{f_2$ on the first toss$\} = 200{,}027$ times. We inspect the 200,027 outcomes having f_2 on the first toss, and we find of these

200,027 outcomes, f_2 occurred on the second toss $n(\mathcal{A} \cap \mathcal{B}) = n\{f_2$ on the first toss, *and* f_2 on the second toss$\} =$ 199,990 times.

If we use $N = 1,200,000$ instead of the limit on the right-hand side of Eq. (23), we would estimate the conditional probability $P\{f_2$ on second toss$|f_2$ on the first toss$\}$ as:

$$P\{f_2 \text{ on second toss}|f_2 \text{ on the first toss}\} \approx \frac{n\{f_2 \text{ on the first toss } and \text{ } f_2 \text{ on the second toss}\}}{n\{f_2 \text{ on first toss}\}} = \frac{199,990}{200,027}$$

If we divide the numerator and denominator in Eq. (23) by N and apply the relative frequency definitions of $P\{\mathcal{A} \cap \mathcal{B}\}$ and $P\{\mathcal{A}\}$, we find that

$$P\{\mathcal{B}|\mathcal{A}\} = \lim_{N \to \infty} \frac{[\frac{n(\mathcal{A} \cap \mathcal{B})}{N}]}{[\frac{n(\mathcal{A})}{N}]} \qquad (24)$$

In the limit $N \to \infty$, the numerator and denominator on the right-hand side become, by definition, $P\{\mathcal{A} \cap \mathcal{B}\}$ and $P\{\mathcal{A}\}$, respectively. This motivates us to define the conditional probability of $\mathcal{A}$ given $\mathcal{B}$ as follows:

Conditional Probability of $\mathcal{B}$ Given $\mathcal{A}$

$$P\{\mathcal{B}|\mathcal{A}\} \triangleq \frac{P\{\mathcal{A} \cap \mathcal{B}\}}{P\{\mathcal{A}\}} \qquad (25)$$

EXAMPLE 8.6   Conditional Probability

There are seven pencils in a box. Each pencil is either red (R), green (G), or black (B) and either hard (H) or soft (S). There are 2 RH pencils, 1 RS pencil, 1 GH pencil, 2 BS pencils and 1 BH pencil. A list of the pencils is given in Table 8.1, where "pj" means pencil j for $j = 1, 2, \ldots, 7$. Consider an experiment consisting of selecting a pencil at random from the box. By selecting "at random" we mean that the probability of selecting pencil pj is 1/7 for $j = 1, 2, \ldots, 7$. Assume that someone selects a pencil and tells you that it is red. What is the probability that this selected pencil is hard, given that it is red?

We can answer this question using the formula for conditional probabilities, we find

$$P(H|R) = \frac{P(R \cap H)}{P(R)} = \frac{P\{p1, p2\}}{P\{p1, p2, p3\}} \qquad (26)$$

Table 8.1
Pencils in a Box

Pencil	R	G	B	H	S
p1	✓			✓	
p2	✓			✓	
p3	✓				✓
p4		✓		✓	
p5			✓	✓	
p6			✓		✓
p7			✓		✓

It is easy to find the probability of selecting a red pencil.

$$P(R) = P\{p1, p2, p3\} = P\{p1\} + P\{p2\} + P\{p3\} = \frac{3}{7} \tag{27}$$

In the preceding equation, we have used axiom 3. Notice that the result is consistent with that obtained using the classical definition of probability because there are three outcomes favorable to R and seven possible outcomes. Similarly it is easy to find $P(R \cap H)$.

$$P(R \cap H) = P\{p1, p2\} = P\{p1\} + P\{p2\} = \frac{2}{7} \tag{28}$$

Therefore

$$P(H|R) = \frac{2/7}{3/7} = \frac{2}{3} \tag{29}$$

Notice that this result is consistent with that obtained using the classical definition of probability. If we are *given* that the selected pencil is red, then there are only three possible outcomes, namely p1, p2, and p3. Of these three, only two are favorable to H.

Let us consider what happens to the formula $P\{\mathcal{B}|\mathcal{A}\} = P\{\mathcal{A} \cap \mathcal{B}\}/P\{\mathcal{A}\}$ when $\mathcal{A}$ and $\mathcal{B}$ are statistically independent. Recall that $\mathcal{A}$ and $\mathcal{B}$ are statistically independent if and only if $P\{\mathcal{A} \cap \mathcal{B}\} = P\{\mathcal{A}\}P\{\mathcal{B}\}$. It follows that when $\mathcal{A}$ and $\mathcal{B}$ are statistically independent

$$P\{\mathcal{B}|\mathcal{A}\} = \frac{P\{\mathcal{A} \cap \mathcal{B}\}}{P\{\mathcal{A}\}} = \frac{P\{\mathcal{A}\}P\{\mathcal{B}\}}{P\{\mathcal{A}\}} = P\{\mathcal{B}\} \tag{30}$$

Thus, $\mathcal{A}$ and $\mathcal{B}$ are statistically independent if and only if $P\{\mathcal{B}|\mathcal{A}\} = P\{\mathcal{B}\}$. Similarly, $\mathcal{A}$ and $\mathcal{B}$ are statistically independent if and only if $P\{\mathcal{A}|\mathcal{B}\} = P\{\mathcal{A}\}$.

In Problem 8.3.2 you are asked to show that in general

$$P(\mathcal{A}|\mathcal{B}) = \frac{P\{\mathcal{B}|\mathcal{A}\}P(\mathcal{A})}{P(\mathcal{B})} \tag{31}$$

where $\mathcal{A}$ and $\mathcal{B}$ are arbitrary events. Eq. (31) is sometimes called the formula for *inverse probability*. It applies whether or not $\mathcal{A}$ and $\mathcal{B}$ are statistically dependent.

8.3.2 "Total Probability" The term "total probability" can be understood by referring to Example 8.6. There we see that the event R can be partitioned into two mutually exclusive parts, namely $R \cap H$ and $R \cap S$. (You should write out the elements in the sets R, $R \cap H$, and $R \cap S$ to make sure you understand.) So we have $R = [R \cap H] \cup [R \cap S]$. The pencils that are both red and hard are clearly distinct from the pencils that are both red and soft, so $R \cap H$ and $R \cap S$ are mutually exclusive events. This means that we can use axiom 3 to write

$$P\{R\} = P\{[R \cap H] \cup [R \cap S]\} = P\{R \cap H\} + P\{R \cap S\} \tag{32}$$

which, by definition of conditional probability can be written as

$$P\{R\} = P\{R|H\}P\{H\} + P\{R|S\}P\{S\} \tag{33}$$

So we can say that the "total probability" of selecting a red pencil is given by the weighted sum of the conditional probabilities $P\{R|H\}$ and $P\{R|S\}$, as shown. The preceding formula can be generalized to three or more conditions.

"Total Probability" Formula

Suppose we can partition an event $\mathcal{A}$ into $\mathcal{M}$ parts: $\mathcal{A}_1, \mathcal{A}_2, \ldots, \mathcal{A}_M$. A partition of $\mathcal{A}$ is defined by the properties

$$\mathcal{A} = \mathcal{A}_1 \cup \mathcal{A}_2 \cup \cdots \cup \mathcal{A}_M \tag{34}$$

and

$$\mathcal{A}_i \cap \mathcal{A}_j = \phi \qquad \text{for } i \neq j \tag{35}$$

The "total probability" of the event $\mathcal{A}$ is given by

$$P\{\mathcal{A}\} = P\{\mathcal{A}|\mathcal{A}_1\}P\{\mathcal{A}_1\} + P\{\mathcal{A}|\mathcal{A}_2\}P\{\mathcal{A}_2\} + \cdots + P\{\mathcal{A}|\mathcal{A}_M\}P\{\mathcal{A}_M\} \tag{36}$$

8.4   Random Variables

A real random variable is a real number assigned to each outcome of a probabilistic experiment. For example, in the die toss, we can assign the number $x(f_1) = 1$ to outcome f_1, the number $x(f_2) = 2$ to outcome $f_2, \ldots,$ the number $x(f_6) = 6$ to outcome f_6. For another example, the experiment might be tossing a coin. Here we might assign the number $x(H) = -10$ to the outcome H (heads shows) and the number $x(T) = \pi$ to the outcome T (tails shows). In each example, the number x is called a random variable. Note that the number assigned to each outcome is a *known* number. The term x is called a *random* variable because the outcome of the experiment (and hence the value of x) is random. Note also that we use boldface to denote a random variable. Therefore we make a distinction between a random variable x and an ordinary variable x.

In each of the preceding two examples, the random variable had a discrete set of possible values. Naturally, we call such random variables *discrete random variables*. We can also have *continuous random variables*, which are random variables for which the set of possible values occupies some continuous range.

8.4.1 Probability Distribution Functions (PDFs)

The *probability distribution function* (PDF) $F_x(x)$ of a random variable x is defined as the probability that x is less than or equal to some number x.

Probability Distribution Function

$$F_x(x) = P\{x \leq x\} \tag{37}$$

There is nothing random about $F_x(x)$: It is a known function. We often plot $F_x(x)$ versus x. $F_x(x)$ is also called a *cumulative probability distribution function*.

The probability distribution function has the following properties:

> **1.** $F_x(-\infty) = 0$.
> **2.** $F_x(\infty) = 1$.
> **3.** $F_x(x)$ is never negative.
> **4.** $F_x(x)$ is a monotonically increasing function of x.

Hand led proofs of these properties are found in the problems. The following example helps you understand how the properties arise.

EXAMPLE 8.7 PDF for a Fair Die Toss

Consider tossing a fair die. Define the random variable $x(f_i) = i = 1, 2, \ldots, 6$. Find and plot the PDF $F_x(x)$.

We can start by considering specific values for x. For example, let's first consider, say, $x = 10$. $F_x(10)$ is the probability that x is less than or equal to 10.

$$F_x(10) = P\{x \le 10\} \tag{38}$$

The event $\{x \le 10\}$ is the set of outcomes for which x is less than or equal to 10. We can see from the definition of $x(f_i)$ that, no matter which face shows, x is less than or equal to 10. Thus

$$\{x \le 10\} = \{f_1, f_2, f_3, f_4, f_5, f_6\} = \mathcal{S} \tag{39}$$

and

$$F_x(10) = P\{x \le 10\} = P\{\mathcal{S}\} = 1 \tag{40}$$

Now by the same reasoning, we can conclude that

$$F_x(x) = 1 \qquad \text{for all } x \ge 6 \tag{41}$$

Let's next consider $F_x(x)$ for $x = 5.5$. The event $\{x \le 5.5\}$ is the set of outcomes for which x is less than or equal to 5.5. We can see from the definition that $x(f_i)$ is less than or equal to 5.5 for $i = 1, 2, 3, 4$, and 5. Thus

$$\{x \le 5.5\} = \{f_1, f_2, f_3, f_4, f_5\} \tag{42}$$

and

$$F_x(5.5) = P\{f_1, f_2, f_3, f_4, f_5\} = \frac{5}{6} \tag{43}$$

Similarly, we can conclude that

$$F_x(x) = \frac{5}{6} \qquad \text{for } 5 \le x < 6 \tag{44}$$

and

$$F_x(x) = \frac{i}{6} \qquad \text{for } i \le x < i+1 \tag{45}$$

for $i = 1, 2, 3, 4, 5$. Finally, there are no outcomes for which $x < 1$, i.e., for all $x < 1$

$$\{x \le x\} = \phi \tag{46}$$

Figure 8.4
PDF for a fair die toss
experiment.

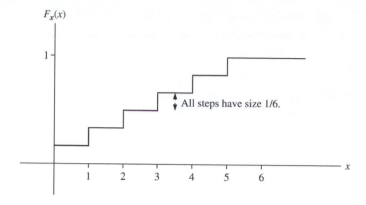

Thus

$$F_x(x) = P(\phi) = 0 \qquad \text{for } x < 1 \tag{47}$$

A plot of $F_x(x)$ is shown in Figure 8.4

EXAMPLE 8.8   PDF for Unfair Coin Toss

Consider tossing an unfair coin. Define the random variable $x(H) = -10$ and $x(T) = \pi$. Let $P(H) = p$ and $P(T) = q = 1 - p$. By using a method similar to that in Example 8.7 (see Problem 8.4.6), we obtain the PDF $F_x(x)$, shown in Figure 8.5.

Figure 8.5
PDF for an unfair coin
toss experiment.

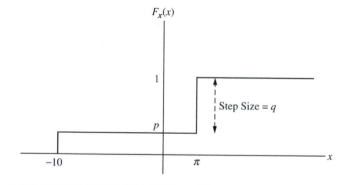

Notice that the PDFs we obtained in these examples look like staircases. Staircaselike PDFs result for every discrete random variable. The height of a step is equal to the probability that the random variable has a value equal to the x coordinate of that step. For continuous type random variables, $F_x(x)$ is continuous (no steps).

8.4.2
Probability
Density Functions
(pdfs)

The *probability density function* (pdf) $f_x(x)$ of a random variable x is defined as the derivative of the distribution function $F_x(x)$.

Probability Density Function (pdf)

$$f_{\boldsymbol{x}}(x) = \frac{d}{dx}F_{\boldsymbol{x}}(x) \tag{48}$$

It follows from the definition that $f_{\boldsymbol{x}}(x)$ is never negative and the area under $f_{\boldsymbol{x}}(x)$ equals 1 (Problem 8.4.8).

It also follows from the definition that

$$F_{\boldsymbol{x}}(x) = P\{\boldsymbol{x} \le x\} = \int_{-\infty}^{x} f_{\boldsymbol{x}}(x)\,dx \tag{49}$$

One of the most important uses of a pdf is to find the probability that $\boldsymbol{x}$ is in some interval $P\{x_1 < \boldsymbol{x} \le x_2\}$. This probability is given by

$$P\{x_1 < \boldsymbol{x} \le x_2\} = \int_{x_1}^{x_2} f_{\boldsymbol{x}}(x)\,dx \tag{50}$$

We can establish Eq. (50) by noticing that

$$\{\boldsymbol{x} \le x_2\} = \{\boldsymbol{x} \le x_1\} \cup \{x_1 < \boldsymbol{x} \le x_2\} \tag{51}$$

where

$$\{\boldsymbol{x} \le x_1\} \cap \{x_1 < \boldsymbol{x} \le x_2\} = \phi \tag{52}$$

Axiom 3 then tells us that

$$P\{\boldsymbol{x} \le x_2\} = P\{\boldsymbol{x} \le x_1\} + P\{x_1 < \boldsymbol{x} \le x_2\} \tag{53}$$

which is

$$\int_{-\infty}^{x_2} f_{\boldsymbol{x}}(x)\,dx = \int_{-\infty}^{x_1} f_{\boldsymbol{x}}(x)\,dx + P\{x_1 < \boldsymbol{x} \le x_2\} \tag{54}$$

and Eq. (50) follows.

We can understand how the name *probability density function* arises if we set $x_2 = x$ and $x_1 = x - \Delta x$ in Eq. (50), where Δx is small. The result is

$$P\{x - \Delta x < \boldsymbol{x} \le x\} = \int_{x-\Delta x}^{x} f_{\boldsymbol{x}}(\psi)\,d\psi \approx f_{\boldsymbol{x}}(x)\,\Delta x \tag{55}$$

where we assume that $f_{\boldsymbol{x}}(\psi)$ is approximately constant for $x - \Delta x \le \psi \le x$. The preceding approximation rearranges to

$$f_{\boldsymbol{x}}(x) \approx \frac{P\{x - \Delta x < \boldsymbol{x} \le x\}}{\Delta x} \tag{56}$$

It follows that $f_{\boldsymbol{x}}(x)$ is approximately equal to the probability that $\boldsymbol{x}$ is in the interval $x - \Delta x < \boldsymbol{x} \le x$ divided by the length of that interval. If we take the limit $\Delta x \to 0$, we obtain

$$f_{\boldsymbol{x}}(x) = \lim_{\Delta x \to 0} \frac{P\{x - \Delta x < \boldsymbol{x} \le x\}}{\Delta x} \tag{57}$$

Because $f_{\boldsymbol{x}}(x)$ equals probability per unit length, it is a probability density.

We saw in Examples 8.7 and 8.8 that a discrete random variable has a staircaselike PDF where the height of each step equals the probability that the value of the random variable equals the x position of the step. For a discrete random variable, $f_x(x)$ consists of impulses located at the discrete possible values of x. The impulse areas equal the probabilities of the corresponding discrete values of x. For example, for the random variable associated with the experiment described in Example 8.8, we have

$$f_x(x) = p\delta(x + 10) + q\delta(x - \pi) \tag{58}$$

If x is a continuous random variable, then $f_x(x)$ does not contain any impulses. Many continuous random variables encountered in engineering are well described by a gaussian pdf

$$f_x(x) = \frac{1}{\sqrt{2\pi\sigma^2}}e^{-\frac{(x-\eta)^2}{2\sigma^2}} \tag{59}$$

The constants η and σ^2 are the *mean* and the *variance* of x. The term σ is called the *standard deviation* of x. Recall that a plot of a gaussian function appeared in Figure 2CT.3 for $\eta = 0$. Its peak occurred at $x = 0$. In general, the peak of $f_x(x)$ occurs at $x = \eta$ and the width of $f_x(x)$ is proportional to σ. In the next section we define the mean, variance, and standard deviation for an arbitrary random variable and explain what these terms signify in general.

8.5 Stochastic Averages (Expectation)

Expectation is a mathematical model for arithmetic averages. To understand how the formulas of expectation are motivated, we consider an experiment in which we toss a die N times. The outcome of each toss is an element from $S = \{f_1, f_2, f_3, f_4, f_5, f_6\}$. Define $x(i)$ as the value on the face that shows on the ith toss. (For example the value of f_2 is 2.) We tabulate the results of the N tosses, as illustrated in Table 8.2. The first column of the table is the number i of the toss: $i = 1, 2, \ldots, N$. and the second column of our table is the value

Table 8.2
List of Outcomes

i	$x(i)$
1	5
2	3
3	6
4	1
5	1
6	5
7	4
8	6
9	2
10	1
11	3
12	4
.	.
.	.
.	.
N	2

of the random variable $x(i)$. The arithmetic average of the entries in the table is given by the formula

$$\text{Ave}\{x\} = \frac{1}{N}\sum_{i=1}^{N} x(i) \tag{60}$$

Let's do some algebra on this formula. There are six possible values for $x(i)$, namely 1, 2, 3, 4, 5, and 6. Let $n(f_k)$ denote the number of times the value k appears in the second column of the table: $k = 1, 2, 3, 4, 5$, and 6. If we add the $n(f_k)$ terms of the sum where $x(i) = k$, the result is $n(f_k) \times k$. This is true for each $k = 1, 2, 3, 4, 5$, and 6. This means that we can rewrite the arithmetic average as

$$\text{Ave}\{x\} = \frac{1}{N}\sum_{i=1}^{N} x(i) = \frac{1}{N}[n(f_1)1 + n(f_2)2 + \cdots + n(f_6)6] \tag{61}$$

or

$$\text{Ave}\{x\} = \sum_{k=1}^{6} \frac{n(f_k)}{N} k \tag{62}$$

Notice that the ratio $n(f_k)/N$ is the relative frequency of face k in N tosses. In the limit $N \to \infty$, we apply the relative frequency definition of probability to obtain

$$\lim_{N\to\infty}\text{Ave}\{x\} = \lim_{N\to\infty}\sum_{k=1}^{6}\frac{n(f_k)}{N}k = \sum_{k=1}^{6}k \times \lim_{N\to\infty}\frac{n(f_k)}{N} = \sum_{k=1}^{6}kP\{f_k\} \tag{63}$$

This result motivates us to define the *expected* die value as

$$E\{x\} = \sum_{k=1}^{6} kP\{f_k\} \tag{64}$$

We can generalize the preceding definition as follows.

Expected Value of a Discrete Random Variable

Let x be a random variable with possible values $x_1, x_2, \ldots, x_M$, where $P\{x = x_k\} = p_k$, $k = 1, 2, \ldots, M$. The *expected value* of x is defined as

$$E\{x\} = \sum_{k=1}^{M} x_k p_k \tag{65}$$

Let's go back to the die toss experiment where we tabulated the face values for the N tosses. Imagine inserting a third column in which the face values of the second column are squared. The first row would be 1, 5, 25, the second row would be 2, 3, 9, and so on. By using algebra similar to that in the preceding paragraph, we can show that the arithmetic average of the numbers in the second column can be written as

$$\text{Ave}\{x^2\} = \sum_{k=1}^{6} \frac{n(f_k)}{N} k^2 \tag{66}$$

which in the limit $N \to \infty$ becomes

$$\lim_{N \to \infty} \text{Ave}\{x^2\} = \sum_{k=1}^{6} k^2 P\{f_k\} \tag{67}$$

and thus

$$E\{x^2\} = \sum_{k=1}^{6} k^2 P\{f_k\} \tag{68}$$

More generally, if x is any random variable that can have possible values $x_1, x_2, \ldots,$ x_M, with probabilities $P\{x = x_k\} = p_k$, $k = 1, 2, \ldots, M$, we define the *expected value* of x^2 as

---

Expected Value of x^2

$$E\{x^2\} = \sum_{k=1}^{M} x_k^2 p_k \tag{69}$$

Generalizing along the same lines, we have for any function $g(x)$ of the discrete random variable x

---

Expected Value of a Function of a Random Variable

$$E\{g(x)\} = \sum_{k=1}^{M} g(x_k) p_k \tag{70}$$

If x is a continuous random variable, these formulas become

Expectation Formulas for a Continuous Random Variable

$$E\{x\} = \int_{-\infty}^{\infty} x f_x(x)\, dx \tag{71}$$

$$E\{x^2\} = \int_{-\infty}^{\infty} x^2 f_x(x)\, dx \tag{72}$$

and

$$E\{g(x)\} = \int_{-\infty}^{\infty} g(x) f_x(x)\, dx \tag{73}$$

The *mean* η of a random variable x is another name for the expected value of x.

$$\eta = E\{x\} \tag{74}$$

The expectation $E\{x^2\}$ is also called the *mean square value* of x^2. The *variance* σ^2 of a random variable x is defined as

---

Variance of a Random Variable

$$\sigma^2 = E\{(x - \eta)^2\}$$ (75)

It is easy to show (Problem 8.5.1) that

$$\sigma^2 = E\{x^2\} - \eta^2$$ (76)

8.6   Random DT Processes

A DT *random process* (also called a *random sequence*) is a sequence of random variables.

EXAMPLE 8.9   IID Binary Random Sequence

An experiment consists of tossing a coin repeatedly. On the nth toss, either a heads H or a tails T appears. A discrete-time stochastic process (a random sequence) is generated by making the nth element of the sequence $x[n]$ equal to 1 if H appears on the nth toss and equal to -1 if T appears on the nth toss. We can imagine a random sequence $x[n]$ generated this way for $-\infty < n < \infty$. It consists of unit impulses having random amplitudes ± 1.

To describe the random sequence statistically, we need to know the probability of heads. Let $P(H) = p$ and $P(T) = q$, where $q = 1 - p$. We also need to know if the outcome of one coin toss affects the outcome of

Figure 8.6
IID binary random
sequence.

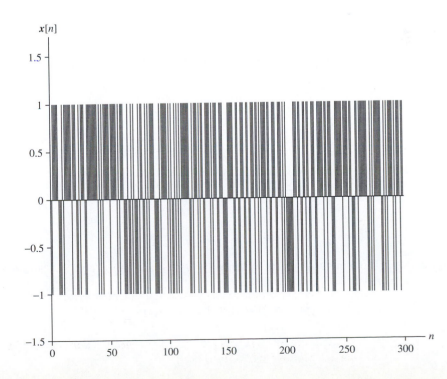

another. Let's assume that the outcomes of the coin tosses (and hence the values in the sequence $x[n]$) are statistically independent. For statistically independent outcomes, the probability of getting heads on both the nth and the mth tosses ($m \neq n$) is p^2. Similarly, the probability of getting a heads on the nth toss and a tails on the mth toss is pq, and the probability of getting a tails on both the nth and the mth tosses is q^2. Thus, for the random processes under consideration, we have

$$P\{x[n] = 1, x[m] = 1\} = \begin{cases} p^2 & m \neq n \\ p & m = n \end{cases} \tag{77}$$

$$P\{x[n] = 1, x[m] = -1\} = Px[n] = -1, x[m] = 1 = pq \quad m \neq n \tag{78}$$

$$P\{x[n] = -1, x[m] = -1\} = \begin{cases} q^2 & m \neq n \\ q & m = n \end{cases} \tag{79}$$

Figure 8.6 shows a typical $x[n]$ for $p = 0.6$.

We can see from the preceding example that for a fixed index, say, n_1, the nth sample $x[n_1]$ in a random sequence is just a random variable. The sequence in the preceding example is called an *IID binary random sequence* because the random binary values of the impulses in the sequence are independent and identically distributed. The term *identically distributed* refers to the fact that the probabilities of 1 or -1 are the same, p and q, respectively, for each impulse.

In general, the impulse values in a random sequence can be any real numbers. The values can be statistically independent of one another or statistically dependent on one another. And the probabilities that govern their values can change with sample position n.

The next example illustrates a ternary (three-valued) random sequence composed of dependent consecutive random variables.

EXAMPLE 8.10   A Ternary Random Sequence

Suppose that the random sequence generated in the preceding example with $P(H) = p = 0.5$ is applied to a LTI system having the impulse response

$$h[n] = \delta[n] + \delta[n-1] \tag{80}$$

The output is a random sequence

$$y[n] = x[n] + x[n-1] \tag{81}$$

The possible values for $y[n]$ for a fixed n are 2, 0, and -2. The probabilities are

$$P\{y[n] = 2\} = P\{y[n] = -2\} = 0.25 \tag{82}$$

and

$$P\{y[n] = 0\} = 0.5 \tag{83}$$

We obtained Eq. (82) by noticing $y[n] = 2$ if and only if $x[n] = 1$ and $x[n-1] = 1$. Therefore, the events $\{y[n] = 2\}$ and $\{\{x[n] = 1\} \cap \{x[n-1] = 1\}\}$ are the same.

$$\{y[n] = 2\} = \{\{x[n] = 1\} \cap \{x[n-1] = 1\}\} \tag{84}$$

and so on.

Using similar reasoning, we can show that consecutive sample values $y[n]$ and $y[n+1]$ are statistically dependent. More specifically, you are asked to show in Problem 8.6.2 that

$$P\{y[n] = 2|y[n-1] = 2\} = P\{y[n] = -2|y[n-1] = -2\} = 0.5 \tag{85}$$

$$P\{y[n] = 2|y[n-1] = 0\} = P\{y[n] = -2|y[n-1] = 0\} = 0.25 \tag{86}$$

$$P\{y[n] = 2|y[n-1] = -2\} = P\{y[n] = -2|y[n-1] = 2\} = 0 \tag{87}$$

$$P\{y[n] = 0|y[n-1] = -2\} = P\{y[n] = 0|y[n-1] = 2\} = 0.5 \tag{88}$$

Figure 8.7 shows a typical $y[n]$ for $P(H) = p = 0.5$.

Figure 8.7
Ternary random sequence: The system is initially at rest, and the first input arrives at $n = 1$. The first output sample represents a start-up transient.

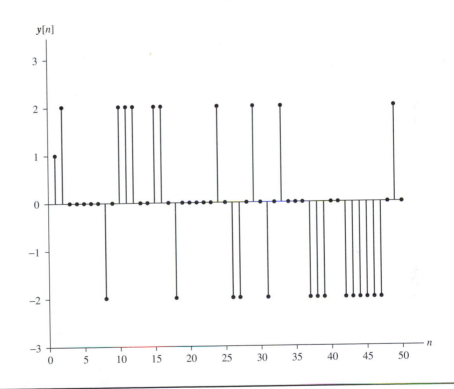

The following example is a generalization of Example 8.10.

EXAMPLE 8.11 Response of N-Point Averager to IID Binary Input

Suppose that an IID binary random sequence, having equally likely sample values ± 1 (Example 8.9), is applied to an N-point averager. The output is a random sequence.

$$y[n] = \frac{1}{N}\sum_{k=0}^{N-1} x[n-k] \tag{89}$$

Figure 8.8
Response of 25-point averager to IID binary input: The system is initially at rest, and the first input arrives at $n = 1$. The first 24 output samples represent a start-up-transient.

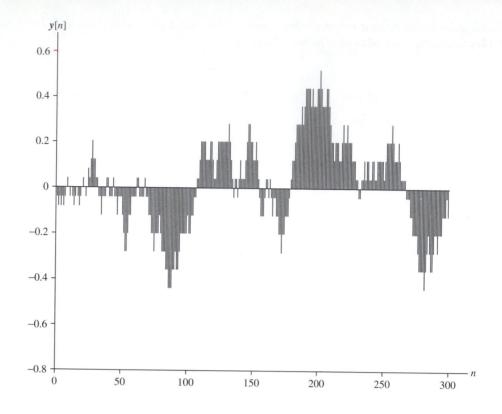

Figure 8.8 shows a typical $y[n]$ for $N = 25$. It can be shown that output values $y[n]$ and $y[n - m]$ are statistically dependent for $m = 0, \pm 1, \pm 2, \ldots, \pm N$ but are independent for $m > N$. Notice and think about how the statistical dependency among the samples is manifest in the plot.

8.6.1 Stationary Random Sequences

A discrete-time stochastic process is called *stationary* if the probabilities that characterize it do not change with sample position n. None of the random sequences in the figures of the preceding examples is stationary because they begin at a specific n ($n = 1$). They would all be stationary, however, if they had begun in the infinite past. For a stationary process, the value of any statistical average is a constant, independent of sample position n. In particular, $E\{x[n]\}$, $E\{x^2[n]\}$, and $E\{g(x[n])\}$ do not depend on n.

A nonstationary sequence is a random sequence that is not stationary.

EXAMPLE 8.12   A Nonstationary Binary Random Sequence

Assume, as in Example 8.9 that a coin is tossed repeatedly and that the outcomes are statistically independent. Assume that the coin tossing began in the remote past (ideally the infinite past). On the nth toss, either a heads H or a tails T appears. A random sequence is generated by making the nth element of the sequence $x[n]$ equal to 1 if H appears on the nth toss and equal to -1 if T appears on the nth toss. As in Example 8.9 this random sequence is characterized by statistically independent impulses having amplitudes ± 1.

Assume that, at index $n = 0$, the coin is switched. The first coin is fair but the second is not. Thus the probabilities of a head and a tail depend on sample position n; so we write them as $P(H, n)$ and $P(T, n)$. Specifically, let's assume that

$$P(H, n) = \begin{cases} 0.5 & n < 0 \\ 0.1 & n \geq 0 \end{cases} \qquad (90)$$

Of course, $P(T, n) = 1 - P(H, n)$.

Figure 8.9 shows a typical $x[n]$. The random sequence $x[n]$ is nonstationary because the coin was changed to the unfair one at $n = 0$. If the second coin had been fair like the first one, the process would have been stationary.

Figure 8.9
Nonstationary
sequence.

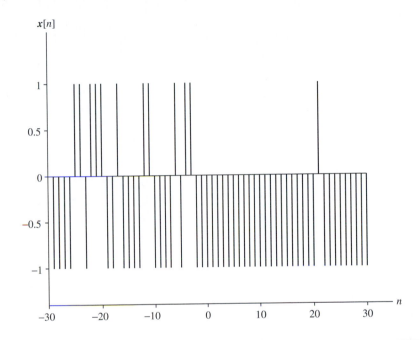

8.6.2
Ergodic
Random
Sequences

The engineering definition of an *ergodic* random process is a stationary random process for which statistical averages are equal to time averages. In this definition, the "time average" involves the values of $x[n]$ from $n = -\infty$ to ∞. For example

$$<x[n]> = \lim_{N \to \infty} \frac{1}{2N + 1} \sum_{n=-N}^{N} x[n] \qquad (91)$$

Similarly

$$<x^2[n]> = \lim_{N \to \infty} \frac{1}{2N + 1} \sum_{n=-N}^{N} x^2[n] \qquad (92)$$

and

$$<g(x[n])> = \lim_{N \to \infty} \frac{1}{2N + 1} \sum_{n=-N}^{N} g(x[n]) \qquad (93)$$

Figure 8.10
Ensemble of stationary ternary random sequences.

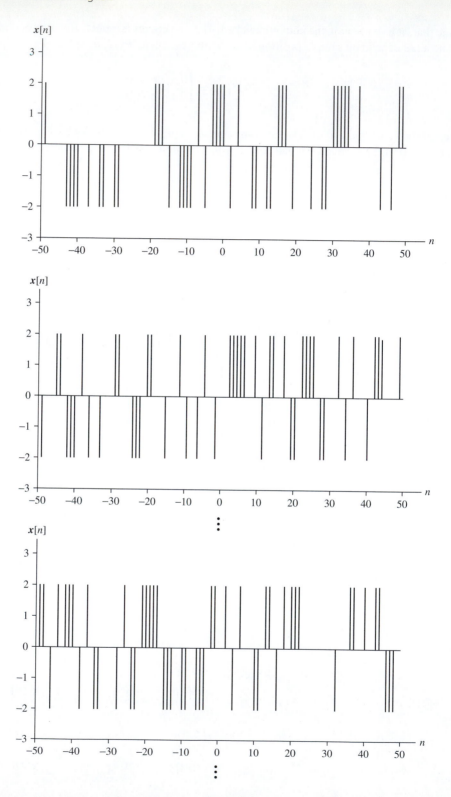

For an ergodic processes, the time averages are equal to the expected values.

$$<x[n]> = E\{x[n]\} \tag{94}$$

$$<x^2[n]> = E\{x^2[n]\} \tag{95}$$

$$<g(x[n])> = E\{g(x[n])\} \tag{96}$$

A good way to look at Eqs. (93)–(95) is in terms of *time averages* and *ensemble averages*. When we look at the sequences drawn and the preceding examples, we realize that each one is only one of many possibilities. We can visualize the many possibilities listed in a table or a figure. For example, consider Figure 8.10, which depicts possible sequences for a ternary random process. These were generated as described in Example 8.10 but beginning in the infinite past. The figure necessarily shows only a few sequences because there is not enough room to list them all. The list of the possible ternary sequences is called an *ensemble*. Now, we can view the time averages in Eqs. (93)–(95) as taken *along the ensemble* (along the coordinate n) where the sequence selected is fixed. We can view the expectations in Eqs. (93)–(95) as taken *down the ensemble* where the value of n is fixed. For an ergodic process, the time averages equal the ensemble averages.

EXAMPLE 8.13   Time Averages and Ensemble Averages

Assume that the random sequence of Example 8.9 starts in the remote past and is ergodic. Find a) $<x[n]>$ and b) $<x^2[n]>$.

a) $<x[n]>$ is the "DC" value of the random sequence. Notice that even though $x[n]$ is random, we can find its DC value!

$$<x[n]> = E\{x[n]\} = 1 \times p - 1 \times q = p - q \tag{97}$$

b) $<x^2[n]>$ is the average power in the random sequence. Even though $x[n]$ is random, we can find its average power!

$$<x^2[n]> = E\{x^2[n]\} = 1^2 \times p + (-1)^2 \times q = p + q = 1 \tag{98}$$

The preceding example illustrated the fact that the DC value and average power of a random sequence are deterministic, not random. We were able to derive the DC value and the average power using the tools of probability theory. Similarly, the power spectrum of a random sequence is a deterministic descriptor of a random sequence that we can derive analytically. The route to obtaining the power spectrum is through the autocorrelation sequence as described in the next section.

8.7   Autocorrelation Sequences and Power Spectrums

8.7.1 Definitions   Recall that in Chapter 4DT we defined the autocorrelation sequence of a real energy sequence as

$$\phi_x[k] = \sum_{n=-\infty}^{\infty} x[n]x[n-k] \tag{99}$$

The autocorrelation sequence $R_x(n, m)$ of a real ergodic random sequence is defined in a similar way.

Autocorrelation Sequence of a Random Sequence

$$R_x[k] = \lim_{N \to \infty} \frac{1}{2N+1} \sum_{n=-N}^{N} x[n]x[n-k] = E\{x[n]x[n-k]\} \qquad (100)$$

Recall that the energy density function of an energy sequence is given by the DTFT of $\phi_x[k]$. Similarly, the *power spectrum* of a real ergodic random sequence is the DTFT of $R_x[k]$.

Power Spectrum of a Random Sequence

$$S_x(e^{j2\pi f}) = \mathcal{DTFT}\{R_x[k]\} = \sum_{k=-\infty}^{\infty} R_x[k]e^{-j2\pi fk} \qquad (101)$$

It follows from the preceding definition that

$$R_x[k] = E\{x[n]x[n-k]\} = \int_{-0.5}^{0.5} S_x(e^{j2\pi f})e^{j2\pi fk} df \qquad (102)$$

The area under a power spectrum is the power (the mean square value) of $x[n]$.

$$R_x(0) = E\{x^2[n]\} = \int_{-0.5}^{0.5} S_x(e^{j2\pi f}) df \qquad (103)$$

It can be shown that the power spectrum of a real random sequence is a real, even, and nonnegative function of f (Problem 8.7.4). The term $S_x(e^{j2\pi f})$ is sometimes called a *power spectral density* because it describes the distribution of signal power with respect to the continuous variable f.

EXAMPLE 8.14   IID Binary Random Sequence

Consider a stochastic process obtained by tossing a coin independently, where we assign $x[n] = 1$ for a head on toss n and $x[n] = -1$ for a tail on toss n. The product $x[n]x[n-k]$ can have only two values, namely -1 and $+1$. If the coin is fair, then

$$R_x[0] = E\{x[n]x[n]\} = E\{x^2[n]\} = (1)^2 \times \tfrac{1}{2} + (-1)^2 \times \tfrac{1}{2} = 1 \qquad (104)$$

and

$$R_x[k] = E\{x[n]x[n-k]\} = (1)^2 \times \tfrac{1}{4} + (1)(-1) \times \tfrac{1}{4} + (-1)(1) \times \tfrac{1}{4} + (-1)^2 \times \tfrac{1}{4} = 0 \qquad (105)$$

for $k \neq 0$. Therefore, the autocorrelation sequence is an impulse

$$R_x[k] = \delta[k] \qquad (106)$$

and the power spectrum is a constant.

$$S_x(e^{j2\pi f}) = \mathcal{DTFT}\{\delta[k]\} = 1 \quad 0.5 \leq f < 0.5 \qquad (107)$$

The random sequence in the preceding example had a constant power spectrum. In general, a random sequence is called *white* if its power spectrum is a constant.

White Random Sequence

Power spectrum:

$$S_x(e^{j2\pi f}) = \frac{\mathcal{N}_o}{2} \tag{108}$$

Autocorrelation sequence:

$$R_x[k] = \frac{\mathcal{N}_o}{2}\delta[n] \tag{109}$$

where $\mathcal{N}_o$ is a constant.

**8.7.2
Power
Spectrum of LTI
System Output**

Assume that a random sequence $x[n]$ is applied to an LTI system having the impulse response $h[n]$. The output is a random sequence

$$y[n] = \sum_{m=-\infty}^{\infty} x[m]h[n-m] = x[n] * h[n] \tag{110}$$

The power spectrum of $y[n]$ is given by

$$S_y(e^{j2\pi f}) = S_x(e^{j2\pi f})|H(e^{j2\pi f})|^2 \tag{111}$$

We can justify the preceding relation by recalling that the amplitude response characteristic of a filter $|H(e^{j2\pi f})|$ equals the ratio of the amplitudes of the output sinusoidal components to the input sinusoidal components. Because power is proportional to amplitude squared, a squared amplitude response characteristic $|H(e^{j2\pi f})|^2$ relates $S_y(e^{j2\pi f})$ to $S_x(e^{j2\pi f})$.

The inverse DTFT of $S_y(e^{j2\pi f})$ gives us the autocorrelation function of $y[n]$:

$$R_y[k] = E\{x[n]x[n-k]\} = \int_{-0.5}^{0.5} S_y(e^{j2\pi f})\,e^{j2\pi fk}df$$

$$= \int_{-0.5}^{0.5} S_x(e^{j2\pi f})|H(e^{j2\pi f})|^2 e^{j2\pi fk}df \tag{112}$$

The area under $S_y(e^{j2\pi f})$ is the power (the mean square value) of $y[n]$.

$$R_y(0) = E\{y^2[n]\} = \int_{-0.5}^{0.5} S_x(e^{j2\pi f})|H(e^{j2\pi f})|^2 df \tag{113}$$

EXAMPLE 8.15   A First-Order DT Stochastic Sequence

We can generate a first-order random sequence by applying the white IID binary sequence of Example 8.14 to the first-order DT filter from Figure 2DT.20, where $|a| < 1$. The power spectrum of the input is

$$S_x(e^{j2\pi f}) = \frac{\mathcal{N}_o}{2} = 1 \tag{114}$$

and the impulse response of the filter is

$$h[n] = (1 - a) * a^n u[n] \tag{115}$$

where the factor $(1 - a)$ normalizes the filter for unity DC gain. (See Figure 3DT.10).) Figure 8.11 illustrates a typical output obtained for $a = 0.95$. The output shown was computed recursively from the system's difference equation starting with $n = 1$. Notice that the output includes a start-up transient introduced by applying the input at $n = 1$.

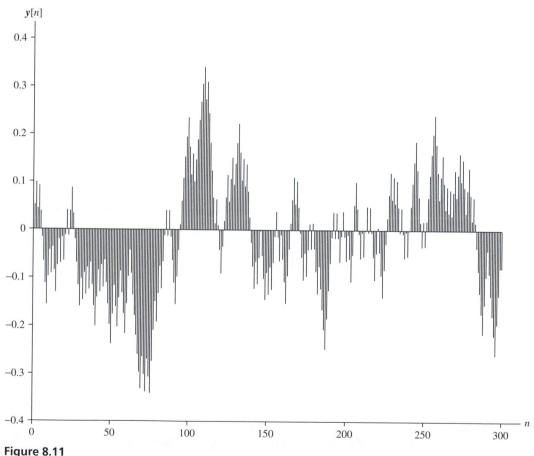

Figure 8.11
Response of first-order LTI system with $a = 0.95$ to IID binary input: The system was initially at rest, and the input sequence was applied at $n = 1$. The first $m \approx 58$ output samples represent a start-up transient.

Figure 8.12
$S_y(e^{j2\pi f})$ power spectrum of a first-order stochastic DT process.

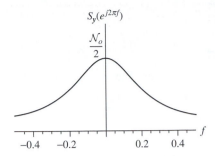

As n increases, the start-up transient dies out and processes approaches a statistical steady state.[4] If we had applied the input at $n = -\infty$ instead of $n = 1$, the statistical steady state would have applied for all n.

Assume that the input is applied at $n = -\infty$ and find

a) The average input power.
b) The power spectrum of the output.
c) The average output power.

Solution

a) The input power is, from Eq. (103)

$$R_x(0) = E\{x^2[n]\} = \int_{-0.5}^{0.5} \frac{N_o}{2} \, df = \frac{N_o}{2} = 1 \tag{116}$$

b) The frequency response characteristic of the filter is, from Eq. (3DT.34)

$$H(e^{j2\pi f}) = \frac{1-a}{1-ae^{-j2\pi f}} \tag{117}$$

The output power spectrum is given by Eq. (111).

$$S_y(e^{j2\pi f}) = \frac{N_o}{2}\left|\frac{1-a}{1-ae^{-j2\pi f}}\right|^2 = \frac{N_o}{2}\left(\frac{1+a^2-2a}{1+a^2-2a\cos(2\pi f)}\right) \tag{118}$$

$S_y(e^{j2\pi f})$ is plotted in Figure 8.12 for $a = 0.5$.

c) The total average output power can be obtained using Eq. (113), which reduces to

$$R_y(0) = E\{y^2[n]\} = \frac{N_o}{2}\int_{-0.5}^{0.5}|H(e^{j2\pi f})|^2 \, df$$

$$= \frac{N_o}{2}\int_{-0.5}^{0.5}\frac{1+a^2-2a}{1+a^2-2a\cos(2\pi f)} \, df \tag{119}$$

This integral can be evaluated using tables. We can obtain a simpler expression for the output power by noticing that $h[n]$ is an energy sequence. Therefore we can apply Parseval's theorem for energy sequences (4DT.90), with $X(e^{j2\pi f})$ replaced with $H(e^{j2\pi f})$ and $x[n]$ replaced with $h[n]$.

$$\int_{-0.5}^{0.5}|H(e^{j2\pi f})|^2 df = \sum_{n=-\infty}^{\infty} h^2[n] \tag{120}$$

[4]The duration of the start-up transient is determined by the "DT memory span" of the filter. We can define it as the value of n for which $|a|^n = 0.05$ ($0.95^n \approx 0.05 \Rightarrow n \approx 58$).

so that Eq. (119) becomes

$$E\{y^2[n]\} = \frac{N_o}{2} \sum_{n=-\infty}^{\infty} h^2[n] = \frac{N_o}{2}(1-a)^2 \sum_{n=0}^{\infty} a^{2n} \tag{121}$$

Identity (B.4) of Appendix B provides a closed form for the series. The result is

$$E\{y^2[n]\} = \frac{N_o}{2} \frac{(1-a)^2}{1-a^2} = \frac{1-a}{1+a} \tag{122}$$

where $|a| < 1$.

8.8   Random CT Processes

A development of random CT processes would bring us beyond the scope of this text. However, we can say that the basic concepts and results developed in Sections 8.6 and 8.7 for random sequences generalize to random waveforms.

Like DT processes, one of the most important of concepts for CT processes is that of an ensemble. Figure 8.13 illustrates a hypothetical ensemble of the waveforms that could be produced by a CT source. This ensemble has an unlimited number of sample waveforms. One of these waveforms (a *sample function*) occurs as the result of an experiment, but we do not know which. The expression $x(t)$ is called a random CT process if we can describe this selection probabilistically. The concepts of time and ensemble (statistical) averages we described in connection with random sequences extend to CT processes. For example, the process $x(t)$ is called *ergodic* if time averages are equal to ensemble averages (expectations). In analogy with random sequences, $x(t)$ is just a random variable when t is a fixed number. Thus, for an ergodic process, the DC value of a sample function equals the expected value of $x(t)$.

$$<x(t)> = \lim_{T\to\infty} \frac{1}{2T} \int_{-T}^{T} x(t)\, dt = E\{x(t)\} \tag{123}$$

and the average power equals the expected value of $x^2(t)$.

$$<x^2(t)> = \lim_{T\to\infty} \frac{1}{2T} \int_{-T}^{T} x^2(t)\, dt = E\{x^2(t)\} \tag{124}$$

Similarly, the product $x(t_1)x(t_2)$ is a random variable for fixed t_1 and t_2. The *autocorrelation function* of an ergodic process given by

$$R_x(\tau) = \lim_{T\to\infty} \frac{1}{2T} \int_{-T}^{T} x(t)x(t-\tau)\, dt = E\{x(t)x(t-\tau)\} \tag{125}$$

The *power spectrum* of $x(t)$ is the Fourier transform of $R_x(\tau)$.

$$S_x(F) = \int_{-\infty}^{\infty} R_x(\tau)\, e^{-j2\pi F\tau}\, d\tau \tag{126}$$

It follows that

$$R_x(\tau) = \int_{-\infty}^{\infty} S_x(F)\, e^{-j2\pi F\tau}\, dF \tag{127}$$

Figure 8.13
Ensemble of
waveforms.

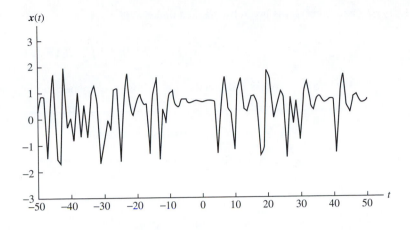

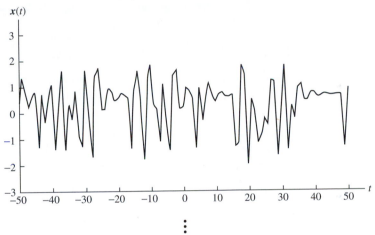

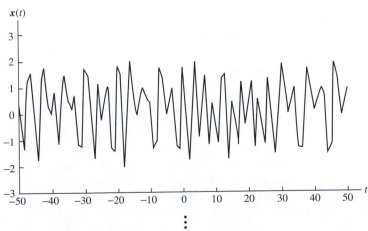

and $E\{x^2(t)\}$ is given by the area under $S_x(F)$.

$$E\{x^2(t)\} = R_x(0) = \int_{-\infty}^{\infty} S_x(F)\,dF \tag{128}$$

A white CT process is defined as one whose power spectrum is constant for all F.

A White CT Random Process

Has the power spectrum:

$$S_x(F) = \frac{\mathcal{N}_o}{2} \tag{129}$$

And autocorrelation function:

$$R_x(\tau) = \frac{\mathcal{N}_o}{2}\delta(\tau) \tag{130}$$

where $\mathcal{N}_o$ is a constant.

Many physical noise sources can be well modeled as white random processes. Perhaps the most famous example of a white noise model is given by Johnson noise.

EXAMPLE 8.16   Johnson Noise

Johnson noise is generated by the thermal agitation of electrons in resistors. It manifests as a random voltage $x(t)$ across the resistor terminals. The power spectrum is given by

$$S_x(F) = \frac{\mathcal{N}_o}{2} = 2Rk\mathrm{T} \tag{131}$$

where R is the resistance in ohms (Ω), k is Boltzmann's constant 1.38×10^{-23} joules per kelvin, and T is the temperature in kelvins. Johnson noise is also called *thermal noise*. Johnson noise is too small to be detected in resistors without amplification, but it can be an important source of noise in high-gain amplifiers (Problem 8.8.1).

The white CT process is a mathematical model. No physical source can produce a truly flat power spectrum for all F. If that were the case, then it would follow that the source would produce a waveform having both infinite bandwidth and infinite power. The mathematical idealization is a useful model for any random process whose power spectrum is approximately flat over the bandwidth of the system it is applied to.

In analogy with DT random processes, we can show that, if any stationary CT process $x(t)$ is applied to an LTI system having impulse response $h(t)$, then the output

$$y(t) = x(t) * h(t) \tag{132}$$

is a random process having the power spectrum

$$S_y(F) = S_x(F)|H(F)|^2 \tag{133}$$

where $H(F)$ is the frequency response characteristic of the system. The autocorrelation function of $y(t)$, $R_y(\tau) = E\{y(t)y(t-\tau)\}$, can be obtained by taking the inverse DTFT of $S_y(F)$.

$$R_y(\tau) = \int_{-\infty}^{\infty} S_y(F) e^{j2\pi F\tau} dF = \int_{-\infty}^{\infty} S_x(F)|H(F)|^2 e^{j2\pi F\tau} dF \qquad (134)$$

If we set $\tau = 0$ in the preceding equation, we obtain

$$R_y(0) = E\{y^2[n]\} = \int_{-\infty}^{\infty} S_x(F)|H(F)|^2 dF \qquad (135)$$

EXAMPLE 8.17   First-Order CT Stochastic Process

A first-order random waveform can be obtained by applying a white random waveform to a first-order CT filter. The power spectrum of the white input waveform is

$$S_x(F) = \frac{N_o}{2} \qquad (136)$$

and the impulse response of the filter is

$$h(t) = \frac{1}{\tau} e^{-\frac{t}{\tau}} u(t) \qquad (137)$$

Determine:

a) The average input power.
b) The power spectrum of the output.
c) The total average output power.

Solution

a) Equation (128) yields

$$R_x(0) = E\{x^2(t)\} = \int_{-\infty}^{\infty} \frac{N_o}{2} df = \infty \qquad (138)$$

This result implies that, like the unit impulse, a white CT process is an idealization.

b) The frequency response characteristic of the filter is

$$H(F) = \frac{1}{1 + j2\pi F\tau} \qquad (139)$$

Equation (133) yields

$$S_y(F) = \frac{N_o}{2} \left| \frac{1}{1 + j2\pi F\tau} \right|^2 = \frac{N_o/2}{1 + (2\pi F\tau)^2} \qquad (140)$$

$S_y(F)$ is plotted in Figure 8.14

c) The total average output power can be obtained using (135) which reduces to

$$R_y(0) = E\{y^2[n]\} = \frac{N_o}{2} \int_{-\infty}^{\infty} |H(F)|^2 dF$$

$$= \frac{N_o}{2} \int_{-\infty}^{\infty} \frac{1}{1 + (2\pi F\tau)^2} dF \qquad (141)$$

Figure 8.14
Power spectrum of a
first-order stochastic
CT process.

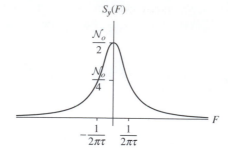

The above integral can be evaluated using tables. Alternatively, we can obtain a simpler expression for the output power by the use of Parseval's theorem for energy waveforms.

$$E\{y^2[n]\} = \frac{N_o}{2} \int_{-\infty}^{\infty} h^2(t)\,dt = \frac{N_o}{2\tau^2} \int_0^{\infty} e^{-\frac{2t}{\tau}}\,dt = \frac{N_o}{4\tau} \tag{142}$$

8.9   Summary

Most real-life waveforms and sequences are not known before they are observed. They are, in principle, different from deterministic signals such as $6.2e^{-0.01t}u(t)$ and $(0.5)^n r[n]$. This chapter was a brief introduction to the probabilistic description of real-life signals. The probabilistic description enables us to derive average signal properties needed in the design of processors. For example, the probabilistic description enables us to derive a signal's average power, its power spectrum, and the percentage of its values that fall below some threshold.

Summary of Definitions and Formulas

Axioms of Probability Theory

1. $P(A) \geq 0$.
2. $P(S) = 1$, where S is the certain event, $S = \{$all possible outcomes$\}$.
3. If A and B are mutually exclusive events, then
$$P(A \cup B) = P(A) + P(B)$$

Useful Formula

$$P(A \cup B) = P(A) + P(B) - P(A \cap B)$$

Statistically Independent Events

Two events A and B are statistically independent if the probability that they both occur $P(A \cap B)$ equals the product of the probabilities of each event.
$$P(A \cap B) = P(A)P(B)$$
Three events, A_1, A_2, and A_3, are statistically independent if
$$P(A_i \cap A_j) = P(A_i)P(A_j) \quad i < j = 1, 2, 3$$
and
$$P(A_1 \cap A_2 \cap A_3) = P(A_1)P(A_2)P(A_3)$$

Conditional Probability of $\mathcal{B}$ Given $\mathcal{A}$

$$P\{\mathcal{B}|\mathcal{A}\} \triangleq \frac{P\{\mathcal{A} \cap \mathcal{B}\}}{P\{\mathcal{A}\}}$$

Inverse Probability

$$P\{\mathcal{A}|\mathcal{B}\} \triangleq \frac{P\{\mathcal{B}|\mathcal{A}\}P\{\mathcal{A}\}}{P\{\mathcal{B}\}}$$

Total Probability

If $\mathcal{A}_1, \mathcal{A}_2, \ldots, \mathcal{A}_M$ partition $\mathcal{A}$, then

$$P\{\mathcal{A}\} = P\{\mathcal{A}|\mathcal{A}_1\}P\{\mathcal{A}_1\} + P\{\mathcal{A}|\mathcal{A}_2\}P\{\mathcal{A}_2\} + \cdots + P\{\mathcal{A}|\mathcal{A}_M\}P\{\mathcal{A}_M\}$$

Random Variable

A real random variable is a real number corresponding to each outcome of a probabilistic experiment.

Probability Distribution Function (PDF)

$$F_x(x) = P\{x \le x\}$$

$F_x(-\infty) = 0$
$F_x(\infty) = 1$
$F_x(x) \ge 0$
$F_x(x)$ is a monotonically increasing function of x.

Probability Density Function (pdf)

$$f_x(x) = \frac{d}{dx}F_x(x) = \lim_{\Delta x \to 0} \frac{P\{x - \Delta x < x \le x\}}{\Delta x}$$

$f_x(x) \ge 0$
$\int_{-\infty}^{\infty} f_x(x)\,dx = 1$
$$P\{x_1 < x \le x_2\} = \int_{x_1}^{x_2} f_x(x)\,dx$$

Expectation

For a discrete random variable:
   Mean:

$$\eta = E\{x\} = \sum_{k=1}^{M} x_k P\{x = x_k\}$$

Mean square:

$$E\{x^2\} = \sum_{k=1}^{M} x_k^2 P\{x = x_k\}$$

Of a function:

$$E\{g(x)\} = \sum_{k=1}^{M} g(x_k)P\{x = x_k\}$$

Variance:

$$\sigma^2 = E\{(x - \eta)^2\} = E\{x^2\} - \eta^2$$

For a continuous random variable:

Mean:

$$\eta = E\{x\} = \int_{-\infty}^{\infty} x f_x(x) \, dx$$

Mean square:

$$E\{x^2\} = \int_{-\infty}^{\infty} x^2 f_x(x) \, dx\}$$

Of a function:

$$E\{g(x)\} = \int_{-\infty}^{\infty} g(x) f_x(x) \, dx$$

Variance:

$$\sigma^2 = E\{(x - \eta)^2\} = E\{x^2\} - \eta^2$$

Random Sequence

A real random sequence is a real sequence assigned to each outcome of a probabilistic experiment. It is a sequence of random variables.

A discrete-time stochastic process is called *stationary* if the probabilities that characterize it do not change with sample position n.

An *ergodic* random process is a random process for which statistical averages are equal to time averages.

Autocorrelation Function (DT)

$$R_x(n, m) = E\{x[n]x[m]\}$$

$$R_x[k] = \lim_{N \to \infty} \frac{1}{2N + 1} \sum_{n=-N}^{N} x[n]x[n - k]$$

$$= E\{x[n]x[n - k]\} \text{ for an ergodic DT process}$$

Power Spectrum (DT)

$$S_x(e^{j2\pi f}) = \sum_{k=-\infty}^{\infty} R_x[k]e^{-j2\pi f k}$$

$$E\{x^2[n]\} = \int_{-0.5}^{0.5} S_x(e^{j2\pi f}) \, df = R_x[0]$$

$$S_y(e^{j2\pi f}) = S_x(e^{j2\pi f})|H(e^{j2\pi f})|^2$$

A white sequence has the power spectrum:

$$S_x(e^{j2\pi f}) = \frac{N_o}{2}$$

Autocorrelation Function (CT)

$$R_x(t_1, t_2) = E\{x(t_1)x(t_2)\}$$

$$R_x(\tau) = \lim_{T \to \infty} \frac{1}{2T} \int_{-T}^{T} x(t)x(t - \tau) \, dt = E\{x(t)x(t - \tau)\} \text{ for an ergodic CT process}$$

Power Spectrum (CT)

$$S_x(f) = \int_{-\infty}^{\infty} R_x(\tau)e^{-j2\pi f\tau}\,d\tau$$

$$E\{x^2(t)\} = \int_{-\infty}^{\infty} S_x(f)\,df = R_x(0)$$

$$S_y(f) = S_x(f)|H(f)|^2$$

A white CT process has the power spectrum:

$$S_x(f) = \frac{N_o}{2}$$

8.10 Problems

8.2 Probability

8.2.1 Use the classical definition of probability to find the probability that:
a) Either face 2 or face 3 or face 4 shows when you toss a fair die.
b) The face numbers sum to 2 when you toss 2 fair dice.
c) The face numbers sum to 6 when you toss 2 fair dice.

8.2.2 One side of the coin is marked with a 0 and the other with a 1. The coin is tossed 5 times. Let n denote the number of the toss and $x[n]$ denote the number that shows (0 or 1) on the nth toss. After 5 tosses, we will have generated a binary word having length 5. Use the classical definition of probability to find the probability that:
a) The word is 00000.
b) The word is 01101.
c) The word contains exactly 3 ones.

8.2.3 The numbers 0, 1, ... 15 are written in binary as 4-bit words: 0000, 0001, ..., 1111. One of the 4-bit words is selected at random. Each 4-bit word is equally likely to be selected. Use the classical definition of probability to find the probability that:
a) The first bit is 1.
b) The first two bits are 11.
c) The first three bits are 011.
d) The selected word is 1011.

8.2.4 A die is tossed with $P\{f_i\} = p_i, i = 1, 2, \ldots, 6$. Let $\mathcal{A} = \{f_1, f_3, f_5\}$ and $\mathcal{B} = \{f_2, f_3\}$. Find $P\{\mathcal{A} \cup \mathcal{B}\}$.

8.2.5 Explain the meaning of the words "event $\mathcal{A}$ occurs." Give an example.

8.2.6 Explain the meaning of the words "event $\mathcal{A} \cup \mathcal{B}$ occurs." Give an example.

8.2.7 Explain the meaning of the words "event $\mathcal{A} \cap \mathcal{B}$ occurs." Give an example.

8.2.8 Show that $P(\mathcal{A} \cup \bar{\mathcal{A}}) = 1$ for every event $\mathcal{A}$.

8.2.9 Show that $P(\phi) = 0$, where ϕ is the empty set.

8.2.10 Use Venn diagrams to illustrate the equality $\mathcal{B} = (\mathcal{B} \cap \mathcal{A}) \cup (\mathcal{B} \cap \bar{\mathcal{A}})$.

8.2.11 An important result in probability theory is

$$P(\mathcal{B}) \geq P(\mathcal{A}) \text{ if } \mathcal{A} \subset \mathcal{B} \tag{143}$$

a) Demonstrate this result with an example from the die toss experiment with $\mathcal{A} = \{f_1, f_2\}$ and $\mathcal{B} = \{f_1, f_2, f_3, f_4\}$.
b) Use Axioms 2 and 3 to prove the inequality. Hint: $\mathcal{B} = (\mathcal{B} \cap \mathcal{A}) \cup (\mathcal{B} \cap \bar{\mathcal{A}})$.

8.2.12 De Morgan introduced the following rules:

$$\overline{(A \cap B)} = \bar{A} \cup \bar{B} \tag{144}$$

and

$$\overline{(A \cup B)} = \bar{A} \cap \bar{B} \tag{145}$$

Use Venn diagrams that illustrate each side of these equalities.

8.3   Statistical Dependence and Independence

8.3.1 Find $P(A)$, $P(B)$, and $P(A \cap B)$ for the die toss experiment for the somewhat less powerful magnetic field defined by Eq. (20) of Example 8.4. Does $P(A \cap B) = P(A)P(B)$?

8.3.2 Derive the inverse probability formula.

8.3.3 The numbers 1, 2, ... 16 are written in binary as 4-bit words. Each 4-bit word has probability 1/16 of being selected. One of the 4-bit words is selected.
a) Find the probability that the first and second bits in the word are both 1s.
b) Find the probability the first or the second bit in the word is 1.
c) Are the values of the first and the second bits statistically independent? Justify your answer.

8.3.4 The following MATLAB m-file simulates the generation of a sequence of statistically independent binary numbers 0 and 1 with $P\{1\} = p$ and $P\{0\} = 1 - p$.

```
function x=cointoss
len=input('Enter length:')
p=input('Enter p:')
r=rand(1,len);
x=zeros(1,len);
  for n=1:len
    if r(n)< p
    x(n)=1;
    else
    end
  end
end
```

Read the MATLAB help for the function `rand`. Then explain the statements in the preceding m-file.

8.3.5 In this problem you should use the program in Problem 8.3.4 to generate binary random numbers generated by tossing an unfair coin whose sides are labeled 1 and 0. Assume that $P\{1\} = 0.6$ and $P\{0\} = 0.4$.
a) Let $x[i]$ denote the ith number that is generated and let $n_m\{1\}$ denote the number of 1s that occur in the sequence: $x[1], x[2], \ldots, x[m]$. The relative frequency of 1s in m trials is given by

$$\rho(m) = \frac{n_m\{1\}}{m} = \frac{1}{m} \sum_{i=1}^{m} x[i] \tag{146}$$

Show that $\rho(m)$ can be computed recursively using the formula

$$\rho(m) = \frac{m-1}{m} \rho(m-1) + \frac{1}{m} x[m] \quad m = 1, 2, \ldots \tag{147}$$

Use this recursion in your program to generate $\rho(m)$. Hint: There is no need to create a vector x in your program because this recursion can updated by the binary numbers as they are generated.
b) Plot $\rho(m)$ versus m for $1 \leq m \leq 10,000$. Use a logarithmic scale for the horizontal axis. Comment on your results.

8.3.6 Consider transmitting binary digits 0 and 1 over a communications channel where errors can occur. The possible inputs to the channel are 0_t and 1_t and the possible outputs are 0_r and 1_r where subscripts "t" and "r" denote "transmitted" and "received," respectively. Before the channel is used, the probabilities that a 0 or a 1 is sent are $P\{0_t\}$ and $P\{1_t\}$, respectively. Assume that $P\{0_r|1_t\} = P\{1_r|0_t\} \equiv p$. A channel having these symmetric error probabilities is called a binary symmetric channel (BSC).

    a) Find the probability that 1_t was transmitted given that 0_r was received. Hint: Use the formula for inverse probability.

    b) Find the probability of an error when 0_t is transmitted.

    c) Find the total probability of an error.

8.4 Random Variables

8.4.1 Show that $F_x(-\infty) = 0$: Hint a random variable is defined to have a value from the set of real numbers. Therefore, what is the event $\{x \le -\infty\}$?

8.4.2 Show that $F_x(\infty) = 1$. Hint: What is the event $\{x \le \infty\}$?

8.4.3 Consider a square law device defined by the I-O relation $y = x^2$. If we apply the random variable x of Example 8.8 to this device the output will be a random variable $y = x^2$. Find $F_y(y)$.

8.4.4 Show that $F_x(x)$ is never negative. Hint: Refer to the axioms of probability.

8.4.5 Show that $F_x(x)$ is a monotonically increasing function of x. Hint: Compare the sets $\{x \le x_1\}$ and $\{x \le x_2\}$, where $x_1 \le x_2$. See Problem 8.2.11.

8.4.6 Derive the PDF shown in Example 8.8.

8.4.7 Let x_1 and x_2 denote the face numbers that show on a pair of dice. Find and plot the PDF of $y = x_1 + x_2$ assuming that the dice are fair and land independently of one another.

8.4.8 Use the definition of a pdf $f_x(x)$ and the properties of a PDF $F_x(x)$ to show that:

    a) $f_x(x)$ is never negative.

    b) The area under $f_x(x)$ equals 1.

8.4.9 A random variable x has pdf $f_x(x)$. Find an expression for the probability that x is either in the interval $1 < x \le 2$ or in the interval $5 < x \le 6$. Write your answer in terms of $f_x(x)$.

8.4.10 A random variable y is defined by the equation $y = u(x)$, where the pdf of x is a known function $f_x(x)$. Find the pdf of y. Hint: It might be easiest to solve this problem by finding the PDF of y first.

8.5 Stochastic Averages (Expectation)

8.5.1 Show that, for an arbitrary random variable x

$$\sigma^2 = E\{x^2\} - \eta^2 \tag{148}$$

8.6 Random Sequences

8.6.1 Write a MATLAB program to plot an IID. binary random sequence $x[n]$ having sample values ± 1, where $P\{1\} = 0.5$. Hint: Modify the MATLAB code of Problem 8.3.4.

8.6.2 Derive the conditional probabilities given in Example 8.10.

8.6.3 Show that, for the random sequence $y[n]$ of Example 8.10, sample values $y[n_1]$ and $y[n_1 + m]$ are statistically independent for $m \ge 2$, where n_1 is any fixed integer.

8.6.4 Write a MATLAB program to generate and plot the random sequence of Example 8.11.

8.6.5 **a)** Find and plot $E\{x[n]\}$ for the nonstationary random process of Example 8.12.
b) Find and plot $E\{x^2[n]\}$ for the nonstationary random process of Example 8.12.
c) Comment on your answers to parts (a) and (b).

8.7   Autocorrelation Sequences and Power Spectrums

8.7.1 Extend the derivation of $R_x[k]$ and $S_x(e^{j2\pi f})$ in Example 8.14 to an unfair coin. Write your answer in terms of p and q, where $p + q = 1$ and p is the probability of a head. Comment on your result.

8.7.2 Derive $R_x[k]$ and $S_x(e^{j2\pi f})$ for a sequence generated by an unfair coin, where $x[n] = 1$ for heads and $x[n] = 0$ for tails. Write your answer in terms of p and q, where $p + q = 1$ and p is the probability of a head. Comment on your result.

8.7.3 Show that the autocorrelation function $R_x[k]$ of a stationary real sequence is a real and even sequence.

8.7.4 Refer to Problem 8.7.3 and show that the power spectrum of a real random sequence is:
a) Real.
b) An even function of f.
c) Nonnegative.

8.7.5 Find the power spectrum and autocorrelation function for the sequence $y[n]$ of Example 8.11.

8.7.6 *The noise equivalent bandwidth* of a DT LTI low-pass filter is defined as

$$w_n = \frac{1}{2|H(1)|^2} \int_{-0.5}^{0.5} |H(e^{j2\pi f})|^2 df \qquad (149)$$

a) Write down the expression for the average output power when white noise having power spectral density $S_x(f) = \mathcal{N}_o/2$ is applied to a low-pass filter having the amplitude response characteristic $|H(e^{j2\pi f})|$.
b) Find the average output power when white noise having the power spectral density $S_x(f) = \mathcal{N}_o/2$ is applied to an ideal low-pass filter having the amplitude response characteristic $|H(1)| \sqcap (f/2w_n)$ for $-0.5 \le f < 0.5$.
c) Equate your answers to parts (a) and (b) and solve for w_n. Then define the phrase "noise equivalent bandwidth" in your own words, using the ideal low-pass filter as a key concept in your definition.

8.7.7 A random sequence $x[n]$ is applied to an ideal band-pass filter of Table 2DT.1 with center frequency f_o, narrow bandwidth $2w = \frac{1}{2}\Delta f$, and unity passband gain. The output of the filter is denoted $y[n] = x_o[n]$ to emphasize that this output is composed of the unaltered frequency components in $x[n]$ that are passed by the filter.
a) Show that, if the power spectrum of $x[n]$, $S_x(e^{j2\pi f})$ is approximately constant in the narrow band-pass of the filter, then the average power of the output of the filter is

$$E\{x_o^2[n]\} \approx S_x(e^{j2\pi f_o})\Delta f \qquad (150)$$

b) Rearrange Eq. (150) and take the limit $\Delta f \to 0$ to show that

$$S_x(e^{j2\pi f_o}) = \lim_{\Delta \to 0} \frac{E\{x_o^2[n]\}}{\Delta f} \qquad (151)$$

Comment on this result.

8.7.8 Figure 1.17 depicts a uniform quantizer. The quantizer converts a continuous amplitude input sequence $x[n]$ to a discrete amplitude output sequence $y[n]$. Details are given in the introduction to miscellaneous Problems 1 and 2 of Chapter 1.
a) Draw the quantizer I-O characteristic for a 16-level uniform quantizer. Show that each level can be uniquely identified by a 4-bit word: $0000, 0001, 0010, \ldots, 1111$.

b) Quantization causes an irreversible degradation in signal quality. The quantizer output is given by

$$y[n] = x[n] + e[n] \tag{152}$$

where

$$e[n] = x[n] - y[n] \tag{153}$$

is the quantization error. Equation (152) shows that $y[n]$ is the sum of the original data $x[n]$ and *quantization noise $e[n]$*. Quantization noise is often modeled as a stationary white random sequence.

$$R_e[k] = \frac{\Delta^2}{12} \delta[k] \tag{154}$$

$$S_e(e^{j2\pi f}) = \frac{\Delta^2}{12} \tag{155}$$

For each n, $e[n]$ is generally modeled as a random variable whose pdf is *uniformly distributed* from $-\Delta/2$ to $\Delta/2$. The uniform pdf is given by

$$f_e(E) = \frac{1}{\Delta} \sqcap \left(\frac{E}{\Delta}\right) \tag{156}$$

The following m-file simulates a uniform quantizer:

```
function y=quantizer(x,delta,nlev)
%x is the input data sequence
%delta is the step size
%nleve is the number of levels. It must be an even number.
len=length(x);
for n=1:len

    if x(n)==0
    x(n)=delta/2;
    end

    if x(n)>=0

    y(n)=delta*sign(x(n))*fix(abs(x(n))/delta+1)-delta/2;

        if y(n)>(nlev/2)*delta-delta/2
        y(n)=(nlev/2)*delta-delta/2;
        end
    else
    y(n)=delta*sign(x(n))*fix(abs(x(n))/delta+1)+delta/2;

        if y(n)<-((nlev/2)*delta-delta/2)
        y(n)=-((nlev/2)*delta-delta/2);
        end
    end
end
```

c) Describe how the m-file works by adding helpful comments in the code. Begin each line of comments with a percentage sign (%).

d) Use the preceding m-file to quantize input data x of your choice. Experiment with various `delta` and `nlev`. Plot x, y, and e = y−x.

e) Read the MATLAB help file for the MATLAB function `xcorr`. Then use `xcorr` to estimate the autocorrelation function of the quantization error. Is your result consistent with a white noise model for e?

f) Read the MATLAB help filter for the MATLAB function `hist`. Then use `hist` to estimate the pdf of the quantization error. Is your result consistent with a uniform model for $f_e(E)$?

g) Show that Eqs. (154) and (156) are consistent. Specifically, show that the uniform model for $f_e(E)$ implies that $R_e[0] = \Delta^2/12$, which is consistent with Eq. (154).

8.8   Random CT Processes

8.8.1 If a the input terminals of a high-gain voltage amplifier are short circuited, there is still an output produced by Johnson noise. Usually, most of the noise comes from the resistors in the input stage. We can model the short-circuited input stage as a Johnson noise voltage source, $x(t)$, a nosieless resistor, R, and a capacitor, C, connected in a closed loop. The voltage across the capacitor is amplified by $\mathcal{A}$ to produce the output.

a) Show that the frequency response characteristic from the voltage source to the amplifier output is

$$H(F) = \frac{A}{1 + j\,2\pi RCF} \tag{157}$$

b) Find $E\{y^2(t)\}$, where $y(t)$ is the amplifier output voltage.

c) Assume that $R = 1$ MΩ (i.e., $R = 10^6$), and $C = 10$ pF (i.e., $C = 10^{-11}$). Find $E\{y^2(t)\}$ if the amplifier is operating at room temperature T = 298 kelvins.

d) Find the value for $\mathcal{A}$ that results in $E\{y^2(t)\} = 1$ mV squared.

8.8.2 The *noise equivalent bandwidth* of a CT LTI low-pass filter is defined as

$$W_n = \frac{1}{2|H(0)|^2} \int_{-\infty}^{\infty} |H(F)|^2 dF \tag{158}$$

a) Write down the expression for the average output power when white noise having the power spectral density $S_x(F) = \mathcal{N}_o/2$ is applied to a low-pass filter having the amplitude response characteristic $|H(F)|$.

b) Find the average output power when white noise having the power spectral density $S_x(F) = \mathcal{N}_o/2$ is applied to an ideal low-pass filter having the amplitude response characteristic $|H(0)| \sqcap (F/2W_n)$.

c) Equate your answers to (a) and (b) and solve for W_n. Then define the term "noise equivalent bandwidth" in your own words, using the ideal low-pass filter as a key concept in your definition.

A Review of Complex Arithmetic

A.1 Complex Numbers

The development of complex numbers began with the recognition that algebraic equations do not always have solutions in the set of real numbers. For example, the equation

$$az^2 + bz + c = 0 \tag{1}$$

does not have real solutions when the argument of the square root in the quadratic formula

$$z = \frac{-b}{2a} \pm \frac{\sqrt{b^2 - 4ac}}{2a} \tag{2}$$

is negative. When real solutions do not exist to an algebraic equation, *complex solutions* can be specified in terms of the *imaginary unit*. The imaginary unit j and its negative $-j$ are defined as the solutions to

$$z^2 = -1 \tag{3}$$

Imaginary Unit

$$j = \sqrt{-1} \tag{4}$$

An example of a complex solution is given by Eq. (2) when the argument of the square root is negative. The complex solutions are

$$z = \frac{-b}{2a} \pm j\frac{\sqrt{4ac - b^2}}{2a} \tag{5}$$

531

Figure A.1
Complex plane as defined in engineering: (a) z as a point in the complex plane; (b) z as a vector in the complex plane.

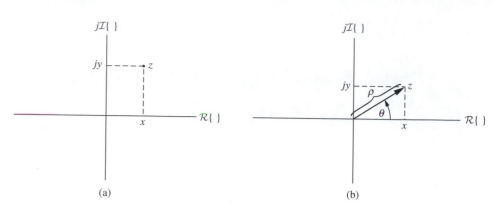

$$(a) \qquad\qquad (b)$$

A.2 Rectangular Form

We can write every complex number in *rectangular form*.

Rectangular Form for a Complex Number

$$z = x + jy \tag{6}$$

where x and y are real numbers.
x is the real part of z.

$$x = \mathcal{R}\{z\} \tag{7}$$

y is the imaginary part of z.

$$y = \mathcal{I}\{z\} \tag{8}$$

There is a one-to-one relationship between a complex number z and a pair of real numbers (x, y).

A.3 The Complex Plane

Complex numbers can be plotted in the *complex plane*. Engineers define this plane in terms of the coordinates $\mathcal{R}\{\cdot\}$ and $jIm\{\cdot\}$, as shown in Figure A.1. The horizontal axis is x and the vertical axis is jy. Mathematicians prefer to define the vertical axis as simply y. A complex number can be drawn as either a point or a vector from the origin to that point, as illustrated in Figures A.1(a) and (b), respectively.

A.4 Polar Form

There is also a shorthand notation for a complex number called the *polar form*. The polar form is based on the vector representation of z in Figure A.1(b).

Polar Form for a Complex Number

$$z = \rho \angle \theta \tag{9}$$

or

$$z = |z| \angle z \tag{10}$$

where

$$\rho = |z| = \sqrt{x^2 + y^2} \tag{11}$$

is the *absolute value* of z and

$$\theta = arg\{z\} = \tan^{-1}\left(\frac{y}{x}\right) \tag{12}$$

is the angle, or *argument*, of z. The argument of a complex number is not unique: θ can be replaced by $\theta + 2k\pi$, where k is an integer, without changing the value of z. The *principal argument* of z is the value of θ in the range $-\pi \le \theta < \pi$.

It follows from Eqs. (11) and (12) or the figure that

$$x = \rho \cos \theta \tag{13}$$

and

$$y = \rho \sin \theta \tag{14}$$

A.5   Euler's Formula and the Exponential Form

Euler's formula is one of the most important results in complex number theory. We can motivate Euler's formula by substituting Eq. (13) and Eq. (14) into Eq. (6). The result of this substitution is

$$z = \rho \cos \theta + j\rho \sin \theta = \rho(\cos \theta + j \sin \theta) \tag{15}$$

Euler's formula lets us express $\cos\theta + j\sin\theta$ as a *complex exponential*.

Euler's Formula

$$e^{j\theta} = \cos \theta + j \sin \theta \tag{16}$$

It enables us to write z in Eq. (15) in *exponential form*.

Exponential Form for a Complex Number

$$z = \rho e^{j\theta} \tag{17}$$

The factor $e^{j\theta}$ is called a *complex exponential*. It is very convenient to use the exponential form when we multiply or divide complex numbers.

We can derive Euler's formula by considering the factor $\cos\theta + j\sin$ in Eq. (15). If we expand the cosine and sine function in Taylor series and do a little algebra, we obtain

$$\cos\theta + j\sin\theta = 1 - \frac{1}{2}\theta^2 + \frac{1}{4!}\theta^4 + \cdots + j\left(\theta - \frac{1}{3!}\theta^3 + \frac{1}{5!}\theta^5\cdots\right)$$

$$= 1 + j\theta - \frac{1}{2}\theta^2 - j\frac{1}{3!}\theta^3 + \frac{1}{4!}\theta^4 + j\frac{1}{5!}\theta^5\cdots$$

$$= 1 + j\theta + \frac{1}{2}(j\theta)^2 + \frac{1}{3!}(j\theta)^3 + \frac{1}{4!}(j\theta)^4 + \frac{1}{5!}(j\theta)^5\cdots \quad (18)$$

The right-hand side of this equation is the Taylor series expansion of $e^{j\theta}$. This observation give us Euler's formula.

We can express $\cos\theta$ (or $\sin\theta$) directly in terms of complex exponentials by adding (or subtracting) the equations

$$e^{j\theta} = \cos\theta + j\sin\theta \quad (19a)$$

$$e^{-j\theta} = \cos\theta - j\sin\theta \quad (19b)$$

The result is the Euler formulas

$$\cos\theta = \frac{1}{2}(e^{j\theta} + e^{-j\theta}) \quad (20a)$$

$$\sin\theta = \frac{1}{2j}(e^{j\theta} - e^{-j\theta}) \quad (20b)$$

A.6 Addition, Subtraction, Multiplication, and Division of Complex Numbers

The imaginary unit j can be treated like any other constant when we work with complex numbers. For example, the sum

$$S = \sum_{i=1}^{n} z_i \quad (21)$$

where

$$z_i = x_i + jy_i \quad i = 1, 2, \ldots, n \quad (22)$$

can be put in the form

$$S = \sum_{i=1}^{n} z_i = \sum_{i=1}^{n}(x_i + jy_i) = \sum_{i=1}^{n} x_i + j\sum_{i=1}^{n} y_i \quad (23)$$

The right-hand side is expressed as a sum in rectangular form. It shows us that the real part of the sum of complex numbers equals the sum of the real parts of the complex numbers, i.e.

$$\mathcal{R}\{S\} = \mathcal{R}\left\{\sum_{i=1}^{n} z_i\right\} = \sum_{i=1}^{n}\mathcal{R}\{z_i\} \quad (24)$$

Similarly, it shows us that the imaginary part of the sum of complex numbers equals the sum of the imaginary parts of the complex numbers, i.e.

$$\mathcal{I}\{S\} = \mathcal{I}\left\{\sum_{i=1}^{n} z_i\right\} = \sum_{i=1}^{n} \mathcal{I}\{z_i\} \tag{25}$$

Similar results hold for subtraction.

The exponential form is convenient when we multiply or divide complex numbers. For example, the product

$$z_1 z_2 \ldots z_n = \rho_1 e^{j\theta_1} \rho_2 e^{j\theta_2} \cdots \rho_n e^{j\theta_n} \tag{26}$$

is given, in exponential form, by

$$z_1 z_2 \ldots z_n = \rho_1 \rho_2 \ldots \rho_n e^{j(\theta_1 + \cdots + \theta_2 + \theta_n)} \tag{27}$$

This result shows us that the absolute value of a product of complex numbers equals the product of the absolute values of the complex numbers

$$|z_1 z_2 \ldots z_n| = |z_1||z_2||z_3| \ldots |z_n| \tag{28}$$

It similarly shows us that the angle of a product of a complex numbers equals the sum of the angles of the complex numbers

$$\angle(z_1 z_2 \ldots z_n) = \angle z_1 + \angle z_2 + \cdots + \angle z_n \tag{29}$$

In a similar way, the exponential form gives us

$$\frac{z_1 z_2 \ldots z_n}{z'_a z'_2 \ldots z'_m} = \frac{|z_1||z_2||z_3| \ldots |z_n|}{|z'_a||z'_2| \ldots |z'_m|} e^{j(\theta_1 + \theta_2 + \cdots + \theta_n - \theta'_1 - \theta'_2 - \cdots - \theta'_n)} \tag{30}$$

Therefore,

$$\left|\frac{z_1 z_2 \ldots z_n}{z'_a z'_2 \ldots z'_m}\right| = \frac{|z_1||z_2||z_3| \ldots |z_n|}{|z'_a||z'_2| \ldots |z'_m|} \tag{31}$$

and

$$\angle\left(\frac{z_1 z_2 \ldots z_n}{z'_a z'_2 \ldots z'_m}\right) = \angle z_1 + \angle z_2 + \cdots + \angle z_n - \angle z'_1 - \angle z'_2 - \cdots - \angle z'_m \tag{32}$$

A.7 The Complex Conjugate

The complex conjugate of z, denoted by z^*, is obtained by replacing j with $-j$.

$$z^* = x - jy = \rho e^{-j\theta} = \rho\angle(-\theta) \tag{33}$$

We can see from (6) and (33) that

$$|z^*| = |z| \tag{34}$$

and

$$\angle z^* = -\angle z \tag{35}$$

Some useful identities involving the complex conjugate are as follows.

Useful Identities Involving the Complex Conjugate

$$(z_1 + z_2)^* = z_1^* + z_2^* \tag{36}$$

$$(z_1 z_2)^* = z_1^* z_2^* \tag{37}$$

$$\left(\frac{z_1}{z_2}\right)^* = \frac{z_1^*}{z_2^*} \tag{38}$$

$$|z| = \sqrt{z \cdot z^*} \tag{39}$$

$$\mathcal{R}\{z\} = \frac{1}{2}(z + z^*) \tag{40}$$

$$\mathcal{I}\{z\} = \frac{1}{2j}(z - z^*) \tag{41}$$

The complex conjugate provides an alternative technique for dividing complex numbers. In this technique we *rationalize the denominator* by multiplying the numerator and the denominator by the complex conjugate of the denominator.

$$\frac{x_1 + jy_1}{x_2 + jy_2} = \frac{x_1 + jy_1}{x_2 + jy_2} \times \frac{x_2 - jy_2}{x_2 - jy_2} = \frac{x_1 x_2 + y_1 y_2 + j(x_2 y_1 - x_1 y_2)}{x_2^2 + y_2^2}$$

$$= \frac{x_1 x_2 + y_1 y_2}{x_2^2 + y_2^2} + j\frac{x_2 y_1 - x_1 y_2}{x_2^2 + y_2^2} \tag{42}$$

A.8 Problems

A.1 Complete the table.

Expression	Rectangular Form	Polar Form	Exponential Form
/////////////////	$3 + j4$		
/////////////////		$6\angle 45°$	
/////////////////			$3e^{j0.1\pi}$
$\dfrac{3 + j4}{5 + j6}$			
$\dfrac{1}{5\angle(-\pi/6)}$			
/////////////////	$R + j\Omega L$		
/////////////////		$\sqrt{R^2 + \left(\dfrac{1}{\Omega C}\right)^2} \angle\tan^{-1}\left(\dfrac{-1}{\Omega RC}\right)$	
$\left(\dfrac{3 + j4}{5 + j6}\right)\left(\dfrac{7 + j84}{9 + j10}\right)$			
$\dfrac{R + j\Omega L}{R + j\Omega L + \dfrac{1}{j\Omega C}}$			
$\dfrac{1 - e^{j\theta}}{1 + e^{-j\theta}}$			

A.2 Prove the identities given for the complex conjugate.

A.3 Find and plot the five solutions to the equation $z = (-1)^{0.2}$. Hint: Express -1 in exponential form.

A.4 Find and plot the roots to the equation $z = (1 + j)^{0.5}$.

A.5 Derive the trigonometric identities

$$\cos(A + B) = \cos A \cos B - \sin A \sin B \tag{43}$$
$$\sin(A + B) = \sin A \cos B + \cos A \sin B$$

Hint: $e^{j(A+B)} = e^{jA}e^{jB}$

A.6 *De Moivre's theorem* states that

$$(\cos\theta + j\sin\theta)^n = \cos(n\theta) + j\sin(n\theta) \tag{44}$$

a) Prove De Moivre's theorem.
b) Use De Moivre's theorem to show that

$$\cos(3\theta) = \cos^3\theta - 3\cos\theta\sin^2\theta \tag{45}$$

and

$$\sin(3\theta) = 3\cos^2\theta\sin\theta - \sin^3\theta \tag{46}$$

A.7 Show that

$$\mathcal{R}\{H(F)e^{j2\pi Ft}\} = |H(F)|\cos(2\pi Ft + \angle H(F)) \tag{47}$$

Power Series

In this appendix we establish the following closed forms for various power series.

1.

$$1 + \xi + \cdots + \xi^{(K-1)} + \xi^K = \frac{1 - \xi^{K+1}}{1 - \xi} \quad \xi \neq 1 \tag{1}$$

2.

$$\xi^{-K} + \xi^{-(K-1)} + \cdots + \xi^{-1} + 1 + \xi + \cdots + \xi^{(K-1)} + \xi^K = \frac{\xi^{K+\frac{1}{2}} - \xi^{-(K+\frac{1}{2})}}{(\xi^{\frac{1}{2}} - \xi^{-\frac{1}{2}})} \quad \xi \neq 1 \tag{2}$$

3.

$$\sum_{n=-K}^{K} e^{-j2\pi\lambda n} = \frac{\sin((2K+1)\pi\lambda)}{\sin(\pi\lambda)} \triangleq (2K+1)\,\text{cinc}(\lambda; K) \tag{3a}$$

where we have defined the *circular sinc* function

$$\text{cinc}(\alpha; \beta) \triangleq \frac{\sin((2\beta+1)\pi\alpha)}{(2\beta+1)\sin(\pi\alpha)} \tag{3b}$$

4.

$$\sum_{n=0}^{\infty} \xi^n = \frac{1}{1-\xi} \quad |\xi| < 1 \tag{4}$$

5.

$$\sum_{n=0}^{\infty} n\xi^n = \frac{\xi}{(1-\xi)^2} \quad |\xi| < 1 \tag{5}$$

6.

$$\sum_{n=0}^{N-1} W_N^{kn} = N\Delta_N[k] \tag{6}$$

where

$$W_N \triangleq e^{-j\frac{2\pi}{N}} \tag{7}$$

and

$$\Delta[k; N] \triangleq \begin{cases} 1 & \text{when } k \text{ is an integer multiple of } N \\ 0 & \text{otherwise} \end{cases} \tag{8}$$

B.1   Derivations

B.1.1   To find a closed form for

$$s(\xi) = 1 + \xi + \cdots + \xi^{(K-1)} + \xi^K \tag{9}$$

we multiply both sides by ξ

$$\xi s(\xi) = \xi + \xi^2 + \cdots + \xi^K + \xi^{K+1} \tag{10}$$

By subtracting Eq. (10) from Eq. (9) and solving for $\xi \neq 1$, we obtain

$$s(\xi) = \frac{1 - \xi^{K+1}}{1 - \xi} \tag{11}$$

B.1.2   Similarly, to find a closed form for

$$S(\xi) = \xi^{-K} + \xi^{-(K-1)} + \cdots + \xi^{-1} + 1 + \xi + \cdots + \xi^{(K-1)} + \xi^K \tag{12}$$

we multiply both sides by ξ

$$\xi S(\xi) = \xi^{-(K-1)} + \cdots + \xi^{-1} + 1 + \xi + \cdots + \xi^K + \xi^{K+1} \tag{13}$$

By subtracting Eq. (13) from Eq. (12) and solving for $S(\xi)$, we obtain

$$S(\xi) = \frac{\xi^{K+1} - \xi^{-K}}{(\xi - 1)} = \frac{\xi^{K+\frac{1}{2}} - \xi^{-(K+\frac{1}{2})}}{(\xi^{\frac{1}{2}} - \xi^{-\frac{1}{2}})} \tag{14}$$

You can confirm the validity of the last step by multiplying numerator and denominator of the right-hand side by $\xi^{\frac{1}{2}}$.

B.1.3   Set

$$\xi = e^{j\theta} \tag{15}$$

Then Eq. (14) becomes

$$S(e^{j\theta}) = \frac{e^{j\theta(K+\frac{1}{2})} - e^{-j\theta(K+\frac{1}{2})}}{(e^{j\theta\frac{1}{2}} - e^{-j\theta\frac{1}{2}})} = \frac{\sin\left[\theta(K+\frac{1}{2})\right]}{\sin\left[\theta\frac{1}{2}\right]} = \frac{\sin\left[\frac{2K+1}{2}\theta\right]}{\sin\left[\frac{\theta}{2}\right]} \tag{16}$$

If we set $\theta = 2\pi\lambda$, we obtain

$$S(e^{j2\pi\lambda}) = \sum_{n=-K}^{K} e^{-j2\pi\lambda n} = \frac{\sin((2K+1)\pi\lambda)}{\sin(\pi\lambda)} \triangleq (2K+1)\,\mathrm{cinc}(\lambda; K) \tag{17}$$

where we have the defined the *circular sinc* function

$$\mathrm{cinc}(\alpha; \beta) = \frac{\sin((2\beta+1)\pi\alpha)}{(2\beta+1)\sin(\pi\alpha)} \tag{18}$$

B.1.4   If we let $K \to \infty$ in Eq. (1), we obtain

$$\sum_{n=0}^{\infty} \xi^n = \frac{1}{1-\xi} \quad |\xi| < 1 \tag{19}$$

B.1.5   We can sum the series

$$\sum_{n=0}^{\infty} n\xi^n \tag{20}$$

by differentiating

$$\sum_{n=0}^{\infty} \xi^{(n+1)} = \frac{\xi}{1-\xi} \quad |\xi| < 1 \tag{21}$$

with respect to ξ. The steps involved in the differentiation are

$$\frac{d}{d\xi}\sum_{n=0}^{\infty} \xi^{(n+1)} = \sum_{n=0}^{\infty} \frac{d}{d\xi}\xi^{(n+1)} = \sum_{n=0}^{\infty}(n+1)\xi^n = \frac{d}{d\xi}\left(\frac{\xi}{1-\xi}\right) = \frac{1}{(1-\xi)^2} \quad |\xi| < 1 \tag{22}$$

This shows us that

$$\sum_{n=0}^{\infty}(n+1)\xi^n = \frac{1}{(1-\xi)^2} \quad |\xi| < 1 \tag{23}$$

On the other hand

$$\sum_{n=0}^{\infty}(n+1)\xi^n = \sum_{n=0}^{\infty} n\xi^n + \sum_{n=0}^{\infty} \xi^n = \sum_{n=0}^{\infty} n\xi^n + \frac{1}{1-\xi} \quad |\xi| < 1 \tag{24}$$

If we combine these two equations and do a little algebra, we obtain the identity sought.

$$\sum_{n=0}^{\infty} n\xi^n = \frac{\xi}{(1-\xi)^2} \quad |\xi| < 1 \tag{25}$$

B.1.6   We can establish Eq. (6). by substituting $W_N = e^{-j\frac{2\pi}{N}}$, $K = (N-1)/2$, and $\lambda = k/N$ into identity 3. This substitution yields

$$\sum_{n=-(N-1)/2}^{(N-1)/2} W_N^{kn} = \frac{\sin(\pi k)}{\sin(\frac{\pi k}{N})} = N \operatorname{cinc}\left(\frac{k}{N}; K\right) \tag{26}$$

The sum equals N when $k =$ is an integer multiple, say lN, of N because $W_N^{lNn} = 1$. The sum equals 0 when k is an odd integer because $\sin(\pi k) = 0$. Therefore

$$\sum_{n=-(N-1)/2}^{(N-1)/2} W_N^{kn} = N\Delta[k; N] \tag{27}$$

where

$$\Delta[k; N] \triangleq \begin{cases} 1 & \text{when } k \text{ is an integer multiple of } N \\ 0 & \text{otherwise} \end{cases} \tag{28}$$

Notice that W_N^{kn} is a periodic sequence in n having period N: $W_N^{k(n+N)} = W_N^{kn}$. Because the terms W_N^{kn} in the sum (27) repeat their values every period as n increases, the sum is the same if we sum from $n = 0$ to $N - 1$. The result is

$$\sum_{n=-(N-1)/2}^{(N-1)/2} W_N^{kn} = \sum_{n=0}^{N-1} W_N^{kn} = N\Delta[k; N] \tag{29}$$

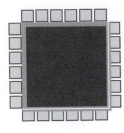

Bibliography

The following bibliography is organized into the headings, *Signal and System Theory, Applications, and Mathematical Foundations*. The field of Signals and Systems is vast. Many valuable contributions have necessarily been omitted.

Signal and System Theory

R.C. Agarwal and J.W. Cooley, "New algorithms for digital convolution." *IEEE Trans. on Acoustics, Speech, and Signal Processing*, ASSP-25:392–410, October 1977.

M. Bellanger, *Digital Processing of Signals*. John Wiley and Sons, New York NY, 1984.

Bennett, W. R., "Spectra of quantized signals." *Bell System Technical J.*, Vol. 27, pp. 446–472, 1948.

R.B. Blackman, *Linear Data Smoothing and Prediction in Theory and Practice*. Addison-Wesley, Reading MA, 1965.

G.E.P. Box and G.M. Jenkins, *Time Series Analysis: Forcasting and Control*. Holden-Day, San Francisco CA, 1970.

J.A. Cadzow and A-N. Huynh, *Discrete-Time Systems*. Prentice-Hall, Englewood Cliffs NJ, 1973.

J.A. Cadzow, *Foundations of Digital Signal Processing and Data Analysis*. Macmillan, New York NY, 1987.

Gordon E. Carlson, *Signal and Linear System Analysis*, John Wiley and Sons, 1998.

Chen, C., *Introduction to Linear Systems Theory*, Holt, Rinehart and Winston, New York, 1970.

J.W. Cooley and J.W. Tukey, "An algorithm for the machine calculation of complex Fourier series." *Math. Computation*, 19:297–301, 1965.

Cooley, J.W., Lewis, P.A.W., and Welch, P.D., "Historical notes on the fast fourier transform." *IEEE Trans. Audio Electroacoustics*, Vol. AU-15, pp. 76–79, June 1967.

Chow, Y., and Cassignol. E., *Linear Signal Flow Graphs and Applications*, John Wiley and Sons, New York, 1962.

DSP Committee, IEEE ASSP Eds., *Selected Papers in Digital Signal Processing II*, IEEE Press, New York, 1976.

DSP Committee, IEEE ASSP Eds., *Programs for Digital Signal Processing*, IEEE Press, New York, 1979.

J.P. Dugre, A.A.L. Beex, and L.L. Scharf. "Generating covariance sequences and the calculation of quantization and roundoff error variances in digital filters." *IEEE Trans. on Acoustice, Speech, and Signal Processing*, ASSP-28:102–104, 1980.

A. Fettweis, "A simple design method of maximally flat delay digital filters." *IEEE Trans. on Audion and Electroacoustics*, AU-20:112–114, June 1972.

R.A. Gabel and R.A.Robert, *Signals and Linear Systems*, John Wiley and Sons, New York NY, 1987.

G. Goertzel, "An algorithm for evaluation of finite trigonometric series." *American Mathematical Monthly*, 65:34–35, January 1958.

B. Gold and C.M. Radar, *Digital Processing of Signals*, McGraw-Hill, New York NY, 1969.

Guillemin, E.A., *Theory of Linear Physical Systems*, John Wiley and Sons, New York, 1963.

R.A. Haddad and T.W. Parsons. *Digital Signal Processing: Theory, Applications, and Hardware*, Computer Science Press, New York NY, 1991.

R.W. Hamming. *Digital Filters*, 3rd edition, Prentice Hall, Englewood Cliffs NY, 1989.

H.D. Helms, "Fast fourier transform method of computing difference equations and simulating filters." *IEEE Trans. Audio Electroacoustics*, Vol. 15, No. 2, pp. 85–90, 1967.

O. Herrman, L.R. Rabiner, and D.S.K. Chan, "Practical design rules for optimum finite impulse response lowpass digital filters." *Bell System Tech. J.*, 52:769–799, 1973.

E.B. Hogenauer, "An economical class of digital filters for decimation and interpolation." *IEEE Trans. Acousitcs, Speech, and Signal Processing*, ASSP-29:155–162, April 1981.

E.C. Ifeachor and B.W. Jervis, *Digital Signal Processing: A Practical Approach*, Addison-Wesley, Reading MA, 1993.

L.B. Jackson, *Signals, Systems, and Transforms*, Addison-Wesley, Reading MA, 1991.

L.B. Jackson, *Digital Filters and Signal Processing*, 3rd edition, Kluwer, Boston MA, 1996.

J.R. Johnson, *Introduction to Digital Signal Processing*, Prentice-Hall, Englewood Cliffs NJ, 1989.

J.F. Kaiser, "On a simple algorithm to calculate the 'energy' of a signal." *Proc. IEEE international Conference on Acoustics, Speech, and Signal Processing*, pp. 381–384, Albuquerque NM, April 1980.

N. Kingsbury, "Second-order recursive digital filter element for poles near the unit circle and the real z axis." *Electronics Letters*, 8:155–156, March 1972.

B.P. Lathi, *Signal Processing and Linear Systems*. Berkley-Cambridge, Carmichael CA, 1998.

McGillem, C.D. and Cooper, G.R., *Continuous and Discrete Signal and System Analysis*, 3rd edition, Holt, Rinehart and Winston, New York NY.

J. Makhoul, "Linear prediction: A tutorial review." *Proc. IEEE*, 62:561–580, April 1975.

S.J. Mason and H.J. Zimmerman, *Electronic Circuits, Signals and Systems*, pp. 251–269, John Wiley and Sons, New York NY, 1960.

S.K. Mitra and J.F. Kaiser, Editors, *Handbook for Digital Signal Processing*, Wiley–Interscience, New York, NY, 1993.

S.K. Mitra and H. Babic, "Partial-fraction expansion of rational z-transforms." *Electronics Letters*, 34: 1726, September 3, 1998.

Nyquist, H. "Certain topics in telegraph transmission theory." *AIEE Trans.*, pp. 617–644, 1928.

A.V. Oppenheim and R.W. Schafer, *Digital Signal Processing*, Prentice-Hall, Englewood Cliffs NJ, 1975.

A.V. Oppenheim, Editor, *Applications of Digital Signal Processing*, Prentice-Hall, Englewood Cliffs NJ, 1978.

A.V. Oppenheim and A.S. Willsky, *Signals and Systems*, Prentice-Hall, Englewood Cliffs NJ, 1983.

A.V. Oppenheim and R.W. Schafer, *Discrete-Time Signal Processing*, Prentice-Hall, Englewood Cliffs, NJ, 1989.

S.J. Orfanids, *Introduction to Signal Processing*, Prentice-Hall, Englewood Cliffs NJ, 1996.

B. Porat, *A Course in Digital Signal Principles*. John Wiley and Sons, New York NY, 1997.

J.G. Proakis and D.G. Manolakis, *Digital Signal Processing: Principles, Algorithms and Applications*, Prentice-Hall, Englewood Cliffs NJ, 2nd edition, 1992.

L.R. Rabiner and B. Gold, *Theory and Application of Digital Signal Processing*, Prentice-Hall, Englewood Cliffs NJ, 1975.

M. Schwartz and L. Shaw, *Signal Processing: Discrete Spectral Analysis, Detection and Estimation*, McGraw-Hill, New York NY, 1975.

Siebert, W. M., *Circuits, Signals, and Systems*, The MIT Press, 1986, Cambridge, MA.

Shannon, C.E., "Communication in the presence of noise." *Proc. IRE*, pp. 10–12, Jan. 1949.

William D. Stanley, *Digital Signal Processing*, Reston Publishing Company, Inc., a Prentice-Hall Company, Reston, VA, 1975.

Steiglitz, K., "The equivalence of analog and digital signal processing." *Information and Control*, Vol. 8, No. 5, pp. 455–467, Oct. 1965.

T.G. Stockham Jr., "High speed convolution and correlation." *166 Spring Joint Computer Conference, AFIPS Proc*, 28:229–233, 1966.

P.P. Vaidyanathan, *Multirate Systems and Filter Banks*, Prentice-Hall, Englewood Cliffs NJ, 1993.

M. Vetterli, "Running FIR and IIR filtering using multirate filter banks." *IEEE Trans. on Acoustics, Speech and Signal Processing*, ASSP-36:381–391, May 1988.

M. Vetterli and D. Lagall, "Perfect reconstruction filter banks: Some properties and factorization." *IEEE Trans. on Acoustics, Speech and Signal Processing*, ASSP-37:1057–1071, July 1989.

P.D. Welch, "The use of the fast Fourier transform for the estimation of power spectra: A method based on time averaging over short modified periodograms." *IEEE Trans. on Audio and Electroacoustics*. AU-15:7073, 1967.

B. Widrow, "A study of rough amplitude quantization by means of Nyquist sampling theory." *IRE Trans. on Circuit Theory*, CT-3:266–276, December 1956.

B. Widrow, "Statustucal analysis of amplitude-quantized sampled-data systems." *AIEE Trans. (Appl. Industry)*, 81:555–568, January 1961.

Sanjit K. Mitra, *Digital Signal Processing A Computer Based Approach*, 2nd Edition, McGraw-Hill, 1998.

P. M. Woodward, *Probability and Information Theory, With Applications to Radar*. 2nd Edition, Pergamon Press Ltd., 1964.

Selected Applications

Audio

Atal, B.S., and Hanauer, S.L., "Speech analysis and synthesis by linear prediction of the speech wave." *J. Acoustical Society of America*, Vol. 50, pp. 637–655, Aug. 1971.

D.L. Duttweiler, "Bell's echo-killer chip." *IEEE Spectrum*, 17:34–37, October 1980.

J. Eargle, *Sound Recording*, Van Nostrand Reinhold, New York, NY, 1976.

J.M. Eargle, *Handbook of Recording Engineering*, Van Nostrand Reinhold, New York NY, 1986.

D. Goedhard, R.J. Van de Plassche, and E.F. Stikvorrs, "Digital-to-analog-conversion in playing a compact disc." *Philips Technical Review*, 40(6):174–179, 1982.

J.P.J. Heemskerk and K.A.S. Immink, "Compact disc: System aspects and modulation." *Philips Technical Review*, 40(6):157–165, 1982.

D.M. Huber and R.A. Runstein, *Modern Recording Techniques*, 3rd edition, Howard W. Sams, Indianapolis IN, 1989.

K. Karplus and A. Strong, "Digital synthesis of plucked-string and drum timbres." *Computer Music Journal*, 7:43–44, Summer 1983.

J.L. Kelly, Jr. and C. Lochbaum, "Speech synthesis. In Proc." *Stockholm, Speech Communication Seminar*, Stockholm, Sweden, September, 1962. Royal Institute of Technology.

E.L. Lerner, "Electronically synthesized music." *IEEE Spectrum*, 17:46–51, June 1983.

Mareky, J.D., and Gray, A.H., *Linear Prediction of Speech*, Springer-Verlag, 1976, New York, NY.

J.A. Moorer, "Signal processing aspects of computer music: A survey." *Proc. IEEE*, 65:1108–1137, August 1977.

J.A. Moorer, "About this reverberatio business." *Computer Music Journal*, 3(2):13–28, 1979.

L.R. Rabiner and R.W. Schafer, *Digital Processing of Speech Signals*. Prentice-Hall, Englewood Cliffs NJ, 1978.

M.R. Schroeder, "Natural sounding aritficial reverberation." *Journal of the Audio Engineering Society*, 10:219–223, 1962.

John R. Pierce, *The Science of Musical Sound*, Scientific American Library, An imprint of Scientific American Books, Inc. New York,1983.

J.M. Worham, *Sound Recording Handbook*, Howard W. Sams, Indianapolis IN, 1989.

Video

R.C. Gonzalez and P. Winz, *Digital Image Processing*, Addison-Wesley, Reading MA, 1987.

Gonzalez and Woods, *Digital Image Processing*, 2nd Edition, Prentice-Hall, Englewood Cliffs NJ, 2002.

T.S. Huang, W.F. Schreiber, and O.J. Tretiak, "Image processing." *Proc. IEEE*, Vol. 59, No. 11, pp. 1586 ff, November 1971 (invited paper).

A.J. Jerri, "The shannon sampling theorem—its various extensions and applications: A tutorial review." *Proceedings of the IEEE*, Vol. 65, No. 11, November 1977, 1565, pp. 1565–1596.

J.S. Lim, *Two-Dimensional Signal and Image Processing*, Prentice-Hall, Englewood Cliffs, NJ, 1990.

Lyon, Richard F., "A brief history of 'pixel'." *Digital Photography II*. Edited by Sampat, Nitin; DiCarlo, Jeffrey M.; Martin, Russel A. *Proceedings of the SPIE*, Volume 6069, pp. 1–15 (2006). February 2006.

Pratt, William K., *Digital Image Processing*, 3rd Edition, Published Online: 19 Feb. 2002, Print ISBN: 0471374075. Online ISBN: 0471221325 Copyright © 2001 John Wiley and Sons, Inc.

Netravali, A.N., and Haskell, B.G., *Digital Pictures: Representation, Compression, and Standards*, 2nd edition, Plenum Press, 1995, New York, NY.

D.P. Sandbank, Editor, *Digital Television*, John Wiley and Sons, New York NY, 1990.

William F. Schreiber, *Fundamentals of Electronic Imaging Systems: Some Aspects of Image Processing*, Springer-Verlag, New York NY, 1986, 1991, 1993.

William F. Schreiber, "Wirephoto quality improvement by unsharp masking." *Pattern Recognition*, Vol. 2, pp. 117–121, March 1970.

Schreiber, W.F., "Image processing for quality improvement." *Proceedings of the IEEE*, Dec. 1978, Volume: 66, Issue: 12, pp. 1640–1651.

William F. Schreiber, "The FCC digital television standards decision." *Prometheus, Special Issue: The Future of Digital TV*, 16, 2, June 1998, pp. 155–172.

William F. Schreiber and R.R. Buckley, "Introduction to 'color television—part I.'" *Proc. IEEE*, 87, 1, January 1999, pp. 173–179 (invited paper).

Other

G.C. Carter, Editor, *Coherence and Time Delay Estimation: An Applied Tutorial for Research, Development, Test, and Evaluation Engineers*. IEEE Press, 1993.

A.C. Dubey, J.F. Harvey, J.T. Broach, and R.E. Dugan, Editors, "Detection and remediation technologies for mines and minelike targets IV." *Proceedings of SPIE*, Vol. 3710, 1999.

A. Cohen, *Biomedical Signal Processing*. Vol. II, CRC Press, Boca Raton FL, 1986.

E.A. Robinson and S. Treitel. *Geophysical Signal Analysis*, Prentice-Hall, Englewood Cliffs NJ, 1980.

K. Shenoi. *Digital Signal Processing in Telecommunication*, Prentice-Hall, Englewood Cliffs NJ 1995.

Mathematical Foundations

R.N. Bracewell, *The Fourier Transform and Its Application*, McGraw-Hill, 1986.

R.S. Burrington, *Handbook of Mathematical Tables and Formulas*, 5th edition, McGraw-Hill, New York NY, 1973.

W.B. Davenport, Jr., *Probability and Random Processes: An Introduction for Applied Scientists and Engineers*, McGraw-Hill, New York, NY, 1970.

R.V. Churchill, and J.W., Brown, *Introduction to Complex Variables and Applications*. 4th edition, McGraw-Hill, New York NY, 1984.

W. Davenport, Jr., and W. Root: *Random Signals and Noise*, McGraw-Hill, New York, 1958.

G. Doetsch, *Introduction to the Theory and Applications of the Laplace Transformation with a Table of Laplace Transformations*, Springler Verlag, 1974, New York NY.

Jury, E.I., *Theory and Application of the z-Transformation Method*, R.E. Krieger, 1982, Malabar FL.

A. Papoulis, *Probability, Random Variables, and Stochastic Processes*, McGraw-Hill, New York NY, 1965.

H. Stark and J.W. Woods, *Probability, Random Processes, and Estimation Theory for Engineers*, Prentice-Hall, Englewood Cliffs NJ 1994.

I. S. Sokolnikoff and R. M. Redheffer, *Mathematics of Physics and Modern Engineering*, McGraw-Hill, New York, NY, 1958.

G.B. Thomas, Jr., and R.L. Finney, *Calculus and Analytic Geometry*, John Wiley and Sons, New York NY, 1969.

A.D. Poularikas, Editor in Chief, *Transforms and Applications Handbook*, CRC Press, Inc. 1996.

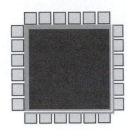

Index